Soft Ferrites

Properties and Applications

Soft Ferrites
Properties and Applications

Second Edition

E C Snelling, B Sc (Eng), C Eng, FIEE

Butterworths
London Boston Singapore Sydney Toronto Wellington

All rights reserved. No part of this publication may be reproduced or transmitted in any
form or by any means, including photocopying and recording, without the written
permission of the copyright holder, application for which should be addressed to the
Publishers, or in accordance with the provisions of the Copyright Act 1956 (as amended),
or under the terms of any licence permitting limited copying issued by the Copyright
Licensing Agency, 7 Ridgmount Street, London WC1E 7AE, England. Such written
permission must also be obtained before any part of this publication is stored in a
retrieval system of any nature.

Any person who does any unauthorized act in relation to this publication may be liable
to criminal prosecution and civil claims for damages.

This book is sold subject to the Standard Conditions of Sale of Net Books and may not
be re-sold in the UK below the net price given by the Publishers in their current price list.

First published 1969
Second edition 1988

© Butterworth & Co. (Publishers) Ltd, 1988

British Library Cataloguing in Publication Data

Snelling, E. C. (Eric Charles), *1923–*
   Soft ferrites.—2nd ed.
   1. Electronic equipment. Magnetically soft ferrites
   I. Title
   621.381

   ISBN 0–408–02760–6

Library of Congress Cataloging-in-Publication Data

Snelling, E. C. (Eric Charles), 1923–
   Soft ferrites : properties and applications / E. C. Snelling.—
2nd ed.
    p. cm.
   Bibliography: p.
   Includes index.
   ISBN 0–408–02760–6 :
   1. Ferrites (Magnetic materials)   I. Title.
TK7871.15.F4S53 1988
620.1′1297—dc19   88–14480

Filmset by Latimer Trend & Company Ltd, Plymouth
Printed and bound in Great Britain by Anchor Brendon Ltd, Tiptree, Essex

Preface

The first edition of *Soft Ferrites* was published in 1969. By that time, ferrites had become firmly established as an important class of magnetic materials and were widely used in linear, digital and microwave applications in electronic engineering. At that time there was no textbook specifically dealing with the properties and applications of the largest and most important group of ferrites, i.e. those having high permeabilities and low losses and generally referred to as magnetically soft ferrites. The first edition was written to fill that gap. It set out to present in easily assessible form the theory, data and procedures required for the design of ferrite cored devices and it soon became established as a widely recognized reference work.

By 1980 the first edition had gone out of print and, while retaining an active interest in ferrites, I had moved on to the area of integrated circuits. The non-availability of the book became a matter of concern; many requests were received from workers in the field who could not obtain a copy or find an alternative. In the latter half of 1985 the opportunity arose for the preparation of a new and completely revised version and work on the second edition was started. The revised text contains important new design approaches and relates to the latest ferrite materials and components. In the interest of continuity the original structure has been retained. The following paragraphs briefly describe the contents and the more significant areas of revision.

The book is concerned with the technically important properties of magnetically soft ferrites at frequencies up to about 100 MHz, and the application of these ferrites to inductors, transformers and related devices. It is primarily intended for electronic engineers and physicists whose work involves design or development using these ferrites. The emphasis is on optimum design, so the theory, the data and the design or selection procedures have been presented in the most objective and useful form. Wherever possible the labour of design or core selection has been minimized by the computation of the relevant functions and the presentation of the results in the form of tables or graphs. A comprehensive list of references and bibliography follows each chapter; these take account of literature published since the first edition.

The four opening chapters deal with the properties of ferrites and magnetic circuits in general. The first gives a broad introduction to ferrite materials and a qualitative description of the origin of the magnetic properties. It also describes those aspects of manufacture and processing that may be of interest to the user; these sections have been extensively revised to embrace the considerable progress that has occurred in this area. Chapter 2 considers in detail the expression of the electrical and magnetic properties of ferrites, ranging from the still-relevant relations introduced by Lord Rayleigh in 1887 to the current recommendations of the International Electrotechnical Commission.

Chapter 3 is devoted to a wide-ranging survey of the properties of currently available manganese zinc and nickel zinc ferrites. It starts with a tabulation of all the ferrite manufacturers known to me; entered against each are the current material code numbers of that manufacturer's products classified according to the principal application. Then follows an extensive presentation of material properties containing over 170 separate graphs based mainly on manufacturers' current data sheets. Clearly, most of the first-edition contents of this chapter have been replaced with new material. Every effort has been made to ensure that the selection of data is as comprehensive and representative as possible. I am grateful to the many manufacturers who have contributed data to this chapter, and apologize for any omissions or imbalance that may become apparent. Chapter 4 considers the relations between the material properties and the geometry of the practical magnetic circuit, e.g. the effect of non-uniform cross-section, air gaps, etc.

The next six chapters each deal with a basic type of ferrite-cored device. To avoid repetition, the aim has been to treat the fundamentals of each basic type of device in some detail, omitting discussion of the more specific applications of the general principles. Thus particular applications such as telephone loading coils, IF transformers, TV line output transformers etc. have not been separately described. However, in as much as the design of such specialized devices is concerned with ferrite cores, the essential design information will be found in these chapters. In turn, they deal with inductors (mainly for low-power resonant circuits), high-frequency transductors, wide-band transformers, pulse transformers, power transformers and inductors, and finally magnetic antennas. Each chapter is a study of the basic theory and design concepts with special reference to core properties and geometry. The extent of the revision varies from one chapter to another; the theory of wide-band transformer design has changed little since the first edition, whereas the progress in power transformer applications and design has been so extensive Chapter 9 has been virtually rewritten and now contains new and improved design procedures.

Finally all design aspects that relate particularly to the windings and are common to a number of the applications, e.g. winding resistance, eddy currents in conductors, self capacitance and leakage inductance, are grouped together in Chapter 11.

The quantities and units used in this book are primarily in accordance with the International System (SI), and the symbols are as far as possible consistent with IEC Publication 27. This representation has now virtually replaced the CGS practical system. However, as many readers will still be more familiar with the CGS system its inclusion as an alternative representation has been retained from the first edition. The format is as follows. In the main text the equations are expressed in SI units; if the CGS version of a numbered equation differs from the SI version, the CGS version is given in a footnote. Unless multiples or sub-multiples are specifically indicated, the symbols refer to the basic units of the respective systems. If a numbered equation does not have a footnote version this generally indicates that the SI and CGS versions are identical, provided the appropriate units are used. In graphs, scales have been given in both systems of units where appropriate. Lastly a small editorial detail. In the first edition the gyromagnetic (spin) resonance was referred to as *ferrimagnetic resonance* to emphasize the ferrimagnetic nature of ferrites. However, following the trend elsewhere, this has been changed in the second edition to *ferromagnetic resonance*, since this term is being generally applied to all types of gyromagnetic resonance unless the context requires the distinction to be made.

ECS
March 1988

Acknowledgements

In preparing the second edition of *Soft Ferrites* for publication I have had the benefit of assistance and co-operation of a number of people and organizations. I wish to express my appreciation to all of them and make the following specific acknowledgements.

Dr B. G. Street (formerly with Mullard Magnetic Components), Mr J. E. Knowles (formerly with Philips Research Laboratories) and Ir T. G. W. Stijntjes of Philips Components; each contributed valuable information and comment relating to Chapter 1, and Dr Street kindly read and commented on the complete draft of this chapter. Drs J. W. Waanders of Philips Components contributed general information on the ferrite industry and in particular, with the assistance of Ir E. J. Pateer of the Central Development Laboratory, supplied specially requested data on particular ferrites for inclusion in Chapter 3. Mr J. A. Houldsworth (formerly with the Mullard Application Laboratory) and his colleagues Mr D. J. Harper and Mr G. M. Fry provided useful background information on power applications during the early stages of the preparation of Chapter 9. Mr L. E. Jansson (also formerly of that laboratory) read the draft of Chapter 9 and made valuable comments. Mr J. A. Garters of the Mullard Applications Laboratory advised on the up-dating of the transistor information in Chapter 10. Dr R. F. Milsom and Dr G. Clark of Philips Research Laboratories kindly allowed me to use their computer network facilities for the generation of the magnetic field plots in Figure 11.11. Mr C. Barrow and Mr D. Wareham of Mullard Magnetic Components supplied data on several aspects of ferrite properties, in particular the performance with magnetic polarization and disaccommodation data. Mr Jan van der Poel of Ferroxcube Division of Amperex, USA, provided contact with the ferrite scene in North America and responded promptly to numerous requests for information. There was frequent need for up-to-date information on the standardization aspects of ferrite components, and Mr N. Crevis, Technical Officer at BSI, was always ready to help. Mrs M. T. Paterson and the staff of the library of Philips Research Laboratories were a constant source of assistance especially in the matter of the literature search. Grateful acknowledgement is also due to the Director of Research at Philips Research Laboratories, Mr K. L. Fuller, for generously allowing the author access to the laboratory facilities. Finally, on a more personal note, I wish to recognize the encouragement and help of my wife, Rita, particularly in the preparation of the typescript and the checking of data.

As indicated in the preface, Chapter 3 is a massive compilation of ferrite material data and this could not have been achieved without the co-operation of the manufacturers. It is therefore appropriate to record my appreciation to all the manufacturers who supplied data on their products and who subsequently checked and approved the detailed representation of their contributions. The manufacturers are:

Western Europe Kaschke KG GmbH & Co., Göttingen, FGR. Krupp Widia-Fabrik, Essen, FGR. Neosid Ltd, Letchworth, UK. Philips Components, Eindhoven, The Netherlands. Salford Electrical Instruments Ltd, Heywood, UK. Siemens AG, Munich, FGR. Thomson CSF, Beaune Cedex, France. Vogt Electronic, Erlau bei Passau, FGR.

Eastern Europe Ei-Feriti, Zemun, Yugoslavia. EMO Elektromodul, Budapest, Hungary. Iskra Elementi, Ljubijana, Yugoslavia. Kombinat VEB, Hermsdorf, GDR. Zakład Materiałow Magnetycznych, Polfer, Warsaw, Poland. Pramet n.p., Šumperk, Czechoslovakia.

USA Allen-Bradley, Eatontown, NJ. Ceramic Magnetics Inc., Fairfield, NJ. Fair-Rite Products Corp., Wallkill, NY. Feronics Inc., Fairport, NY. Ferroxcube, Division of Amperex Electronic Corp., Saugerties, NY. Krystinel Corp., Paterson, NJ. National Magnetics Group, Newark, NJ. Magnetics, Division of Spang & Co., Butler, Pa. D. M. Steward Mfg. Co., Chattanooga, TN.

Japan Nippon Ferrite Ltd, Tokyo. Sumitomo Special Metals Co. Ltd, Osaka. TDK Corporation, Tokyo. Tokin Corporation, Tokyo. Tomita Electric Co. Ltd, Tottori City.

Contents

Preface   v

Acknowledgements   vii

1 Ferrites: their nature, preparation and processing
1.1 Introduction   1
1.2 Magnetism in ferrites   3
1.3 Manufacture   7
1.4 Finishing processes available to the user   18
1.5 References and bibliography   21

2 The expression of electrical and magnetic properties
2.1 Magnetization   26
2.2 The expression of low-amplitude properties   28
2.3 The expression of high-amplitude properties   40
2.4 References and bibliography   42

3 Properties of some manganese zinc and nickel zinc ferrites
3.1 Introduction   44
3.2 Survey and classification   45
3.3 Mechanical and thermal properties   54
3.4 Magnetic and electrical properties   55
3.5 References and bibliography   133

4 Magnetic circuit theory
4.1 Introduction   136
4.2 Closed magnetic cores   136
4.3 Open magnetic cores   149
4.4 References and bibliography   157

5 Inductors
5.1 Introduction   158
5.2 Form of core   160
5.3 Air gap and the calculation of inductance   163
5.4 Inductance adjustment   165
5.5 Constancy of inductance   169
5.6 Self capacitance   173
5.7 Q-factor   175
5.8 Waveform distortion and intermodulation   187
5.9 References and bibliography   188

6 High-frequency transductors
6.1 Introduction   189
6.2 General mode of operation   189
6.3 Applications and performance   192

6.4 Design and operating techniques 196
6.5 References and bibliography 199

7 Wide-band transformers
7.1 Introduction 201
7.2 The transmission and reflection characteristics in terms of the equivalent circuit elements 201
7.3 Low and medium frequency wide-band transformers 213
7.4 High frequency wide-band transformers 224
7.5 References and bibliography 231

8 Pulse transformers
8.1 Introduction 232
8.2 Pulse distortion in terms of the equivalent circuit elements 233
8.3 Determination of the element values 239
8.4 Pulse distortion in relation to bandwidth 247
8.5 Pulse permeability 248
8.6 Practical considerations 252
8.7 References and bibliography 254

9 Power transformers and inductors
9.1 Introduction 255
9.2 Heat transfer and temperature rise 256
9.3 General electrical design considerations 264
9.4 Low-frequency power transformers and chokes 268
9.5 Higher frequency power transformers 291
9.6 References and bibliography 301

10 Ferrite antennas
10.1 Introduction 303
10.2 Antenna circuit theory 303
10.3 Practical aspects of design 308
10.4 References and bibliography 312

11 Properties of windings
11.1 Introduction 313
11.2 Number of turns in a given winding area 313
11.3 The d.c. winding resistance 315
11.4 Power loss due to eddy currents in the winding 317
11.5 The inductance of an air-cored coil 330
11.6 The self capacitance of a winding 330
11.7 Leakage inductance in transformers 334
11.8 References and bibliography 338

Appendix A Calculation of the core factors of a pot core 339
Appendix B Wire tables 341
Appendix C Publications relevant to ferrite cores prepared by Technical Committee No. 51 of the International Electrotechnical Commission 352
Appendix D Symbols 354

Index 363

1 Ferrites: their nature, preparation and processing

1.1 Introduction

Ferrites may be defined[1] as magnetic materials composed of oxides containing ferric ions as the main constituent.* The term is often restricted to such materials having the cubic crystal structure of the mineral spinel, but it is also loosely applied to magnetic oxides in general irrespective of their crystal structure. The scope of this book is restricted to the properties of those ferrites that are magnetically soft (i.e. having low coercivity) and of technical importance, and to the application of such ferrites in devices that, in the broadest sense, may be described as inductors or transformers. Before embarking on a text constrained to this scope, it is perhaps appropriate that this introduction should put these particular ferrites into a proper perspective by surveying the wider field.

Magnetite, or ferrous ferrite, is an example of a naturally-occurring ferrite. It has been known since ancient times and its weak permanent magnetism found application in the lodestone of the early navigators. Hilpert[2] in 1909 published the first systematic study of the relation between the chemical and magnetic properties of a number of binary iron oxides but experienced difficulty in identifying the magnetic phases of his preparations. Around 1928 Forestier[3] in France and Hilpert and Wille[4] in Germany made quantitative investigations into the relation between the chemical composition, the saturation magnetization and the Curie temperature. Magnetic oxides were also studied by Japanese workers between 1932 and 1935.[5,6]

This research was motivated by the realization that if the high electrical resistivity of the oxides could be combined with useful magnetic properties then the resultant materials might extend the frequency range of inductors and transformers. Metallic cores are limited to relatively low frequencies unless the material is divided into thin laminations or powder; this is expensive and not wholly effective.

In 1936 Snoek[7] was studying magnetic oxides in The Netherlands; similar studies in Japan were reported by Takei[8] in 1937 and 1939. Snoek and his co-worker, Six, realized that the most important property of a material intended as a core for an inductor is the loss tangent divided by the permeability, the so-called loss factor. This is because the loss can always be reduced by the introduction of an air gap provided the resultant permeability remains adequate. This led Snoek to the development of manganese-zinc-ferrous ferrite in which low loss and high permeability were combined by minimizing the magnetocrystalline anisotropy and the magnetostriction. By 1945 Snoek[9] had laid the foundations of the physics and technology of practical ferrites and a new industry came into being.

Since that time the use of ferrites has become established in many branches of telecommunication and electronic engineering and they now embrace a very wide diversity of compositions, properties and applications.

A more detailed account of the history of ferrite research and development has been given by Wijn.[10]

Ferrites are ceramic materials, dark grey or black in appearance and very hard and brittle. A ferrite core is made by pressing a mixture of powders containing the constituent raw materials to obtain the required shape and then converting it into a ceramic component by sintering. The magnetic properties arise from interactions between metallic ions occupying particular positions relative to the oxygen ions in the crystal structure of the oxide. In magnetite, in the first synthetic magnetic oxides and indeed in the majority of present-day magnetically soft ferrites, the crystal structure is cubic; it has the form of the mineral spinel. The general formula of the spinel ferrite is $MeFe_2O_4$ where Me usually represents one or, in mixed ferrites, more than one of the divalent transition metals Mn, Fe, Co, Ni, Cu and Zn, or Mg and Cd. Other combinations, of equivalent valency, are possible and it is also possible to replace some or all of the trivalent iron ions with other trivalent metal ions.

In the early practical ferrites Me represented Cu + Zn + Fe, Mn + Zn + Fe or Ni + Zn + Fe. The first of these compounds was soon discarded and the other two, generally referred to as manganese zinc

---

*In metallurgy and mineralogy, the term *ferrite* has other meanings.

ferrite and nickel zinc ferrite respectively (often abbreviated to MnZn ferrite and NiZn ferrite) were developed for a wide range of applications where high permeability and low loss were the main requirements. These two basic compounds are still by far the most important ferrites for such applications and constitute the great majority of present-day soft ferrite production. By varying the ratio of Zn to Mn or Ni, or by other means, both types of ferrite may be made in a variety of grades, each having properties that suit it to a particular class of application. The range of permeabilities available extends from about 15 for nickel ferrite to more than 10 000 for some manganese zinc ferrite grades.

The applications started in the field of telephony transmission where carrier (Frequency Division Multiplex) systems, being essentially analogue, required large numbers of high-performance inductors and transformers mainly operating in the frequency range from about 40 kHz to 500 kHz. The combination of good magnetic properties and high resistivity made these ferrites very suitable as core materials in this application. Since the resistivities could be at least a million times greater than the values for metallic magnetic materials, laminated or powder cores could be replaced with solid ferrite cores. These could be made in more functional shapes than their metallic counterparts and their better performance enabled Frequency Division Multiplex (FDM) telephony to be extended to higher frequencies than had hitherto been possible.

The first extension to applications outside telephony was to domestic television receivers where MnZn ferrite became the undisputed core material for the line time-base/e.h.t. transformer and the field-shaping yoke used in the picture tube beam-deflection system. Later, other ferrite compositions, such as LiZn, NiZn or MgZn, ferrites were introduced for deflection yokes requiring high resistivity to facilitate the winding of the coils directly onto the ferrite. However, most high-performance television deflection yokes are made of MnZn ferrite. In domestic radio receivers, ferrite rods or plates enabled compact a.m. antennas to be made. Other applications proliferated.[11] Of these, the most important were the use of ferrites for recording heads[12] for audio, data and, later, video recorders, their use in inductive elements for radio interference suppression, and their application as cores for transformers in switched-mode power supplies.[13,14] In recent years the latter application has increased very strongly and is now comparable to television ferrite applications in terms of production quantities.

Meanwhile, in 1952, a new class of magnetic oxides having permanent magnet properties was announced.[15] They have a hexagonal crystal structure and a general chemical formula $MeFe_{12}O_{19}$, where Me represents barium, strontium or lead. They are characterized by having very large coercivities, e.g. greater than 160 kA/m (2000 Oe). Because of this high coercivity, their relatively low mass density and their composition of relatively inexpensive and readily available raw materials, these magnetic oxides, often referred to as hexaferrites, found ready application. As an alternative to being made in isotropic form, the crystal axes may be orientated by the application of a magnetic field during the pressing operation. The anisotropic properties so induced enable much higher magnetic energies to be stored.[16,17] In both isotropic and anisotropic form, but mainly the latter, these hexaferrites are used in a wide variety of permanent magnet applications. The main products are loudspeaker magnets and field magnets in d.c. motors, particularly for the automobile industry. Other hexaferrites were developed[18,19] having useful soft magnetic properties that are maintained up to frequencies approaching 1000 MHz.

Another class of application arose when it was found possible to prepare spinel ferrites, such as MnMg, MnCu and LiNi ferrite, having substantially rectangular hysteresis loops.[20] These offered the prospect of the rapid storage and retrieval of digital information by the switching of small toroidal cores from one remanent state to the other.[21] Their appearance coincided with the rapid development of the digital computer and the attendant, almost insatiable, demand for mass data storage.[22,23] Many billions of tiny toroidal cores of these ferrites were woven into memory planes for computer main frame storage before the advent of semiconductor memories led to their being superseded in all but a few specialized applications.

The application of magnetic oxides at microwave frequencies resulted from the discovery that the aligned magnetic moments of the electron spins within the crystal lattice may be made to precess at a frequency that depends on the strength of the static internal magnetic field. An incident circularly-polarized electromagnetic wave will stimulate precessional resonance (and the accompanying absorption) only if the rotation of the polarization is in the right sense with respect to the direction of the static field. This gives rise to non-reciprocal devices in which the transmission properties in one direction are quite different to those in the other.[24,25] A variety of microwave devices have been developed which depend on this principle. They include wave guide and strip line isolators, switches, circulators, modulators, limiters, etc. The performance of these devices depends among other things on the saturation magnetization of the material and the width of the resonance absorption peak. Many different magnetic oxides have been developed to meet the diversity of requirements that occur in microwave engineering. Among them are spinel ferrites such as Ni ferrite, MnMg ferrite and NiCuCoMnAl ferrite, hexagonal ferrite such as $BaFe_{12}O_{19}$ and a variety of mixed oxides based on the structure of the garnet, notably yttrium iron garnet with various substitutions.

There are, of course, many miscellaneous applications of ferrites. Although a catalogue of these would be out of place in the present context, it may be of interest to mention a few of the more significant ones to fill in the broader perspective. Massive toroidal and frame cores are used in particle accelerators, e.g. synchrotrons, to couple the driving energy to the particle beam. The development of these machines has been very dependent on the availability of suitable ferrites in sufficient sizes.[26] MnZn ferrite with a suitably low Mn/Zn ratio can have a Curie temperature at or below room temperature and such ferrites can in principle be used in a variety of temperature-sensitive or temperature-activated devices.[27,28] Very high-frequency ferrite has been used in powder form dispersed in a plastic matrix to provide cladding on buildings or vehicles to suppress reflection of radio waves.[29,30] Finally, garnet materials found a new potential rôle when it was discovered that thin wafers of the single crystal of suitably substituted yttrium iron garnet could support isolated microscopic domains of reverse magnetization. These could be switched on or off and moved along a sequence of locations defined by surface electrodes thus creating a sequential data store having very high data density, the so-called bubble memory.[31]

In terms of production volume there are five principal magnetic oxide products. In the permanent magnet sector there are magnets for motors and magnets for loudspeakers, and in the soft ferrite sector there are television deflection yokes and cores for television line output/e.h.t. transformers and switched mode power supplies. It is estimated that together these five applications account for about 85% of the total weight of magnetic oxides produced. According to Ruthner[32] the world production figures for ferrite powder have been estimated as shown in Table 1.1.

While the conversion from analogue to digital systems and the massive increase in the use of microelectronic circuits based on silicon has certainly resulted in a large reduction in the demand for high-performance inductors and analogue transformers, the overall trend in soft ferrite production remains a steady increase. Digital electronic systems, particularly in the area of information technology and computing have generated an increasing demand for ferrite components for:

- power supplies, e.g. transformer cores,
- display, e.g. line time base transformers and deflection yokes,
- recording,
- interference suppression.

Having briefly reviewed ferrites in the wider context, the scope now narrows somewhat to the sector defined at the beginning of this section, namely the properties and applications of magnetically soft MnZn and NiZn ferrites. In a book primarily concerned with the engineering aspects of the subject, it is not appropriate to consider in any detail the physics of magnetism in general or of ferrites in particular. This is extensively covered in the literature. However, a short qualitative description of the spinel lattice, the magnetic domain and the magnetization process, as they relate to ferrites, will provide a useful basis for the discussion of the properties of these materials. The reader is referred to standard text books for a more complete treatment.[33-36]

1.2 Magnetism in ferrites

1.2.1 The spinel lattice

Figure 1.1 shows a unit cell of the spinel lattice and the sites of the various ions. The small cubic diagram shows how the unit cell is composed of octants of alternate kind: in the large diagram only the four nearest octants are shown complete, the remainder having the symmetry shown in the smaller diagram. All the octants contain the same tetrahedral arrangement of oxygen ions (anions), the sites being defined by four corners of a smaller cube, which in practice is usually slightly distorted. In the octants corresponding to the unshaded parts of the small diagram, the remaining corners of the smaller cube are occupied by metal ions (cations). In the alternate octants these corners are not occupied; instead there is a site in the centre of the octant. This site, being surrounded by a tetrahedral arrangement of oxygen ions, is called a tetrahedral site or A site. A tetrahedral site is shown separately at the top of the diagram. All the black spheres are in tetrahedral sites although this is not obvious when considering only one isolated unit cell.

Table 1.1 Production of ferrite powder

Year	Estimated world production in kilotonnes/yr	
	Soft ferrites	Permanent magnet hexaferrites
1975	70	172
1980	102	226
1985	108	285
1990	150	400

The remaining metal ion sites are surrounded by six oxygen ions in the form of an octahedron. These are referred to as octahedral sites or *B* sites; an isolated octahedral site is shown at the bottom of the diagram.

In the unit cell there are 64 possible tetrahedral sites and 32 possible octahedral sites. Of these only 8 tetrahedral and 16 octahedral sites are occupied in a full unit cell. If, in the preparation, the ratio of metal ions to oxygen ions is too small, i.e. there is excess oxygen then some of the normally occupied metal ion sites may be unoccupied. These sites are then referred to as vacancies. If the ratio is correct, i.e. if the composition is stoichiometric, then the unit cell contains 24 metal ions and 32 oxygen ions, i.e. there are 3 metal ions for every 4 oxygen ions. Thus the spinel unit cell may be considered from the chemical point of view to consist of 8 molecules having the formula $MeFe_2O_4$. Of the three metal ions, one is on a tetrahedral site and two are on octahedral sites. If the spinel were 'normal', the divalent Me ion would occupy a tetrahedral (*A*) site while the trivalent Fe ions would occupy the octahedral (*B*) sites. In an 'inverse' spinel the divalent Me ion occupies one of the *B* sites while the trivalent Fe ions occupy the other *B* site and the *A* site. In terms of a unit cell:

Number of ions
$\begin{cases} & \text{Me} & \text{Fe}_2 & \text{O}_4 \\ A \text{ site} & - & 8 \\ B \text{ site} & 8 & 8 \end{cases}$ 32

In practice spinel ferrites have an ion distribution somewhere between 'normal' and 'inverse'.

The spinel structure consists of a number of interlaced face-centred cubic lattices. The most obvious one in Figure 1.1 is that formed by the *A* sites on the cell corners and face centres. The remaining *A* sites (octant centres) form another face-centred cubic lattice displaced from the first along the cube diagonal.

The positions of the oxygen ions are also defined by a set of interlaced face-centred cubic lattices. Any oxygen ion may be taken as occupying a corner of a face-centred cube having the same dimensions as the unit cell; all other sites in this face-centred cube are also occupied by oxygen ions. Again, the octahedral (*B*) ions occupy sites on four face-centred cubic lattices. Each of these lattices has the same dimensions as the unit cell and they are displaced from one another along the edge of the smaller cube in the unshaded octants.

These interlaced lattices are called sub-lattices and they play an important part in the magnetism of ferrites.

1.2.2 Magnetization

Electrons spin about an axis and, by virtue of this spin

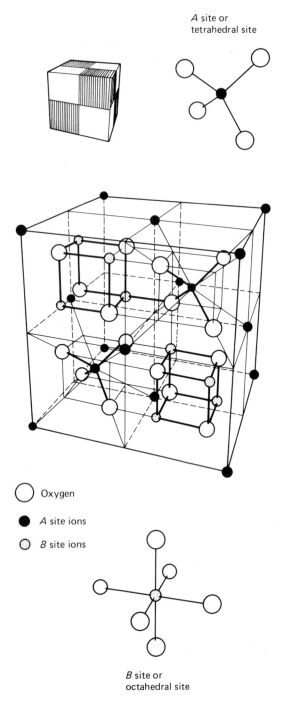

Figure 1.1 Unit cell of the spinel lattice

and their electrostatic charge, exhibit a magnetic moment. Normally, in an ion with an even number of electrons, the spins or moments cancel, and when the number of electrons is odd there will be one uncompensated spin. For the transition metals the number of uncompensated spins is larger, e.g. the trivalent Fe ion has a moment equivalent to five uncompensated spins.

When the atoms of these transition metals are combined in metallic crystals, as they are, for example, in iron, the atomic moments are spontaneously held in parallel alignment over regions within each crystallite. The net number of uncompensated spins will be less than for the isolated ion due to the band character of the electron energies in a metal. The regions in which alignment occurs are called domains and may extend over many thousands of unit cells. The spin orientation is along a direction of minimum energy, i.e. external energy is required to deflect the magnetization from this direction and if the external constraint is removed the magnetization will return to a preferred direction. This directional or anisotropic behaviour may arise from a number of factors. Magnetocrystalline anisotropy is inherent in the lattice structure, the magnetization lying preferentially in the direction of the cube edge or cube diagonal. Figure 1.2 represents the magnetocrystalline anisotropy energy surface[37] for a MnZn ferrite which has a [100] preferred direction. The length of the radius vector from the origin to the surface is a measure of the anisotropy energy when the magnetization lies along that direction. Clearly the [100] direction is one of minimum energy. Mechanical strain also causes anisotropy and so can the shape of the grain boundary. The result is that the magnetization is held to a certain direction, or to one of a number of directions, as if by a spring. The greater the anisotropy, the stiffer the spring and the more difficult it is to deflect the magnetization by an external magnetic field, i.e. the lower the permeability (see Chapter 2 for definitions of permeability, etc.)

The parallel spin alignment implies that the material within the domain is magnetically saturated. The magnetization is defined as the magnetic moment per unit volume and is therefore proportional to the density of magnetic ions and to their magnetic moments. This magnetism arising from parallel alignment is called ferromagnetism.

In a ferrite the metal ions are separated by oxygen ions. As a result of this the ions in the A sub-lattice (tetrahedral sites) are orientated antiparallel to those in the B sublattice (octahedral sites). If these sub-lattices were identical the net magnetization would be zero in spite of the alignment and the ferrite would be classified as antiferromagnetic. In the majority of practical ferrites the two sub-lattices are different in number and in the type of ions so that there is a resultant magnetization. Such materials are classified as ferrimagnetic. For example, in the foregoing section it was stated that in the general spinel molecule $MeFe_2O_4$ one metal ion occupies an A site while two occupy B sites; thus in the case of $MnFe_2O_4$ where both metal ions have 5 uncompensated spins the net magnetization is 5 spins per molecule. This compares with a net moment of 2.2 spins per *atom* in the case of metallic iron. For this reason, a ferrite has a much lower saturation magnetization ($\mu_o M_s = J \simeq 0.5$ tesla) than metallic iron (about 2.0 tesla). However, in spite of the partial cancellation of the spin moments, ferrites possess sufficient saturation magnetization to make them useful in a wide range of applications.

The crystallite is normally divided into a number of domains of various spin orientations, e.g. opposite (180°) and orthogonal (90°), so that the crystallite has very little external field arising from the internal magnetization, i.e. the demagnetizing fields are small. The domain boundaries (Bloch walls) consist of regions many unit cells in thickness in which there is a gradual transition of spin orientation, see Figure 1.3. This transition must act against the magnetocrystal-

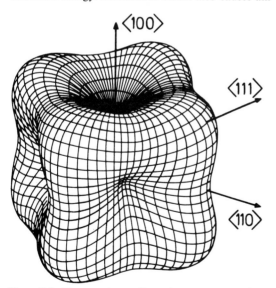

Figure 1.2 The magnetocrystalline anisotropy energy surface for a MnZn ferrite having a positive anisotropy. (After Pearson,[37] courtesy John Wiley)

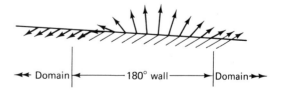

Figure 1.3 Transition of spin direction at a 180° Bloch wall (domain boundary)

line anisotropy which is tending to hold the spins in the preferred direction, thus the formation of a domain boundary involves the storage of energy. The number and arrangement of domains in a crystallite is such that the sum of the energies, mainly the wall energy and the demagnetizing field energy, is a minimum. Figure 1.4 shows an idealized arrangement of domains. If an external field is applied, the domain walls experience a pressure which tends to make those domains having a component of magnetization in the direction of the field grow at the expense of the unfavourably orientated domains.

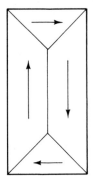

Figure 1.4 Idealized domain pattern

In practice it is energetically favourable for domain walls to pass through certain imperfections such as voids, stressed regions, non-magnetic inclusions, etc. Figure 1.5 is a simplified representation of the situation. In the absence of an applied field the walls are straight and might occupy the positions shown in (a). The dots represent imperfections. If a small field is applied in the direction shown (b) the walls remain pinned by the imperfections and the crystallite boundaries bulge as would a membrane under pressure. These movements are reversible. The change in magnetization is restricted by the stiffness of the walls. Under these circumstances the lower part of the magnetization curve is traced. As the field increases, the pressure on the walls overcomes the pinning effect and the walls move by a series of jumps (c, d, e). These movements are irreversible, i.e. if a certain field change is required to produce a jump, the reversal of that change, i.e. the restoration of the field, will not in general cause the wall to jump back. During this part of the process the magnetization curve rises steeply. Finally, when all the domains have been swept away, further increases in field strength cause the magnetization vector to rotate reversibly towards the external field direction until complete alignment is approached (f). No further increase in magnetization is then possible and the material is said to be saturated. Normally in a polycrystalline material there is a wide distribution of grain sizes, domain sizes, orientations, etc., and irreversible and reversible processes merge together. However the above illustration represents the main stages in the magnetization process (see also Reference 38).

If the magnetic field, having reached the maximum value corresponding to Figure 1.5(f), is made to alternate cyclically about zero at the same maximum

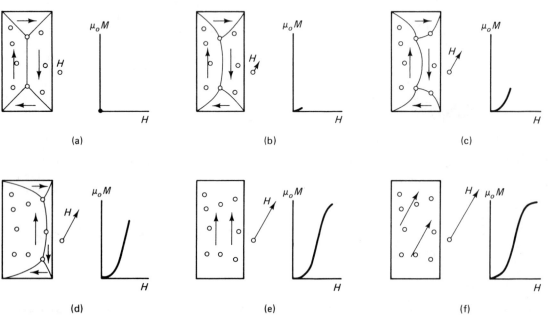

Figure 1.5 A simplified representation of the part played by domain boundaries in the process of magnetization

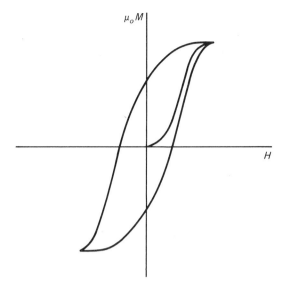

Figure 1.6 Hysteresis loop

amplitude, the initial magnetization curve will not be retraced. Due to the irreversible domain wall movements the magnetization will always lag behind the field and an open loop will be traced. This phenomenon is known as magnetic hysteresis and the loop is called a hysteresis loop, see Figure 1.6.

The ease with which the magnetization may be changed by a given magnetic field depends on the anisotropy, i.e. magnetic stiffness, whether the change is due to reversible or irreversible wall movements or rotations. A low anisotropy leads to a large induced magnetization for a given magnetic field and therefore to a large value of susceptibility and permeability (see Chapter 2).

1.3 Manufacture

1.3.1 Manufacturing processes

The processes used in ferrite manufacture on an industrial scale are similar to those used in the manufacture of other ceramics. The description of these processes given in this chapter is intended mainly for the information of the user, so that the possibilities and limitations of manufacturing may be taken into account when a particular ferrite core design or application is being considered.

Figure 1.7 shows a multipath flow diagram that represents the main processes in the commercial production of sintered (polycrystalline) ferrites. Another commercially important process, that must by definition be excluded from this diagram, is the preparation of single-crystal ferrites mainly for the manufacture of magnetic heads for video and other recorders. For this purpose, boules of single-crystal ferrite, usually MnZn ferrite and weighing up to 6 kg, are grown by the Bridgman method.[39-41] The individual heads are fabricated from the boule by cutting. Recording heads are also made from high-density sintered ferrites.[42] As recording heads are outside the scope of this book the single crystal process will not be pursued further.

Returning to Figure 1.7, the route through this flow diagram chosen by a manufacturer depends on a number of factors, such as the type of product to be made, the type of equipment installed in the plant and on the characteristics and price of the available raw materials. In the following sections the various process stages will be described, but first a few general remarks can be made.

The effects of the process variables on the magnetic and electrical properties of the finished ferrite pieces have always been a subject of great importance.[43-50] The factors that have most influence on these properties are the purity of the constituent oxides, their proportions and homogeneity in the powder mix and the control of temperature and atmosphere during sintering. The effect of process variables will be considered in a little more detail under the appropriate section headings.

Another important factor is process economics. Innovative formulations or process stages, that in research and development can produce ferrite of superior magnetic performance, may prove to be uneconomic in production.[51] Usually the electronic equipment manufacturer will accept improved component quality only if the premium to be paid is more than offset by the reduction of the cost of the end product. This generally means that the development of improved ferrite properties involving more expensive materials or processes must be accompanied by economies elsewhere in production. The trend is towards better control and less expensive processing by the use of more automation and computer-aided manufacture.[52]

1.3.2 Raw materials

In current industrial ferrite manufacture, there are two, essentially different, starting processes. In the conventional and most widely used method, the raw materials are normally oxides or carbonates of the constituent metals and are delivered to the factory as powders. This is the l.h. start in Figure 1.7. The other technique, that has been pioneered over many years and is now in use by some manufacturers, is the so-called wet method.

In the conventional process, the important characteristics of the raw materials are the chemical analysis, the shape and the size distribution of the particles, the

8 Ferrites: their nature, preparation and processing

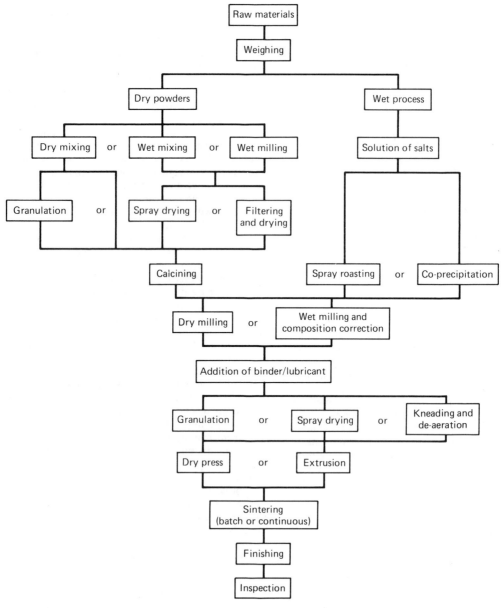

Figure 1.7 Flow diagram for typical processes of ferrite manufacture

consistency of supply and the cost. Because minor constituents or impurities have a great influence on the properties of the finished ferrite it is essential to analyse the incoming materials in an attempt either to ensure that the composition does not vary significantly from batch to batch or that the variations can be compensated. Particle size of the powders has a profound effect on the behaviour of the product during manufacture.[53] The ease of mixing, the compressibility, and the shrinkage and reactivity during sintering all depend on the particle size so it is important to keep a check on this parameter to ensure uniformity between batches. Refined materials of controlled uniformity are usually expensive, and successful large-scale manufacture depends on the skill with which lower priced materials may be used to produce, consistently, ferrites having the required magnetic properties.

The iron oxide is particularly important, and increasingly the ferrite industry is using a by-product of the steel industry, the so-called Ruthner[54] spray roasted ferric oxide. It originates from the regeneration of the steel industry's spent hydrochloric acid pickle liquors. It is possible to control this process, in certain steel production units, to produce iron oxide with good chemical analysis and physical properties and at a competitive price. This makes it particularly suitable as a raw material for ferrite production, especially for products made in very large quantities.

The constituent raw materials are weighed into batches to give the proportions required to characterize the type of ferrite being made. In mixed ferrites there are three or more different oxides or carbonates. Ignoring the influence of the minor constituents, the relative proportions of the principal metal ions and the oxygen ions have a dominant effect on the resultant magnetic properties, so accuracy is important.

In the wet process[54-59] the metals, in the correct proportions, are dissolved into an aqueous solution with sulphuric acid and are then co-precipitated by combination with an alkaline solution, e.g. a solution of ammonium hydroxide which may also contain ammonium bicarbonate. The resulting precipitate, generally the metal hydroxides or oxides, can, with oxygenation, be converted to spinel ferrite having a particle size of the order of 1 µm. This is filtered and dried. Alternatively, in the Ruthner process,[54,59,60] manganese metal or chloride is added to steel works pickle liquor (hydrochloric acid containing ferrous chloride). The resultant slurry is spray roasted to produce manganese iron oxide powder and then finely milled zinc oxide is added. These methods provide powder with fine particles and very good homogeneity.

Figures 3.3.1 and 3.3.2 give an indication of the effect on the saturation flux density of changes in the ratio of zinc to the other non-ferrous metal. Figures 3.8.1 and 3.8.2 give the corresponding effects on the permeability–temperature relation. Zinc ferrite is paramagnetic at room temperatures; it is seen that an increasing proportion of zinc has the effect of reducing the Curie temperature and, except at the higher temperatures, it gives rise to a maximum in the saturation flux density.

If in the final sintered ferrite there is an excess of iron over that required by the stoichiometric composition $MeFe_2O_4$, due to its excess in the starting powder or by appropriate control of the partial pressure of oxygen during sintering, then the excess iron appears in the form of divalent ions. Then Me, representing the divalent ions, includes some ferrous ions. The effects of this ferrous iron are rather fundamental to the attainment of the required properties of MnZn ferrite. The Fe^{2+} ion contributes a strong positive anisotropy which can at a given temperature compensate the small negative anisotropy of the stoichiometric formulation.[61,62] The compensation temperature depends on the divalent iron content. The divalent iron results in a higher permeability and lower loss at the compensation temperature (e.g. see Figures 3.8.4 and 3.13). The higher permeability often appears as a somewhat suppressed secondary peak in the permeability–temperature curve and can be used to adjust the temperature coefficient of permeability over the working range. On the debit side, the divalent iron provides conduction electrons and this lowers the resistivity; for this reason the amount of iron is not normally allowed to exceed about 10% of the amount needed to supply the trivalent iron.

In addition to the main constituents, small amounts (a few mol%) of cationic substitutions or additions are often used in ferrites, particularly MnZn ferrites, to improve the magnetic and electric properties.[63,64] There are three categories.

First, there are additions to promote sintering by introducing a liquid phase; borates and alkali fluorides are examples. In the second category are additions that appear at the grain boundaries as a second phase having very high resistivity. The effect is to increase the bulk resistivity of the ferrite and so reduce eddy currents.[65-67] For this purpose the usual additions are SiO_2 and CaO. Because these grain boundary phases are non-magnetic they reduce the permeability of the ferrite so the amounts added are a compromise depending on the required properties. They also have an important influence on the grain growth.[68]

The third category are cations that are soluble in the host lattice and enter regular positions on the tetrahedral or octahedral sites. They have a fundamental influence on the intrinsic magnetic properties such as magnetization, anisotropy and the stability of the properties with time. In principle most of the cations having ionic radii between 0.05 and 0.1 nm and charges between 1^+ and 4^+ are soluble in the MnZn ferrite lattice. In practice, the most successful have been substitutions of titanium or tin.[63,64,69-73]

The amount of these additional constituents to be added into the powder mix depends, of course, on the amounts already present as impurities in the raw materials and so must be adjusted from batch to batch depending their chemical analysis.

1.3.3 Mixing

It will be clear from the foregoing section that ferrite powder prepared by the wet chemical process does not require a mixing operation.

In the conventional process, i.e. the l.h. start in Figure 1.7, a mixing operation is necessary to combine the starting materials into a thoroughly homogeneous mixture.[74] If crystallites of uniform composition and properties are to be formed at the sintering stage then the constituents must be present in the correct proportions in any microscopic sample of the mixed powder.[75]

The mixing may be carried out dry, or the constituents may have water added to form a slurry which may be stirred with a turbine blade. Another method is wet ball-milling. The constituents are placed in a rotating steel-lined drum with steel balls and a medium such as water. Steel is used because any iron picked up by the mixture due to wear of the lining and balls may be allowed for in the initial composition of the powder. The fluid is mainly for cooling and mixing purposes. Wet milling usually continues for periods up to about twelve hours. For most of the raw materials used, particle size reduction is not a primary aim of this process at this stage.

In the mixing process it is important to avoid demixing, i.e. one or more of the constituents becoming slightly more concentrated in certain regions of the mass, by virtue of the particle size or density. This may be due to agitation, either during mixing or transport, or in the wet process, sedimentation or selective filtration.

In the earlier manufacturing processes, the powder, after wet mixing, was poured off as a slurry into a filter press where the water was squeezed out. The resulting masses were dried in an oven and then pressed or broken into blocks ready for calcining.

The most usual current practice is to calcine in rotary kilns, and for this the material is usually granulated. Wet milled material is prepared in the form of a slurry suitable to be processed in a spray drier, which as the name implies, breaks the slurry down into airborne droplets and dries them rapidly into granules. This process is described in a little more detail in Section 1.3.5. In some plants the slurry is pumped directly into the rotary kiln. Dry mixed powders are roll-compacted or pelletized by rolling through a perforated grating.

1.3.4 Calcining

Calcining, also called pre-firing or pre-sintering, is a process in which the temperature of the powder is raised to the region of 1000°C. Calcining is not essential to the preparation of ferrites, indeed in the wet process, which produces spinel ferrite powder, it is not required. However, in the dry process it provides an important means of obtaining the necessary degree of control over the properties of the finished product. The main functions attributed to calcining[76] are:

(1) It decomposes the carbonates or higher oxides, thereby reducing the evolution of gases in the final sintering.
(2) It assists in homogenizing the material.
(3) It reduces the effects of variations in the raw materials.
(4) It reduces or controls the shrinkage occurring during the final sintering.

The equipment most commonly used for calcining in large-scale manufacture is the rotary kiln. This is a cylindrical chamber the axis of which is inclined at about 5° to the horizontal. The powder enters at the higher end of the chamber which, by virtue of its slow rotation, conveys the powder along its length by a tumbling action. The chamber is heated, e.g. by gas, and maintained at a temperature of typically 1000°C. The older method was to use a tunnel kiln through which the charge of compacted powder in the form of blocks or pellets was conveyed on trolleys or similar carriers. Whatever the method, calcining is almost invariably carried out in an air atmosphere.

During the calcining process the constituents combine by solid-state reaction to form, partially or completely, spinel ferrite. The extent to which spinel is formed depends on the reactivity of the constituents and the oxidation during the cooling part of the cycle.

1.3.5 Processing of the calcined powder

The calcined powder is milled to reduce it to small, uniformly sized particles (crystallites). The particle size depends on the application, e.g. typical mean particle sizes are 1.7 μm for TV deflection yokes and 1.1 μm for inductor and transformer cores. The process is carried out in a wet ball mill, a vibratory mill or an attritor (in which the powder and a charge of steel balls are stirred in water in a stationary vessel). As previously stated, steel balls are used because allowance can be made for the inevitable wear by adjusting the amount of iron in the starting composition; balls of other materials might cause contamination.

Milling times of up to 12 h are commonly used. After an initial period of rapid breakdown the particle size decreases in proportion to the milling time and then approaches a limiting value. The extent of the milling affects the forming characteristics of the powder, the sintered density and the magnetic properties such as permeability and losses.[77] Since the extent of the milling depends on milling efficiency as well as the milling time it is preferable to mill to a certain particle size rather than a certain time. Particle size may be controlled by measuring the aggregate surface area of the particles in a sample of the powder or by using a Fisher Sub-Sieve analyzer.

After milling, water content is adjusted to the amount required in the subsequent process. In the case of wet milling, this may involve decanting or filtering off the surplus water; if dry milling was used then water may need to be added.

At this stage, the appropriate binder and lubricants are added. The choice of these additives depends on the subsequent granulation process, the method of forming (pressing or extrusion), the required strength of the formed piece-part before firing, and the avoidance of undesirable residues after burning-out during sintering. Commonly used binders are gum arabic, ammonium alginate, acrylates, polyvinyl alcohol, while waxes and wax emulsions, zinc or ammonium stearate may be used as lubricants to improve powder flow during forming. The quantities involved are quite small; using water as a vehicle these additives are blended with the powder to give the right consistency. Sometimes the lubricant is added later, when the powder has been granulated.

For the extrusion process the consistency must be that of a stiff dough. This is obtained by kneading. The batch is then divided into separate charges of the right size and shape for the cylinders of the extrusion machines. They are de-aerated to remove voids which would be detrimental to the extrusion process and to prevent excessive porosity of the sintered extrusion.

For dry pressing the powder must be in a form in which it will flow readily and uniformly into the die. This may be achieved by granulation. The most common method of granulation used in large-scale production is spray drying. This uses a slurry having a high solid content of ferrite powder together with the appropriate binder and lubricants. In some plants this may be pumped directly from the milling equipment. The spray drier consists of a vertical cylindrical vessel into which the slurry is sprayed and atomized into droplets. The droplets travel through currents of hot air before dropping into the collecting zone as dry granules.

In other production processes, the granulation may be carried out by forming the powder into tablets and then breaking these down by sieving or tumbling.

The optimum size of granule is dictated mainly by the die in which the part is to be pressed. If the granules are too small or contain too many fines, the powder will work its way in between the sliding faces of the tool and cause excessive wear and friction. If they are too large they may not adequately fill the die. Good flow of granules is important; this is improved if very fine particles can be eliminated from the mass. To ensure a dense compact, the granules must readily deform under the pressure in the die.

1.3.6 Forming

For the ferrites that are the subject of this book, the usual forming method is dry-pressing or extrusion. Normally the main purpose of this operation is, of course, to form the powder into a shape which is as near as possible to the final shape required. The dimensions of the forming tools must be larger than the final part dimensions by a factor which allows for shrinkage during sintering. The product when sintered is very hard, and shape modification at that stage is usually expensive. Another, equally important, purpose of forming is to force the particles into close proximity so that during sintering, they may densify and/or grow into a low porosity polycrystalline ceramic. There are occasions when the compacting is more important than the shaping, e.g. in the isostatic technique described among the miscellaneous methods at the end of this section, the powder is pressed into a simple shape of high density from which the required final shape is cut after sintering, the quality of the product justifying the cost of the technique.

1.3.6.1 Dry pressing

In its simplest form this process consists of pouring the correct quantity of granulated powder into the die and then closing the die with a prescribed pressure, which is usually in the range 1.6 to 16 kg mm^{-2} (1 to 10 ton in^{-2}). In practice the flow properties of the powder must be taken into account and therein lies the art of dry pressing. As the powder is compressed it builds up friction between its outer surface and the adjacent die walls. Due to this friction and the viscosity of the powder, the pressure (and therefore the density) in the pressed powder is not uniform.[78-82] Figure 1.8(a) shows the simple pressing of a cylinder and indicates by shading the higher pressure zones caused by friction. If the length l were long enough, the friction alone would be enough to balance the applied force and the powder at the bottom of the die would not be compressed at all. This difficulty may be somewhat alleviated by compressing the powder from both ends as in Figure 1.8(b). In practice it is usual for the die to move in a vertical line during pressing, the movements of die and punches being so related that the most uniform pressed density is obtained. Even so, shapes having l/d ratios of greater than about 5 are difficult to press successfully.

The design, construction and finish of the die is of great importance. In most cases, dies used in mass production are made from tungsten carbide using spark erosion machining; the internal surfaces are finished to a high polish and in use these surfaces may periodically be given a very thin coating of lubricant.

Inhomogeneous pressed density leads to inhomogeneous magnetic properties and so degrades the product. It also leads to non-uniform shrinkage during the sintering and this causes shape distortion as shown in exaggerated form in Figure 1.8(c). In more complicated shapes, non-uniform pressed density may result from some zones being compressed more than others, see Figure 1.8(d) and (e). The simpler the shape the more successful is the pressing and the

12 Ferrites: their nature, preparation and processing

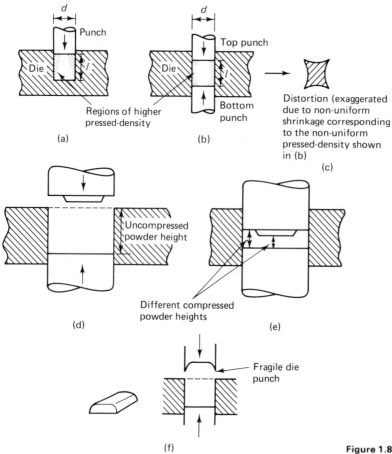

Figure 1.8 Some aspects of pressing techniques

better is the control of the electrical and mechanical parameters. On the other hand, as the understanding of the flow properties of the powder improves and pressing technology advances, more complicated shapes may be successfully pressed. However, some shapes will always be inherently difficult to press and control. Such shapes are those that include:

(1) zones of widely unequal powder compression
(2) zones having large aspect ratios (l/d) in the direction of pressing
(3) abrupt change of cross-section
(4) features that lead to fragile die punches (see Figure 1.8(f))

Shapes having cavities extending perpendicular to the line of pressing cannot normally be pressed as they would require split dies. Such dies are rare in ferrite pressing technology.

The pressed parts have the consistency of drawing chalk. They are said to be in the green state. As the parts leave the press a proportion of them are weighed to check the density and they are then stacked on trays to await sintering.

1.3.6.2 Extrusion

The extrusion process is used mainly for rods and tubes. Such shapes cannot easily be dry-pressed, because if the pressing were axial the length/diameter ratio would be too great whereas if they were pressed transversely the die punches would have feather edges.

The extrusion process is very simple. The de-aerated charge of plasticized powder is placed in the cylinder of an extrusion machine. A ram then forces the charge through a suitable orifice to form a length of rod or tube. When a convenient length, e.g. about 0.7 m, has emerged, the extrusion is stopped and the length is detached from the machine to be laid on a tray. The tray has a number of parallel grooves in which the still-plastic extrusions may rest and be held straight while they dry. When the extrusions are dry and hard the ends, which have been deformed by handling, are removed and the remainder may be cut into pieces of the required length (allowing for shrinkage during sintering). Where a large production is required it is possible for the extrusion, cutting and drying to be done automatically and this leads to a more consistent product. Sometimes, particularly for shorter lengths or where greater length accuracy is necessary, the final cutting is done after sintering.

1.3.6.3 Miscellaneous forming methods

Isostatic pressing The inhomogeneity arising from the friction and flow of granulated or plasticized powders may be avoided by isostatic pressing. In this method the finely milled powder (particle size 0.1–0.3 µm) is placed, either loose or after being compacted approximately to the required shape, into a latex envelope itself approximating to the required shape, e.g. a cylindrical compact may be formed and placed into a cylindrical sleeve. The air is evacuated and the sleeve is sealed. It is placed in a hydraulic pressure vessel and the pressure is raised to about 1000 atmospheres. This pressure, acting on all sides of the envelope, compresses the powder evenly and with the minimum interference from friction.

This process may contribute to the preparation of high density ferrites, e.g. having porosities of 0.1% or less but the shapes obtainable from isostatic pressing are relatively simple and this normally leads to cutting and shaping. The cutting and shaping may either be carried out before sintering or, if greater final accuracy is required, after sintering. For items such as ferrite recording heads for industrial or professional equipment, high density is a prerequisite to good wear characteristics, and production cost is less important than good performance. Accordingly the ferrite parts for these heads are often cut from blocks that have been isostatically pressed and sintered to a high density.

Hot pressing[42] This is a process in which the pressing and sintering occur simultaneously. The die is made of a refractory material heated to the sintering temperature by a surrounding electric furnace and pressure is applied by die punches which do not enter the sintering zone. This process, which is virtually confined to cylindrical shapes, enables dense ferrites to be obtained at relatively low sintering temperatures. It is an important means of making ferrite for recording heads.

Hot isostatic pressing In this process, sintered pieces are reheated in a pressurized argon atmosphere. It has been used for high density products such as recording heads.

Injection moulding This method is similar to that used for moulding plastics. A dough of ferrite powder and thermosetting binder is transferred under pressure into the mould which, after a suitable interval to allow the binder to set, is opened to release the moulded piece-part. The binder is burnt-off in the early stages of sintering in the usual way. The process is particularly suitable for the mass-production of small parts having complex shapes since some of the constraints mentioned in connection with Figure 1.8 do not apply.

Slip casting This process is widely used in the pottery industry for making cup shaped items and it has been adapted by some ferrite manufacturers for making TV deflection yokes. A two-part mould is made of some porous material such as gypsum and a slurry of ferrite powder is poured in. The water is rapidly removed through the porous mould which is then opened to release the moulded component.

Slurry pressing This technique is used when it is necessary to induce orientation of the particles during forming. The powder is made into a slurry and when the required amount has been admitted to the die a strong magnetic field is applied so that the orientation is obtained. With the field maintained, the die closes and the fluid is expelled by the pressure through filters at the bottom of the die. This process is mainly used to make anisotropic hexaferrite permanent magnets.

1.3.7 Sintering

When the powder has been compacted into the required shape, the partially or fully reacted particles press against each other over part of their surface, the remaining surface forming the walls of the interstitial voids or pores. At temperatures in the region of 1000°C and above, crystal growth proceeds where the particles are in contact, the free surface containing the voids decreases and the particles grow together to form crystallite grains with an accompanying rise in density. If the process is not too rapid, the residual pores and any non-soluble constituents are carried to the grain boundaries. If grain growth proceeds too rapidly the pores and impurities are trapped within the grain where they can provide pinning centres for the domain walls, lowering the permeability and increasing the losses. The rate of grain growth and the resultant micro-structure depend in a complex way on many factors, such as the sintering temperature, the oxygen partial pressure, stoichiometry, the amount and nature of the non-soluble constituents and the presence or otherwise of a liquid phase.[83–94] Typical grain size is between 5 and 40 µm.

During this sintering process the linear dimensions

of the piece part shrink between 10 and 25% depending on the powder and the pressing technique and this must be allowed for in the design of the forming tool or mould.

The ferrite properties are, of course, also strongly dependent on the chemistry occurring within the grain during sintering and many of the references cited above also consider this aspect.

A mixed ferrite such as Mn-Zn-Fe ferrite can assume a wide variety of compositions within the limitations of the spinel structure. Even when the proportions of the metal oxides have been fixed during the powder preparation, the valency states will depend on the amount of oxygen in the structure and this will in turn depend on the oxygen equilibrium between the ferrite and the surrounding atmosphere during sintering and the subsequent cooling.[95-97]

If, for instance, during the cooling period the partial pressure of oxygen in the kiln is very low, any excess iron in the composition would be reduced to ferrous iron and would take its place with the other divalent ions, e.g. Mn and Zn. As explained in Section 1.3.2, a small proportion of ferrous ions provides compensation of the magnetocrystalline anisotropy and results in increased permeability and lower losses near the compensation temperature. However, the co-existence of trivalent and divalent ions of the same element reduces the electrical resistivity, so the ferrous iron content must be carefully controlled. On the other hand, excessive oxygen could prevent the formation of ferrous iron; in addition to higher losses and resistivity this would result in the formation of cation vacancies since the ratio of oxygen ions (anions) to metal ions (cations) would be excessive. The presence of vacancies is to be avoided as far as possible because they give rise to after-effects such as change of permeability with time, known as disaccommodation (see Figure 3.10).

It is beyond the scope of this chapter to discuss further the chemistry and physics of ferrites as they are affected by the sintering process; the subject is well covered in the references. The above brief excursion into this field is intended to illustrate that the control of atmosphere during sintering has a direct bearing on the composition of the ferrite and therefore on its properties. The control is complicated because equilibrium oxygen partial pressure varies with temperature and at any one temperature the optimum with respect to iron may not be the optimum with respect to the other metals, e.g. manganese.[95] In practice a compromise has to be found empirically. Another difficulty is that, under normal production conditions, uniform equilibrium may not be attainable because, as the sintering proceeds and the porosity decreases, the inner regions of the material may become less susceptible to the kiln atmosphere and inhomogeneity may result.

Figure 1.9 shows a typical sintering cycle for ferrite production. During the heating period the oxygen partial pressure is usually quite high to promote the burn-out of the binder. As the maximum temperature is approached the oxygen pressure is reduced rather abruptly and continuously adjusted to maintain the required equilibrium as the temperature falls. On the success of this operation depends the properties of a high quality ferrite. However, for some ferrites where the properties are not critical this atmosphere control is not necessary and it is possible to carry out the whole sintering process in air.

There are many types of kiln that may be used in the manufacture of ferrite. They may be classified according to the method of heating and to whether the charge remains stationary (static kiln) or is passed through the kiln (continuous kiln).

Originally ferrites were fired in rather simple electrically-heated box kilns, in which the temperature and atmosphere programme was set and controlled by mechanical means. They had three main disadvantages; limited capacity, difficulty in maintaining uniform conditions, particularly temperature, throughout the working volume, and difficulty in controlling the process because there was no output to monitor until the sintering process had been completed.

As the demand for greater production capacity and better control grew, large continuous kilns were installed and the box kilns were relegated to small scale production and development. Later, as described below, they were to reappear, their shortcomings having been overcome by the application of new technology.

The continuous kiln is in the form of a long tunnel through which the ferrite parts travel. The temperature and atmosphere vary along the length of the kiln in accordance with the required programme, the spatial distribution being typically as in Figure 1.9 except that the time scale is replaced by one of distance along the kiln. Control is maintained by suitable temperature sensors and gas analyzers stationed at intervals along the length and these provide the necessary data for operation of controls that hold

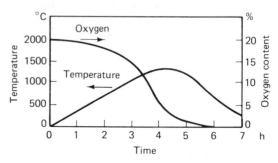

Figure 1.9 Typical sintering cycle

the conditions at any position constant. The continuous output of ferrite cores is sampled for quality measurements to provide additional process control.

Continuous kilns fired by town gas are extensively used where a large throughput is required and where sintering in air is permissible, e.g. where the finished product is not required to have high performance. Such kilns might be up to 43 m (140 ft) long and produce up to 250 kg ($\frac{1}{4}$ ton) of sintered parts per hour. The parts are stacked on refractory kiln furniture that is carried through the kiln on trolleys.

The majority of higher quality ferrite cores have been sintered in electrically-heated continuous kilns. These are usually of smaller cross-section than the gas-fired kilns but are of similar length. Figure 1.10 shows in diagrammatic form the side elevation and alternative cross-sections of such a kiln. The charge is stacked on refractory pusher plates which are pushed through the kiln in procession by means of a hydraulic ram. After the ram has completed a stroke equal to the length of the pusher plate it returns and another loaded plate is introduced. The pusher plates either slide along a track formed of tiles (a) or along grooves formed in the lower parts of the tunnel walls (b). Care is necessary to avoid the plates being obstructed and causing a catastrophic hold-up or pile-up inside the kiln.

The kiln is heated by elements, usually of wire such as Kanthal or of silicon carbide rods. They may be located either in the side walls of the tunnel as in section (a) of Figure 1.10 or along the roof and floor as in section (b). The temperature may be kept more uniform if the aperture of the kiln is shallow; for this reason the roof and floor offer better accommodation for the elements.

As described earlier, the temperature control in the various zones of the kiln uses thermocouples placed in the walls. Electrically screened against leakage currents from the heating elements, they operate potentiometric recorders and the means of controlling the current to the elements. Atmosphere control is normally required to produce a pre-programmed transition from air or oxygen in the rising-temperature zones to a low and controlled oxygen partial pressure at or about the zone of maximum temperature. This may be obtained by feeding air in at the entrance of the kiln and nitrogen in at the exit. Inlets for air or nitrogen are also provided at intermediate points. The flow of gas past the charge is impeded by baffles. The nitrogen and air pass along the kiln towards the maximum temperature region where they are allowed to leave the kiln through vents. An air-lock is necessary at the exit to prevent air leaking into the cooling zone. Again, instruments are used to measure and record the oxygen pressure at a number of points in the kiln and by means of these instruments the oxygen profile is automatically controlled.

As ferrite applications developed, particularly in the field of high-quality professional components, two trends became apparent. The size of the parts decreased and the degree of kiln control required to achieve the necessary quality increased to a level difficult to attain in a continuous kiln. This led to the reintroduction of the static box kiln by some manufacturers. The modern box kiln is a very sophisticated equipment in which advanced technology has been used to overcome the difficulties presented by the early versions.

The charge, stacked on the kiln furniture in the usual way, is placed on the rectangular base of the kiln and the cover (sides and top assembly) is lowered onto the base (or the base is raised into the cover assembly) to complete the enclosure. The joint is hermetically sealed by a water-cooled gasket. The load volume is between one and five cubic metres and the firing cycle can be completed in about 24 hours or, in some installations, as little as 12 hours. The complete temperature and gas cycle is programmed and controlled by computer. The temperature control allows for the gradual burn-out of the organic additives, a $\pm 2°C$ total temperature spread throughout the load at maximum temperature (in the larger installations this figure may increase to $\pm 10°C$) and provides for programmed cooling, including rapid cooling if required. The atmosphere can be varied smoothly from 100% oxygen to 0.001% oxygen in nitrogen, either predetermined by time throughout the cycle or by the load temperature. The accuracy of the oxygen partial pressure can be better than $\pm 5\%$

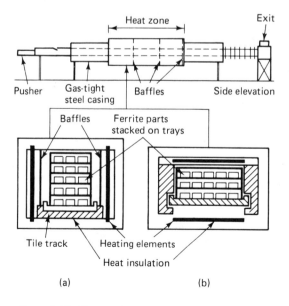

Figure 1.10 Diagram of a pusher type kiln; (a) and (b) show alternative cross-sections of heat zone

16 Ferrites: their nature, preparation and processing

of the nominal set point value. Figure 1.11 is a photograph of such a computer-controlled box kiln. The trend is for the installation of a battery of kilns of this type within a production unit to provide for the sintering of all high-quality professional ferrite cores.

1.3.8 Finishing

When the sintered part leaves the kiln it is 10–25% smaller in linear dimensions than its pressed size. If the pressed density and the firing cycle have been correctly controlled it is normally possible to hold the dimensions of the sintered part within a tolerance of $\pm 1\%$ to $\pm 2\%$ for pressed parts (depending on their geometry) and $\pm 3\%$ for extruded parts. As explained in the section on pressing, non-uniform pressed density can lead to distortion during sintering, and where the geometric form of the part (e.g. straightness, roundness) is important it may be necessary to specify geometric tolerances or form gauges.

If the specified tolerances of dimensions or form are smaller than those which may easily be achieved by sintering alone then some form of finishing process will be required. Such a process usually consists of grinding or cutting. As the sintered product is hard and abrasive, the grinding and cutting will significantly increase the production cost. For this reason it is usual to avoid, where possible, tolerances smaller than those obtained by sintering.

The most usual, and often unavoidable, finishing process is the grinding of surfaces which are required to give low-reluctance butt-joints in the assembled magnetic circuit. Figure 1.12 shows typical mating parts. The surfaces $A-A$ must be made flat and smooth so that the residual gap obtained on assembly is as small and as stable as possible. Surface grinding

Figure 1.11 A production computer-controlled box kiln. (Courtesy Philips Components, and Ferroxcube Division of Amperex Electronic Corp, USA)

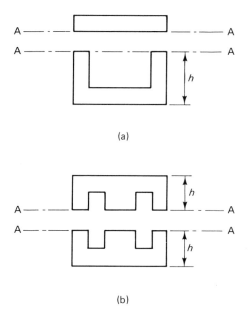

Figure 1.12 Typical mating parts

may also be required to control a dimension, such as h, to close tolerances.

As far as mating surfaces are concerned the degree of flatness and smoothness obtainable depends to a large extent on the acceptable cost of the operation. In principle a reasonable surface may be obtained by the use of a normal surface grinder and this is often used when only a few parts are involved. In mass production it is usual to use a vertical spindle batch grinder or a continuous grinder. In the case of the batch grinder, the work, perhaps a thousand pieces or more, is attached magnetically to a large circular surface plate which rotates slowly in a horizontal plane. Above this chuck plate is a horizontal spinning grinding plate which carries an abrasive annulus of silicon carbide (green) grit or in some cases it could be diamond plated or diamond impregnated. The work is flooded with coolant, usually water. The vertical clearance between the lower plate and the abrasive surface is gradually reduced until the ferrite parts have been ground to the correct height. It is usually necessary to grind the back faces of the work first to establish a stable base. In the case of a continuous grinder a similar horizontal abrasive plate is used; it is usually diamond plated or impregnated. The cutting regions are arranged as a coarse outer annulus and a fine inner annulus. The vertical clearance to the inner annulus is set to the required finished height of the work while the clearance to the coarser outer annulus is greater and increases towards the outer radius. The work is carried through the grinder on a continuous belt and is ground to size in a single pass.

Since the grinding operation represents a significant production cost the physics of material removal by grinding has been the subject of special study.[98,99] It is observed that grinding generally results in some plastic deformation in the ferrite surface and this causes stresses that affect the magnetic properties of the core. Knowles[100,101] has shown that grinding can cause the surface to be left in a state of compressive stress, the value of which may vary from the ultimate compressive strength of the ferrite at the extreme surface to zero at a depth of about 5 µm. In this stressed surface the permeability is very low and the losses are very high; this gives rise to an effective gap of the order of 5 µm at mating surfaces even if there is no physical air gap. A surface compressive stress causes a significant tensile stress in the bulk of the ferrite core and such a stress affects the magnetic properties by disturbing the anisotropy balance.

Notably, the temperature dependence of the permeability is strongly influenced in the region of the compensation temperature. The stresses can be relieved by lapping, using a suitably mild abrasive such as a slurry of colloidal silica, or by acid etching.

When it is required to obtain the smallest possible effective gap in production cores, e.g. high permeability transformer cores, lapping is used. This is a relatively expensive operation; initially it was carried out in small batches on conventional lapping machines but more recently the technique has been adapted to mass production.

Where it is necessary to remove sharp edges, e.g. the edges of a ring core on which a conductor is to be wound, this is done by tumbling the cores in a barrel tumbler together with a suitable abrasive aggregate. Another widely used finishing process is the cutting to length of extruded rods or tubes. This is done with diamond impregnated metal-bonded saws.

Some modern colour television deflection yokes are required to have a precisely contoured internal flare and they are assembled in one piece with saddle coils having precise geometry. This arrangement eliminates many of the subsequent adjustments that are otherwise necessary during picture tube alignment. In this case the internal flare is ground to the required profile on automatic machines using diamond wheels.

The more conventional deflection yoke does not need profile grinding but it is generally required to be divided into two halves along a diametral plane to enable the windings to be assembled. This is done during manufacture; the yoke is cracked along a diametral plane by the heat of two fish tail gas burners applied along the required dividing lines. Sometimes the required dividing lines are defined by notches or grooves formed during pressing. Due to the somewhat random nature of the cracks, the cracked parts must remain paired so that when incorporated in the deflection assembly they may be restored to an almost perfect magnetic circuit.

1.3.9 Inspection

The mechanical dimensions are checked by conventional means, e.g. gauges. The magnetic properties pose greater problems. For a typical ferrite core it is necessary to control about seven magnetic parameters and some of these present appreciable measuring difficulty, e.g. the measurement of very low magnetic losses, temperature coefficients of permeability, etc. Some measurements cannot easily be done on the production line and usually require the specialized resources of a quality control laboratory. Although such facilities, used on a sampling basis, give adequate inspection of the manufactured product, the inevitable time delay makes the effective feedback of process control information rather difficult. This latter aspect of inspection is probably the more important because it can ensure that the process will be maintained in a high-yield condition. In practice adequate inspection and control is attained by a combination of on-line and laboratory testing backed up by the accumulated inspection records from which control relations may be derived.

Most of the widely-used ranges of inductor and transformer cores are now inspected in accordance with Assessed Quality schemes in which a nationally or internationally agreed schedule of tests and associated acceptable quality levels are specified. The manufacturer must have qualification or capability approval and the tests are carried out by an approved test laboratory, usually that of the manufacturer, the whole operation being subject to the surveillance and authority of a National Supervising Inspectorate. The ferrite components are released with a certificate of conformance that declares that they have been tested and proved acceptable in accordance with the assessed quality specification, thus facilitating interchange of equivalent cores made by different manufacturers operating under the scheme. Test records are retained on all batches to provide cumulative evidence of the manufacturer's quality performance. The system was developed in the United Kingdom under the direction of the British Standards Institution, where it is known as the BS9000 System. It has subsequently been developed and adopted by the European Cenelec Electronic Components Committee and more recently by the International Electrotechnical Commission as the IECQ System. The three levels of the system are now harmonized.[102]

1.4 Finishing processes available to the user

Usually a ferrite part leaves the factory in its final form and the user has only to combine it with the required winding and hardware to complete the wound assembly. There are, however, instances where the user must carry out some finishing process on the ferrite itself: (a) in the development of a prototype design it may be necessary to carry out machining operations, (b) in the production of inductors or transformers the design often requires core halves to be cemented together and (c) the encapsulation of the complete wound assembly is sometimes necessary to provide adequate environmental protection of the winding. In many cases users have developed their own techniques; in this section some brief notes on these topics will be given for the guidance of those encountering these problems for the first time.

1.4.1 Machining

Normal ceramics machining techniques and standard machine shop equipment may be used for most shaping operations on ferrites; see Section 3.3 for data on hardness, etc.

For surface (and cylindrical) grinding, 'green' silicon carbide wheels may be used. Roughing operations require a fairly hard grade of wheel with a coarse texture while for finishing a softer wheel with finer grains may be used. A peripheral speed of 1500 to 1800 m min^{-1} (5000 to 6000 ft min^{-1}) is suitable. The cut should be limited to about 0.025 mm (0.001 in) and the traverse should be rapid so that local heating is restricted. The work should be flooded with coolant; water or a dilute soluble oil solution is suitable. The wheel must be dressed true so that 'hammering' of the work due to eccentricity is avoided. The dressing must be repeated at frequent intervals because ferrite rapidly dulls the wheel.

Diamond machining is a much more efficient technique. Diamond plated or diamond impregnated metal-bonded tools may be used for surface grinding, drilling, reaming and sawing. Because the cutting is much more efficient, less heat is generated and larger cuts and slower traverse may be used. For surface grinding, wheels having diamond particles of 150 to 180 μm (100 to 80 mesh approx.) are normally used and the peripheral speeds should be similar to those quoted for silicon carbide wheels. Cuts of 0.6 mm (0.025 in) may be taken but it should be noted that diamond grinding can cause higher residual stresses than silicon carbide grinding especially when the cut is deep. For sawing, the size of the diamond particles may vary from 150 to 90 μm (100 to 170 mesh approx.) depending on the thickness of the wheel. As before, peripheral speeds of 1500 to 1800 m min^{-1} (5000 to 6000 ft min^{-1}) are recommended. The cut may be as deep as the available radius of the wheel; a 25 mm (1 in) cut is quite common. Again an ample quantity of coolant is essential; water or dilute soluble oil solution may be used.

If the magnetic properties of the core are critical then care must be taken to avoid excessive residual stress; see Section 1.3.8.

A valuable technique, which cannot be classed as normal machine shop routine, is ultrasonic drilling. A tool, having a cross-section identical to that of a required hole, is attached vertically to the end of a piezoelectric transducer which is resonated longitudinally at ultrasonic frequency, e.g. 20 kHz. The end of the tool is brought into contact with the ferrite and it delivers a rapid succession of minute blows. An abrasive slurry, such as 45 µm (325 mesh approx.) boron carbide in water, submerges the end of the tool and the result is that a downward pressure on the tool causes it to cut its way through the ferrite. In this way holes or patterns of holes of any cross-section may be drilled. Large holes may be trepanned by using hollow tools thus reducing the amount of ferrite cutting to a minimum. The tool is usually made of mild steel or brass and an ultrasonic power of 300 W is suitable for most purposes.

1.4.2 Cementing

Ferrite-ferrite bonds may be made with thermosetting resin adhesives, in particular the epoxy resins, but if the bond is to be strong and permanent a number of precautions must be observed. The main cause of the deterioration of bond strength between ferrite parts appears to be the ingress of water vapour between the ferrite and the resin. Some resins are more resistant to this deterioration than others and the porosity of the ferrite is probably also a factor. The test of a good bond is its strength after it has been subjected to a number of damp heat cycles. Often a wound component, e.g. an inductor, must be protected from dampness in order to prevent corrosion or degradation of the winding during its service life. If such protection includes the ferrite core, the deterioration of joint strength due to dampness ceases to be a problem. Before subjection to dampness, a good bond has a tensile strength that is greater than that of the ferrite. Not all the factors contributing to a good bond are understood, but experience has shown that the following points are significant. See also References 103–105.

1.4.2.1 Choice of adhesive

The available range is very large. The most important factors influencing the choice are the required curing temperature and the viscosity. The economic curing temperature must not be above the maximum temperature to which the assembly may be safely raised. If the assembly consists only of ferrite and other inorganic parts it will not usually impose a limit to the curing temperature. To bond the parts of such an assembly a hot setting resin may be used. Such resins cure quickly at about 180°C to give very strong bonds. They are available in solid, powder or liquid form. As solids or powders they present some difficulties in application on a production scale. The liquid form usually consists of two components, the resin and the hardener, and these are mixed together by the user. Usually one or both components contain a solvent and in that case the manufacturer generally recommends that as much of the solvent as possible be driven off by warming the mixture in free air conditions. This may be done after the adhesive has been applied but before the mating surfaces have been brought together, or alternatively the process may be carried out on the adhesive in bulk before application. While this appears to be essential for a good ferrite-metal bond experience has shown that solvent removal is not necessary for ferrite-ferrite bonds.

If the assembly includes temperature limiting materials such as thermoplastic or other organic insulating parts then resins with lower curing temperatures, e.g. 100 to 120°C, will have to be used. Cold or warm setting resins are also available but they are usually inferior to the hot setting resins in bond durability and stability. Flexible adhesives have advantages under some circumstances, but they tend to creep under load and may have poor stability even when unstressed. If the bond is required to withstand adverse environmental conditions, particularly damp heat, the strength of sample joints should be determined experimentally after exposure to such conditions.

The other important factor is the viscosity. If it is too high application is difficult; if it is too low the resin may run out of a poorly-fitting joint or, in a well-fitting joint, it may be absorbed by the porosity of the ferrite. It is sometimes an advantage to incorporate a filler such as finely powdered chalk or silica. This will reduce the temperature coefficient of linear expansion, reduce the shrinkage occurring during curing and increase the viscosity. On the other hand it makes the exclusion of air or water vapour from the mixture more difficult and it is usually advisable to remove the water vapour from the filler before mixing.

The selected adhesive should be easy to prepare. It should have an adequate pot life and a simple and reasonably short curing cycle. Needless to say, the manufacturer's instructions for a particular resin must be meticulously followed. The constituents must not have exceeded the permitted shelf life and must not have been exposed to excessive atmospheric moisture before or after mixing. Quite often bond weakness may be traced to insufficient care in the preparation of the resin mixture.

1.4.2.2 Preparation of surfaces

The surfaces to be bonded must be perfectly dry and free from contamination. It is advisable to clean the surfaces before bonding. Degreasing in solvents is widely employed but it has been found that ultrasonic washing in hot water containing a small amount of detergent is the most effective process. After cleaning, the surfaces should not be touched by hand.

1.4.2.3 Application

For maximum bond strength, the resin manufacturers usually specify the thickness of the adhesive in the joint; figures between 0.025 and 0.25 mm (0.001 and 0.01 in) are often given. However, in a ferrite-ferrite bond it is usually necessary to make the joint as thin as possible in order to minimize the reluctance of the residual gap in the magnetic circuit. In practice adequate joint strengths are possible between flat ferrite surfaces from which all the surplus resin has been expelled by the application of pressure or by rubbing the surfaces together. It is obviously preferable to use resins that have an optimum joint thickness that is near the lower end of the above-quoted range. Joints prepared in this way have effective gaps in the magnetic circuit that are not significantly larger than those obtained when the joints are dry.

1.4.2.4 Curing

After application of the adhesive, the mating surfaces should be clamped lightly together to provide contact pressure, e.g. a pressure of 0.02 kg mm^{-2} (27 lb in^{-2}) has been found satisfactory. The manufacturer's curing instructions should then be followed. If a warm or cold setting resin has been used it is usually preferable to cure at a suitable elevated temperature (e.g. 60–80°C) as this will result in improved strength. However if the viscosity becomes very low at the curing temperature so that there is a risk of excessive absorption by the ferrite, it is recommended that partial curing be allowed at room temperature and that this is followed by a suitable heat cycle to complete the cure.

Bonds between ferrite and other materials are more difficult insofar as there is a difference in the temperature coefficients of linear expansion. This coefficient, α, for ferrites is about $10 \times 10^{-6} \text{°C}^{-1}$. Thus steel ($\alpha = 12 \times 10^{-6} \text{°C}^{-1}$) may be successfully bonded (although steel, having large magnetic losses, is not recommended for intimate connection to a ferrite core), while brass ($\alpha = 19 \times 10^{-6} \text{°C}^{-1}$) presents the difficulty of accommodating the differential expansion or contraction. In practice the interfaces are locked together as the resin cures at a temperature that is usually well above room temperature. On subsequent cooling a stress is created. The attachment is then prone to failure due to the resin or the ferrite fracturing under the stress. The best remedy is to alleviate the stress by reducing the shear strain in the adhesive. This may be done either by using a flexible resin or by making the adhesive layer as thick as possible, e.g. by including an open glass cloth scrim wetted with adhesive. As stated earlier, if the resin mixture contains solvent it is usually advisable to remove as much of it as possible by warming the mixture in free air conditions. This may be done after application but before the mating surfaces have been brought together or it may be done before the adhesive has been applied.

Apart from the additional aspects, the procedures described for the ferrite-to-ferrite bonds apply equally well to these dissimilar bonds.

1.4.3. Encapsulation

Encapsulation is a technique of protection whereby an electrical component or assembly is embedded in a casting of insulating material, e.g. synthetic resin. The component or assembly is suitably mounted in a mould and the encapsulant is poured in until it is covered. Usually all air and water vapour is removed by evacuation, thus ensuring that the finished casting is without voids. The encapsulant is then allowed to set or cure into a solid or elastic mass. The main protection is against mechanical shock and vibration, and against hostile atmospheric constituents such as water vapour, sulphur dioxide and salt. Ferrites, being inert to atmospheric conditions, do not normally require such protection, but in most wound assemblies it is difficult to provide the essential protection of the winding without including the core.

There are many encapsulants available, the main ones being epoxy resins and silicone compounds. To perform its task successfully an encapsulant should be tough, impervious to the atmosphere (particularly to water vapour) and capable of adhering to any parts of the assembly, e.g. terminal connections, which break the surface of the casting. The latter point is of importance because lack of adhesion could provide capillary paths by which water vapour may reach the inside of the assembly.

The encapsulation of a wound ferrite assembly presents special difficulties. It has been observed that except for mechanical protection the ferrite core does not need to be encapsulated. In practice the core must usually be included and it is the resultant stresses that give rise to the difficulties. The main factors are:

(1) Most resins shrink during the curing process and exert a compressive stress on the core. If the curing has been done at an elevated temperature the differential temperature coefficient will increase the compressive stress. It is shown in Chapter 3 that mechanical stress can appreciably change the initial permeability of a ferrite, so the process of encapsulation may cause a change of inductance. This difficulty may be overcome by using a non-rigid resin, e.g. a silicone elastomer or an epoxy resin which has been made flexible by the addition of liquid polyamide resin or polysulphide rubber. Such non-rigid resins must be selected with great care because they may lack toughness, or the ability to adhere to the terminal connections or they may not be a sufficient barrier to water vapour.

Another possibility is to load the resin with a filler in an attempt to equate the temperature coefficients of the encapsulant and the ferrite. However, because filled resin sets hard and exact temperature coefficient compensation is unlikely, appreciable residual stress must be expected.

(2) If the resin completely fills the winding space, and has been cured at room temperature, a rise of temperature will cause a bursting stress on the core as the differential temperature coefficient reverses any stress due to curing shrinkage. This effect can easily break any ferrite core which, like a pot core, encloses the winding, even if the encapsulant is quite flexible. For this reason it is undesirable to allow the encapsulant to fill spaces enclosed by ferrite.

A particular case in this category is the effect of a hard resin in the air gap of an inductor. Differential expansion will cause the gap to increase with temperature and this will tend to make the temperature coefficient of inductance negative. Even if fracture does not occur the resultant temperature coefficient of inductance will be unpredictable. The stresses due to volumetric expansion of the encapsulant in ferrite cavities may be made negligible by the use of foam encapsulants since the gas in the foam is easily compressible. The foam encapsulant must of course be satisfactory in all other respects.

(3) If the resin sets hard and adheres to the surface of the ferrite, the differential temperature coefficient will cause shear stresses in the surface of the ferrite, an action similar to that of a bimetal strip. Such stresses could cause fracture, so it is good practice to prevent adhesion to the ferrite.

(4) Most resins have fairly high permittivity and dielectric loss. Such resins, if allowed to enter the winding or the winding space, could increase the losses of most medium or high frequency inductors. This may be another reason for preventing the resin from entering the winding space.

Summing up, the encapsulant should be:

- sufficiently tough,
- a sufficient barrier to adverse environmental conditions,
- sufficiently flexible to prevent compressive stresses in the ferrite,
- capable of good adhesion to parts of the assembly which break the surface, e.g. terminal connections.

The encapsulation technique should:

- avoid solid filling ferrite-enclosed spaces, e.g. winding spaces and air gaps,
- avoid adhesion to the ferrite if a hard encapsulant is used.

It is not easy to select an encapsulant and develop a technique that fulfils these requirements. For this reason encapsulation is not widely employed in the production of high-performance inductors and transformers using ferrite cores. See also References 106–109.

1.5 References and bibliography

General

A number of the entries given below refer to papers presented at the meetings of the International Conference on Ferrites. In the interests of brevity the reference is given in the form: ICFx, page, (year of conf.), where x indicates the serial number of the conference. The books or journal supplements containing the proceedings of these important conferences provide a valuable source of literature on all aspects of ferrites and the reader is referred to them for more general reading. The full references to these proceedings are as follows:

ICF1 (1970), *Ferrites, Proc. of Int. Conf. on Ferrites*, edited by Y. Hoshino, S. Iida, and M. Sugimoto, University Park Press, Tokyo (1971)

ICF2 (1976), *J. Phys. (France)*, Colloque C1, Suppl. to No. 4, **4** (1977)

ICF3 (1980), *Ferrites, Proc. of Int. Conf. on Ferrites*, edited by H. Watanabe, S. Iida, and M. Sugimoto, Centre for Academic Publications, Japan (1982)

ICF4 (1985), *Fourth Int. Conf. on Ferrites*, pt. I, edited by F. F. Y. Wang, Advances in Ceramics, **15** (1985) and pt. II, edited by F. F. Y. Wang, Advances in Ceramics, **16** (1985)

Section 1.1

1. The International Electrotechnical Vocabulary, *International Electrotechnical Commission, Publication 50*, Chap. 221, item 221-01-17, Geneva (1988)

2. HILPERT, S. Genetische und konstitutive Zusammenhänge in den magnetischen Eigenschaften bei Ferriten und Eisenoxyden, *Ber. dtsch. chem. Ges.*, **42**, 2248 (1909)
3. FORESTIER, H. Transformations magnétiques du sesquioxide de fer, de ses solutions solides, et des ses combinaisons ferromagnétiques, *Ann. Chim., Xe Série*, **IX**, 316 (1928)
4. HILPERT, S. and WILLE, A. Zusammenhänge zwischen Ferromagnetismus und Aufbau der Ferrite, *Z. Phys. Chem.*, **B18**, 291 (1932). See also *ibid.*, **B22**, 395 (1933) and **B31**, 1 (1935)
5. KATO, V. and TAKEI, T. Permanent oxide magnet and its characteristics, *J. Instn elect. Engrs Japan*, **53**, 408 (1933)
6. KAWAI, N. Formation of a solid solution between some ferrites, *J. Soc. chem. Ind. Japan*, **37**, 392 (1934)
7. SNOEK, J. L. Magnetic and electrical properties of the binary systems $MO.Fe_2O_3$, *Physica, Amsterdam*, **3**, 463 (1936)
8. TAKEI, T. *J. Electrochem. Ass. Japan*, **5**, 411 (1937)
9. SNOEK, J. L. *New developments in ferromagnetic materials*, Elsevier Publishing Co., New York-Amsterdam (1947)
10. WIJN, H. P. J. Some remarks on the history of ferrite research in Europe, *ICF1**, xix (1970)
11. OWENS, C. D. A survey of the properties and applications of ferrites below microwave frequencies, *Proc. I.R.E.*, **44**, 1234 (1956)
12. HIROTA, E., HIROTA, K. and KUGIMIYA, K. Recent development of ferrite heads and their materials, *ICF3**, 667 (1980)
13. ROESS, E. Soft magnetic ferrites and applications in telecommunications and power converters, *IEEE Trans. Magn.*, **MAG-18**, 1529 (1982)
14. BRACKE, L. P. M. Progress in SMPS magnetic component optimisation, *Electron. Components and Appl.*, **5**, 171, June (1983)
15. WENT, J. J., RATHENAU, G. W., GORTER, E. W. and VAN OOSTERHOUT, G. W. Ferroxdure, a new class of permanent magnet materials, *Philips tech. Rev.*, **13**, 194 (1951–52)
16. STUIJTS, A. L., RATHENAU, G. W. and WEBER, G. H. Ferroxdure II and III, anisotropic permanent magnet materials, *Philips tech. Rev.*, **16**, 141 (1954–55)
17. VAN DEN BROEK, C. A. M. and STUIJTS, A. L. Ferroxdure, *Philips tech. Rev.*, **37**, 157 (1977)
18. JONKER, G. H., WIJN, H. P. J. and BRAUN, P. B. Ferroxplana, hexagonal ferromagnetic iron-oxide compounds for very high frequencies, *Philips tech. Rev.*, **18**, 145 (1956–57)
19. NECKENBURGER, E., SEVERIN, H., VOGEL, J. K. and WINKLER, G. Ferrite hexagonaler Kristallstructur mit hoher Grenzfrequenz, *Z. Angew. Phys.*, **18**, 65 (1964)
20. ALBERS-SCHONBERG, E. Ferrites for microwave circuits and digital computers, *J. appl. Phys.*, **25**, 152 (1964)
21. PELOSCHEK, H. P. Square loop ferrites and their applications, in *Progress in Dielectrics*, **5**, edited by J. B. Birks and J. Hart, Heywood & Co., London (1963)
22. RAJCHMAN, J. A. A survey of magnetic memories, *ICF1**, 409 (1970)
23. PUGH, E. W. Ferrite core memories that shaped an industry, *IEEE Trans. Magn.*, **MAG-20**, 1499 (1984)
24. LAX, B. and BUTTON, K. J. *Microwave ferrites and ferrimagnetics*, McGraw-Hill Book Co (1962)
25. DIONNE, G. F. A review of ferrites for microwave applications, *Proc. IEEE*, **63**, 777 (1975)
26. BROCKMAN, F. G., VAN DER HEIDE, H. and LOUWERSE, M. W. Ferroxcube for proton synchrotrons, *Philips tech. Rev.*, **30**, 312 (1969)
27. PASNICU, C. and CONDURACHE, D. On some electric and magnetic properties of the Mn-Zn ferrites with Curie point about 0°C, *Buletinul Institutului Politechnic Din Iasi*, **XXI (XXV)**, Sectia I, 101 (1975)
28. KATO, U. and ENDO, M. The thermal reedswitch, *Proc. 22nd Nat. and 3rd Int. Relay Conf. Sillwater, Okla.*, 13-1, Publ. by Nat. Assoc. Relay Manufacturers, Scottsdale, Arizona (1974)
29. AKITA, K. Countermeasures against TV ghost interference using ferrite, *ICF3**, 885 (1980)
30. UENO, R., OGASAWARA, N. and INUI, T. Ferrites- or Iron-oxides-impregnated plastics serving as radiowave scattering suppressors, *ICF3**, 890 (1980)
31. PULLIAM, G. R., MEE, J. E. and HEINZ, D. M. Bubble domain materials for high density devices, *ICF3**, 449 (1980)
32. RUTHNER, M. J. The importance of hydrochloric acid regeneration processes for the industrial production of ferric oxides and ferrite powders, *ICF3**, 64 (1980)
33. SMIT, J. and WIJN, H. P. J. *Ferrites*, Philips Technical Library, Eindhoven (1959)
34. CHIKAZUMI, S. *Physics of Magnetism*, John Wiley & Sons, New York (1964)
35. CRAIK, D. J. (Ed.) *Magnetic oxides, Pts. 1 & 2*, John Wiley and Sons, London (1975)
36. WOHLFARTH, E. P. (Ed.), *Ferromagnetic materials*, North-Holland Publishing Co., Amsterdam-New York, Vol. 1 (1980), Vol. 2 (1980) and Vol. 3 (1982)

GORTER, E. W. Some properties of ferrites in connection with their chemistry, *Proc. Inst. Radio Engrs*, **43**, 1945 (1955)

STUIJTS, A. L., VERWEEL, J. and PELOSCHEK, H. P. Dense ferrites and their applications, *IEEE Trans. Commun. Electronics*, No. 75, 726 (1964)

BROESE VAN GROENOU, A., BONGERS, P. F. and STUIJTS, A. L. Magnetism, microstructure and crystal chemistry of spinel ferrites, *Mater. Sci. Eng.*, **3**, 317 (1968–69)

SUGIMOTO, M. Research and technological development of ferrites in Japan, *Japan Elect. Engrg*, No. 65, 25 (1972)

HECK, C. *Magnetic materials and their applications*, Butterworths, London (1974)

KRUPIČKA, S. Soft-ferrites, achievements and problems, *J. Magn. and Magn. Mater.*, **19**, 88 (1980)

ENZ, U. Magnetism and magnetic materials: Historical developments and present rôle in industry and technology, in *Ferromagnetic materials Vol. 3*, Chap. 1, edited by E. P. Wohlfarth, North-Holland Publishing Co., Amsterdam-New York (1982)

GOLDMAN, A. Understanding ferrites, *Am. Ceram. Soc. Bul.*, **63**, 582 (1984)

KULIKOWSKI, J. Soft magnetic ferrites—development or stagnation?, *J. Magn. and Magn. Mater.*, **41**, 56 (1984)

*See *General* (page 21)

POSTUPOLSKI, T. Soft ferrites, *J. Phys. (France)*, Col. C6, Suppl. to No. 9, **46**, C6-159 (1985)

Section 1.2.2
37. PEARSON, R. F. Magnetic anisotropy, in *Experimental magnetism*, edited by G. M. Kalvius and R. S. Tebble, 153, John Wiley & Sons, Chichester, England (1979)
38. SNELLING, E. C. and GILES, A. D. *Ferrites for inductors and transformers*, Chaps. 1 & 2, Research Studies Press, Letchworth, Hertfordshire, England; distrib. by John Wiley & Sons, New York (1983)
    KNOWLES, J. E. The simulation of domain wall motion in square loop ferrites, and other aspects of wall behaviour, *Brit. J. Appl. Phys. (J. Phys. D)*, Ser. 2, **1**, 821 (1968)
    KNOWLES, J. E. The magnetostatic energy associated with polycrystalline ferrite, *Brit. J. Appl. Phys. (J. Phys. D)*, Ser. 2, **1**, 987 (1968)
    DEMBIŃSKA, M. Domain structure of lithium ferrite, *Acta Phys. Polonica*, **A45**, 33 (1974)
    GLOBUS, A. Some physical considerations about the domain wall size theory of magnetization mechanisms, *ICF2**, C1-1(1976)
    TSUNEKAWA, H. *et al.* Microstructure and properties of commercial grade manganese zinc ferrite, *IEEE Trans. Magn.*, **MAG-15**, 1855 (1979)

Section 1.3.1
39. WATANABE, H. and TAKEDA, S. Growth and properties of manganese zinc tin ferrite single crystals, *ICF2**, C1-51 (1976)
40. STOPPELS, D., BOONEN, P. G. T., ENZ, U. and VAN HOOF, L. A. H. Monocrystalline high-saturation magnetization ferrites for video recording head application, *J. Magn. and Magn. Mater.*, **37**, Pt. I, 116, Pt. II, 123 and Pt. III, 131 (1983)
41. KOBAYASHI, T. and TAKAGI, K. Crystal growth of Mn-Zn ferrite by the travelling solvent zone melting method, *J. Crystal Growth*, **62**, 189 (1983)
42. HIROTA, K., SUGIMURA, M. and HIROTA, E. Hot-press ferrites for magnetic recording heads, *Ind. Eng. Chem. Prod. Res. Dev.*, **23**, 323 (1984)
43. GORTER, E. W. Some properties of ferrite in connection with their chemistry, *Proc. Inst. Radio Engrs*, **43**, 1945 (1955)
44. NATANSOHN, S. and BAIRD, D. H. Effect of synthesis parameters on the magnetic properties of manganese zinc ferrites, *J. Am. Ceram. Soc.*, **52**, 127 (1969)
45. AURADON, J-P., DAMAY, F. and CHOL, G. Modern investigation methods for the optimization of high-quality soft-ferrite manufacture, *IEEE Trans Magn.*, **MAG-5**, 276 (1969)
46. MAGEE, J. H., MORTON, V., FISHER, R. D. and LOWE, I. J. Factors affecting magnetic properties of linear ferrites, *ICF1**, 217 (1970)
47. DE LAU, J. G. M. and BROESE VAN GROENOU, A. High-frequency properties of Ni-Zn-Co ferrites in relation to iron content and microstructure, *ICF2**, C1-17 (1976)
48. BONGERS, P. F., DEN BROEDER, F. J. A., DAMEN, J. P. M., FRANKEN, P. E. C. and STACY, W. T. Defects, grain boundary segregation, and secondary phases of ferrites in relation to the magnetic properties, *ICF3**, 265 (1980)
49. JOHNSON, D. W. and GHATE, B. B. Scientific approach to processing of ferrites, *ICF4**, Part I, 27 (1985)
50. ROESS, E. Manufacturing problems and quality assurance in the production of soft ferrites, *ICF4**, Part I, 39 (1985)
51. ROESS, E. Soft magnetic ferrites and applications in telecommunications and power converters, *IEEE Trans. Magn.*, **MAG-18**, 1529 (1982)
52. STREET, B. G. Ferrite component manufacture, *Powder Metallurgy*, **25**, 173 (1982)

Section 1.3.2
53. ERZBERGER, P. Correlation of particle size with other physical properties of iron oxides for ferrite synthesis, *Proc. Brit. Ceram. Soc.*, **2**, 19 (1964)
54. RUTHNER, M. J. The importance of hydrochloric acid regeneration processes for the industrial production of ferric oxides and ferrite powders, *ICF3**, 64 (1980)
55. GOLDMAN, A. and LAING, A. M. A new process for coprecipitation of ferrites, *ICF2**, C1-297 (1976)
56. ROBBINS, H. The preparation of Mn-Zn ferrites by coprecipitation, *ICF3**, 7 (1980)
57. YU, B. B. and GOLDMAN, A. Effect of processing parameters on morphology of Mn-Zn ferrite particles produced by hydroxide-carbonate coprecipitation, *ICF3**, 68 (1980)
58. TAKADA, T. Development and application of synthesizing technique of spinel ferrites by the wet method, *ICF3**, 3 (1980)
59. WAGNER, U. Spray firing for preparation of presintered powder for soft ferrites, *J. Magn. and Magn. Mater.*, **19**, 99 (1980)
60. RUTHNER, M. J. Spray-roasted iron oxides: the nature of minor impurities, *ICF4**, Part I, 103 (1985)
61. OHTA, K. Magnetocrystalline anisotropy and magnetic permeability of Mn-Zn-Fe ferrites, *J. Phys. Soc. Japan*, **18**, 685 (1963)
62. BROESE VAN GROENOU, A., BONGERS, P. F. and STUIJTS, A. L. Magnetism, microstructure and crystal chemistry of spinel ferrites, *Mater. Sci. Eng.*, **3**, 317 (1968-9)
63. STIJNTJES, T. G. W., BROESE VAN GROENOU, A., PEARSON, R. F., KNOWLES, J. E. and RANKIN, P. Effects of various substitutions in Mn-Zn-Fe ferrites, *ICF1**, 194 (1970)
64. KOENIG, U. Substitutions in manganese zinc ferrites, *Appl. Phys.*, **4**, 237 (1974)
65. GILES, A. D. and WESTENDORP, F. F. The effect of silica on the microstructure of MnZn ferrites, *ICF2**, C1-317 (1976)
66. WAGNER, U. Aspects of the correlation between raw material and ferrite properties, *J. Magn. and Magn. Mater.*, Pt I, **4**, 116 (1977) and Pt II, **23**, 73 (1981)
67. NISHIYAMA, T., KANAI, K., IIMURA, T. and HARADA, H. Analysis of loss on Mn-Zn ferrites containing CaO, *ICF4**, Part I, 491 (1985)
68. HIROTA, K., FUJIMOTO, Y., WATANABE, K. and SUGIMURA, M. High-B and -μ′ Mn-Zn ferrite with improved mechanical strength, *ICF4**, Part I, 385 (1985)

*See *General* (page 21)

69. STIJNTJES, T. G. W., KLERK, J. and BROESE VAN GROENOU, A. Permeability and conductivity of Ti-substituted MnZn ferrites, *Philips Res. Rep.*, **25**, 95 (1970)
70. FOMENKO, G. V. and BASHKIROV, L. A. Effect of replacing iron with titanium, zirconium, and hafnium on the intitial magnetic permeability of manganese-zinc ferrites, *Izv. Akad. Nauk. SSSR Neorg. Mater.*, **18**, 1886 (1982), *Trans. in Inorg. Mater.*, **18**, 1622 (1982)
71. BUTHKER, C., ROELOFSMA, J. J. and STIJNTJES, T. G. W. Low loss power material with improved temperature behaviour, *Fall Mtg. Am. Cer. Soc.*, 58-BE-82F (Sept. 1982)
72. STIJNTJES, T. G. W. and ROELOFSMA, J. J. Low-loss power ferrites for frequencies up to 500 kHz, *ICF4**, Part II, 493 (1984)
73. MULLIN, J. T. and WILLEY, R. J. Grain growth of Ti-substituted Mn-Zn ferrites, *ICF4**, Part I, 187 (1985) FODOR, L., HIDASI, B. and VECSEY, B. The effects of some additives and impurities on the magnetic properties of manganese-zinc ferrites, *Periodica Polytechnica*, **28**, 15, Budapest (1983)

Section 1.3.3
74. CHOL, G. and AUBAILE, J. P. Influence of the mixing methods and duration on the physical and electrical properties of high quality Mn-Zn ferrites, *ICF1**, 243 (1970)
75. REYNOLDS, T. G. and JONES, E. W. Comparison of analytical results of manganese zinc ferrites, *ICF3**, 74 (1980)

Section 1.3.4
76. SWALLOW, D. and JORDAN, A. K. The fabrication of ferrites, *Proc. Br. Ceram. Soc.*, **2**, 1 (1964).

Section 1.3.5
77. ALAM, M. I., JALEEL, S. A. and VENUGOPALAN, R. Effect of dispersant on grinding efficiency of a ferrite powder, *ICF4**, Part I, 109 (1985)

Section 1.3.6
78. STRIJBOS, S., RANKIN, P. J., KLEIN WASSINK, R. J., BANNINK, J. and OUDEMANS, G. J., Stresses occurring during one-sided die compaction of powders, *Powder Technology*, **18**, 187 (1977)
79. STRIJBOS, S. Powder-wall friction: the effects of orientation of wall grooves and wall lubricants, *Powder Technology*, **18**, 209 (1977)
80. BROESE VAN GROENOU, A. Pressing of ceramic powders: a review of recent work. *Powder Metallurgy Int.*, **10**, 206 (1978)
81. STRIJBOS, S., BROESE VAN GROENOU, A. and VERMEER, P. A. Recent progress in understanding die compaction of powders, *J. Am. Ceram. Soc.*, **62**, 57 (1979)
82. STRIJBOS, S. Phenomena at the powder-wall boundary during die compaction of a fine oxide powder, *Ceramurgia Int.*, **6**, 119 (1980)

Section 1.3.7
83. HECK, C. and WEBER, J. How firing atmospheres influence ferrite properties, *Ceramic Ind.*, **77**, Part 1, No. 5, 75, Part 2, No. 6, 66 (1961)
84. GUILLAUD, C. and PAULUS, M. Perméabilité initiale et grosse des grains dans les ferrites de manganèse-zinc, *C. R. Acad. Sci. Paris*, **242**, 2525 (1956)

85. STUIJTS, A. L. Microstructural considerations in ferromagnetic ceramics, *Proc. 3rd Int. Mater. Symp. on Ceramic Microstructures*, Univ. Calif., Berkeley (June 1966)
86. REIJNEN, P. *Equilibria, reactions and sintering in systems with iron oxide as one of the components*, Thesis, Delft (1969)
87. CHOL, G., AURADON, J-P. and DAMAY, F. Influence of the sintering conditions on the densification of manganese-zinc ferrites, *IEEE Trans. Magn.*, **MAG-5**, 281 (1969)
88. STUIJTS, A. L. Control of microstructures in ferrites, *ICF1**, 108 (1970)
89. ROESS, E. Magnetic properties and microstructure of high permeability Mn-Zn ferrites, *ICF1**, 203 (1970)
90. BROESE VAN GROENOU, A., BONGERS, P. F. and STUIJTS, A. L. Magnetism, microstructure and crystal chemistry of spinel ferrites, *Mater. Sci. Eng.*, **3**, 317 (1968–69)
91. FRANKEN, P. E. C. and VAN DOVEREN, H. Determination of the grain boundary composition of soft ferrites by Auger electron spectroscopy, *Ber. Dtsch. Keram. Ges.*, **55**, 287 (1978)
92. NORMIRA, T., OKUTANI, K., KITAGAWA, T. and OCHIAI, T. Sintering of MnZn ferrites for power materials, *Fall Mtg. Am. Cer. Soc.*, 57-BE-82F (1982)
93. KIMURA, O. and CHIBA, A. Formation of commercial Mn-Zn ferrites, *ICF4**, Part I, 115 (1985)
94. DROFENIK, M., BESENICAR, S., LIMPEL, M. and GARDASEVIC, V. Influence of the dimensions of MnZn-Ferrite samples on their microstructural and magnetic properties, *ICF4**, Part I, 229 (1985)
95. MORINEAU, R. and PAULUS, M. Chart of PO_2 versus temperature and oxidation degree for Mn-Zn ferrites in the composition range: $50 \leqslant Fe_2O_3 \leqslant 54$; $20 \leqslant MnO \leqslant 35$; $11 \leqslant ZnO \leqslant 30$ (mole %), *IEEE Trans. Magn.*, **MAG-11**, 1312 (1975)
96. MULLIN, J. T. and WILLEY, R. J. Effects of post-sinter cooling on Ti-substituted Mn-Zn ferrites, *J. Magn. and Magn. Mater.*, **41**, 66 (1984)
97. RIKUKAWA, H. and SASAKI, I. On the sintering atmosphere of Mn-Zn ferrites, *ICF4**, Part I, 215 (1985)

Section 1.3.8
98. BROESE VAN GROENOU, A., VELDKAMP, J. D. B. and SNIP, D. Scratching and grinding parameters of various ferrites, *ICF2**, C1-285 (1976)
99. BROESE VAN GROENOU, A. Grinding of ferrites, some mechanical and magnetic aspects, *IEEE Trans. Magn.*, **MAG-11**, 1446 (1975)
100. KNOWLES, J. E., The effect of surface grinding upon the permeability of manganese-zinc ferrites, *J. Phys. D: Appl. Phys.*, **3**, 1346 (1970)
101. KNOWLES, J. E. The origin of the increase in magnetic loss induced by machining ferrites, *IEEE Trans. Magn.*, **MAG-11**, 44 (1975)

Section 1.3.9
102. Harmonized system of quality assessment for electronic components. Inductor and transformer cores for telecommunication, Part 0. Generic specification, *British Standards Inst.*, BS9925: Part 0, Lon-

*See *General* (page 21)

don (1984), also *Cenelec Electronic Components Committee, CECC 25000* (1982) and *Int. Electrotechnical Commission, IEC Publication 723-1*, Geneva (1982) (Note: these Generic Specifications form the basis of a structure of Sectional and Detail Specifications relating to specific classes of cores and individual core types respectively)

Section 1.4.2
103. HOUWINK, R. and SALOMON, G. *Adhesion and adhesives*, Elsevier Publishing Co., Vol. 1, 2nd edition (1965), Vol. 2. 2nd Edn. (1967)
104. SHIELDS, J. *Adhesives handbook*, Butterworths, London, 3rd Edn. (1985)
105. SKEIST, I. *Handbook of adhesives*, Van Nostrand Reinhold Co., New York, 2nd Edn. (1977)

Section 1.4.3
106. MAGEE, J. H. and FISHER, R.D. Polymer encapsulation of linear ferrite cores, *IEEE Trans. Magn.*, **MAG-6,** 34 (1970)
107. KIRK, W. J. *Ferrite-core transformer bonding and potting*, Issued by Bendix Corp., Kansas City, Mo., USA, Report No. BDX-613-992. Availability: NTIS, Springfield, Va., 22151, USA (1974)
108. ARNETT, J. R. and BUNKER, E. R. Low stress potting for stress sensitive magnetic cores, *Coil Winding Chicago '80 Proc.*, **42** (1980)
109. VAN DER POEL, J. M. Vacuum impregnation of wound ferrite components: potential problems and pitfalls, *Insul./Circuits*, **27,** 35 (1981)

2 The expression of electrical and magnetic properties

2.1 Magnetization

The magnetic field strength, H, inside a very long uniform solenoid having N_1 turns per axial length l and carrying I amperes is given by

$$H = \frac{N_1 I}{l} \quad \text{A m}^{-1} \qquad (2.1)*$$

Its direction is parallel to the axis of the solenoid and it is uniform across the internal cross section.

The associated flux density, B, is given by

$$B = \mu_0 H \quad \text{tesla† (T)} \qquad (2.2)$$

where μ_0 is the magnetic constant or the permeability of free space. It has the numerical value $4\pi \times 10^{-7}$ and has the dimensions henries/metre or $[LMT^{-2}I^{-2}]$. Thus in the SI units, flux density is dimensionally different from field strength.

---

*Many of the quantities used in this section are vector quantities and therefore the equations involving them are vector equations. However, in the present limited treatment, the relative directions of the vectors are implied in the text so no special symbols will be used to denote vector quantities or vector operations. For a more general treatment the reader is referred to textbooks on electromagnetic theory.

The International Electrotechnical Vocabulary (IEV)[1] of the IEC, particularly Chapters 121, Electromagnetism, and 221, Magnetic Materials and Components, contains internationally agreed terms and definitions for most of the quantities referred to in Chapters 2 and 4 of this book.

†1 T = 1 weber per square metre = 10^4 Gs
1 mT = 10 Gs

In the CGS system of units H and B have the same dimensions and therefore the oersted and the gauss are strictly the same units. The CGS equations equivalent to the text equations are given as footnotes below and on the following pages.

(2.1) $H = \dfrac{4\pi N_1 I}{10 l}$ Oe (1 Oe $\approx$ 80 A m^{-1})

(2.2) $B = H$ Gs

If the solenoid is now filled with a magnetic material the applied magnetic field will act upon the magnetic moments of the ions composing the material. This process has been described qualitatively in Section 1.2.2. The ions, by virtue of the spinning electrons, behave as microscopic current loops each having a magnetic moment. These moments may, in general, be considered to be aligned parallel to each other over small regions, or domains, within the material. In the demagnetized state the domains are distributed so that the vector sum of the magnetization of the domains is zero. Under the influence of an applied field the ion moments are re-orientated, either by the growth and contraction of the various domains or by the rotation of the magnetization within them, so that the ionic moments effectively augment the applied field. This increase in magnetic field is called the magnetization, M, and it is expressed in A m^{-1}. It is the vector sum of the magnetic area moments* of all the microscopic currents in a given volume of material, divided by that volume. The internal magnetic field, H_i, becomes

$$H_i = \frac{N_1 I}{l} + M \quad \text{A m}^{-1} \qquad (2.3)$$

and the flux density becomes:

$$B = \mu_0 H_i = \mu_0 (H + M) \quad \text{T} \qquad (2.4)$$

or $B = \mu_0 H + J$ \qquad T \qquad (2.5)

---

*Magnetic area moment, m, is the product of a current and the area of the loop in which it flows, the direction is normal to the plane of the loop and when viewed in this direction the current has clockwise rotation.

(2.3) $H_i = \dfrac{4\pi N_1 I}{10 l} + 4\pi M$ Oe

(2.4)
(2.5) $\Big\}$ $B = H_i = H + 4\pi M$ \quad Gs

where J is the magnetic polarization in teslas; it is sometimes referred to as intrinsic flux density

$$J = \mu_o M \quad \text{T} \tag{2.6}$$

Thus M is the increase in the field strength due to the magnetic material and J is the corresponding increase in flux density. The ratio of the magnetization and the applied field strength is called the susceptibility, κ; it is dimensionless.

From Eqn (2.4)

$$\frac{B}{H} = \mu_o \left(1 + \frac{M}{H}\right) = \mu_o(1 + \kappa) \tag{2.7}$$

This quotient of flux density and applied field strength is called the absolute permeability and is sometimes denoted by μ. However it is more usual to show it as the product of the magnetic constant and a dimensionless constant called the relative permeability, μ_r. In the chapters that follow, the relative permeability is such a widely used parameter and is given such a variety of qualifying subscripts that it is convenient to drop the adjective 'relative'. Thus permeability will refer to the dimensionless ratio and in equations it will normally be associated with the magnetic constant, μ_o. The absolute permeability, as such, will not be used.

$$\frac{B}{H} = \mu_o \mu \tag{2.8}$$

from which it follows that

$$\kappa = \mu - 1 \tag{2.9}$$

The applied field strength may be determined by measuring the current and using Eqn (2.1). The measurement of flux density depends on the law of induction, i.e.

$$e = -d\varphi/dt \quad \text{V} \tag{2.10}$$

where φ is the magnetic flux i.e. the area integral of the flux density; it is expressed in webers (Wb).

In the ideal solenoid $\varphi = BA$ where A is the cross-sectional area of the magnetic material. If N_2 turns are wound tightly round the magnetic material the e.m.f. induced will be

$$e = -N_2 A\, dB/dt \quad \text{V} \tag{2.11}$$

By integration, the average e.m.f. during a change of flux density, ΔB, is given by

$$\bar{E}\, \Delta t = -N_2 A\, \Delta B \quad \text{V s} \tag{2.12}$$

The negative sign indicates that the e.m.f. is in such a direction that it would produce current opposing the change of flux. If the flux density is sinusoidal, e.g. if $B = \hat{B} \sin \omega t$, then from Eqn (2.11), dropping the sign:

$$e = N_2 A \hat{B} \omega \cos \omega t = \hat{E} \cos \omega t$$
$$\therefore \hat{E} = \omega \hat{B} A N_2 \quad \text{V}$$

or $E = \dfrac{\omega \hat{B} A N_2}{\sqrt{2}} \quad \text{V} \tag{2.13}$

If the current in the ideal solenoid is increased from zero, the field strength increases and the magnetization will increase non-linearly by the processes illustrated in Figure 1.5. It is more usual to consider the dependence of the flux density on field strength. Such a B–H curve is shown in Figure 2.1. Starting with the

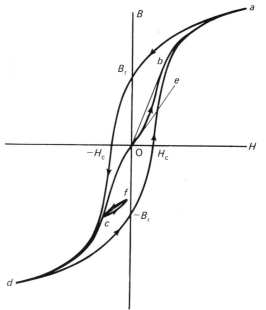

Figure 2.1 Hysteresis loop

(2.7) $\dfrac{B}{H} = 1 + 4\pi \dfrac{M}{H} = 1 + 4\pi\kappa$

(2.8) $\dfrac{B}{H} = \mu$

(2.9) $4\pi\kappa = \mu - 1$

(2.10) $e = -\dfrac{d\varphi}{dt} \times 10^{-8} \quad \text{V}$

(2.11) $e = -N_2 A \dfrac{dB}{dt} \times 10^{-8} \quad \text{V}$

(2.12) $\bar{E}\, \Delta t = -N_2 A \Delta B \times 10^{-8} \quad \text{V s}$

(2.13) $E = \dfrac{\omega \hat{B} A N_2}{\sqrt{2}} \times 10^{-8} \quad \text{V}$

magnetic material in an unmagnetized or neutralized state the B–H curve will follow the path **oba**. If, on reaching the point **a**, the field strength is decreased, the B–H curve will follow the upper limb of the open loop. As explained in Chapter 1, a reason for this lag in the change of flux density is the irreversible movement of the domain walls. If, on reaching point **d**, symmetrical with **a**, the field strength is increased again towards its former positive maximum value, the lower limb of the loop will be traced. The loop is called a hysteresis or B–H loop. Although, on the first cycle, the loop may not close exactly, a number of cycles will result in a closed loop. If the excursions of H are symmetrical about the origin the material is then said to be in a symmetrical cyclic magnetic state.

If the field strength is large enough to take the material substantially into saturation, i.e. to a point where the magnetization M cannot be significantly increased, then the intercept of the hysteresis loop with the B-axis, B_r (or $-B_r$), is referred to as the remanence of the material and the intercept with the H-axis, H_c (or $-H_c$), is referred to as the coercivity. The tips of loops for smaller excursions of H lie very close to the initial magnetization curve **oba**.

Since B is a two-valued function of H, the instantaneous ratio B/H depends on the magnetic history. However in alternating magnetization it is usually relevant to consider only the peak amplitudes of B and H, i.e. the tips of the loops. If the material is in a symmetrical cyclic state and H is vanishingly small, the permeability is designated μ_i, the initial permeability. It is $1/\mu_o$ times the slope of the line **oe**. If H is not vanishingly small, then the permeability is referred to as the amplitude permeability designated μ_a. It is $1/\mu_o$ times the slope of the line connecting the origin to the tip of the loop produced by that particular value of H. As H increases, μ_a increases, until the tip of the loop reaches **b** and the slope of the line **ob** is the maximum value of $\mu_o\mu_a$.

A non-symmetrical or minor loop is traced if, on reaching a point such as **c**, the field variation is reversed and the material is cycled between **c** and **f**. The slope of **cf** divided by μ_o is called the incremental permeability, μ_Δ, and if the amplitude of the excursion is made vanishingly small it becomes the reversible permeability, μ_{rev}. Finally, the slope at any point on a hysteresis loop or curve is referred to as the differential permeability.

Of course, B, H and μ are physical quantities and may be observed without the need for windings, e.g. in a waveguide or a cavity. However, in the present context, it is by means of windings on magnetic cores that magnetic properties may be observed and put to useful purposes. The ideal solenoid is not a practical arrangement but it may be simulated by a uniformly wound toroidal core. An ideal toroidal core is radially thin so that it approaches an elementary ring and consequently has a uniform distribution of field strength and flux density across its cross-section. Such an ideal toroid is assumed in the remainder of this chapter. It is further assumed that the windings fit closely to the core, the flux is contained entirely in the core and that the effects of stray inductance, capacitance or resistance due to the winding are negligible so that the impedance measured across a winding is due entirely to the properties of the core material.

The methods of expressing core properties will be related to such a core. In a practical device the core departs from the ideal, having, in general, a relatively large non-uniform cross-section and perhaps an air gap in the magnetic circuit. The modifications to the theoretical relations required to cover such a core are considered in Chapter 4.

2.2 The expression of the low-amplitude properties

In the majority of applications of magnetic materials in inductors and transformers, the field strength or flux density is varying in a more-or-less complex way. For the purpose of expressing material properties and design relations it is convenient to consider only sinusoidal wave forms. Ferrite cores are often used at quite low amplitudes. At these low amplitudes the non-linearity between B and H is small so that, to a first order, the waveform distortion may usually be neglected. Under these conditions, if the field strength is sinusoidal then the flux density and the e.m.f (proportional to dB/dt) may be taken as sinusoidal. Thus simple a.c. theory may be used to describe the influence of a magnetic material on an electrical circuit. When distortion must be taken into account this may be done by specifically introducing non-linearity into the relations.

For the present purpose, low-amplitude may be taken as corresponding to flux densities of the order of, or less than, 1% of the saturation flux density.

2.2.1 Complex permeability

The inductance of a circuit may be defined as the flux linkage per unit current, i.e. for an alternating current of peak amplitude $\hat{I}$,

$$L = \frac{N\hat{\Phi}}{\hat{I}} \quad \text{H}$$

For a winding of N turns on an ideal toroid of magnetic length l and cross-sectional area A

$$L = \frac{N\hat{B}A}{\hat{I}} = \frac{NA}{\hat{I}}\mu_o\mu\frac{N\hat{I}}{l}$$

$$\therefore L = \frac{\mu_o\mu N^2 A}{l} \quad \text{H} \qquad (2.14)$$

$$= L_o \mu$$

where

$$L_o = \frac{\mu_o N^2 A}{l}$$

= the inductance that would be measured if the core had unity permeability, the flux distribution remaining unaltered.

In general the impedance of the winding will not be a pure reactance; there will be a resistive component due to the loss of energy incurred as the magnetization alternates. The impedance may be expressed in terms of a complex permeability.

Whereas a loss-free core will present a reactance $X = j\omega L$, a core having magnetic loss may be represented by an impedance:

$$\begin{aligned} Z &= j\omega L_s + R_s \quad \Omega \\ &= j\omega L_o(\mu'_s - j\mu''_s) \quad \Omega \end{aligned} \quad (2.15)$$

where R_s is the series loss resistance
L_s is the series inductance
μ'_s is the real component of the series complex permeability
μ''_s is the imaginary component of the series complex permeability.

Then

$$\left. \begin{aligned} \omega L_s &= \omega L_o \mu'_s \quad \Omega \\ R_s &= \omega L_o \mu''_s \quad \Omega \end{aligned} \right\}$$

and the magnetic loss tangent: (2.16)

$$\left. \begin{aligned} \tan \delta_m &= \frac{R_s}{\omega L_s} \\ &= \frac{\mu''_s}{\mu'_s} \end{aligned} \right\}$$

where δ_m is the loss angle, i.e. the phase angle between B and H.

Figure 2.2(a) shows the vector diagram corresponding to the series circuit.

An alternative approach is to represent the same impedance in terms of parallel components. Figure 2.2(b) shows the equivalent vector diagram, the applied voltage U and current I in each case corre-

(2.14) $\quad L = 4\pi\mu \frac{N^2 A}{l} \times 10^{-9} \quad H$

$\quad = L_o \mu$ where $L_o = 4\pi \frac{N^2 A}{l} \times 10^{-9}$

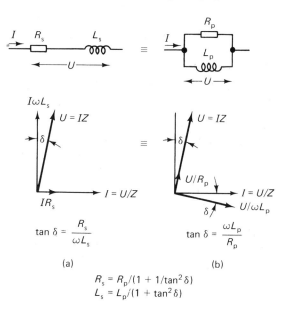

$$\tan \delta = \frac{R_s}{\omega L_s} \qquad \tan \delta = \frac{\omega L_p}{R_p}$$

(a)   (b)

$$R_s = R_p/(1 + 1/\tan^2 \delta)$$
$$L_s = L_p/(1 + \tan^2 \delta)$$

Figure 2.2 Equivalence of series and parallel inductive circuits

sponding exactly in magnitude and phase. Then the admittance,

$$Y = \frac{1}{j\omega L_p} + \frac{1}{R_p} = \frac{1}{j\omega L_o}\left(\frac{1}{\mu'_p} - \frac{1}{j\mu''_p}\right) \quad \Omega^{-1} \quad (2.17)$$

where R_p is the parallel loss resistance
L_p is the parallel inductance
μ'_p is the real component of the parallel complex permeability
μ''_p is the imaginary component of the parallel complex permeability

Then

$$\left. \begin{aligned} \omega L_p &= \omega L_o \mu'_p \quad \Omega \\ R_p &= \omega L_o \mu''_p \quad \Omega \end{aligned} \right\}$$

and the magnetic loss tangent: (2.18)

$$\tan \delta_m = \frac{\omega L_p}{R_p} = \frac{\mu'_p}{\mu''_p}$$

By analogy with the conversion of series to parallel impedances quoted in Figure 2.2 the following relations exist between the series and parallel components of the permeability.

$$\left. \begin{aligned} \mu'_p &= \mu'_s(1 + \tan^2 \delta_m) \\ \mu''_p &= \mu''_s(1 + 1/\tan^2 \delta_m) \end{aligned} \right\} \quad (2.19)$$

Graphical representation of the components of the complex permeability and of the values of $\tan \delta_m$ as functions of frequency are common methods of indicating the performance of ferrites at low field strengths. Where the term 'permeability' or the symbol μ is used without any identification of the complex components, the real component is assumed.

2.2.2 Variability

The permeability of a magnetic material may change for a variety of reasons. When such changes occur as the result of unavoidable variations in the operating conditions, they are usually detrimental to the performance of the magnetic material and must be minimized or their magnitude controlled. These changes are referred to as variability and the quantitative expression of the types of variability will now be considered briefly. Experimental observations and some discussion of the phenomena are given in Chapter 3.

The most obvious cause of variation is the change of temperature. Over a limited temperature range the reversible variation of permeability with temperature may be described by a temperature coefficient defined by:

$$a_\mu = \frac{\Delta\mu}{\mu\,\Delta\theta} \qquad (2.20)$$

It is usually expressed in parts per million per degree centigrade. In principle this and the following expressions may be applied to any of the previously described forms of permeability, including the loss components; in practice they usually refer only to the initial permeability.

If the range of temperature is small the above expression is satisfactory but, if $\Delta\mu/\mu$ becomes appreciable, it is necessary to define the value of μ used in the denominator. Further, if the relation is non-linear or has a turning point within the temperature range considered, further ambiguity may arise. In present practice the above expression is interpreted as:

$$a_\mu = \frac{(\mu_2 - \mu_1)}{\mu_1(\theta_2 - \theta_1)} \qquad (2.21)$$

where μ_1 is the permeability observed at θ_1 and μ_2 at θ_2. If the permeability change is large, the value in the denominator should be a mean value. It has been suggested that the geometric mean gives a more satisfactory result.

Then

$$a_\mu = \frac{(\mu_2 - \mu_1)}{\sqrt{\mu_1\mu_2}(\theta_2 - \theta_1)} \qquad (2.22)$$

If the relation is very non-linear or has a turning point between θ_2 and θ_1 then it is better to abandon the concept of temperature coefficient and to specify that the curve shall lie within a prescribed area, bounded on either side by θ_1 and θ_2 and at the top and bottom by lines giving the permissible change of fractional permeability. Such an area might be rectangular or, if some compensation of another temperature coefficient, e.g. that of a capacitor, is to be attempted, the shape may be that of a parallelogram.

In Chapter 4 it is shown that when an air gap is inserted into a magnetic circuit so that its permeability is reduced to an effective value, μ_e, the effect of permeability variations are reduced in the ratio μ_e/μ_i. It is therefore convenient to divide the temperature coefficient by μ_i. This gives a parameter generally referred to as temperature factor and having the symbol a_F. Corresponding to Eqn (2.20)

$$a_F = \frac{\Delta\mu_i}{\mu_i^2\,\Delta\theta} \qquad (2.23)$$

while corresponding to Eqn (2.22)

$$a_F = \frac{(\mu_2 - \mu_1)}{\mu_1\mu_2(\theta_2 - \theta_1)} \qquad (2.24)$$

To obtain the temperature coefficient of the effective permeability it is only necessary to multiply the temperature factor by μ_e. Temperature factor as defined by Eqn (2.23) is at present the usual method of expressing the temperature dependence of the initial permeability of a ferrite material. Since Eqns (2.23) and (2.24) represent the change in $1/\mu$ (reluctivity) with temperature, this parameter is now designated temperature factor of reluctivity in the IEV.[2]

Normally the initial permeability rises monotonically with temperature until it reaches a peak just below the Curie temperature. Then, as the forces that hold the spins in alignment are overcome by thermal agitation the permeability falls abruptly to values approaching unity and the material becomes paramagnetic. It is possible to introduce secondary peaks and even minima in the μ/θ characteristic in order to produce a low temperature factor within a limited operating temperature range.

A second form of variability is the variation of permeability with time, or disaccommodation as it is sometimes called. This phenomenon is described in some detail in the introduction to Figure 3.10. It is sufficient here to state that if a magnetic material is given some form of disturbance, the initial permeability is, in general, instantaneously raised to an unstable value from which it returns as a function of time. Let t_1 and t_2 represent two time intervals measured from the time of the disturbance, and let μ_1 and μ_2 be the corresponding values of the permeability observed at these times, then in general the disaccommodation (of permeability) is defined as

$$D = \frac{\mu_1 - \mu_2}{\mu_1} \qquad (2.25)$$

and is usually expressed in percent.

The times must also be defined. Usually the fractional change in permeability is small so there is no need to use a mean value of permeability in the denominator.

In the discussion of Figure 3.10, specific types of disturbance are referred to and specific measuring techniques are described. These lead to particular forms of the above expression. The one most commonly used in the past, and still sometimes quoted, expresses the time change of permeability, at constant temperature, between 1 min and 24 h after the material has been disturbed by an alternating field which decreases to zero from a value corresponding to saturation, i.e. μ_1 is measured at $t = 1$ min and μ_2 is measured at $t = 24$ h after the disturbance. More recently time intervals of 10 min and 100 min have come into use.

Because it is observed that the change of permeability is approximately proportional to the logarithm of time, the IEV[3] defines a disaccommodation coefficient (of permeability) as

$$d = \frac{\mu_1 - \mu_2}{\mu_1 \log_{10}(t_2/t_1)} \quad (2.26)$$

where t_1 and t_2 are arbitrary but defined time intervals after the disturbance.

Within the validity of the assumed relations this expression will yield a value which is independent of the times of the measurement. By a similar reasoning to that used for deriving temperature factor, a disaccommodation factor (of permeability) is defined:

$$D_F = \frac{d}{\mu_1} \quad (2.27)$$

2.2.3 Hysteresis

There are a variety of coefficients that have been used to express hysteresis loss.[4] In this section only the Rayleigh and Peterson coefficients will be introduced because they are derived from consideration of ideal hysteresis loop shapes and have formed the basis for the more recent methods of expression.

2.2.3.1 The Rayleigh relations

Rayleigh,[5] in 1885, during experiments on specimens of iron, observed that at low and decreasing values of magnetization the amplitude permeability linearly approached the initial permeability,

i.e. $\mu_a = \mu_i + v\hat{H}$ (2.28)

where v is a hysteresis coefficient having the units $(A/m)^{-1}$.

Multiplying by $\mu_0 \hat{H}$ this relation becomes

$$\hat{B} = \mu_0(\mu_i \hat{H} + v\hat{H}^2) \quad T \quad (2.29)$$

He further observed that, for the specimen examined, the sides of the hysteresis loop could be represented by parabolic curves such that

$$B = \mu_0\{(\mu_i + v\hat{H})H \pm \frac{v}{2}(\hat{H}^2 - H^2)\} \quad T \quad (2.30)$$

where B and H are instantaneous values and $\hat{H}$ is the maximum value of H.

It should be noted that Rayleigh used only one constant to express the change of permeability with field strength and to express the width of the hysteresis loop, two properties of a magnetic material which are not obviously related.

The hysteresis loop is essentially a d.c. loop. With alternating magnetization the non-hysteresis losses will result in a phase angle between B and H. This will give rise to an elliptical B–H relation which will be superimposed on the hysteresis loop. If hysteresis predominates in the range of amplitudes for which the Rayleigh loop applies then the overall loop will tend to be parabolic. If the other losses predominate the loop will tend to be elliptical, but the parabolic component and the associated hysteresis will still be present.

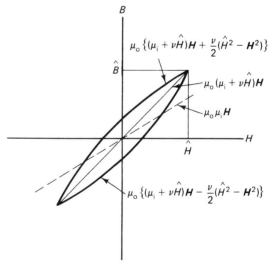

Figure 2.3 The Rayleigh loop

(2.28) $\mu_a = \mu_i + v\hat{H}$

v is a hysteresis coefficient having the units Oe^{-1}

$v_{CGS} = \frac{10^3}{4\pi} v_{SI}$

(2.29) $\hat{B} = \mu_i \hat{H} + v\hat{H}^2$ Gs

(2.30) $B = \{(\mu_i + v\hat{H})H \pm \frac{v}{2}(\hat{H}^2 - H^2)\}$ Gs

32   The expression of electrical and magnetic properties

The Rayleigh loop is shown in Figure 2.3. From Eqn (2.30), the vertical width of the loop at any given field strength is $\mu_o v(\hat{H}^2 - H^2)$ and, by integration of this width from $-\hat{H}$ to $+\hat{H}$, the area of the loop is $4\mu_o v\hat{H}^3/3$. The energy lost during each cycle in a unit volume of the material due to hysteresis is equal to the loop area. It is called the hysteresis energy loss (volume) density and is denoted by w_h

$$w_h = \oint B dH = \frac{4\mu_o v\hat{H}^3}{3} = \frac{4v\hat{B}^3}{3\mu_o^2\mu_a^3} \quad \text{J m}^{-3}\text{cycle}^{-1} \tag{2.31}$$

The hysteresis power dissipated in a core of volume Al which is subjected to an alternating field of amplitude $\hat{H}$ and frequency f is

$$P_h Al = w_h Alf$$
$$= I^2 R_h \quad \text{W}$$

where R_h is the series hysteresis loss resistance and I is the r.m.s. current $= \hat{H}l/\sqrt{2}N$.

Combining these equations with Eqn (2.31)

$$R_h = \frac{4v\hat{B}}{3\pi\mu_o\mu_a^2} 2\pi f \cdot \frac{N^2 A \mu_o \mu_a}{l} \quad \Omega$$

$$= \frac{4v\hat{B}}{3\pi\mu_o\mu_a^2} \omega L \quad \Omega \tag{2.32}$$

∴ The hysteresis loss tangent:

$$\tan \delta_h = \frac{R_h}{\omega L} = \frac{4v\hat{B}}{3\pi\mu_o\mu_a^2}$$

$$= \frac{4v\hat{H}}{3\pi\mu_a} \tag{2.33}$$

As far as a.c. bridge measurements are concerned, the Rayleigh relations are embodied in Eqns (2.28) and (2.33). For a magnetic material to obey the Rayleigh relations the measured properties should follow each of these equations with the same value of constant, or, combining the equations to eliminate the constant, it should obey the following equation:

$$\mu_a = \mu_i + \frac{3\pi\mu_a \tan \delta_h}{4}$$

or $\quad \mu_a = \dfrac{\mu_i}{1 - \dfrac{3\pi \tan \delta_h}{4}} \tag{2.34}$

There is no obvious reason why this relation should be obeyed however low the field strength is. The Rayleigh region is a term loosely applied to the low amplitude region in which $\tan \delta_h$ is approximately proportional to $\hat{B}$ or $\hat{H}$, the value of μ_a in Eqn (2.33) being assumed equal to μ_i; this is equivalent to saying that the hysteresis energy loss is proportional to $\hat{B}^3$ (see Eqn 2.31).

Figure 3.5 shows relations between μ_a and low-amplitude flux density measured on typical ferrites. Values of v have been determined from the tangents of the curve at zero flux density. Figure 3.15 gives the corresponding $\tan \delta_h$ relations measured on the same specimens and from these the values of v have also been derived. Comparison shows that, for the materials considered, the two independently derived values of v approximately agree. However, in both figures, the relations are not linear but tend to fall away from the tangent at $\hat{B} = 0$ as the flux density increases.

2.2.3.2 The Peterson relations

Peterson,[6] in 1928, expressed the flux density as a double power series of the instantaneous and maximum field strengths $(H, \hat{H})$. Making certain assumptions about the loop symmetry and ignoring terms higher than those of the third degree, he obtained for the magnetization curve (not loop)

$$\left. \begin{array}{l} B = \mu_o[a_{10}H + a_{11}H^2 + (a_{12} + a_{30})H^3 \ldots] \quad \text{T} \\ \text{or } \mu_a = a_{10} + a_{11}H + (a_{12} + a_{30})H^2 \ldots \end{array} \right\} \tag{2.35}$$

where a_{10}, etc., are coefficients in the power series.

In particular $a_{10} =$ the initial permeability μ_i

$$a_{11} = d\mu/dH \text{ and has the units } (\text{A/m})^{-1}$$

From the power series he also deduced that the hysteresis energy (volume) density is

$$w_h = \mu_o \frac{8}{3}(a_{02}\hat{H}^3 + a_{03}\hat{H}^4 + \ldots)$$

$$\text{J m}^{-3}\text{cycle}^{-1} \tag{2.36}$$

where a_{02} has the units $(\text{A/m})^{-1}$.

(2.31) $\quad w_h = \dfrac{1}{4\pi} \times \text{loop area} = \dfrac{1}{3\pi}v\hat{H}^3 = \dfrac{v\hat{B}^3}{3\pi\mu_a^3}$
$\qquad\qquad\qquad\qquad\qquad\qquad\qquad\qquad\qquad \text{ergs cm}^{-3}\text{cycle}^{-1}$

(2.32) $\quad R_h = \dfrac{4v\hat{B}}{3\pi\mu_a^2}\omega L \quad \Omega$

(2.33) $\quad \dfrac{R_h}{\omega L} = \tan \delta_h = \dfrac{4v\hat{B}}{3\pi\mu_a^2} = \dfrac{4v\hat{H}}{3\pi\mu_a}$

(2.35) $\left\{ \begin{array}{l} B = a_{10}H + a_{11}H^2 + (a_{12} + a_{30})H^3 \ldots \quad \text{Gs} \\ \mu_a = a_{10} + a_{11}H + (a_{12} + a_{30})H^2 \end{array} \right.$

where $a_{10} =$ the initial permeability $= \mu_i$

$$a_{11} = d\mu/dH \text{ and has the units Oe}^{-1}$$

$$(a_{11})_{\text{CGS}} = \frac{10^3}{4\pi}(a_{11})_{\text{SI}}$$

Therefore at low field strength, using the same reasoning as that used to obtain Eqn (2.33) and taking only the first term of Eqn (2.36),

$$\frac{R_\mathrm{h}}{\omega L} = \tan \delta_\mathrm{h} = \frac{8 a_{02} \hat{B}}{3 \pi \mu_\mathrm{o} \mu_\mathrm{a}^2} \quad (2.37)$$

Whereas Rayleigh started with limited experimental observations and found relations to express them, Peterson assumed only that the loop is symmetrical about the origin, that the locus of its tips is the initial magnetization curve and that it disappeared when $H = 0$. Thus Peterson's approach was more general; provided the hysteresis loop is regarded as a continuous function and enough terms are taken, the hysteresis loop can always be expressed.

Comparing Eqns (2.28) and (2.35) and ignoring Peterson's third degree term

$$a_{10} = \mu_\mathrm{i} \quad \text{and} \quad a_{11} = \nu \quad (2.38)^*$$

Similarly comparing Eqns (2.33) and (2.37)

$$a_{02} = \frac{\nu}{2} \quad (2.39)^*$$

Thus for a material which satisfies the Rayleigh relations the following correspondence must exist in the Peterson coefficients:

$$a_{11} = 2 a_{02} \quad (2.40)$$

Both Rayleigh and Peterson considered only the loss of energy that is proportional to the area of the low frequency or static B–H loop. This is hysteresis loss by definition. Also, they were only concerned with the low field strength region. More recently expressions have been derived that give the B–H loops and μ–H curves from low field strengths up to saturation.[7] Such expressions are useful in the Computer Aided Design of systems involving magnetic components.

---

*The discrepancy between these equations and those of Peterson's original paper is due to his use of 2ν in place of ν used here.

(2.36) $\quad w_\mathrm{h} = \dfrac{2}{3\pi}(a_{02}\hat{H}^3 + a_{03}\hat{H}^4 + \ldots)10^{-7}$ J cm^{-3} cycle^{-1}

where a_{02} is a Peterson hysteresis coefficient having the units Oe^{-1}

$$(a_{02})_\mathrm{CGS} = \frac{10^3}{4\pi}(a_{02})_\mathrm{SI}$$

(2.37) $\quad \dfrac{R_\mathrm{h}}{\omega L} = \tan \delta_h = \dfrac{8 a_{02} \hat{B}}{3\pi \mu_\mathrm{a}^2}$

2.2.4 Eddy current phenomena

2.2.4.1 Resistivity and permittivity

Ferrites are semiconductors and commercial grades have resistivities ranging from $0.1\,\Omega\,\mathrm{m}$ to greater than $10^6\,\Omega\,\mathrm{m}$ ($10\,\Omega\,\mathrm{cm}$ to $> 10^8\,\Omega\,\mathrm{m}$) at room temperatures. Associated with the resistivity there is an effective permittivity which in the case of the low resistivity ferrites can be as high as 100 000. A ferrite rarely forms a part of a physical electrical circuit but nevertheless the dielectric properties are important. Significant eddy currents may flow in the bulk of a ferrite core and give rise to energy losses or dispersion of permeability. It is also possible that a ferrite core may contribute to the stray capacitance of a winding. In this section the theory of ferrite dielectric properties will be considered.[8]

The bulk resistivity may be measured on a block of ferrite by the four-probe method, i.e. a known direct current is passed through the block by means of electrodes at each end of the block and the voltage drop across two intermediate electrodes of known separation is measured with a high resistance voltmeter. The ratio of the voltage to the current gives the resistance of the section between the voltage probes. The result is independent of the contact resistance of the probes. Alternatively metallic contacts may be deposited on each end of the block. If the contact resistance can be made negligible the resistivity, ρ, and the relative parallel permittivity, ε, of the block may be measured in terms of the equivalent two terminal impedance (or admittance). On an admittance bridge the equivalent admittance is expressed in the parallel components of Figure 2.4.

If A is the cross-sectional area of the block and l is the length separating the contacts then

$$\left.\begin{array}{l} C_\mathrm{p} = \varepsilon_\mathrm{o} \varepsilon \dfrac{A}{l} \quad \mathrm{F} \\[1em] R_\mathrm{p} = \dfrac{\rho l}{A} \quad \Omega \end{array}\right\} \quad (2.41)$$

where ε_o is the electric constant or permittivity of free space $= 8.854 \times 10^{-12}$ F m^{-1}. It has the dimensions $[\mathrm{L}^{-3}\mathrm{M}^{-1}\mathrm{T}^4\mathrm{I}^2]$.

An alternative presentation is the parallel complex permittivity which is analogous to parallel complex permeability (see Eqn (2.17)).

(2.41) $\left\{\begin{array}{l} C_\mathrm{p} \simeq \dfrac{\varepsilon A 10^{-12}}{3.6\pi l} \quad \mathrm{F} \\[0.5em] \quad = 0.0885 \varepsilon A/l \quad \mathrm{pF} \\[0.5em] R_\mathrm{p} = \rho l / A \quad \Omega \end{array}\right.$

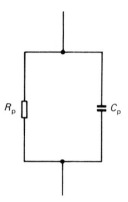

Figure 2.4 Equivalent circuit of a dielectric

Then

$$\varepsilon = \varepsilon'_p - j\varepsilon''_p$$

$$\therefore Y = j\omega C_p + \frac{1}{R_p} = j\omega\varepsilon_o \frac{A}{l}(\varepsilon'_p - j\varepsilon''_p) \ \Omega^{-1}$$ (2.42)

$$\therefore \varepsilon''_p = \frac{1}{\omega\varepsilon_o A R_p} = \frac{1}{\omega\varepsilon_o \rho}$$

The dielectric loss tangent is given by

$$\tan\delta_d = \frac{1}{\omega C_p R_p} = \frac{\varepsilon''_p}{\varepsilon'_p} = \frac{1}{\omega\varepsilon_o \varepsilon'_p \rho}$$ (2.43)

Some writers remove the d.c. resistance contribution from ε''_p thus representing the impedance as a constant d.c. resistance in parallel with a capacitance having dielectric loss. This is not the procedure adopted here.

Consideration of the polycrystalline structure of ferrites and the variation of their resistivity and permittivity with frequency (see Figure 3.22) has led to the view that polycrystalline ferrite may be regarded as a compound dielectric consisting of semi-conducting grains (crystallites) surrounded by thin boundaries having much higher resistivity. These grain boundaries generally consist of non-soluble material phases which are transported on the boundaries as the crystallite grows during sintering. As well as having high resistivity they are generally non-magnetic. They are usually composed of silicon and calcium oxides; indeed, such constituents are often introduced into the initial composition in order to control grain growth and increase the bulk resistivity of the sintered product. Figure 2.5 shows a unit cube of material, of resistivity ρ_1 and real permittivity ε_1, having a boundary layer, of thickness a, with corresponding dielectric properties ρ_2 and ε_2. This represents a single crystallite. The admittance between the faces A and B will characterize the admittance of a material built up of a chain of such blocks, or more generally, a polycrystalline material. Of course, in practice there is a wide distribution of grain size and shape and this representation is very much over-simplified.

Applying Eqn (2.41), the crystallite and boundary may be represented by the equivalent electrical circuit as shown. This circuit may at a given frequency be resolved into a capacitance, $\varepsilon_o \varepsilon$ in parallel with a resistance, ρ; ε and ρ represent the apparent dielectric

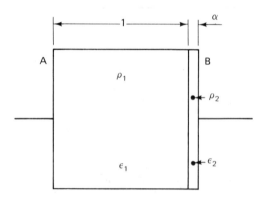

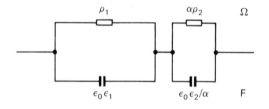

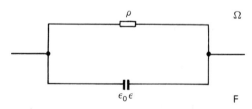

Figure 2.5 Equivalent circuit of a crystallite and boundary

(2.42) $Y = j\omega C_p + \frac{1}{R_p} \simeq \frac{j\omega A 10^{-12}}{3.6\pi l}(\varepsilon'_p - j\varepsilon''_p) \ \Omega^{-1}$

$\therefore \varepsilon''_p \simeq \frac{3.6\pi/10^{12}}{\omega A R_p} = \frac{3.6\pi 10^{12}}{\omega\rho}$

(2.43) $\tan\delta_d = \frac{1}{\omega C_p R_p} = \frac{\varepsilon''_p}{\varepsilon'_p} \simeq \frac{3.6\pi 10^{12}}{\varepsilon'_p \omega\rho}$

properties of the composite dielectric.

It has been found that in polycrystalline ferrites $\alpha \ll 1$, $\varepsilon_2 \approx \varepsilon_1$ and $a\rho_2 \gg \rho_1$. Therefore at low frequencies the impedance of the crystallite is negligible compared to that of the boundary; so the resistivity approximates to $a\rho_2$ while the permittivity approaches ε_2/α. This is analogous to calculating dielectric properties from measurements on a specimen between the plates of a capacitor, using a dielectric length $1/\alpha$ times the actual value.

At very high frequencies the boundary capacitance short circuits the boundary resistance and the bulk dielectric properties approach those of the crystallite.

Summarizing

$$\left. \begin{array}{l} \text{when } f \to 0 \quad \rho \to a\rho_2 \\ \qquad\qquad\quad \varepsilon \to \varepsilon_2/\alpha \\ \text{when } f \to \infty \quad \rho \to \rho_1 \\ \qquad\qquad\quad \varepsilon \to \varepsilon_1 \end{array} \right\} \quad (2.44)$$

It follows that if the boundary layer is relatively thin, i.e. $\alpha \ll 1$, then at low frequencies the bulk resistivity and the permittivity will both be high. The properties will vary from the low-frequency values to the high frequency values in accordance with a relaxation curve.

The relaxation time for both ρ and ε is given by

$$\tau = \varepsilon_0 \frac{\varepsilon_1 + \varepsilon_2/\alpha}{\frac{1}{\rho_1} + \frac{1}{a\rho_2}} \quad \text{s} \quad (2.45)$$

$$\tau = 1/\omega_r$$

where ω_r is $2\pi \times$ the relaxation frequency. This is the frequency at which the values of the parameters are midway between the values at the two extreme frequencies. Figure 3.22 shows some experimental results.

2.2.4.2 Eddy current loss

An alternating magnetic flux in a conductive medium will induce eddy currents in that medium and these will result in an energy loss called eddy current loss.

The magnitude of this loss depends, among other things, on the size and shape of the conductive medium and, as is well known, may be reduced by subdivision of the medium into electrically insulated regions, e.g. laminations or grains. When a material is thus subdivided the magnitude of the eddy current loss per unit volume depends on the size and shape of the insulated regions and not on the shape of the bulk material. Ferrites, on the other hand, are usually regarded as homogeneous materials as far as eddy currents are concerned. Under certain circumstances significant micro-eddy currents may circulate within the crystallites and cause additional loss.[9] Any loss due to this cause is usually included with the residual loss and not separately considered; this is the practice adopted in these chapters.

At any given frequency, the effective dielectric properties discussed in the previous section may be used to calculate the magnitudes of the eddy currents that may flow through the material as a whole. Figure 3.22 shows some measured dielectric properties.

At low frequencies, where the inductive effect of the eddy currents may be neglected, the eddy current power loss (volume) density is given by

$$P_F = \frac{(\pi \hat{B} f d)^2}{\rho \beta} \quad \text{W m}^{-3} \quad (2.46)$$

where $\hat{B}$ is the peak value of the flux density assumed perpendicular to the plane containing the cross-sectional dimension d, ρ is the bulk resistivity in Ω m and
$\beta = 6$ for laminations of thickness d m,
$= 16$ for a cylinder of diameter d m,
$= 20$ for a sphere of diameter d m.

Using a similar reasoning to that which led to Eqn (2.33) the eddy current core loss tangent is:

$$\tan \delta_F = \frac{R_F}{\omega L} = \frac{\pi \mu_0 \mu d^2 f}{\rho \beta} \quad (2.47)$$

where R_F is the eddy current series loss resistance.

These expressions may be used to calculate the eddy current loss in simple core shapes; the result for a square cross-section bar may be taken as approximately equal to that for the cylinder of the same cross-sectional area. For more complicated shapes, expressions must be derived from first principles. An elementary eddy current path is assumed and the flux linkage is expressed in terms of its dimensions. From the corresponding induced e.m.f. and the resistance of the path, an expression for the power loss may be obtained. This may then be integrated over the whole cross-section. An example of such a calculation, albeit for a copper strip with transverse flux, is given in Section 11.4.3.

(2.45) $\quad \tau = \dfrac{10^{-12}}{3.6\pi} \times \dfrac{\varepsilon_1 + \varepsilon_2/\alpha}{\dfrac{1}{\rho_1} + \dfrac{1}{a\rho_2}} \quad$ s

(2.46) $\quad P_F = \dfrac{(\pi \hat{B} f d)^2}{\rho \beta} 10^{-16} \quad \text{W cm}^{-3}$

(2.47) $\quad \tan \delta_F = \dfrac{R_F}{\omega L} = \dfrac{4\pi^2 \mu d^2 f}{\rho \beta} 10^{-9}$

2.2.4.3 Dimensional resonance

The calculation of eddy current loss in the previous section assumed that the flux density is uniformly distributed across the section of the core. At higher frequencies this may not be true because in some ferrites the high values of permeability and permittivity give rise to standing electromagnetic waves within the ferrite. This is called dimensional resonance.

In a loss-free medium the velocity of propagation of electromagnetic waves is given by

$$v = (\mu_o \mu \varepsilon_o \varepsilon)^{-1/2} = f\lambda \quad \text{m s}^{-1} \qquad (2.48)$$

where λ = the wavelength in the medium.

In a typical manganese zinc ferrite, $\mu = 10^3$ and $\varepsilon = 10^5$.
Therefore at a frequency of 1 MHz

$$v = (4\pi \times 10^{-7} \times 10^3 \times 8.854 \times 10^{-12} \times 10^5)^{-1/2}$$

$$\simeq 3 \times 10^4 \quad \text{m s}^{-1}$$

and $\lambda \simeq 0.03$ m

If the smallest cross-sectional dimension of the core (perpendicular to the magnetic field) is half a wavelength, i.e. 15 mm, then a fundamental mode standing wave will be set up across the section. Under these conditions the net reactive flux is zero; the surface flux which is in phase with the surface magnetic field is cancelled by antiphase flux at the centre of the core. Thus dimensional resonance is characterized by the observed permeability dropping to zero. Since the phenomenon is essentially electromagnetic there is a corresponding standing wave in the electric flux density. This gives rise to a similar dispersion in the observed permittivity. It occurs when the smallest dimension of the cross-section perpendicular to the electric field equals half a wavelength. Since the electric and magnetic fields are perpendicular both standing waves appear across the same dimension.

If the medium has some loss then μ and ε in Eqn (2.48) must be replaced by $\mu'_s - j\mu''_s$ and $\varepsilon'_p - j\varepsilon''_p$. If $\tan \delta_m = \mu''_s/\mu'_s$ and $\tan \delta_d = \varepsilon''_p/\varepsilon'_p$, and $\tan \delta_m \tan \delta_d \ll 1$ then dimensional resonance will still be observed but the resonant dimension is now obtained by replacing $\mu\varepsilon$ in Eqn (2.48) by the real part of the complex product. The resonant dimension is then given by[10]

$$(2.48) \quad v = \frac{c_o}{\sqrt{(\mu\varepsilon)}} = f\lambda \quad \text{cm s}^{-1}$$

where c_o = velocity of electromagnetic waves in vacuo

$$\simeq 3 \times 10^{10} \quad \text{cm s}^{-1}$$

λ = wavelength in cm

$$\frac{\lambda}{2} = \frac{\sqrt{2}\pi c_o}{\omega\sqrt{[|\mu||\varepsilon| + \mu'_s\varepsilon'_p(1 - \tan \delta_m \tan \delta_d)]}} \quad \text{m} \quad (2.49)$$

where c_o is the velocity of electromagnetic waves in vacuo = $(\mu_o \varepsilon_o)^{-1/2}$

In such a medium there is a quadrature standing wave which gives rise to a peak in the observed magnetic loss, μ''_s, and dielectric loss ε''_p at resonance.

If $\tan \delta_m \tan \delta_d \gg 1$ the material will not support standing waves and there is no dimensional resonance. Instead the flux will be attenuated as it is propagated through the cross-section and there will be a penetration depth, Δ, as in metal magnetic materials.[10]

$$\Delta = \frac{\sqrt{2}c_o}{\omega\sqrt{[|\mu||\varepsilon| - \mu'_s\varepsilon'_p(1 - \tan \delta_m \tan \delta_d)]}} \quad \text{m} \quad (2.50)$$

As with the resonant dimension, Δ is measured perpendicular to the magnetic field or the electric field. In calculating $\lambda/2$ and Δ, the variation of the permeability and permittivity components with frequency must be taken into account. Relevant data may be found in Figures 3.11 and 3.22. Strictly the above expressions apply only to the thickness of an infinite plate but in practice they may be applied to practical core cross-sections for the purpose of estimation. Using the properties of typical manganese zinc ferrites, $\lambda/2$ has been calculated from Eqn (2.49) and the results are given in Figure 4.7.

An analytical study of dimensional resonance was made by Brockman et al.[11] They give expressions for the real and imaginary parts of the complex permeability and permittivity as functions of frequency and they calculate these functions for a typical manganese zinc ferrite. Their results are reproduced in Figure 2.6.

2.2.5 Residual loss

Two forms of magnetic loss have so far been distinguished, namely hysteresis loss which gives rise to a loss tangent proportional to $\hat{B}$ (see Eqn (2.33)) and eddy current loss which gives rise to a loss tangent proportional to f (see Eqn (2.47), i.e.

$$\tan \delta_h \propto \hat{B}$$

$$\tan \delta_F \propto f$$

If, for a metallic magnetic material, e.g. a laminated or powder core, the total loss tangent is measured as a function of $\hat{B}$ and f at low frequencies where eddy current screening effects are negligible, a set of curves similar to that shown in Figure 2.7 is obtained. It is seen that when $B \to 0$ and $f \to 0$, the total loss tangent is not zero. The remainder is called the residual loss.

The total loss tangent may be written

The expression of electrical and magnetic properties 37

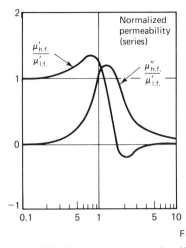
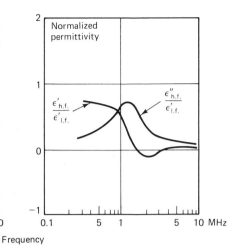

Figure 2.6 Dimensional resonance in a 12.5 mm thick sample of manganese zinc ferrite, calculated from typical complex permeability and permittivity data. The imaginary part of the permittivity has the contribution from the d.c. resistance removed; see Eqn (2.42). The results are normalized with respect to the low-frequency values. (Courtesy Brockman et al.[11])

$$\tan \delta_m = \tan \delta_h + \tan \delta_F + \tan \delta_r \quad (2.51)$$

where $\tan \delta_r$ is the residual loss tangent.

This is the original definition of residual loss. In a ferrite, which has high resistivity, it is very easy to ensure that the bulk eddy current loss is negligible by making measurements on an adequately small sample. On such a sample it is also possible to measure the loss tangent at vanishingly small flux densities. By this procedure the bulk eddy current and hysteresis losses may be eliminated at any frequency and the remaining loss tangent is regarded as residual loss. This is the usual meaning of the term residual loss in ferrites, and the residual loss tangent expressed as a function of frequency is an important loss parameter relating to ferrites.

There are several loss processes that contribute to residual loss and different processes may apply at different parts of the spectrum. A brief discussion of this aspect of residual loss is given in the introduction to Figure 3.11 and Figure 3.12 where typical residual loss spectra are illustrated.

2.2.6 General loss expressions

2.2.6.1 Loss factor

If a loss-free winding having N turns is placed round a closed magnetic core and an alternating voltage U is applied, a power loss (volume) density, $P_m = U^2 G/Al$, will be observed, where G is the conductance appearing across the winding due to the magnetic loss in the core. The induced e.m.f, the frequency and flux density are related by Eqn (2.13). If the loss tangent is small the applied voltage will approximately equal the induced e.m.f. so

$$U = \frac{\hat{B}AN\omega}{\sqrt{2}} \quad \text{V} \quad (2.52)$$

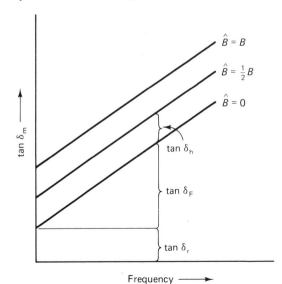

Figure 2.7 Loss tangent as a function of frequency; illustration of Eqn (2.51)

$$(2.52) \quad U = \frac{\hat{B}AN\omega}{\sqrt{2}} \times 10^{-8} \quad \text{V}$$

Thus for a given core cross-sectional area A and number of turns, the voltage and frequency determine the flux density.

The loss tangent due to the magnetic core,

$$\tan \delta_m = \frac{\omega L_p}{R_p} = \omega L_p G$$

It has already been postulated that $\tan \delta_m \ll 1$, therefore the distinction between series and parallel inductance may be dropped.

Substituting $L_p = \mu_0 \mu N^2 A/l$

$$\tan \delta_m = \omega \mu_0 \mu N^2 A G/l \qquad (2.53)$$

Putting $G = P_m Al/U^2$ and substituting for U from Eqn (2.52)

$$\frac{\tan \delta_m}{\mu} = \frac{\mu_0}{\pi f \hat{B}^2} P_m \qquad (2.54)$$

The factor $(\tan \delta_m)/\mu$ depends on the power loss (volume) density and this depends on the frequency and flux density. If the eddy current core loss is negligible the loss density is independent of core shape and is a material property dependent only on f and B. Under these conditions $(\tan \delta_m)/\mu$ is also a material property and a function of only f and B.

The analysis of the effect of an air gap must now be anticipated. If an air gap is introduced into a magnetic circuit the inductance will be reduced so that the core appears to have an effective permeability μ_e. It is shown in Chapter 4 that the loss tangent of a gapped core is related to the material loss tangent by the following expression (see Eqn 4.49):

$$\frac{(\tan \delta_m)_{gapped}}{\mu_e - 1} = \frac{\tan \delta_m}{\mu - 1} \qquad (2.55)$$

These expressions are referred to as loss factors. Because μ_e and μ are usually much greater than unity the above equation is usually approximated to

$$\frac{(\tan \delta)_{gapped}}{\mu_e} = \frac{\tan \delta}{\mu} \qquad (2.56)$$

The subscript m has been dropped because clearly the loss tangent need not be due to the total loss: it may refer to any specific form of loss. Within the conditions stated the loss factor is independent of the gap and therefore for those ferrites that are normally gapped in use it is an important material property. Residual loss of a ferrite material is frequently expressed by the corresponding loss factor.

An important practical consequence of Eqn (2.56) is that the loss tangent due to the magnetic loss of a gapped core is given by

$$(\tan \delta)_{gapped} = \left(\frac{\tan \delta}{\mu}\right) \mu_e \qquad (2.57)$$

i.e. loss factor × effective permeability

Considering again the total loss factor, Eqn (2.51) may be written in terms of loss factors

$$\frac{\tan \delta_m}{\mu} = \frac{\tan \delta_h}{\mu} + \frac{\tan \delta_F}{\mu} + \frac{\tan \delta_r}{\mu} \qquad (2.58)$$

Substituting for the loss tangent expressions from Eqns (2.33) and (2.47) and letting $\mu_a \to \mu$:

$$\frac{\tan \delta_m}{\mu} = \frac{4v\hat{B}}{3\pi \mu_0 \mu^3} + \frac{\pi \mu_0 d^2 f}{\rho \beta} + \frac{\tan \delta_r}{\mu} \qquad (2.59)$$

$$= k_1 \hat{B} + k_2 f + k_3 \qquad (2.60)$$

where k_1 and k_3 are hysteresis and residual loss coefficients characterizing the material and k_2 is an eddy current loss coefficient that depends on the material and its shape. Eqn (2.60) is the basis of a number of loss expressions that have been used in the past.[12,13,14] As a result of the activity of IEC Technical Committee 51, for example Publication 401,[15] most of these have now been discarded in favour of internationally agreed standard expressions. Before quoting these, it is of interest to mention the expression introduced by Legg[13] in 1936 to characterize loss measurements made on magnetic laminations:

$$\frac{R_s}{\mu f L} = a\hat{B} + ef + c \qquad (2.61)$$

where a is the hysteresis loss coefficient in T^{-1},
e is the eddy current loss coefficient in seconds,
c is the residual loss coefficient and is a number.
[In CGS units a has the units Gs^{-1}
$\therefore (a)_{CGS} = 10^{-4}(a)_{SI}$]
This expression is directly equivalent to Eqn (2.60). Since $R_s/\mu f L$ equals $(2\pi \tan \delta_m)/\mu$ it may be written in terms of loss factor:

$$\frac{\tan \delta_m}{\mu} = \frac{1}{2\pi}(a\hat{B} + ef + c) \qquad (2.62)$$

Although this expression corresponds to that proposed by Legg it should be noted that strictly the loss factor has $\mu - 1$ in the denominator, see Eqn (2.55).

(2.53) $\tan \delta_m = \omega 4\pi \mu N^2 AG \, 10^{-9}/l$

(2.54) $\dfrac{\tan \delta_m}{\mu} = \dfrac{4P_m 10^7}{f\hat{B}^2}$

(2.59) $\dfrac{\tan \delta_m}{\mu} = \dfrac{4v\hat{B}}{3\pi\mu^3} + \dfrac{4\pi^2 d^2 f \times 10^{-9}}{\rho \beta} + \dfrac{\tan \delta_r}{\mu}$

The IEC uses $(\tan\delta_r)/\mu$ and $(\tan\delta_F)/\mu$ for the residual and eddy current loss factors respectively, and the International Electrotechnical Vocabulary[16] defines a hysteresis constant for the material, η_B, by the following expression:

$$\frac{\tan\delta_h}{\mu} = \eta_B \hat{B} \qquad (2.63)$$

where η_B is in T^{-1} if $\hat{B}$ is in T.

There is a corresponding hysteresis constant for complete cores; this will be introduced in Chapter 4, Eqn (4.44).

The IEV also defines a parameter quantifying the μ–H dependence expressed in the Rayleigh relation given in Eqn (2.28). It is the permeability rise factor, δ_H:

$$\delta_H = \frac{\mu_{a2} - \mu_{a1}}{\mu_{a1}(\hat{H}_2 - \hat{H}_1)} \qquad (2.64)$$

where μ_{a1} and μ_{a2} are the amplitude permeabilities measured at field strengths $\hat{H}_1$ and $\hat{H}_2$ respectively.

2.2.6.2 Discussion of loss expressions

The losses in magnetic materials are fundamentally energy losses and may be expressed as such, e.g. in terms of $J\,m^{-3}\,cycle^{-1}$, but at low amplitudes it is more convenient to express them in terms of loss angle or loss resistance. This is because at these low amplitudes the influence of the core on the associated electrical circuit is best studied by expressing the core properties in terms of resistances and inductances. These may then be readily combined with other circuit elements by the normal processes of circuit theory. In particular, time constants and Q-factors may easily be derived.

It has been previously observed that the power loss per unit volume is a function of flux density and frequency. A double power series in f and B may easily be set up to express the loss per unit volume. Common observation will lead to the discarding of some of the lower order terms. If all the higher order terms are also discarded one obtains an expression identical in form with Eqn (2.60). Such an exercise does not introduce such concepts as hysteresis or eddy current loss and yet for $(\tan\delta_m)/\mu$ it yields one term proportional to B, one proportional to f and another that is independent of either. The point of this digression is that although it is appropriate to describe these terms as due to hysteresis, eddy current and residual loss respectively, in practice the adequate expression of the behaviour of a material requires higher terms and the distinction between the phenomena can become obscure.

It is not usual to invoke these higher power terms. It is more convenient to express the losses in general by three coefficients and to regard these coefficients as empirical functions of frequency and flux density. Thus the hysteresis coefficient is usually specified at a low frequency and at a low flux density, under which conditions it may be regarded as a material constant. If the frequency or flux density is raised appreciably the measured value of the hysteresis coefficient will change and it is convenient to express this change by means of a graph which is characteristic of the particular material.

The eddy current loss coefficient depends on the degree of sub-division in the case of laminated or powdered cores, or, in the case of homogeneous cores, on the dimensions of the core cross-section. Ferrite cores, although not entirely homogeneous, come into the latter category so the eddy current loss coefficient is not a material property but depends on the core size and shape (see Eqns (2.59) and (2.60)). For this reason the eddy current loss coefficient is rarely used in a ferrite material specification; the value of the bulk resistivity is a more useful parameter.

So the loss coefficients most commonly used to express losses in ferrites are the hysteresis and the residual loss coefficients. In this context residual loss is taken as referring to the loss angle or loss resistance remaining when the eddy current loss is negligible and the flux density is vanishingly small. In practice it depends on frequency but by definition does not depend on flux density.

Some experimental data on these losses are given in Chapter 3.

2.2.7 Waveform distortion and intermodulation

At very low field strengths the low frequency hysteresis loop approximates to the parabolic form considered by Rayleigh and Peterson; it is illustrated in Figure 2.3. If a sinusoidal field of constant peak amplitude $\hat{H}$ is applied to such a material the flux density will alternate at a peak amplitude $\hat{B}$ but due to the non-linear relation between $\boldsymbol{B}$ and $\boldsymbol{H}$ the waveform will be distorted. In the absence of magnetic bias the loop has symmetry about the origin so this distortion may be expressed in terms of only the odd harmonics of the applied field.

Assuming the parabolic loop equation, an expression may easily be derived for the flux density as a function of time and by Fourier analysis the amplitude of the harmonics may be obtained. Peterson's[6] expression for the flux density when the applied field is $\hat{H}\cos pt$ is

40 The expression of electrical and magnetic properties

$$B = \mu_o \left\{ \hat{H}(a_{10} + a_{11}\hat{H}) \cos pt + \frac{8a_{02}}{3\pi} \hat{H}^2 \sin pt \right.$$

$$\left. + a_{30} \frac{\hat{H}^3}{4} \cos 3pt - \frac{8a_{02}}{15\pi} \hat{H}^2 \sin 3pt \ldots \right\} \quad (2.65)$$

The first term is the fundamental frequency inductance term, the second is the fundamental frequency hysteresis loss term, the third term is the third harmonic resulting from the permeability change and is negligible, and the fourth is the main third harmonic term. The third harmonic term depends on the hysteresis loss coefficient a_{02} (see Eqn (2.36). It usually has much larger amplitude than any of the higher order harmonics and its amplitude is usually taken as a measure of the total distortion. Denoting the fundamental and the third harmonic by subscripts a and $3a$ respectively, the amplitude of the third harmonic flux density is

$$\hat{B}_{3a} = \frac{8}{15\pi} \mu_o a_{02} \hat{H}_a^2 \quad \text{T} \quad (2.66)$$

$$\frac{\hat{B}_{3a}}{\hat{B}_a} = \frac{8a_{02}\hat{B}_a}{15\pi\mu_o\mu^2} \quad (2.67)$$

Since the induced e.m.f. is proportional to frequency

$$\frac{E_{3a}}{E_a} = \frac{3\hat{B}_{3a}}{\hat{B}_a} = \frac{8a_{02}\hat{B}_a}{5\pi\mu_o\mu^2}$$

$$= 0.6 \tan \delta_h \quad (2.68)$$

from Eqn (2.37).

If the amplitude of the applied field is increased, the loop will increase in length and breadth and $\tan \delta_h$ will increase in proportion to $\hat{H}$ (see Eqn (2.33)). At the same time the slope of the major axis of the loop, i.e. the permeability, will increase in accordance with Eqn (2.35). The permeability rise has negligible effect in the production of third harmonic distortion. How-

ever, in the intermodulation of two simultaneously applied currents of different frequency, both hysteresis phenomena play a part.

If two alternating fields of equal amplitude $\hat{H}$ and angular frequencies $2\pi f_a = (p + \omega)$ and $2\pi f_b = (p - \omega)$ respectively are simultaneously applied to a magnetic material the resultant field has the equation:

$$H = \hat{H} \cos 2\pi f_a t + \hat{H} \cos 2\pi f_b t = 2\hat{H} \cos \omega t \cos pt \quad (2.69)$$

The combined r.m.s. amplitude is numerically equal to $\hat{H}$, i.e. it is $\sqrt{2}$ times the r.m.s. amplitude of the single frequency field considered in the analysis of third harmonic. If the difference between f_a and f_b is small, i.e. $\omega \ll p$, then Eqn (2.69) represents a wave of angular frequency p, the peak amplitude of which varies relatively slowly from $2\hat{H}$ to $-2\hat{H}$. Thus the distortion arising in the flux wave will be partly due to the parabolic shape of the loop (characterized by the hysteresis loss coefficient) and partly due to the rise and fall of the slope of the loop axis (permeability) as the amplitude varies. The flux wave will contain components corresponding to the intermodulation of the two applied fields.

Latimer[17] in 1935 and Kalb and Bennett[18] in the same year derived analytically the amplitudes of the third order intermodulation products arising from the application of such simultaneous fields to a material having a B–H loop described by the parabolic loop equation.

The results are set out in tabular form in Reference 19, from which the following expressions may usefully be quoted. The subscripts a and b refer to frequencies f_a and f_b in Eqn (2.69), $3f_a$ refers to the third harmonic of f_a, and $2a \pm b$ refers to the sum or difference products, etc.

$$\frac{E_{2a-b}}{E_a} = \frac{E_{2b-a}}{E_a} = 0.869 \tan \delta_h \quad (2.70)$$

$$\frac{E_{2a+b}}{E_a} = \frac{E_{2b+a}}{E_a} = 1.019 \tan \delta_h \quad (2.71)$$

$$\frac{E_{3a}}{E_a} = \frac{E_{3b}}{E_a} = 0.204 \tan \delta_h \quad (2.72)$$

2.3 The expression of high-amplitude properties

At the beginning of this chapter the B–H loop was described and the amplitude permeability was defined as

$$\mu_a = \frac{\hat{B}}{\mu_o \hat{H}} \quad (2.73)$$

(2.65) $B = \hat{H}(a_{10} + a_{11}\hat{H}) \cos pt + \frac{8a_{02}}{3\pi} \hat{H}^2 \sin pt$

$+ a_{30} \frac{\hat{H}^3}{4} \cos 3pt - \frac{8a_{02}}{15\pi} \hat{H}^2 \sin 3pt$ Gs

(2.66) $\hat{B}_{3a} = \frac{8}{15\pi} a_{02} \hat{H}_a^2$ Gs

(2.67) $\frac{\hat{B}_{3a}}{\hat{B}_a} = \frac{8a_{02}\hat{B}_a}{15\pi\mu^2}$

(2.68) $\frac{E_{3a}}{E_a} = \frac{3\hat{B}_{3a}}{\hat{B}_a} = \frac{8a_{02}\hat{B}_a}{5\pi\mu^2} = 0.6 \tan \delta_h$

(2.73) $\mu_a = \hat{B}/\hat{H}$

As the field strength or flux density increases from zero the amplitude permeability rises until it reaches a maximum value at the knee of the magnetization curve and then it falls progressively as the amplitude is further increased. In high amplitude applications the *B–H* relation is of particular importance as it indicates the level at which the onset of saturation occurs, a condition usually to be avoided in power circuits.

At these high amplitudes the *B–H* relation becomes increasingly non-linear so that there is appreciable waveform distortion. There are two limiting cases, (1) the field strength waveform may be made sinusoidal by driving the magnetizing winding from a high impedance source or (2) the flux waveform may be made sinusoidal by driving from a low impedance source. In the latter case the winding resistance must not be high enough to cause appreciable voltage drop.

The waveforms associated with each case are shown in Figure 2.8.

Due to this appreciable non-linearity it becomes difficult to express the magnetic performance with any accuracy in terms of simple coefficients and a.c. theory, although models based on a number of coefficients have been developed for the purposes of computer-aided design.[20] It is usual therefore to express the high-amplitude loss of a magnetic ferrite in terms of power dissipation per unit volume. It may be noted that normalization with respect to volume contrasts with the practice followed for magnetic alloys and steels where power loss per unit mass is favoured. The power loss may be measured by a suitable a.c. bridge, an electronic wattmeter or a calorimeter and may be expressed as a function of frequency and flux density. For such data to be truly descriptive of the material, i.e. independent of size and shape of the core, the eddy

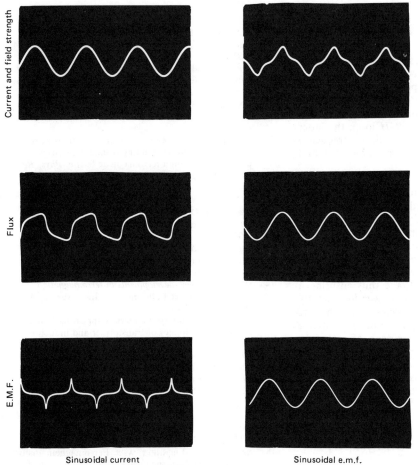

Figure 2.8 High-amplitude waveform distortion in magnetic materials. The oscillograms show current (and field strength), flux and e.m.f. for (a) sinusoidal current and (b) sinusoidal e.m.f.

current loss must be excluded either by using cores suitably dimensioned to make this loss negligible or by calculating its magnitude (e.g. from Eqn (2.46)) and substracting it from the measured loss. This high amplitude loss is loosely called hysteresis loss since it is closely related to the area of the hysteresis loop. However, magnetic hysteresis is defined as the irreversible variation of flux density or magnetization associated with the change of magnetic field strength and is *independent of the rate of change*,[21] that is, it is related to the static *B–H* loop. As the frequency increases, other magnetic loss mechanisms may become significant, e.g. ferromagnetic resonance loss. So the high amplitude magnetic loss, excluding eddy current loss, depends in practice on empirical powers of f and B. In this book the term *magnetic loss* will be used to refer to the total high amplitude losses excluding eddy current loss.

It is seen from Fig. 3.19 that this loss may be expressed in the form:

$$P_M = k f^m \hat{B}^n \tag{2.74}$$

where, for a typical manganese zinc ferrite intended for power applications, m is about 1.3 between 10 and 100 kHz and the Steinmetz exponent, n, lies between 2 and 3. The fact that m is greater than unity indicates that there are loss contributions other than that due to excursions round the static *B–H* loop; the latter loss would simply increase in proportion to frequency as predicted from the Rayleigh relations, see Eqn (2.31).

In power applications a ferrite material may be excited by a variety of repetitive waveforms, often approximately rectangular and often not symmetrical about the origin. However, it has been usual to measure and quote the loss expressed in Eqn (2.74) in terms of a symmetrical sinusoidal flux density variation. This is convenient both for specification and measurement since it is relatively easy to excite the material with a sinusoidal voltage thus ensuring a sinusoidal flux. In practice, the power loss density based on sinusoidal flux density gives results that are sufficiently relevant to a wide range of applications, and in particular for the comparison of materials. This discussion is pursued at the end of Section 4.2.3.

To obtain the total loss in a given core, the eddy current loss must be calculated for that core and, if significant, it must be added to the magnetic loss of the complete core, i.e. $P_M V_e$.

2.4 References and bibliography

Section 2.1
1. 'The International Electrotechnical Vocabulary', *Int. Electrotechnical Commission, Publication 50*, Geneva, particularly Chaps 121, Electromagnetism (1978) and 221, Magnetic Materials and Components (1988)

Section 2.2.2
2. *Ibid.*, Chap. 221, item 221-02-49 (1988)
3. *Ibid.*, Chap. 221, item 221-02-55 (1988)

Section 2.2.3
4. OLSEN, E. *Applied magnetism, a study in quantities*, Philips Technical Library, Eindhoven (1966)
5. RAYLEIGH, LORD Notes on electricity and magnetism, III. On the behaviour of iron and steel under the operation of feeble magnetic forces, *Phil. Mag.*, **23**, 225 (1887)
6. PETERSON, E. Harmonic production in ferromagnetic materials at low frequencies and low flux densities, *Bell Syst. tech. J.*, **7**, 762 (1928)
7. RIVAS, J., ZAMARRO, J. M., MARTIN, E. and PEREIRA, C. Simple approximation for magnetization curves and hysteresis loops, *IEEE Trans. Magn.*, **MAG-17**, 1498 (1981)

Section 2.2.4
8. VAN UITERT, L. G. Dielectric properties of and conductivity in ferrites, *Proc. Inst. Radio Engrs*, **44**, 1294 (1956)
9. ROESS, E. Eddy current and hysteresis losses in manganese zinc ferrites, in *Ferrites, Proc. of Int. Conf. on Ferrites, 1970*, edited by Y. Hoshino, S. Iida and M. Sugimoto, 187, University Park Press, Tokyo (1971)
10. VAN DER BURGT, C. M. and GEVERS, M. unpublished report
11. BROCKMAN, F. G., DOWLING, P. H. and STENECK, W. G. Dimensional effects resulting from a high dielectric constant found in a ferromagnetic ferrite, *Phys. Rev.*, **77**, 85 (1950)

Section 2.2.6
12. JORDAN, H. Die ferromagnetischen Konstanten für schwache Wechselfelder, *Elekt. Nachr Tech.*, **1**, 7 (1924)
13. LEGG, V. E. Magnetic measurements at low flux densities using the alternating current bridge, *Bell Syst. tech. J.*, **15**, 39 (1936)
14. SNOEK, J. L. *New developments in ferromagnetic materials*, Elsevier Publishing Co. Inc., New York-Amsterdam (1947)
15. 'Information on ferrite materials appearing in manufacturers' catalogues of transformer and inductor cores', *Int. Electrotechnical Commission, Publication 401*, Geneva (1972)
16. The International Electrotechnical Vocabulary, *Int. Electrotechnical Commission, Publication 50*, Chap. 221, item 221-03-33, Geneva (1988)

Section 2.2.7
17. LATIMER, K. E. Intermodulation in loaded telephone cables, *Electl Commun.*, **14**, 275 (1935–36)
18. KALB, R. M. and BENNETT, W. R. Ferromagnetic distortion of a two-frequency wave, *Bell Syst. tech. J.*, **14**, 322 (1935)
19. SNELLING, E. C. *Soft Ferrites: properties and applications*, 1st Edn, p. 35, Butterworths, London (1969)

Section 2.3

20. RIVAS, J. *et al.* See Reference 7.
21. The International Electrotechnical Vocabulary, *Int. Electrotechnical Commission, Publication 50*, Chap. 221, item 221-01-19, Geneva (1988)

General

Letter symbols to be used in electrical technology, Part 1: General, *Int. Electrotechnical Commission, Publication 27-1*, Geneva (1971)

BLEANEY, B. I. and BLEANEY, B. *Electricity and magnetism*, 3rd Edn, Oxford University Press, Oxford (1976)

DUFFIN, W. J. *Electricity and Magnetism*, 3rd Edn, McGraw-Hill (1980)

Cores for inductors and transformers for telecommunication, Part 1: Measuring methods, *Int. Electrotechnical Commission, Publication 367-1*, Geneva (1982)

3 Properties of some manganese zinc and nickel zinc ferrites

3.1 Introduction

This chapter presents a wide range of data relating to the technical properties of currently available manganese zinc and nickel zinc ferrites. In the majority of cases the information is intrinsic to the material, i.e. it is not, in the theoretical sense, dependent on the core geometry. The purpose is to provide the designer with the data necessary to predict the performance of a given core under a wide variety of conditions, e.g. frequency, flux density, temperature, etc. The relations between the material properties described here and the core performance are considered in the next chapter: they depend essentially on the core geometry.

In practice the intrinsic material properties may also depend somewhat on the size and shape of the core due to the limitations imposed by the manufacturing processes, e.g. a core with a large cross-section may have a material permeability that differs from that of a smaller cross-section core made from the same powder because the exposure to kiln conditions is different in the two cases (see Section 1.3.7). Thus, although in theory the data presented here may be used to estimate the typical properties of an existing or proposed core design, in practice the core geometry and processing conditions will influence the performance. Therefore the relations between material and core properties set out in Chapter 4 should only be regarded as a guide.[1]

All the data appearing in this chapter refer, as far as possible, to typical properties of the materials indicated. Although the designer usually requires limit values it would not be appropriate to include them here as they are essentially a matter of contract between the supplier and purchaser. In any case, the range of data required by a designer is far larger than the scope of a normal magnetic material specification.

Any survey of material properties must in some respects be inadequate. It will be incomplete because (i) it must be kept to a reasonable size by selection, (ii) data from some manufacturers may be more easily accessible than from others and (iii) some manufacturers, or particular grades of material, may have been overlooked. The data were selected according to the following procedure. The first step was to compile a list describing the electrical and magnetic data that are useful in ferrite applications. These data are usually available as measured functions expressed graphically. An attempt was made to obtain such data from all the principal manufacturers known to the author; he apologises now for any omissions. All available data were correlated to the prepared list and the selection was made on the widest possible basis. The data for some of the figures were specially prepared by the author and his colleagues and consequently relate only to ferrites manufactured by the Company with which the author was associated. However, the range of materials and manufacturers represented has been made as wide as possible, and the emphasis has been on the variety of type and grades currently available rather than on comparisons of competitive products. Indeed quantitative comparisons based on typical data would be quite unreliable.

In cases where particular information on a grade of ferrite is, for some reason, omitted it may often be inferred from the corresponding data of a similar grade of ferrite, the similarity being established by a comparison of those properties that are given for both grades.

All types and grades are referred to by manufacturer's code numbers. They are compiled into Table 3.1, in which they are tabulated against the manufacturer's name and broadly classified according to application. This table provides a general guide to the types and grades currently available at the time of going to press and, together with the accompanying index, enables any code number to be identified with the manufacturer.

The application classification is followed by data

on the mechanical and thermal properties. The next section starts with a table of electrical and magnetic properties in which typical values are given for the parameters that are normally quoted in manufacturers' catalogues. Then follows the main part of the chapter, i.e. the graphical data. This starts with the $B-H$ loops and ranges through all the more important properties.

3.2 Survey and classification

A study has been made of the catalogues and trade literature of all manufacturers of manganese zinc and nickel zinc ferrites known to the author. Only magnetically soft ferrites suitable for inductors and transformers (in the broadest sense of these terms) were considered. It appears that although there are very many grades and types of ferrite, nearly all of them can be placed into relatively few categories according to the principal application for which they are intended. The application classification is as follows:

Manganese zinc ferrites

I Inductors for resonant circuits operating at frequencies up to about 200 kHz.
II Inductors for resonant circuits operating in the approximate frequency range 100 kHz to 2 MHz. Ferrite antennas for medium and long wave broadcast bands.
III High permeability applications, in particular wide-band transformers (lower cut-off frequency up to about 1 MHz depending on the permeability) and low-power pulse transformers.
IV Applications requiring high saturation flux density and low loss at high flux densities in the approximate frequency range 10 kHz to 1 MHz, in particular transformers and chokes for switched mode power supplies and line scanning transformers in television receivers.

Nickel zinc ferrites

V $\mu > 1000$. Wide-band transformers operating in the approximate frequency range 1–300 MHz, pulse transformers for short duration pulses, e.g. $t_d < 0.1$ μs.
VI μ: 500–1000. Wide-band transformers operating in the approximate frequency range 5–300 MHz, pulse transformers, ferrite antennas and cores for electromagnetic interference suppression.
VII μ: 150–500. Ferrite antennas for medium and long wave broadcast bands, power transformers operating in the approximate frequency range 0.5–5 MHz and cores for electromagnetic interference suppression.
VIII μ: 70–150. Inductors for resonant circuits operating in the approximate frequency range 2–20 MHz, ferrite antennas for short wave broadcast bands, power transformers for the approximate frequency range 2–30 MHz and cores for electromagnetic interference suppression.
IX μ: 35–70. Inductors for resonant circuits operating in the approximate frequency range 10–40 MHz and cores for electromagnetic interference suppression.
X μ: 12–35. Inductors for resonant circuits operating in the approximate frequency range 20–60 MHz and cores for electromagnetic interference suppression.
XI $\mu < 12$. Inductors for resonant circuits operating at frequencies above about 30 MHz and cores for electromagnetic interference suppression.

Manganese zinc or nickel zinc ferrites

XII Ferrites dedicated to electromagnetic interference suppression or special applications not listed in other classifications.

The classification of all the grades and types of material which were included in the survey is shown in Table 3.1. Although great care has been taken omissions are inevitable and it must be recognized that in due course such a table becomes out-of-date. Nevertheless it does give a broad picture of the commercial availability of manganese zinc and nickel zinc ferrites that have been developed for inductors and transformers (see also the index to Table 3.1).

Notes: Where a manufacturer has several grades of ferrite in one category this may be due to the introduction of improved grades or because there are several versions of a given grade, each having a special specification, e.g. a particular temperature factor. The relative merit of different manufacturers' ferrites should not be inferred from the relative positions of entries. Relative performance may only be established by reference to manufacturers' specifications; the vertical columns of the table simply indicate application classification.

Sometimes the application of a particular grade is obscure or the catalogue places a given grade in several categories; in such cases the grade may appear several times.

Finally there are a number of special grades of these ferrites which are intended for applications not covered by the scope of this book, e.g. magnetostrictive ferrites and specially dense ferrites used for recording heads; in general these have not been included.

46 Properties of some manganese zinc and nickel zinc ferrites

Table 3.1 Survey of some ferrite grades

| | | | MnZn ferrites | | | APPLICATION |
CLASS		I	II	III	IV	V
Initial permeability		800–4000	500–1000	3000–20 000	2000–5000	>1000
Main applications		L.f. resonant circuits	M.f. resonant circuits, antennas	Low-power wide-band & pulse transformers, e.m.i. supprsn.	Power & high-flux density applications, e.m.i. supprsn.	Wide-band & pulse transformers
Approx. frequency range		<0.2 MHz	0.1–2 MHz	l.f.–200 MHz	0.01–1 MHz	1–300 MHz
Manufacturer, country and brand name						
WESTERN EUROPE						
1	Kaschke (FRG) *Kamafer*	K2005	K300 K600 K700	K4000 K6000 K10000	K2002 K2004 **K2006**	
2	Krupp (FRG) *Hyperox*	DIS2 DIS3 **DIS4**	DIS5	D1 DIS1 **DIS11**	C2 **C21** C23 C24	
3	Neosid (UK)	P10 P11 P12		F9 **F10**	F4 F5 **F3** F6 FP4a FP5a	
4	Phillips Components (The Netherlands) *Ferroxcube*	3B 3B7 3H1 3H3 A5^M	3D3 3D35	3E1 3E2 3E4 3E5 3H2	3C2^Y 2A2^Y 3C6 3C8 3C85 3F3 3B8 A9^M	
5	SEI Magnetic Components (UK)	Q3 Q7	S1	Q5 P T3 T4 T6	L2	
6	Siemens (FRG) *Siferrit*	N22 N28 N48	M33	N26 N30 T35 T38	N41 N27 N47 N67	
7	Thomson CSF (France) *Ferrinox*	**T10 T14 T13**	B10 T31	T22 T4 T6 T4A T6A	T22 B50 B51 B52 B30^Y B31^Y	
8	Vogt Electronic (FRG) *Ferrocarit*	**Fi311 Fi323**	Fi262 Fi291	Fi340 Fi360	Fi323 Fi322 Fi311	
EASTERN EUROPE						
9	Ei Feriti (Yugoslavia)	M42 M58	M31	M42 M48 M56 M51 M52	M46 M48 M32^Y	
10	EMO Elektromodule (Hungary) *Maferrit Niferrit*	**M1 M2 M2F** **M2F-A M4-A**	**M05F-A**	M2 M3 M4 **M4-A** M5 M6	M2T M2TN **M2TN-A M2TN-B**	
11	Iskra (Yugoslavia) *Elvefer*	1G 6G 8G **11G** **14G 16G** 26G	**10G**	8G 9G 19G 22G 12G 25G	5G 15G 25G 35G	
12	Kombinat VEB (GDR)	M163 M183 M150	M143	M195	M194 M196	
13	Polfer (Poland) *Ferroxyd*	F-1001 F-1501 F-2001 F-2002	F-605	F-1501 F-2001 F-2004 F-3001 F-3002 F-6001	F-1502 F-2004 F-3002 **F-807** F-803^Y F-804^Y	
14	Pramet (Czechoslovakia) *Fonox*	H23	H6 H7	H22	H11 H20 **H21**	
USA						
15	Allen-Bradley *Ferramic*	BBR7400 BBR7500 BBR7600 BBR7700 BBR7800 BBR7950		BBR7400 BBR7500 BBR7600 BBR7700 BBR7800	**IR8000 IR8100** **IR8200** IR8500 IR8600 IR8700 05 06	

Properties of some manganese zinc and nickel zinc ferrites

CLASSIFICATION							
	NiZn Ferrites						MnZn or NiZn
VI	VII	VIII	IX	X	XI		XII
500–1000	150–500	70–150	35–70	12–35	<12		–
H.f. wide-band & pulse transformers, antennas, e.m.i. supprsn.	Antennas, h.f. power transformers, e.m.i. supprsn.	Resonant circuits, antennas, h.f. power transformers, e.m.i. supprsn.	Resonant circuits, e.m.i. supprsn.	Resonant circuits, e.m.i. supprsn.	Resonant circuits, e.m.i. supprsn.		Dedicated e.m.i. suppression, special applications
5–300 MHz	0.5–5 MHz	2–20 MHz	10–40 MHz	20–60 MHz	>30		–
K650	K250	K150 K80	K50 K40	K14	K10		
E2	E4	E5*		E7*			
F13	F14	F16	F25*	F28* F29*	F27*		F8
4A4 B1^M	4B1	4C6	4D1	4E1			3H2 3S1 4S2
	K4	K6 K10	K7	K8			
		K1		K12	U60 U17*		N22
H10	H20	H30 H32	H50 H52	H60 H62			T4A T6A
Fi292 Fi293	Fi221 Fi222	Fi212	Fi150	Fi130	Fi110 Fi091		
N32	N31	N23	N22	N25 N21	F11		
	N300 N200	N100	N50	N20	N10		
1C	2C 3C	3F 1F	2F	1E 2E			
	M360	M340 M343		M330 M321	M320		
	F-302 F-201	F-81 F-82		U-31*	U-11* U-6		
	N2	N1 N08P	N05	N02	N01 N01P		
H	G	Q1	Q2	Q3			AR9100 AR9700

48 Properties of some manganese zinc and nickel zinc ferrites

Table 3.1—*contd*

					APPLICATION
	MnZn ferrites				
CLASS	I	II	III	IV	V
Initial permeability	800–4000	500–1000	3000–20 000	2000–5000	>1000
Main applications	L.f. resonant circuits	M.f. resonant circuits, antennas	Low-power wide-band & pulse transformers, e.m.i. supprsn.	Power & high-flux density applications, e.m.i. supprsn.	Wide-band & pulse transformers
Approx. frequency range	<0.2 MHz	0.1–2 MHz	l.f.–200 MHz	0.01–1 MHz	1–300 MHz
Manufacturer, country and brand name					
USA *(contd)*					
16 Ceramic Magnetics	MN30		MN60 MN100	MN80 MN60LL MN67 MN8CX	CMD5005 CMD6
17 Fair-rite		33	73 75	77 83[1]	
18 Ferronics			B		
19 Ferroxcube Div. of Amperex	3B9 3B7	3D3	3E2A 3E5	3C6A 3C8 3C85 3F3	4A6
20 Krystinel	K18	K16 KE1	K82 KT6 KS6 KT7 KTA	KB2 KB3 K72	KF1
21 National Magnetics Group		R P			
22 Magnetics Div. of Spang & Co.	D G V	A	D G J W H	F P	
23 D. M. Steward	33		35 40	33	
JAPAN					
24 Nippon Ferrite			GP-5 GP-7 GP-9 GP-11 6Q-5C	SB-5F SB-5HK SB-5L SB-5LK SB-5S SB-7 SB-7C SB-7E SB-7H SB-9C SB-9H SB-5M[Y]	DL-3 DL-2 DL-8C T-314 RL-5A
25 Sumitomo Ceraxcube	3B2 3B4 3B4A 3B6 3F3 3F4 3F5		3E1 3E1A 3E1B	3F5 3F5B 3F6	4T1
26 TDK	H6A H6A3 H6B H6H3 H6K H6Z	H6F	H5A **H5B H5B2** H5C2 H5D H5E HP3000 HP4000 HP5000 NF H1B H1D	H7A H7B **H7C1** H7C2 H7C4	
27 Tokin	1801F 1000SFP 2001F 2002F 2101F 2300F	801F 1000SFP	4000H 5000H 700H **12001H**	2500B 2500B2 2500B3 3100B 5000B	2000L
28 Tomita	2B3 2B4		2C2 2C3 2E4 2E1 2D1 2D3	2E6 **2E7** 2E4	3A9

Notes

Where a material code number is printed in bold type, some property or properties of that material are presented in the graphical data in Figures 3.1–3.25.

* These materials have properties induced by special heat treatment during manufacture; irreversible changes in properties will be caused if the material is subjected to strong magnetic fields.

Properties of some manganese zinc and nickel zinc ferrites

CLASSIFICATION

	NiZn Ferrites						MnZn or NiZn
VI	VII	VIII	IX	X	XI		XII
500–1000	150–500	70–150	35–70	12–35	<12		–
H.f. wide-band & pulse transformers, antennas, e.m.i. supprsn.	Antennas, h.f. power transformers, e.m.i. supprsn.	Resonant circuits, antennas, h.f. power transformers, e.m.i. supprsn.	Resonant circuits, e.m.i. supprsn.	Resonant circuits, e.m.i. supprsn.	Resonant circuits, e.m.i. supprsn.		Dedicated e.m.i. suppression, special applications
CN20	C2025	C2050	C2075	N40			
	64*	61* 65*	63* 67*	68*			43 73
J	F	K*	P*				
4A	4B	4C4					
K52	KA1 KD1	K01 KU8	K21	K31 KU1			KV1
H		M	M2	M3			
28		18 21	16	15			
DL-7C CL-81 DB-3T DL-6C TH-55	L-82 QL-400 SM-30C DL-5C L-81 KP-4B DM-1N DM-8 QM-051 DL-4C DM-7 DM-6 DM-1	KQ-1 KP-2S KP-3S QM-201 SM-2C SV-5B	DV-6 DV-2A DV-2C SV-1 SV-1D KP-3M VH-50 DV-1A DV-1S DV-1N SV-1AC SV-2C	RM-1K RM-2K SV-2D RV-6 RV-5 SV-3C SV-5D RV-4 SV-3A SV-3H VH-200	RV-2 RV-3 SF-7 SH-6L		
FS1				SMC-2			
	K5	K6A		K7A K8			NF1 NF2 NF3 NF4 NF5
601L 700L	400L 250L	150L 100L	50L 40L	20L	10L		
3A 3A4 3A6	3A3 3D2 3H2 4A 4A2 4A3 4A5 D3 D4A D5A D7E D8E D9E D11A D12A D13A D14A F3 F3L 4B4 4D 4D4 4D8 4H2 4C	DF4A 4D7 5B 5B3 5H2 6D8 6H3	DF7P DF11R DF10 6A 6D2 6D3 6D7	6B2 6B3 7A5 7A6 7B2 7B5 7B5E 7D1	7A3 7B4 7C 7D2 7D3 7D4 7D5C 7D5E 7D6 7D7		

(1) This material has a substantially rectangular hysteresis loop and is intended for use in saturating transformers and chokes.
(M) These materials are discontinued Mullard grades but are retained in this table (under the Philips Components name) because useful properties appear in the following graphical data.
(Y) Materials intended mainly for TV deflection yokes.

Table 3.1.1 Index to ferrite code numbers listed in Table 3.1

Ferrite code number	Manufacturer ref.	Ferrite code number	Manufacturer ref.	Ferrite code number	Manufacturer ref.
A	22	D1	2	F25	3
AR9100	15	D1S1	2	F27	3
AR9700	15	D1S2	2	F28	3
A5	See Table 3.1.2.	D1S3	2	F29	3
A9	See Table 3.1.2.	**D1S4**	2	F-81	13
B	18	**D1S5**	2	F-82	13
BBR7400	15	**D1S11**	2	F-201	13
BBR7500	15	D3	28	F-302	13
BBR7600	15	D4A	28	F-605	13
BBR7700	15	D5A	28	F-803	13
BBR7800	15	D7E	28	F-804	13
BBR7950	15	D8E	28	**F-807**	13
B1	See Table 3.1.2.	D9E	28	F-1001	13
B10	7	D11A	28	F-1501	13
B30	7	D12A	28	F-1502	13
B31	7	D13A	28	F-2001	13
B50	7	D14A	28	F-2002	13
B51	7			F-2004	13
B52	7	E2	2	F-3001	13
		E4	2	F-3002	13
CL-81	24	E5	2	F-6001	13
CMD6	16	E7	2		
CMD5005	16			G	15
CN20	16	F	18	G	18
C2	2	F	22	G	22
C21	2	Fi-091	8	GP-5	24
C23	2	Fi-110	8	**GP-7**	24
C24	2	Fi-130	8	**GP-9**	24
C2025	16	Fi-150	8	**GP-11**	24
C2050	16	Fi-212	8	**GQ-5C**	24
C2075	16	Fi-221	8		
		Fi-222	8	H	15
D	22	**Fi-262**	8	H	21
DB-3T	24	**Fi-291**	8	H	22
DF4A	28	Fi-292	8	**HP3000**	26
DF7P	28	Fi-293	8	**HP4000**	26
DF10	28	**Fi-311**	8	**HP5000**	26
DF11R	28	**Fi-322**	8	H1B	26
DL-2	24	**Fi-323**	8	H1D	26
DL-3	24	**Fi-340**	8	H5A	26
DL-4C	24	**Fi-360**	8	**H5B**	26
DL-5C	24	FP4a	3	**H5B2**	26
DL-6C	24	FP5a	3	**H5C2**	26
DL-7C	24	F3	28	**H5D**	26
DL-8C	24	F3L	28	**H5E**	26
DM-1	24	**F3**	3	H6	14
DM-1N	24	F4	3	H6A	26
DM-6	24	F5	3	H6A3	26
DM-7	24	F6	3	H6B	26
DM-8	24	F8	3	H6F	26
DV-1A	24	F9	3	H6H3	26
DV-1N	24	**F10**	3	H6K	26
DV-1S	24	F11	9	H6Z	26
DV-2A	24	F13	3	H7A	26
DV-2C	24	F14	3	H7B	26
DV-6	24	F16	3	**H7C1**	26

Table 3.1.1—*contd*

Ferrite code number	Manufacturer ref.	Ferrite code number	Manufacturer ref.	Ferrite code number	Manufacturer ref.
H7C2	26	K10	1	M31	9
H7C4	26	**K10**	5	M32	9
H7	14	**K12**	6	**M33**	6
H10	7	K14	1	M42	9
H11	14	K16	20	M46	9
H20	7	K18	20	M48	9
H20	14	K21	20	M51	9
H21	14	K31	20	M52	9
H22	14	K40	1	M56	9
H23	14	K50	1	M58	9
H30	7	K52	20	M143	12
H32	7	K72	20	M150	12
H50	7	K80	1	M163	12
H52	7	K82	20	M183	12
H60	7	K150	1	M194	12
H62	7	K250	1	M195	12
		K300	1	M196	12
IR8000	15	K600	1	M320	12
IR8100	15	K650	1	M321	12
IR8200	15	K700	1	M330	12
IR8500	15	K2002	1	M340	12
IR8600	15	K2004	1	M343	12
IR8700	15	K2005	1	M360	12
		K2006	1		
J	18	K4000	1	**NF**	26
J	22	K6000	1	NF1	26
		K10000	1	NF2	26
K	18			NF3	26
KA1	20	**L2**	5	NF4	26
KB2	20	L-81	24	NF5	26
KB3	20	L-82	24	N01	14
KD1	20			N01P	14
KE1	20	M	21	N02	14
KF1	20	**MN8CX**	16	N05	14
KP-2S	24	**MN30**	16	N08P	14
KP-3S	24	**MN60**	16	N1	14
KP-4B	24	MN60LL	16	N2	14
KP-3M	24	**MN67**	16	N10	10
KQ-1	24	**MN80**	16	N20	10
KS6	20	**MN100**	16	N21	9
KTA	20	**M05F-A**	10	**N22**	6
KT6	20	**M1**	10	N22	9
KT7	20	**M2**	10	N23	9
KU1	20	M2	21	N25	9
KU8	20	**M2F**	10	**N26**	6
KV1	20	**M2F-A**	10	**N27**	6
KO1	20	M2T	10	**N28**	6
K1	6	M2TN	10	**N30**	6
K4	5	**M2TN-A**	10	N31	9
K5	26	**M2TN-B**	10	N32	9
K6	5	M3	10	N40	16
K6A	26	M3	21	N41	6
K7	5	M4	10	**N47**	6
K7A	26	**M4-A**	10	**N48**	6
K8	5	M5	10	N50	10
K8	26	M6	10	**N67**	6

Table 3.1.1—contd

Ferrite code number	Manufacturer ref.	Ferrite code number	Manufacturer ref.	Ferrite code number	Manufacturer ref.
N100	10	SV-5B	24	3A4	28
N200	10	SV-5D	24	3A6	28
N300	10	S1	5	3A9	28
				3B	4
P	5	T3	5	3B2	25
P	18	T4	5	3B4	25
P	21	T4	7	3B4A	25
P	22	T4A	7	3B6	25
P10	3	T6	5	3B7	4, 19
P11	3	T6	7	3B8	4
P12	3	T6A	7	3B9	19
		T10	7	3C	11
QL-400	24	T13	7	3C2	4
QM-051	24	T14	7	3C6	4
QM-201	24	T22	7	3C6A	19
Q1	15	T31	7	3C8	4, 19
Q2	15	T35	6	3C85	4, 19
Q3	15	T38	6	3D2	28
Q3	5	T-314	24	3D3	4, 19
Q5	5	TH-55	24	3D35	4
Q7	5			3E1	4
		U-6	13	3E1	25
R	21	U-11	13	3E1A	25
RL-5A	24	U17	6	3E1B	25
RM-1K	24	U-31	13	3E2	4
RM-2K	24	U60	6	3E2A	19
RV-2	24	V	22	3E4	4
RV-3	24	VH50	24	3E5	4, 19
RV-4	24	VH200	24	3F	11
RV-5	24			3F3	4, 19
RV-6	24	W	22	3F3	25
				3F4	25
SB-5F	24	ZF-7	24	3F5	25
SB-5HK	24			3F5B	25
SB-5L	24	05	15	3F6	25
SB-5LK	24	06	15	3H1	4
SB-5M	24	1C	11	3H2	4
SB-5S	24	1E	11	3H2	28
SB-7	24	1F	11	3H3	4
SB-7C	24	1G	11	3S1	4
SB-7E	24	2A2	4	4A	19
SB-7H	24	2B3	28	4A	28
SB-9C	24	2B4	28	4A2	28
SB-9H	24	2C	11	4A3	28
SH-6L	24	2C2	28	4A4	4
SMC-2	25	2C3	28	4A5	28
SM-2C	24	2D1	28	4A6	19
SM-30C	24	2D3	28	4B	19
SV-1	24	2E	11	4B1	4
SV-1AC	24	2E1	28	4B4	28
SV-1D	24	2E4	28	4C	28
SV-2C	24	2E6	28	4C4	19
SV-2D	24	2E7	28	4C6	4
SV-3A	24	2F	11	4D	28
SV-3C	24	3A	28	4D1	4
SV-3H	24	3A3	28	4D4	28

Properties of some manganese zinc and nickel zinc ferrites

Table 3.1.1—contd

Ferrite code number	Manufacturer ref.	Ferrite code number	Manufacturer ref.	Ferrite code number	Manufacturer ref.
4D7	28	7D5C	28	63	17
4D8	28	7D5E	28	64	17
4E1	4	7D6	28	65	17
4H2	28	7D7	28	67	17
4S1	25	8G	11	68	17
4S2	4	9G	11	73	17
4T1	25	**10G**	11	75	17
5B	28	10L	27	**77**	17
5B3	28	**11G**	11	**83**	17
5G	11	12G	11	100L	27
5H2	28	**14G**	11	150L	27
6A	28	15	23	250L	27
6B2	28	15G	11	400L	27
6B3	28	16	23	601L	27
6D2	28	**16G**	11	700L	27
6D3	28	18	23	801F	27
6D7	28	19G	11	1000SFP	27
6D8	28	20L	27	1801F	27
6G	11	21	23	2000L	27
6H3	28	22G	11	2001F	27
7A3	28	25G	11	2002F	27
7A5	28	26G	11	2101F	27
7A6	28	28	23	2300F	27
7B2	28	33	17	**2500B**	27
7B4	28	**33**	23	**2500B2**	27
7B5	28	**35**	23	**2500B3**	27
7B5E	28	35G	11	3100B	27
7C	28	40	23	**4000H**	27
7D1	28	40L	27	5000B	27
7D2	28	43	17	5000H	27
7D3	28	50L	27	7000H	27
7D4	28	61	17	**12001H**	27

Table 3.1.2 Equivalence of Mullard and Philips grades

Mullard code number	Application classification (see Table 3.1)	Approx. Philips type	
A5 *	I	–	
A13	I	3H1	
A10	II	3D3	
A7	III	3E2	
A9 *	IV	–	
A16	IV	3C8	
B1 *	VI	4A1	(superseded by 4A4)
B2	VII	4B1	
B10	VIII	4C6	
B4	IX	4D1	
B5	X	4E1	

*discontinued

Apart from being a broad survey of ferrite grades Table 3.1 provides a means of identifying the grade numbers in the graphical data which forms the major part of this chapter. In some instances different manufacturers use the same code numbers to designate quite different materials; to avoid ambiguity all grades quoted in the graphical data are printed in bold type in both Table 3.1 and the accompanying index.

The Mullard grades which appeared in Table 3.1 in the first edition have generally been omitted in the revised table because either they have Philips equivalents or they have been discontinued. However, a number of important properties presented in the graphical data in the first edition were specifically measured on the Mullard grades and some of these properties are still relevant. In such cases the graphical data have been carried over to the second edition, using the Mullard material code numbers but assigning them to Philips Components. In Table 3.1, these grades carry the superscript 'M'. To put these grades into context, a short list (Table 3.1.2) giving the approximate Philips equivalents has been appended to the index of code numbers.

3.3 Mechanical and thermal properties

Ferrites, being ceramic materials formed by sintering, have mechanical properties similar to those of pottery. In particular the properties depend on the sintered density. As described in Chapter 1, the pressed core before firing consists of a relatively porous compact of oxides. During sintering the oxides react to form crystallites, or grains, of the required composition, the grains nucleating at discrete centres and growing outwards until the boundaries meet those of neighbouring crystallites. During this process the density of the mass rises; if this process were to yield perfect crystals meeting at perfect boundaries the density would rise to the theoretical maximum, i.e. the X-ray density, ρ_x, which is the mass of material in a perfect unit crystal cell divided by the cell volume. In practice imperfections occur and the sintered mass has microscopic voids both within the grains and at the grain boundaries. The resulting density is referred to as the sintered density, ρ_s.

If the porosity is denoted by p then

$$p = 100(1 - \rho_s/\rho_x) \text{ per cent} \qquad (3.1)$$

In normal production the porosity might range from 1% to 15% depending on the grade of ferrite.

In Table 3.2 typical values of the sintered density and porosity are given for a number of representative polycrystalline ferrites. The porosity has an effect on the mechanical and magnetic properties and low porosities are usually preferable.

For specific application some ferrites have been developed having porosities of less than 1%; indeed values of 0.1% are not uncommon.[2] These are particularly valuable in the manufacture of devices such as recording heads which require intricately formed, highly polished surfaces having the greatest possible wear resistance. High density ferrites having high permeability are used for transformer cores; for this application accurately lapped pole faces are essential.

Some of the principal mechanical properties of a representative selection of ferrite specimens have been determined by Sellwood.[3] The results are given in Table 3.2. Care should be exercised in using these figures, particularly the tensile strength values, since the presence of porosity, voids or hair-line cracks will in practice make the breaking load uncertain. Some additional data are given in the literature.[4-6]

The average hardness of ferrite was measured on a limited number of samples. The results in terms of the Vickers Pyramid Number were 600 to 700 for manganese zinc ferrite and 800 to 900 for nickel zinc ferrite.

Table 3.3 gives typical values of the thermal properties of ferrites. See also Reference 8.

Table 3.2 Mechanical properties of some ferrites

Property	MnZn ferrite		NiZn ferrite		Units
	Pressed	Extruded	Pressed	Extruded	
Sintered density	4800		4600		kg m^{-3}
Porosity	8		13.5		%
Ultimate tensile strength	4.8	3.9	4.9	5.2	kgf mm^{-2}
Ultimate compressive strength	47	55	44	40	kgf mm^{-2}
Modulus of elasticity	12.5	13.6	12.1	11.5	kgf mm^{-2}
Impact strength (Charpy)		0.043		0.043	J

The above figures are the average values measured on a wide range commercial ferrite grades. Considerable varience is to be expected; standard deviations of 30% are not uncommon.

1000 kg m^{-3} = 1 g cm^{-3}
1 kgf = weight of 1 kg $\simeq$ 9.81 newtons
1 kgf mm^{-2} $\simeq$ 9.81 × 10^6 pascals $\simeq$ 1422 lbf in^{-2}
1 J $\simeq$ 0.738 ft lbf

Table 3.3 Typical thermal properties of ferrites

Property	Conditions	Value	Units
Coeff. of linear expansion			
MnZn ferrites	0–50°C	10×10^{-6}	°C^{-1}
	0–200°C	11×10^{-6}	°C^{-1}
NiZn ferrites	0–50°C	7×10^{-6}	°C^{-1}
	0–200°C	8×10^{-6}	°C^{-1}
Specific heat[7]			
MnZn ferrites	25°C	700–800	J kg^{-1} °C^{-1}
		0.16–0.19	cal g^{-1} °C^{-1}
NiZn ferrites	25°C	750	J kg^{-1} °C^{-1}
		0.18	cal g^{-1} °C^{-1}
Thermal conductivity			
MnZn ferrites		3500 to 4300	µW mm^{-1} °C^{-1}
NiZn ferrites	25–85°C	35 to 43	mW cm^{-1} °C^{-1}
		0.0083 to 0.010	cal s^{-1} cm^{-1} °C^{-1}

3.4 Magnetic and electrical properties

Before presenting the graphical data it will be useful to give in tabular form, for each application category, typical values for those parameters which may be quoted in manufacturers' catalogues. These Tables, 3.4 and 3.5, enable comparisons between the categories of ferrite to be readily made. However it must be emphasized that the figures are merely representative of the category and should not be interpreted as being typical of any particular grade.

The rest of this chapter consists of graphical data presented as a series of figures, each dealing with a specific property as a function of an independent variable. Each figure is accompanied by a brief commentary intended to draw attention to the salient features, to indicate in a qualitative way any underlying physical processes that may be relevant, and to provide cross-references and relevant formulae. Each commentary serves as an extended caption for all graphs in one figure. It indicates, after the heading, the values of those quantities, e.g. f, B, θ, etc., that were constant during the variation of the independent variable(s).

In general each figure consists of a number of graphs each displaying typical data for the grade or grades of ferrite indicated in the panel at the top right-hand corner. The legend accompanying the graph number indicates the type of ferrite, the manufacturer and the application classification, as designated in Table 3.1. The order of the graphs appearing under one figure number is approximately in accordance with the application classifications listed in Section 3.2 and Table 3.1, e.g. any data relating to l.f. inductor grades would appear first while the data on higher frequency nickel zinc ferrites would appear in the later graphs. The grade references are the code numbers used by the manufacturers and appear in bold type in the survey in Table 3.1 and the accompanying index.

It is emphasized once again that the data are typical and should not be used to compare the performance of similar grades manufactured by different companies. Where comparison is required reference should be made to the manufacturer's specification.

Table 3.4 Typical values of specification parameters for manganese zinc ferrites

			Measuring conditions*			Principal application categories (see Table 3.1)				
				$\hat{B}$		I	II	III	IV	
Property	Symbol or expression	f (kHz)	(mT)	(Gs)	Misc.	L.f. resonant circuits	M.f. resonant circuits, antennas	Low-power wide-band and pulse transformers, e.m.i. supprsn.	Power and high B_{sat} applns., e.m.i. supprsn.	Units
Approx. application frequency	f					<0.2	0.1–2	l.f.–200	0.01–1	MHz
Initial permeability	μ_i	<10	<0.1	<1		800–4000	500–1000	3000–20 000	2000–5000	
Saturation flux density	B_{sat}				$H = 1\,\text{kA m}^{-1}$ $= 12.5\,\text{Oe}$	350–500 3500–5000	350–400 3500–4000	300–500 3000–5000	450–520 4500–5200	mT Gs
Remanence	B_r				from sat	50–160 500–1600	150–200 1500–2000	40–140 400–1400	90–200 900–2000	mT Gs
Coercivity	H_c				from sat	10–35 0.13–0.44	40–100 0.5–1.2	2.4–10 0.03–0.13	10–35 0.13–0.44	Am^{-1} Oe
Residual loss factor	$(\tan\delta_r)/\mu$	10 30 100 1000	<0.1	<1		0.8–1.8 1–3 1–6	5–12 15–30	2.5–15 3–20 6–75		10^{-6} 10^{-6} 10^{-6} 10^{-6}
Hysteresis coefficient	η_B	10	from 1.5 to 3	from 15 to 30		0.1–1.3	0.5–2	0.3–1.5		mT$^{-1} \times 10^{-6}$
Power loss density	P_m	25 100	200 100	2000 1000	25°C 100°C 25°C 100°C				100–200 90–200 100–250 70–250	μW mm^{-3} = mW cm^{-3}
Curie point	θ_c	<10	<0.1	<1		130–210	200–280	110–200	180–250	°C
Temperature factor	$\dfrac{(\mu_2-\mu_1)}{\mu_1\mu_2(\theta_2-\theta_1)}$	<10	<0.25	<2.5	$\theta_1 = 25°\text{C},$ $\theta_2 = 55°\text{C},$	0.5 to 2.0 or −0.6 to 0.6	0 to 3.0			°C$^{-1} \times 10^{-6}$
Disaccommodation factor	$\dfrac{(\mu_1-\mu_2)}{\mu_1^2\log_{10}(t_2/t_1)}$	<10	<0.25	<2.5	$t_1 = 10\,\text{min}$ $t_2 = 100\,\text{min}$	1.5–5	8–15	1–4		10^{-6}
Resistivity	ρ	d.c.				1–50 100–5000	>1.5 >150	0.02–2 2–200	0.2–10 20–1000	Ωm Ωcm

Table 3.5 Typical values of specification parameters for nickel zinc ferrites

Property	Symbol or expression	Measuring conditions* f (kHz)	Measuring conditions* $\hat{B}$ (mT)	Measuring conditions* $\hat{B}$ (Gs)	V W.b. and pulse trans- formers	VI H.f, w.b. and power trans- formers, antennas, e.m.i. supprsn.	VII Antennas, h.f. power transformers, e.m.i. supprsn.	VIII Resonant circuits, antennas, h.f. power transformers, e.m.i. supprsn.	IX Resonant circuits, e.m.i. supprsn.	X Resonant circuits, e.m.i. supprsn.	XI Resonant circuits, e.m.i. supprsn.	Units
Approx. application frequency	f				1–300	5–300	0.5–5	2–20	10–40	20–60	>30	MHz
Initial permeability	μ_i	<10 kHz	<0.1	<1	>1000	500–1000	150–500	70–150	35–70	12–30	<12	
Saturation flux density	B_{sat}				260 at $H=1$ 2600 at $H=12.5$	280–340 at $H=1$ 2800–3400 at $H=12.5$	270–360 at $H=2$ 2700–3600 at $H=25$	250–420 at $H=4$ 2500–4200 at $H=50$	230–300 at $H=4$ 2300–3000 at $H=50$	150–260 at $H=8$ 1500–2600 at $H=100$	100–270 at $H=8$ 1000–2700 at $H=100$	mT kA m^{-1} Gs Oe
Remanence	B_r				60–90 600–900	140–190 1400–1900	140–250 1400–2500	150–300 1500–3000	140–200 1400–2000	70–160 700–1600	50–170 500–1700	mT Gs
Coercivity	H_c				16–25 0.2–0.3	50–100 0.63–1.3	80–160 1–2	200–500 2.5–6.3	450–750 5.6–9.4	800–1300 10–16	800–1600 10–20	A m^{-1} Oe
Residual loss factor	$(\tan \delta_r)/\mu$	100 kHz 300 kHz 1 MHz 3 MHz 10 MHz 30 MHz 100 MHz	<0.1	<1	30 130	10–30 20–40 40–90	25–90 50–200	20–50 25–70 60–120	50–300 100–500	150–300 200–500	100 130–400 180–2000	10^{-6} 10^{-6} 10^{-6} 10^{-6} 10^{-6} 10^{-6} 10^{-6}
Hysteresis coefficient	η_B	100 kHz	from 1.5 to 3	from 15 to 30	6	1.8–4	8–20	6–40	40–90	40–80	27–40	mT^{-1} × 10^{-6}
Curie point	θ_c	<10 kHz	<0.1	<1	100	90–150	135–300	250–400	300–500	400–500	350–550	°C
Temperature factor	$\dfrac{(\mu_2-\mu_1)}{\mu_1\mu_2(\theta_2-\theta_1)}$	<10 kHz	<0.25	<2.5	3 to 15 $\theta_1=25°C$ $\theta_2=55°C$	2 to 16	0 to 10	0 to 10	0 to 15	−10 to 15		°C^{-1} × 10^{-6}
Resistivity	ρ				10^2 10^4	10^3–10^6 10^5–10^8	10^3–10^6 10^5–10^8	10^3–10^6 10^5–10^8	10^3–10^6 10^5–10^8	10^3–10^6 10^5–10^8	10^4–10^5 10^6–10^7	Ω m Ω cm

*The measuring temperature is 25°C unless otherwise stated.
These tables list the parameters that may be found in a ferrite material specification. In each of the application categories typical values of these parameters are given. They normally refer to toroidal specimens of the material; no limits are implied.

3.4.1 Index to graphical data

Figure 3.1 $B-H$ loops, p. 59
Figure 3.2   Minor $B-H$ loops, p. 67
Figure 3.3   Saturation flux density as a function of temperature, p. 69
Figure 3.4   Permeability as a function of high amplitude of flux density, p. 71
Figure 3.5   Permeability as a function of low amplitude flux density, p. 73
Figure 3.6   Incremental permeability as a function of static field strength, p. 74
Figure 3.7   Hanna curves, p.77
Figure 3.8   Initial permeability as a function of temperature, p. 80
Figure 3.9   Incremental permeability as a function of temperature with static field strength as a parameter, p. 87
Figure 3.10 Disaccommodation, p. 88
Figure 3.11 Complex initial permeability spectrum, p. 90
Figure 3.12 Residual loss factor spectrum, p. 99
Figure 3.13 Residual loss factor as a function of temperature, p. 103
Figure 3.14 Residual loss tangent as a function of frequency with superimposed static field as a parameter, p. 105
Figure 3.15 Hysteresis loss factor as a function of flux density, p. 107
Figure 3.16 Hysteresis loss factor as a function of frequency, p. 109
Figure 3.17 Hysteresis loss factor as a function of temperature, p. 110
Figure 3.18 Magnetic distortion, p. 111
Figure 3.19 Magnetic power loss as a function of flux density and frequency, p. 114
Figure 3.20 Magnetic power loss as a function of temperature, p. 123
Figure 3.21 Resistivity as a function of temperature, p. 126
Figure 3.22 Resistivity and permittivity as functions of frequency, p. 127
Figure 3.23 High frequency resistivity and permittivity as functions of temperature, p. 129
Figure 3.24 Static magnetostriction, p. 130
Figure 3.25 Initial permeability as a function of stress, p. 131

B–H loops, Figure 3.1

$f \to 0$ (ballistic measurement), or
$f \simeq 10$ kHz (dynamic measurement)

A basic property of any magnetic material is the relation between flux density and field strength, the B–H loop. This figure shows the B–H loops for a wide and representative range of ferrites. Most are static (ballistic) loops; those obtained by dynamic measurement are distinguished by the letter D in the bottom r.h. corner of the figure.

The B–H loops for the manganese zinc ferrites are given first and the later loops are for the nickel zinc ferrites. The saturation flux density depends on the composition and decreases as the temperature rises (see also Figure 3.3); its effective value is also decreased by the porosity of the specimen.

At room temperature the saturation flux densities range from 0.30 to 0.5 T (3000 to 5000 Gs) for the manganese zinc ferrites quoted and from 0.12 to 0.41 T (1200 to 4100 Gs) for the nickel zinc ferrites.

The coercivities range from 4 to 100 A m^{-1} (0.05 to 1.25 Oe) for the manganese zinc ferrites and from 16 A m^{-1} (0.2 Oe) for the nickel zinc ferrite having the highest zinc content to about 1400 A m^{-1} (18 Oe) for nickel ferrite having no zinc.

The area of the B–H loop is a measure of the energy loss due to hysteresis in a unit volume during one cycle of magnetization. At high frequencies residual loss and eddy current loss will generally modify the loop shape. Thus the loop area is only an indication of the high amplitude low frequency loss: the losses at small amplitudes, e.g. the loss represented by the hysteresis coefficient, cannot be inferred.

Figure 3.1.29 illustrates B–H loops obtained by special heat treatment of nickel zinc ferrites having small cobalt additions.

See also Section 2.1, References 9, 10, 11.

Field strength	$H = NI/l$	A m^{-1}	(see Eqn (2.1))
Flux density	$B = \mu_o \mu H$	T	(see Eqn (2.8))
Magnetic polarization	$J = B - \mu_o H$		(see Eqn (2.5))
Hysteresis energy loss density	$w_h = \oint B \, dH$	J m^{-3} cycle^{-1} (see Eqn (2.31))	

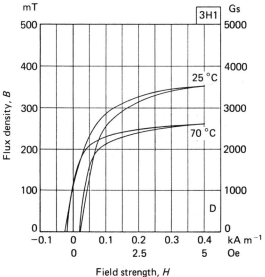

Figure 3.1.1

	Code No.	Manufr	Class
MnZn ferrite:	3H1	4	I

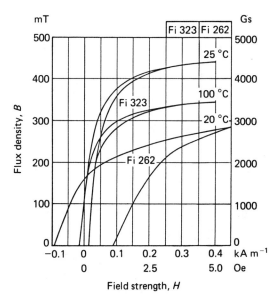

Figure 3.1.2

	Code No.	Manufr	Class
MnZn ferrite:	Fi323	8	I, IV
	Fi262	8	II

60 Properties of some manganese zinc and nickel zinc ferrites

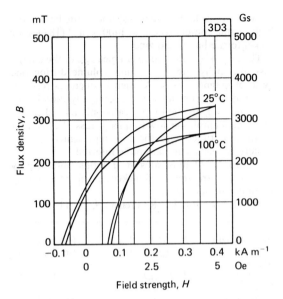

Figure 3.1.3

	Code No.	Manufr	Class
MnZn ferrite:	3D3	4, 19	II

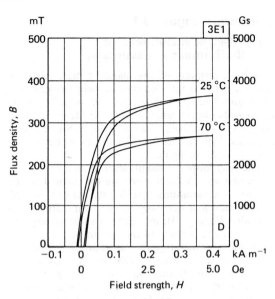

Figure 3.1.5

	Code No.	Manufr	Class
MnZn ferrite:	3E1	4	III

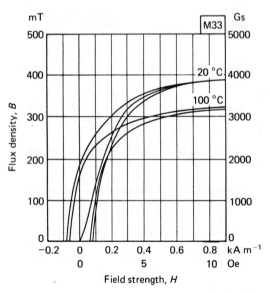

Figure 3.1.4

	Code No.	Manufr	Class
MnZn ferrite:	M33	6	II

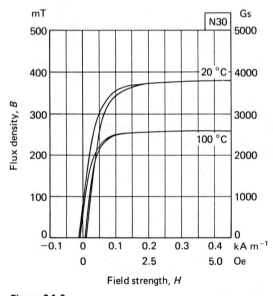

Figure 3.1.6

	Code No.	Manufr	Class
MnZn ferrite:	N30	6	III

Properties of some manganese zinc and nickel zinc ferrites 61

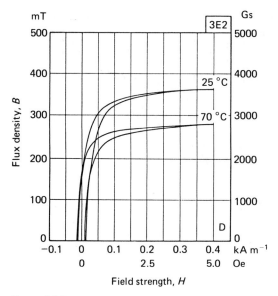

Figure 3.1.7

	Code No.	Manufr	Class
MnZn ferrite:	3E2	4	III

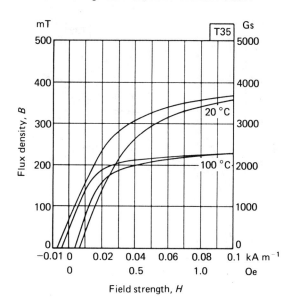

Figure 3.1.9

	Code No.	Manufr	Class
MnZn ferrite:	T35	6	III

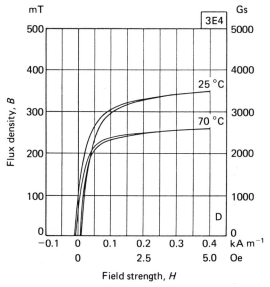

Figure 3.1.8

	Code No.	Manufr	Class
MnZn ferrite:	3E4	4	III

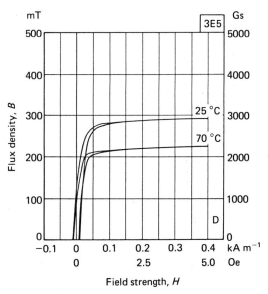

Figure 3.1.10

	Code No.	Manufr	Class
MnZn ferrite:	3E5	4, 19	III

Properties of some manganese zinc and nickel zinc ferrites

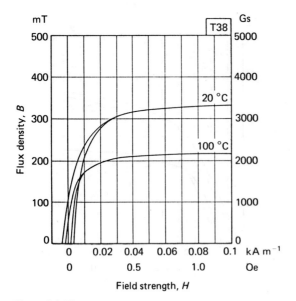

Figure 3.1.11

	Code No.	Manufr	Class
MnZn ferrite:	T38	6	III

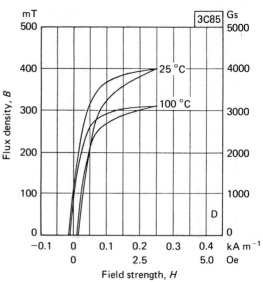

Figure 3.1.13

	Code No.	Manufr	Class
MnZn ferrite:	3C85	4, 19	IV

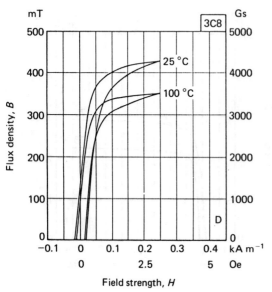

Figure 3.1.12

	Code No.	Manufr	Class
MnZn ferrite:	3C8	4, 19	IV

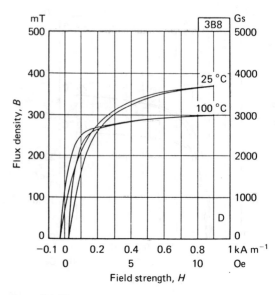

Figure 3.1.14

	Code No.	Manufr	Class
MnZn ferrite:	3B8	4	IV

Properties of some manganese zinc and nickel zinc ferrites 63

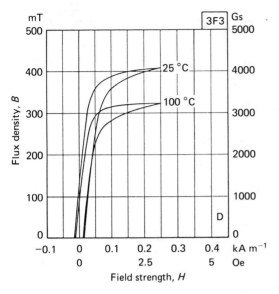

Figure 3.1.15

	Code No.	Manufr	Class
MnZn ferrite:	3F3	4, 19	IV

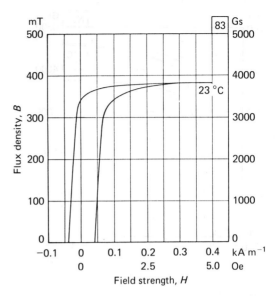

Figure 3.1.17

	Code No.	Manufr	Class
MnZn ferrite:	83	17	IV

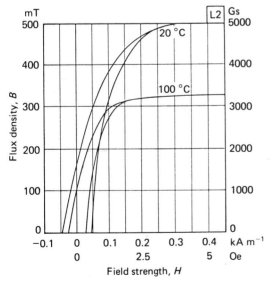

Figure 3.1.16

	Code No.	Manufr	Class
MnZn ferrite:	L2	5	IV

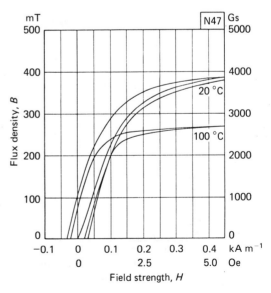

Figure 3.1.18

	Code No.	Manufr	Class
MnZn ferrite:	N47	6	IV

64 Properties of some manganese zinc and nickel zinc ferrites

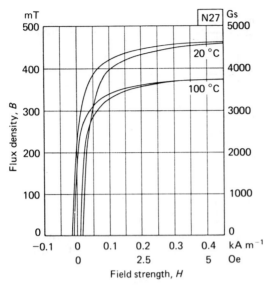

Figure 3.1.19

	Code No.	Manufr	Class
MnZn ferrite:	N27	6	IV

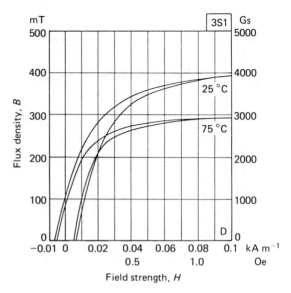

Figure 3.1.21

	Code No.	Manufr	Class
MnZn ferrite:	3S1	4	XII

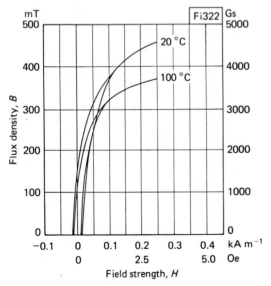

Figure 3.1.20

	Code No.	Manufr	Class
MnZn ferrite:	Fi322	8	IV

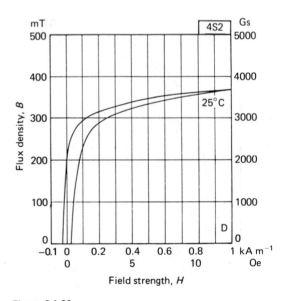

Figure 3.1.22

	Code No.	Manufr	Class
NiZn ferrite:	4S2	4	XII

Properties of some manganese zinc and nickel zinc ferrites 65

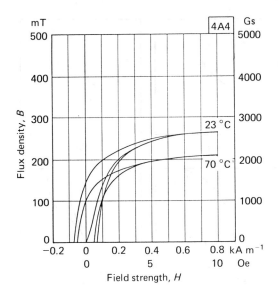

Figure 3.1.23
	Code No.	Manufr	Class
NiZn ferrite:	4A4	4	VI

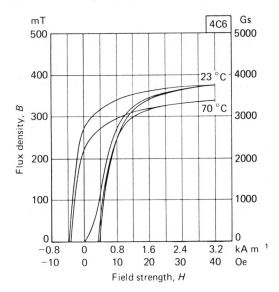

Figure 3.1.25
	Code No.	Manufr	Class
NiZn ferrite:	4C6	4	VIII

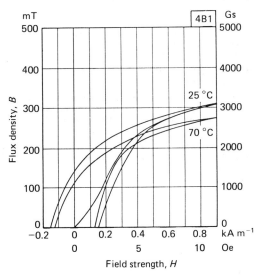

Figure 3.1.24
	Code No.	Manufr	Class
NiZn ferrite:	4B1	4	VII

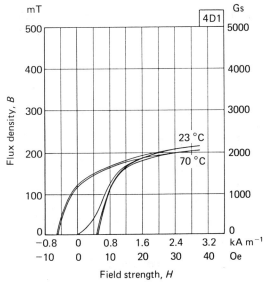

Figure 3.1.26
	Code No.	Manufr	Class
NiZn ferrite:	4D1	4	IX

66 Properties of some manganese zinc and nickel zinc ferrites

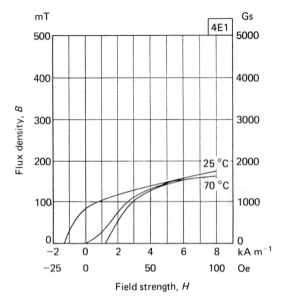

Figure 3.1.27

	Code No.	Manufr	Class
NiZn ferrite:	4E1	4	X

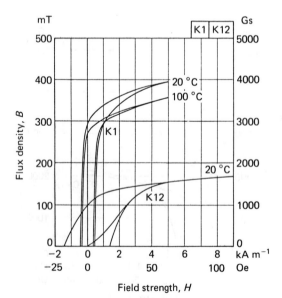

Figure 3.1.28

	Code No.	Manufr	Class
NiZn ferrite:	K1	6	VIII
	K12	6	X

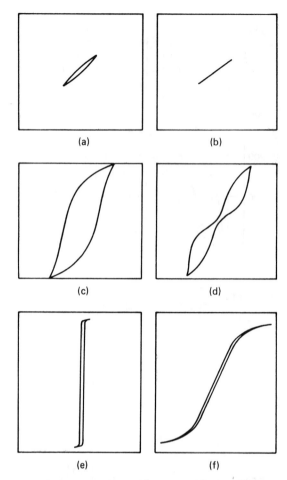

Figure 3.1.29 *B–H* loops obtained by special heat treatment of nickel zinc ferrite having a small addition of cobalt. These materials have been described by Kornetzki[9] and others.[10,11]

Loops (a) and (c): low and high amplitude loops for the ferrite without special treatment.

Loops (b) and (d): corresponding loops obtained by slow cooling after sintering. The loop shows virtually no hysteresis until, at an 'opening field', a butterfly loop is obtained. This opening of the loop is irreversible.

Loops (e) and (f): loops obtained by cooling in a magnetic field (e) parallel to the measuring field and (f) perpendicular to the measuring field

Minor B–H loops, Figure 3.2

$f \rightarrow 0$ (ballistic measurement)

$\theta \simeq 20°C$

The graphs in this figure show the initial magnetization curve for a typical MnZn power ferrite and for two NiZn ferrites having permeabilities of about 500 and 250 respectively. Branching from the magnetization curves at various field strengths are the return B–H curves obtained when the field is reduced from these field strengths. Thus if a number of unidirectional pulses of equal field strength are applied to a core, these graphs will show the corresponding peak flux density and give an indication of the remanent flux density. The first pulse will leave the material at the appropriate remanent flux density. The second will take it, by the lower arm of the minor loop (e.g. as shown by the broken curve in Figure 3.2.1), approximately to the previous value of peak flux density. Subsequent pulses will traverse substantially the same minor loop. Thus the graphs will indicate the total excursion of flux density and the incremental (or pulse) permeability.

If a number of uni-directional pulses, each corresponding to a given peak flux density, are applied to a core, the minor loop that will be traversed may be estimated by finding or interpolating a minor loop having a total flux excursion equal to that of the applied flux density. This use of these graphs is explained in more detail in Chapter 8.

It should be noted that, as these are static loops, their application may become unreliable at high pulse repetition frequencies. However they can provide a useful guide to pulse operation.

In Figure 3.2.4, the pulse permeability μ_p corresponding to the slope of the minor loop is given as a function of the total unidirectional flux excursion.

See also Sections 2.1, 8.5, 9.3, and Figures 8.21 and 8.22.

Permeability, $\mu_p = \dfrac{\Delta B}{\mu_o \Delta H}$ (see Eqn (8.30))

$\Delta B = \dfrac{U t_d}{NA}$ (see Eqn (8.31))

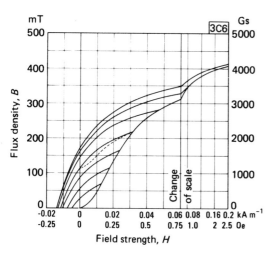

Figure 3.2.1

	Code No.	Manufr	Class
MnZn ferrite:	3C6	4	IV

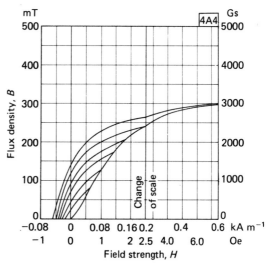

Figure 3.2.2

	Code No.	Manufr	Class
NiZn ferrite:	4A4	4	VI

68 Properties of some manganese zinc and nickel zinc ferrites

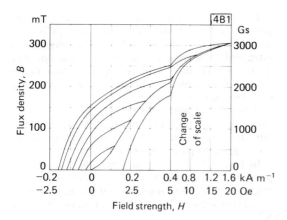

Figure 3.2.3

	Code No.	Manufr	Class
NiZn ferrite:	4B1	4	VII

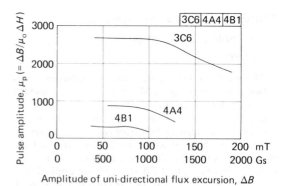

Amplitude of uni-directional flux excursion, ΔB

Figure 3.2.4

	Code No.	Manufr	Class
MnZn ferrite:	3C6	4	IV
NiZn ferrite:	4A4	4	VI
	4B1	4	VII

Saturation flux density as a function of temperature, Figure 3.3

As the temperature rises from 0 K the magnetic alignment within the domains is increasingly disturbed by thermal agitation and as a result the saturation flux density falls until, at the Curie point, magnetic alignment is virtually destroyed and the material becomes paramagnetic. In manganese zinc and nickel zinc ferrites the larger the proportion of zinc the lower the Curie point. At room temperature the saturation flux density reaches a maximum at a particular proportion of zinc. This is illustrated in the first two graphs which show the B_{sat}–temperature relations for an experimental series of ferrite compositions (after Smit and Wijn[12]).

See also Figure 3.1, Reference 12.

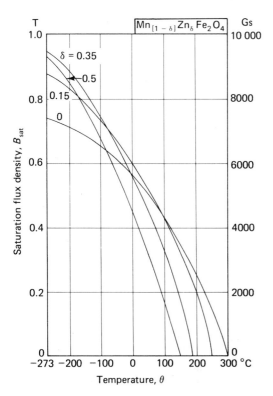

Figure 3.3.1

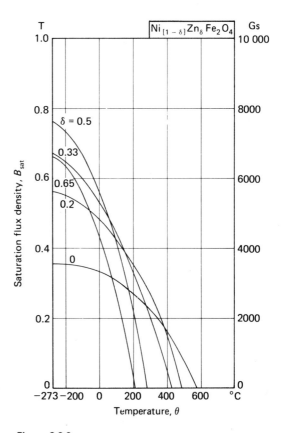

Figure 3.3.2

70   Properties of some manganese zinc and nickel zinc ferrites

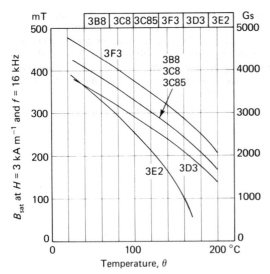

Figure 3.3.3

MnZn ferrite:	Code No.	Manufr	Class
	3B8	4	IV
	3C8, 3C85, 3F3	4, 19	IV
	3D3	4, 19	II
	3E2	4	III

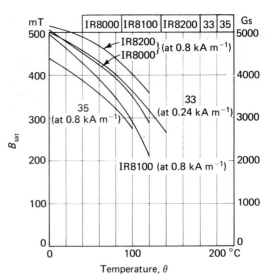

Figure 3.3.5

MnZn ferrite:	Code No.	Manufr	Class
	IR8000, IR8100, IR8200	15	IV
	33	23	I, IV
	35	23	III

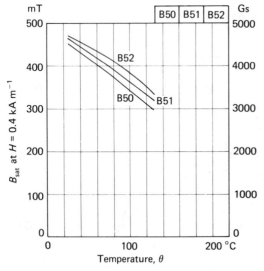

Figure 3.3.4

MnZn ferrite:	Code No.	Manufr	Class
	B50, B51, B52	7	IV

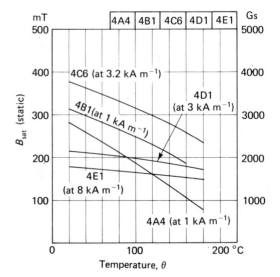

Figure 3.3.6

NiZn ferrite:	Code No.	Manufr	Class
	4A4	4	VI
	4B1	4	VII
	4C6	4	VIII
	4D1	4	IX
	4E1	4	X

Permeability as a function of high amplitude flux density, Figure 3.4

$f < 10$ kHz

As described in Section 2.1 the permeability rises with flux density from the initial value when the flux density is nearly zero to a maximum value corresponding to the slope of a line from the origin, tangential to the knee of the initial magnetization curve. Any further increase in $\hat{B}$ reduces the permeability.

The graphs illustrate a wide variety of behaviour. In some grades the increase of permeability is not great but on the other hand μ_{max}/μ_i ratios exceeding 2 may often be found, and values of μ_{max} between 5000 and 30 000 are observed.

At higher frequencies, e.g. > 50 kHz, the rise in permeability may be expected to be less.

From the practical point of view these amplitude permeabilities are rarely of benefit in design work because a device that must have a given performance at high flux densities must usually operate equally well at low flux densities. The main use of these data is in checking that a given design stays within requirements over the specified range of flux densities. See also Section 2.1.

Amplitude permeability $\mu_a = \dfrac{\hat{B}}{\mu_o \hat{H}}$ (see Eqn (2.8))

Initial permeability $\mu_i = \lim_{\hat{B} \to 0}(\mu_a)$

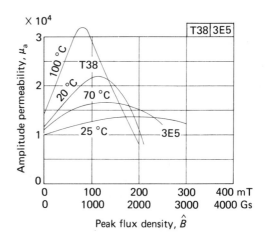

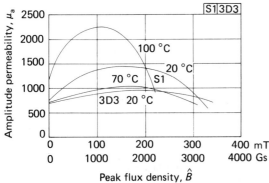

Figure 3.4.1

MnZn ferrite:	Code No.	Manufr	Class
	3H1, 3H3	4	I

Figure 3.4.2

MnZn ferrite:	Code No.	Manufr	Class
	S1	5	II
	3D3	4, 19	II

Figure 3.4.3

MnZn ferrite:	Code No.	Manufr	Class
	T38	6	III
	3E5	4, 19	III

Figure 3.4.4

MnZn ferrite:	Code No.	Manufr	Class
	3B8	4	IV
	3C8	4, 19	IV

72 Properties of some manganese zinc and nickel zinc ferrites

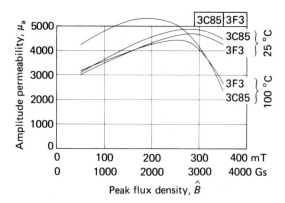

Figure 3.4.5
MnZn ferrite: Code No. 3C85, 3F3   Manufr 4, 19   Class IV

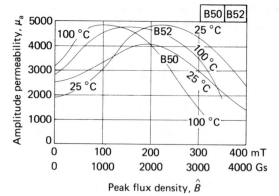

Figure 3.4.8
MnZn ferrite: Code No. B50, B52   Manufr 7   Class IV

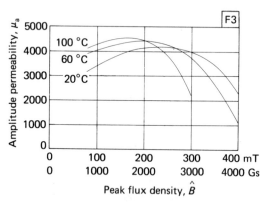

Figure 3.4.6
MnZn ferrite: Code No. F3   Manufr 3   Class IV

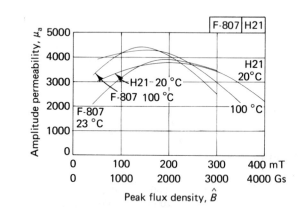

Figure 3.4.9
MnZn ferrite: Code No. F-807, H21   Manufr 13, 14   Class IV, IV

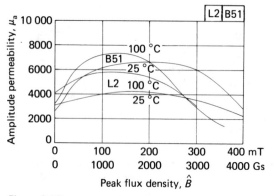

Figure 3.4.7
MnZn ferrite: Code No. L2, B51   Manufr 5, 7   Class IV, IV

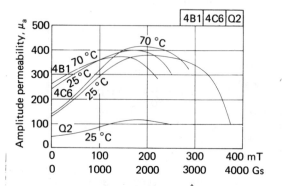

Figure 3.4.10
NiZn ferrite: Code No. 4B1, 4C6, Q2   Manufr 4, 4, 15   Class VII, VIII, IX

Permeability as a function of low amplitude flux density, Figure 3.5

$f = 10$ kHz

$\theta \approx 20°C$

The graphs show the rise in permeability as a function of the alternating flux density at low amplitudes. Such data are useful in estimating the stability of inductors subject to changes of low flux density.

They also give an indication of the permeability rise factor used in wave-form distortion calculations, e.g. the Rayleigh coefficient v and the Peterson coefficient a_{11}. For this purpose it is usual to work in terms of field strength rather than flux density. However, as the equations below indicate, the conversion may be readily obtained. Using the slopes of these graphs at zero flux the value of the above coefficients are as shown in Table 3.6.

See also Sections 2.2.3, 2.2.7, and Figures 3.15 and 3.18.

Table 3.6

Specimen	$d\mu/d\hat{H}$ or v or a_{11}	
	(SI units)	(CGS)
3H1	18	1430
3H3	24	1900
3D3	2.1	170
3E2	700	56000

Compare values obtained in Figure 3.15.

$$\left(\frac{\delta\mu}{\delta H}\right)_{dc} = \left(\frac{\delta\mu}{\delta H}\right)_{ac} = v = a_{11} = \frac{\mu_o \mu_a \delta\mu}{\delta \hat{B}}$$

(see Eqn (2.28))

where μ_a is the amplitude permeability at flux density $\hat{B}$. $\mu_a \to \mu_i$ when $\hat{B} \to 0$

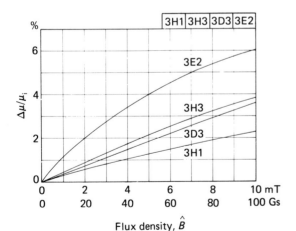

Figure 3.5.1

	Code No.	Manufr	Class
MnZn ferrite:	3H1, 3H3	4	I
	3D3	4, 19	II
	3E2	4	III

Incremental permeability as a function of static field strength, Figure 3.6

$f < 10$ kHz

The permeability of a material measured with an alternating field in the presence of a superimposed static, or steady, field ($= NI_o/l_e$) is referred to as the incremental permeability, μ_Δ.

Apart from a possible small initial increase it becomes progressively less as the static field is increased. If an air gap is introduced into the magnetic circuit there are two effects: in the absence of a static field the effective permeability is reduced (see Eqn (4.26)) and the reluctance of the magnetic path to the static field is increased and this reduces the superimposed flux density. The overall result of the introduction of an air gap is that the value of μ_Δ is reduced at the lower values of the static field but it is increased when the applied static field exceeds a certain value depending on the ratio of gap length to magnetic path length, l_g/l_e. If the incremental permeability is measured as a function of static field strength with a number of different gap lengths a family of curves is obtained. Each curve emerges above the others over a limited range of applied field strength indicating that for a given field strength there is a particular air gap ratio that will give the maximum effective incremental permeability.

These curves are useful in predicting the effect of superimposed d.c. on a given design but they cannot easily be used for designing inductors or transformers carrying d.c. This is because until the effective μ_Δ is known it is not possible to estimate the number of turns for a required inductance; until the number of turns is known the applied d.c. ampere turns cannot be calculated and so the optimum effective μ_Δ is not known. This problem is solved by deriving from these graphs others called Hanna curves. These are presented in Figure 3.7.

See also Sections 2.1, 4.2.2, 6.2.1, and Figure 3.7.

$$\mu_e = \frac{C_1}{l_g/A_g + C_1/\mu}$$ (see Eqn (4.26))

$$\mu_\Delta = \Delta B / \mu_o \Delta H$$ (see Eqn (2.8))

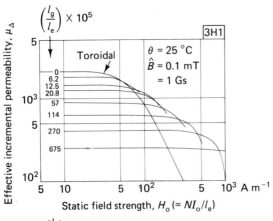

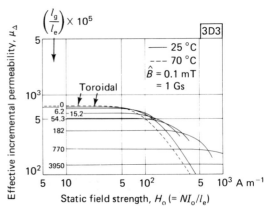

Figure 3.6.2

	Code No.	Manufr	Class
MnZn ferrite:	3D3	4, 19	II

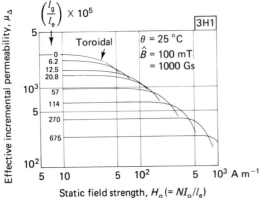

Figure 3.6.1

	Code No.	Manufr	Class
MnZn ferrite:	3H1	4	I

Properties of some manganese zinc and nickel zinc ferrites 75

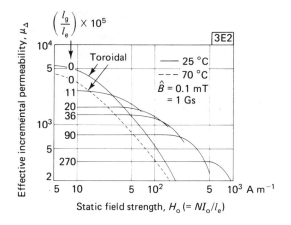

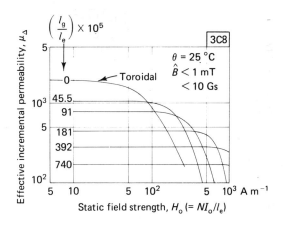

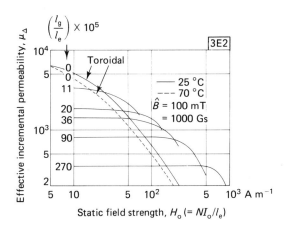

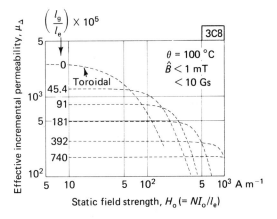

Figure 3.6.3

	Code No.	Manufr	Class
MnZn ferrite:	3E2	4	III

Figure 3.6.4

	Code No.	Manufr	Class
MnZn ferrite:	3C8	4, 19	IV

76 Properties of some manganese zinc and nickel zinc ferrites

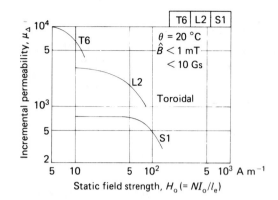

Figure 3.6.5

	Code No.	Manufr	Class
MnZn ferrite:	T6	5	III
	L2	5	IV
	S1	5	II

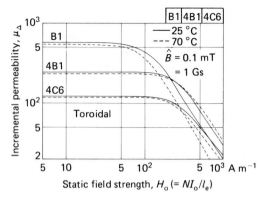

Figure 3.6.7

	Code No.	Manufr	Class
NiZn ferrite:	B1	4	VI
	4B1	4	VII
	4C6	4	VIII

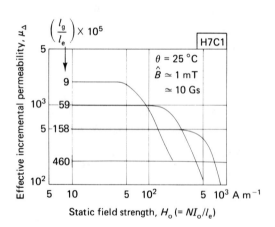

Figure 3.6.6

	Code No.	Manufr	Class
MnZn ferrite:	H7C1	26	IV

Properties of some manganese zinc and nickel zinc ferrites 77

Hanna curves, Figure 3.7

$f < 10$ kHz

In the introduction to the preceding figure (μ_Δ v. static field strength) it was observed that the usefulness of such graphs was confined to predicting the behaviour of a given design rather than preparing an original design. For convenience in the design of inductors and transformers carrying direct current the data are best presented in the form of Hanna curves.[13] From the experimental data the parameter LI_o^2/V_e is calculated as a function of NI_o/l_e for each gap ratio. L is the inductance of N turns on the particular core used in the measurement, l_e and V_e are its magnetic length and volume and I_o is the direct current. It was shown by Hanna that the relation between these quantities is independent of the core geometry provided the cross section is constant and it may in practice be used quite generally. The curve for each given air gap ratio has a region in which it is above all the other curves; it is usual to draw a single envelope curve and mark along its length a scale of optimum air gap ratios.

To use the composite curve the parameter LI_o^2/V_e is calculated from the value of the inductance required at a given direct current through a winding on a given core. From the curve, the corresponding value of NI_o/l_e is found. Since I_o and l_e are known the value of N may be obtained. The point at which the calculated value of LI_o^2/V_e intersects the curve will give the air gap ratio that must be used to meet the requirements.

This will be the optimum design. It should be remembered that in some cases, e.g. E cores or U cores, the air gap so calculated is normally provided by means of spacers. Since there will be two spacers in the magnetic path each spacer thickness is half the required air gap.

The graphs of Figure 3.7 show Hanna curves for a representative selection of ferrites, the majority measured at room temperature and at an elevated temperature. In two cases curves are shown for both low amplitudes and high amplitudes of alternating flux density. The curves for the nickel zinc ferrites are not true Hanna curves because data were available only for the ungapped core, however they have been included because in a limited way they are useful. Some indication of gapped performance may be obtained by analogy with the curves for the manganese zinc ferrites by extrapolating the lower part of the curve upwards, ignoring the falling off at the knee.

In the derivation, l_e and V_e are used as though they referred to an ideal core shape, i.e. a radially thin toroid. In the case of a practical core the effective length and volume may be used without much error. Where a core has a particularly non-uniform cross-section it may be preferable to determine experimentally a special curve for it, expressing LI_o^2 as a function of NI_o with spacer thickness as a parameter so that the core geometry does not figure in either the preparation or the practical use of the curve. This has been done for some specific cores in Chapters 7 and 9.

See also Sections 7.3.1, 9.4.3, Figures 3.6, 7.12, 9.12, and Reference 13.

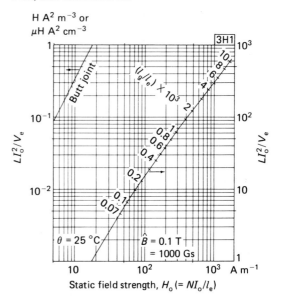

Figure 3.7.1

	Code No.	Manufr	Class
MnZn ferrite:	3H1	4	I

78 Properties of some manganese zinc and nickel zinc ferrites

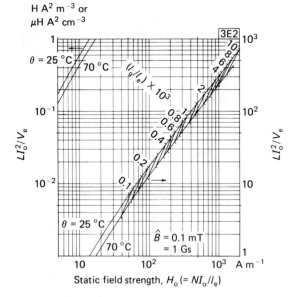

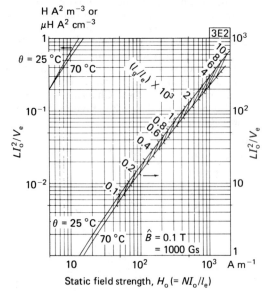

Figure 3.7.2

	Code No.	Manufr	Class
MnZn ferrite:	3E2	4	III

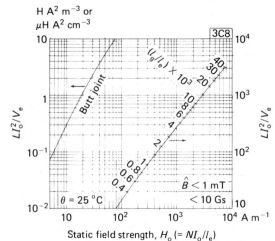

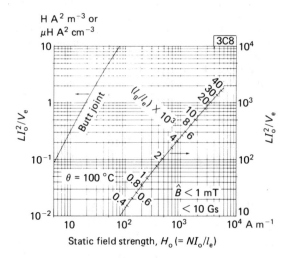

Figure 3.7.3

	Code No.	Manufr	Class
MnZn ferrite:	3C8	4, 19	IV

Properties of some manganese zinc and nickel zinc ferrites 79

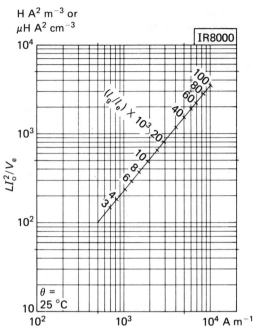

Figure 3.7.4

	Code No.	Manufr	Class
MnZn ferrite:	IR8000	15	IV

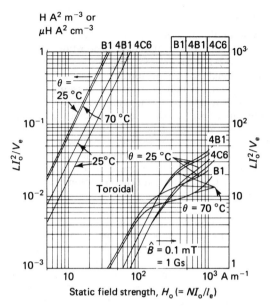

Figure 3.7.6

	Code No.	Manufr	Class
NiZn ferrite:	B1	4	VI
	4B1	4	VII
	4C6	4	VIII

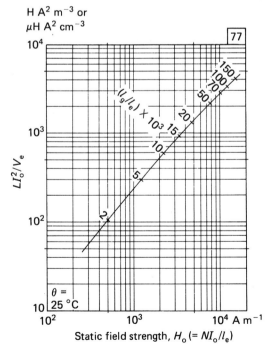

Figure 3.7.5

	Code No.	Manufr	Class
MnZn ferrite:	77	17	IV

Initial permeability as a function of temperature, Figure 3.8

f = low
$\hat{B}$ = low

The rotational component of the initial permeability of a magnetic material is proportional to the square of the saturation magnetization and inversely proportional to the anisotropy energy. Smit and Wijn[14] point out that since both of these parameters vary with temperature, the temperature dependence of the initial permeability can be complicated. As the anisotropy generally depends on the second or higher power of the saturation magnetization, which goes to zero at the Curie point, the permeability will normally increase with temperature, reaching a maximum just below the Curie point and then dropping rapidly to unity.

In mixed ferrites, particularly MnZn ferrites, the anisotropy is often arranged to go through zero at a temperature (the compensation temperature) that is well below the Curie point. As described in Section 1.3.2, this can be achieved by controlling the oxygen partial pressure during sintering in such a way that some of the iron appears in divalent form. This causes the $\mu_i(\theta)$ curve to exhibit a secondary maximum which can in principle be placed at any temperature below the Curie point; this maximum can be a pronounced peak or a rather suppressed undulation of the curve. By suitable process control the $\mu_i(\theta)$ curve may thus be engineered to meet specific requirements, such as a specified temperature coefficient or a high permeability over a particular temperature range, see for example Figures 3.8.3 and 3.8.4.

Other ions, such as cobalt, can have a similar effect and are sometimes introduced in small quantities for this purpose.

In measuring the temperature coefficient care must be taken to avoid time effects. Obviously sufficient time must be allowed for the specimen to achieve a stable and uniform temperature. When this has been done there will in general remain a time change of permeability resulting from the thermal disturbance as described in the introduction to Figure 3.10. If this is ignored the temperature change will appear to have an irreversible component and the measurements will be unreliable. The difficulty may be avoided by allowing sufficient time, e.g. 1 h, after the specimen has reached the required uniform temperature; the time change will usually have become negligible by that time.

The first two graphs show the effect of zinc content on the permeability/temperature relation (after Smit and Wijn[14]). As the zinc content becomes larger the Curie temperature falls and the room-temperature permeability increases. The manganese zinc ferrites, which are the subject of the first graph, contain a small amount of ferrous ions and show the characteristic secondary peak.

In Figures 3.8.16 and 3.8.17 the graphs show the temperature dependence of the amplitude permeability as well as that of the initial permeability.

See also Sections 1.3.2, 1.3.7, 2.2.2, 4.2.2, 5.5.2 and References 14–23.

Temperature coefficient = $\Delta\mu_i/\mu_i\Delta\theta$ (see Eqn (2.20))
Temperature factor = $\Delta\mu_i/\mu_i^2\Delta\theta$ (see Eqn (2.23))

Properties of some manganese zinc and nickel zinc ferrites

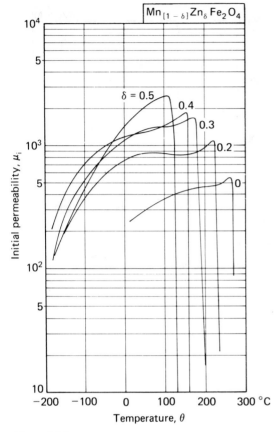

Figure 3.8.1

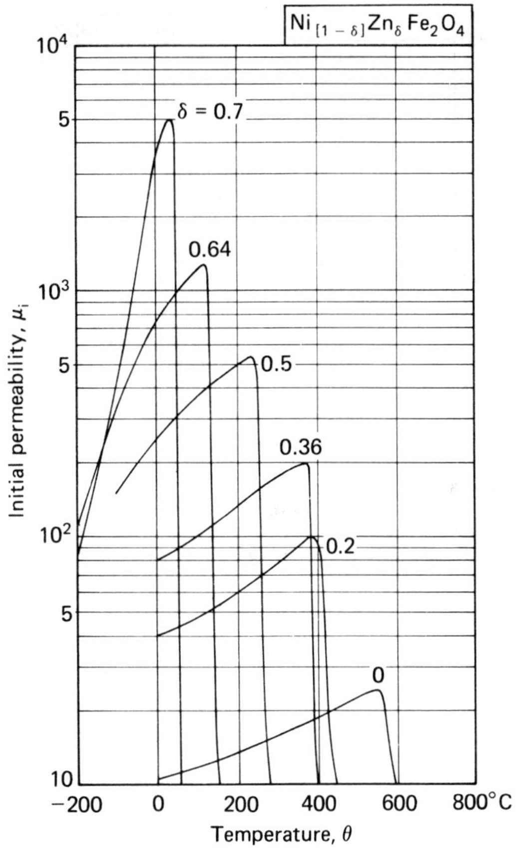

Figure 3.8.2

82   Properties of some manganese zinc and nickel zinc ferrites

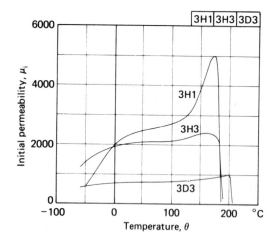

Figure 3.8.3

MnZn ferrite:	Code No.	Manufr	Class
	3H1 3H3	4	I
	3D3	4, 19	II

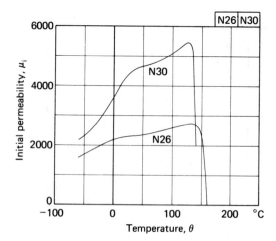

Figure 3.8.5

MnZn ferrite:	Code No.	Manufr	Class
	N26, N30	6	III

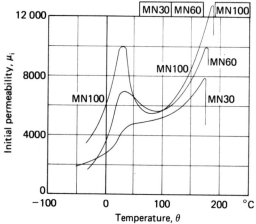

Figure 3.8.4

MnZn ferrite:	Code No.	Manufr	Class
	MN30	16	I
	MN60, MN100	16	III

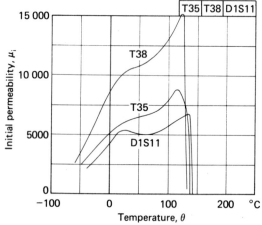

Figure 3.8.6

MnZn ferrite:	Code No.	Manufr	Class
	T35, T38	6	III
	D1S11	2	III

Properties of some manganese zinc and nickel zinc ferrites 83

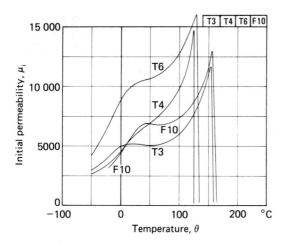

Figure 3.8.7

MnZn ferrite:	Code No.	Manufr	Class
	T4, T3, T6	5	III
	F10	3	III

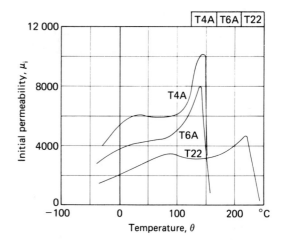

Figure 3.8.9

MnZn ferrite:	Code No.	Manufr	Class
	T4A, T6A	7	III, XII
	T22	7	III, IV

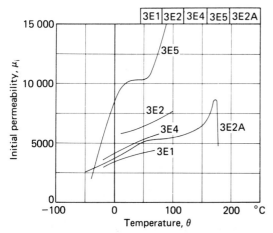

Figure 3.8.8

MnZn ferrite:	Code No.	Manufr	Class
	3E1, 3E2, 3E4	4	III
	3E5	4,19	III
	3E2A	19	III

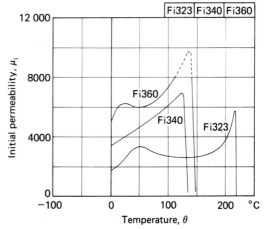

Figure 3.8.10

MnZn ferrite:	Code No.	Manufr	Class
	Fi323	8	I, IV
	Fi340, Fi360	8	III

84 Properties of some manganese zinc and nickel zinc ferrites

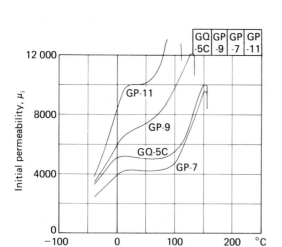

Figure 3.8.11

	Code No.	Manufr	Class
MnZn ferrite:	GQ-5C, GP-9, GP-7, GP-11	24	III

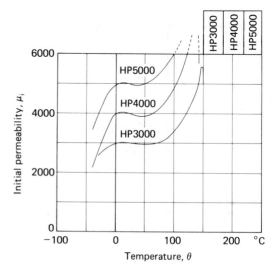

Figure 3.8.13

	Code No.	Manufr	Class
MnZn ferrite:	HP3000, HP4000, HP5000	26	III

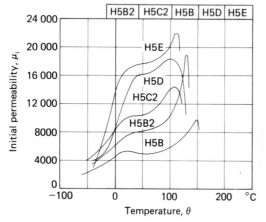

Figure 3.8.12

	Code No.	Manufr	Class
MnZn ferrite:	H5B2, H5C2, H5B, H5D, H5E	26	III

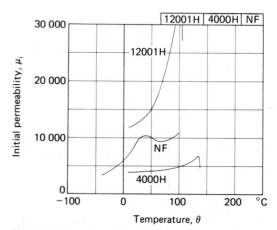

Figure 3.8.14

	Code No.	Manufr	Class
MnZn ferrite:	12001H, 4000H	27	III
	NF	26	III

Properties of some manganese zinc and nickel zinc ferrites 85

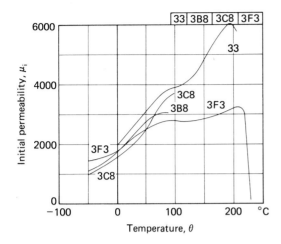

Figure 3.8.15

MnZn ferrite:	Code No.	Manufr	Class
	33	23	I, IV
	3B8	4	IV
	3C8, 3F3	4, 19	IV

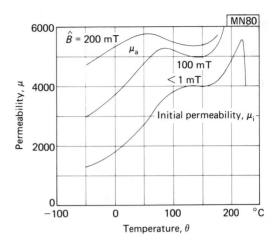

Figure 3.8.17

MnZn ferrite:	Code No.	Manufr	Class
	MN80	16	IV

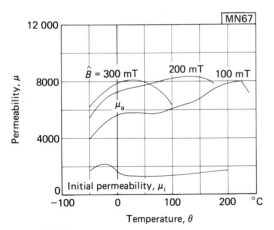

Figure 3.8.16

MnZn ferrite:	Code No.	Manufr	Class
	MN67	16	IV

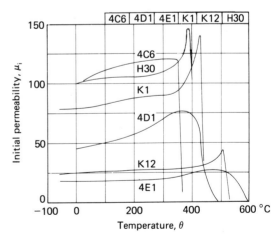

Figure 3.8.18

NiZn ferrite:	Code No.	Manufr	Class
	4C6	4	VIII
	4D1	4	IX
	4E1	4	X
	K1	6	VIII
	K12	6	X
	H30	7	VIII

86 Properties of some manganese zinc and nickel zinc ferrites

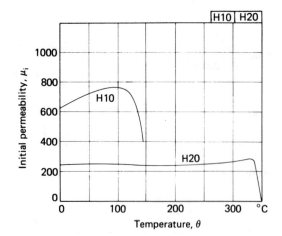

Figure 3.8.19

	Code No.	Manufr	Class
NiZn ferrite:	H10	7	VI
	H20	7	VII

Properties of some manganese zinc and nickel zinc ferrites 87

Incremental permeability as a function of temperature with static field strength as a parameter, Figure 3.9

$f = 5$ kHz

$B = 0.1$ mT (1 Gs)

It has been seen in the previous figure that the temperature coefficient of initial permeability is normally positive unless there is a pronounced secondary peak in the curve. However the incremental permeability is governed to some extent by the saturation flux density, e.g. referring to Figure 3.1.1 it may be deduced that the incremental permeability at a field strength of 0.1 kA m^{-1} would fall as the temperature rises from 25°C to 70°C simply due to the diminishing B–H loop (see also Figure 6.3). So it is to be expected that as the static field strength, H_o, is increased the temperature coefficient of the incremental permeability will decrease and become negative. The graphs that follow illustrate the behaviour of some ferrites that might find application in transductors.

The effect of a static field is to reduce the primary peak of the permeability/temperature curve that occurs just below the Curie point and the whole curve is lowered. Over a small range of temperature, which depends on the value of the static field strength, the temperature coefficient is approximately zero. The first and last graphs show secondary peaks due to anisotropy compensation in the region of room temperature. The static field does not eliminate the compensation point but tends to move it to lower temperatures.

See also Section 6.2.2, and Figure 3.8.

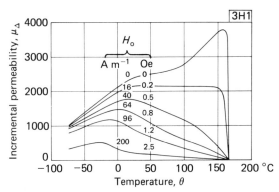

Figure 3.9.1
MnZn ferrite: Code No. 3H1 Manufr 4 Class I

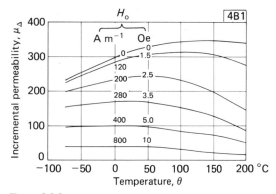

Figure 3.9.3
NiZn ferrite: Code No. 4B1 Manufr 4 Class VII

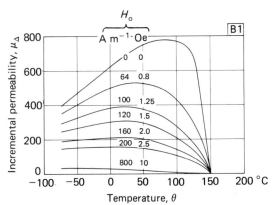

Figure 3.9.2
NiZn ferrite: Code No. B1 Manufr 4 Class VI

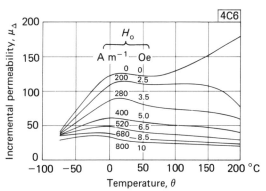

Figure 3.9.4
NiZn ferrite: Code No. 4C6 Manufr 4 Class VIII

Disaccommodation, Figure 3.10

$f = 5$ kHz

$\hat{B} = 0.1$ mT (< 1 Gs)

If a magnetic material is given a disturbance, which may be magnetic, thermal or mechanical, the initial permeability observed immediately after the cessation of the disturbance is normally found to be raised to an unstable value from which it returns, as a function of time, to its undisturbed or stable value. This phenomenon has been called time-change of permeability but is usually referred to as disaccommodation. The process is repeatable indefinitely and in this sense it is different from ageing, which is a permanent change with time and which is not generally observed in ferrites at normal operating temperatures.

There are a number of possible mechanisms of disaccommodation in ferrites and they all depend on migratory processes within the lattice. These processes often involve the anisotropic or preferred distribution of ferrous ions and/or cation vacancies over the four octahedral sublattices of the spinel structure. The preference for a particular sublattice depends on the direction of the domain magnetization and, therefore, on the position of the domain walls. The distribution tends to fix, and be fixed by, the location of the domain walls, giving rise to a more stable or lower permeability state. After the disturbance the vacancies and domain walls are no longer in a mutually low energy state (therefore the initial permeability is higher) but the lower energy state is progressively approached as the ferrous ions and/or vacancies migrate to the new preferred sites conditioned by the new domain wall positions. In other words the disturbed domain wall, having taken up a new position, sinks slowly into an energy trough at that position, losing mobility and reducing domain wall contribution to the permeability as it does so. The rate at which the vacancies are redistributed or diffuse depends very much on the temperature, shorter time constants being observed at the higher temperatures. If the permeability is measured with a large amplitude signal it is clear that domain walls will be constantly jumping to new positions and disaccommodation will not be observed; the maximum disaccommodation is observed at vanishingly small measuring amplitudes.

Perhaps the most fundamental disturbance is an excursion of the temperature of the material above the Curie point and back again. As this is rather a lengthy process it is not a very convenient basis for a quantitative method of measuring disaccommodation. Instead, a disturbance procedure similar to an a.c. demagnetization is usually preferred; it is particularly suitable for measurements on manganese zinc ferrites. On the other hand it is not suitable for field sensitive materials, such as those described in Figure 3.1.29; these suffer permanent change of properties due to high fields and so the Curie point method is perhaps the only means by which their disaccommodation may be assessed.

The a.c. field method consists of subjecting the material to 'magnetic conditioning' by applying a saturating alternating field for a few periods and then reducing the amplitude progressively from just above the knee of the B–H loop to zero. Excessive field amplitude or duration should be avoided because this may cause heating of the core by the winding. An exponential decay of field is convenient and effective; a frequency of about 100 Hz and a time constant of about 0.08 s has been found to give reproducible results. Using such a standard disturbance the time-change of permeability may be measured (at constant temperature) and materials compared.[24] It should be remembered when using disaccommodation data that the test disturbance is artificial and as such is unlikely to occur, for instance, during the manufacture and life of a filter inductor. However, investigations have shown[25] that the time-change following the test disturbance is similar in form and magnitude to that following a temperature change to which a ferrite-cored component may be subjected, e.g. a 50°C step.

The graphs show, for various grades of ferrite and at various temperatures, the time-change of initial permeability following the test disturbance. It may be seen that, measured over the interval from 1 min to 24 h after the test disturbance, the initial permeability of inductor ferrites at room temperature may typically decrease between 0.5 and 2.5%.

For convenience the relation is usually taken to be proportional to the logarithm of time, and based on this assumption standard methods for the numerical expression of disaccommodation have been defined.[24] For the a.c. magnetic conditioning procedure, the disaccommodation, D, is expressed as the fractional change of permeability between 10 and 100 minutes after magnetic conditioning. If other time intervals are used. e.g. t_1 and t_2 respectively, then a disaccommodation coefficient, d is defined as the fractional change divided by $\log_{10}(t_2/t_1)$.

As shown in Section 4.2.2, any change in initial permeability is diluted in its effect on a gapped core by the ratio μ_e/μ_i. Thus, for a given μ_e, a higher value of μ_i will result in a lower value of effective disaccommodation in a gapped core. This consideration gives rise to the concept of normalized disaccommodation or disaccommodation factor, D_F, which is defined[26] as D or d divided by μ_i. This is the parameter which characterizes the time variability of a material, since a higher value of disaccommodation is permissible provided the μ_i is proportionately higher so that the effective value in a gapped core is unchanged:

$$\text{effective disaccommodation} = \frac{d}{\mu_i} \times \mu_e = D_F \times \mu_e$$

The IEC[26] defines the general expression for D_F as

$$\frac{(\mu_1-\mu_2)}{\mu_1^2 \log_{10} t_2/t_1}$$

where the suffices refer to the first and second measurement and t is the appropriate time interval after the test disturbance. If the graph is a straight line then the same value of D_F will be obtained whatever values of t_1 and t_2 are chosen. When $t_1 = 10$ and $t_2 = 100$ minutes then the logarithm is unity.

Returning to the experimental results it is clear that the graphs are not quite straight lines but, at least over the early period, tend to droop below the tangent drawn at $t = 1$ min. The true form of the variation and its nature has been described in the literature.

It has been observed previously that a thermal disturbance such as a rise in temperature (well below the Curie temperature) will cause a variation with time similar in form to that just described. It is important to note that the graphs in this figure depend on the specimens being held at a very constant temperature. Under service conditions the normal temperature fluctuations will prevent the permeability from progressing down the curve towards unnaturally quiescent values. The result is that the mean permeability will take up a pseudo-stable position at some intermediate value, perhaps corresponding approximately to the value at $t = 10^4$ minutes. The only subsequent effect of disaccommodation will then be a small exaggeration of the temperature dependence.

The implications of disaccommodation in the measurement of temperature coefficient have been mentioned in the introduction to Figure 3.8. Sufficient time should be allowed at each temperature so that the rate of change of permeability with time becomes negligible.

In any accurate measurement of permeability or inductance the possibility of time change due to previous disturbance should be taken into account and sufficient recovery time allowed. Apart from its effect on the accuracy of measurements, the principal importance of disaccommodation is in the design of high stability inductors used in wave filters as described in Chapter 5.

The substitution of small amounts of titanium or tin into MnZn ferrite has the effect of reducing the activation energy of the diffusion of the ferrous ions and so speeds the disaccommodation process.[32, 33] The practical effect is that the disaccommodation measured over the usual time intervals is significantly reduced. Figure 3.10.1 shows the time-dependence of permeability for such a ferrite (3H3).

See also Sections 2.2.2, 4.2.2, 5.5.2, and References 24–40.

Disaccommodation:

$$D = \frac{\mu_1 - \mu_2}{\mu_1}, \text{ often expressed in percent.}$$

(see Eqn (2.25))

Disaccommodation coefficient:

$$d = \frac{\mu_1 - \mu_2}{\mu_1 \log_{10}(t_2/t_1)}$$

(see Eqn (2.26))

Disaccommodation factor:

$$D_F = \frac{d}{\mu_1}$$

(see Eqn (2.27))

where μ_1 and μ_2 are the permeabilities at times t_1 and t_2 after the disturbance respectively.

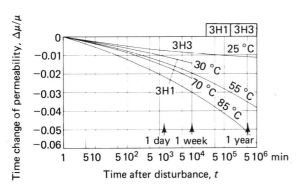

Figure 3.10.1

	Code No.	Manufr	Class
MnZn ferrite:	3H1, 3H3	4	I

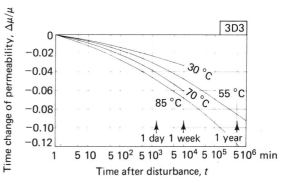

Figure 3.10.2

	Code No.	Manufr	Class
MnZn ferrite:	3D3	4, 19	II

Complex initial permeability spectrum, Figure 3.11

$B < 0.1$ mT (1 Gs)

$\theta \simeq 20°$C

In Section 2.2.1 the well-known concept of complex permeability is introduced. The real and imaginary components represent reactive and loss terms respectively and may be expressed as series components (μ'_s, μ''_s) or parallel components (μ'_p, μ''_p).

The graphs in this figure show the components of the complex initial permeability (i.e. permeability measured with $B \to 0$) as functions of frequency. For each group of materials the series components are given first and the succeeding graph shows the corresponding parallel components. It often happens that the designer requires the product of the angular frequency and the permeability components. As discussed below, these quantities are proportional to the corresponding series and parallel components respectively of the circuit impedance. They are readily obtained by placing a 45° graticule in the appropriate position over the permeability curves; the new graticule may then be scaled in terms of $\omega\mu'_s$ and $\omega\mu''_s$ or $\omega\mu'_p$ and $\omega\mu''_p$. This has been done for each of the graphs in this figure; it should be used in conjunction with the normal frequency scale.

The general form of these relations may be seen by referring to the first pair of graphs as an example. These illustrate the complex permeability of a group of manganese zinc ferrites. The measurements have been made in such a way that the loss due to macroscopic eddy currents and phenomena associated with dimensional resonance (see Section 2.2.4) have been eliminated. It is seen that at low frequencies the real part of the initial permeability, μ'_s, is about 2000 for 3H1 and about 500 for 3D3. As the frequency rises, each curve remains level at first and then rises to a shallow peak before falling rapidly to relatively low values. The loss component, μ''_s, rises to a pronounced peak as μ'_s falls. In high permeability ferrites this dispersion is principally due to the ferromagnetic resonance[41] (spin precession resonance). In a magnetically saturated ferrite this occurs as a sharp resonance at microwave frequencies; in the absence of an externally applied static field there is a distribution of domain magnetizations and the precession resonance is rather broad. Domain wall motion may, in general, also contribute to the magnetization process, so wall resonance or relaxation may contribute to μ'_s and μ''_s in this region.[42]

Snoek[43] observed that the frequency of ferromagnetic resonance varies inversely as the initial permeability. He gave the following relation:

$$f_{res} = \frac{\gamma M_{sat}}{3\pi(\mu_i - 1)} \quad \text{Hz} \quad (3.2)$$

where

f_{res} = frequency at which μ''_s is maximum

γ = gyromagnetic ratio
$\simeq 0.22 \times 10^6$ rad s^{-1} A^{-1} m

M_{sat} = saturation magnetization in A m^{-1}

μ_i = initial permeability, i.e. $\lim_{f \to 0} \mu'_s$

$$\therefore f_{res} = 23.4 \times 10^3 \frac{M_{sat}}{(\mu_i - 1)} \quad \text{Hz} \quad (3.3)$$

The saturation magnetization in most cubic ferrites lies between 250×10^3 and 350×10^3 A m^{-1} (250 and 350 Gs). Therefore if M_{sat} is given a mean value of 300×10^3 the ferromagnetic resonance frequency may be estimated with reasonable accuracy for any value of initial permeability. This treatment is rather oversimplified; for a fuller discussion the references should be consulted.

The useful frequency range of a ferrite is limited by the onset of ferromagnetic resonance, either because the permeability begins to fall or, at a somewhat lower frequency, the losses rise steeply. The maximum frequency for which the ferrite has usefully low loss may be taken as a fraction, e.g. 1/6, of f_{res} depending on the limit of $\tan \delta_r$ that is chosen. In illustration, ferrite grade 4C6 in Figure 3.11.6 has $\mu_i = 130$ and therefore the calculated value of f_{res} is about 54 MHz compared with a measured value of about 60 MHz. For this ferrite $\tan \delta_r = 0.01$ at about 10 MHz, i.e. at 1/6th of f_{res}.

The applications of high permeability manganese zinc ferrites are usually confined to the lower frequency region, e.g. less than about 2 MHz, although it will be seen below that this is not necessarily the case in transformer applications.

Reading the curves against the 45° graticule, the $\omega\mu'_s$ product rises at first in proportion to the frequency and then it either peaks or becomes almost independent of frequency.

The corresponding parallel components are shown in the second of each pair of graphs. The resonance dispersion is of course still apparent but μ'_p does not fall so rapidly with frequency. The products with ω are more revealing because, for a winding on a core, $\omega\mu'_p$ is proportional to the parallel reactance, and $\omega\mu''_p$ is proportional to the parallel resistance, arising from the core properties. These values are of importance in

---

(3.2) $f_{res} = \dfrac{4\gamma M_{sat}}{3(\mu_i - 1)}$ Hz

($\gamma \simeq 17.6 \times 10^6$ rad s^{-1} Oe^{-1})

transformer design. It is seen that $\omega\mu_p''$ ($\propto R_p$) peaks at relatively low frequency and then becomes almost constant with frequency. The product $\omega\mu_p'$ ($\propto \omega L_p$) rises with frequency and, except for the highest permeability material, the rising characteristic extends to frequencies beyond 100 MHz. It is for this reason that a ferrite-cored transformer, having sufficient shunt impedance at the lower end of the pass band, will not usually suffer extra attenuation due to the core at higher frequencies unless it is particularly sensitive to changes of R_p.

The nickel zinc ferrites behave in a similar way. It is interesting to note that in general their values of $\omega\mu_p'$ ($\propto \omega L_p$) remain lower than corresponding values of the manganese zinc ferrites even up to 100 MHz; the advantage of the higher frequency of the resonance being offset by the lower initial permeability.

See also Sections 2.2.1, 7.3.1, 7.4.1, and References 41–48.

Series impedance in terms of complex permeability:

$$Z = j\omega L_s + R_s = j\omega L_o(\mu_s' - j\mu_s'')$$
$$\omega L_s = \omega L_o \mu_s'$$
$$R_s = \omega L_o \mu_s''$$
$$\tan \delta_m = \frac{R_s}{\omega L_s} = \frac{\mu_s''}{\mu_s'}$$

(see Eqns (2.15), (2.16))

Parallel impedance and admittance in terms of complex permeability:

$$Y = \frac{1}{j\omega L_p} + \frac{1}{R_p} = \frac{1}{j\omega L_o}\left(\frac{1}{\mu_p'} - \frac{1}{j\mu_p''}\right)$$
$$\omega L_p = \omega L_o \mu_p'$$
$$R_p = \omega L_o \mu_p''$$
$$\tan \delta_m = \frac{\omega L_p}{R_p} = \frac{\mu_p'}{\mu_p''}$$

(see Eqns (2.17), (2.18))

Conversion:

$$\mu_p' = \mu_s'(1 + \tan^2 \delta_m)$$
$$\mu_p'' = \mu_s''(1 + 1/\tan^2 \delta_m)$$

(see Eqn (2.19))

92 Properties of some manganese zinc and nickel zinc ferrites

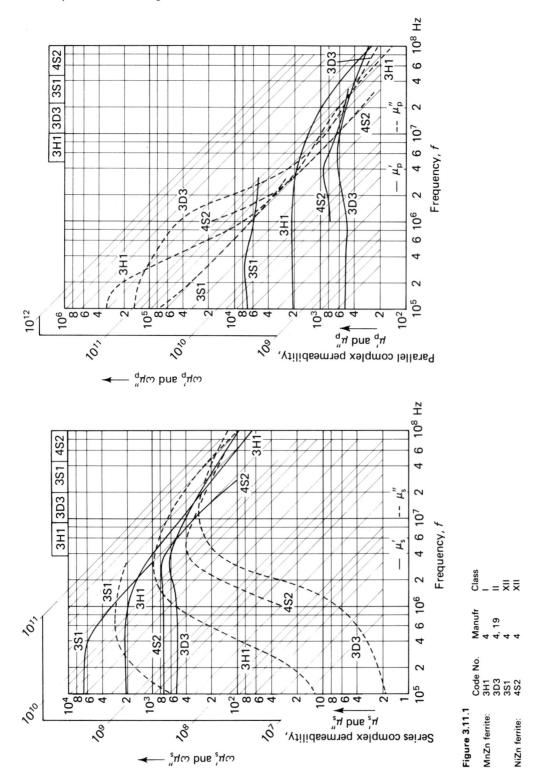

Figure 3.11.1

	Code No.	Manufr	Class
MnZn ferrite:	3H1	4	I
	3D3	4, 19	II
	3S1	4	XII
NiZn ferrite:	4S2	4	XII

Properties of some manganese zinc and nickel zinc ferrites 93

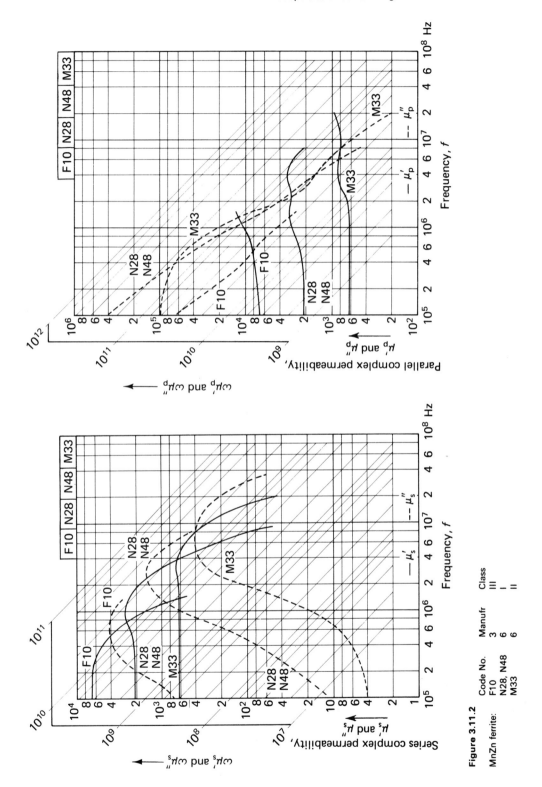

Figure 3.11.2
MnZn ferrite:

Code No.	Manufr	Class
F10	3	III
N28, N48	6	I
M33	6	II

94 Properties of some manganese zinc and nickel zinc ferrites

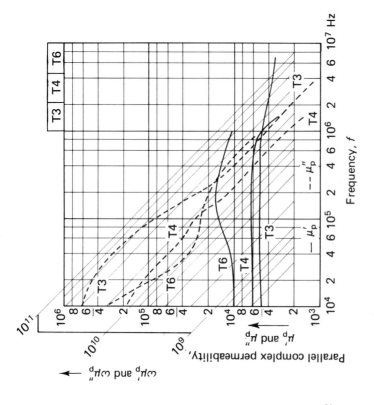

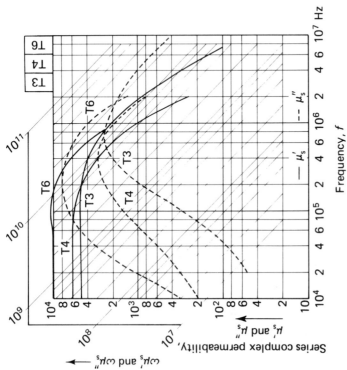

Figure 3.11.3
MnZn ferrite: Code No. T3, T4, T6   Manufr 5   Class III

Properties of some manganese zinc and nickel zinc ferrites 95

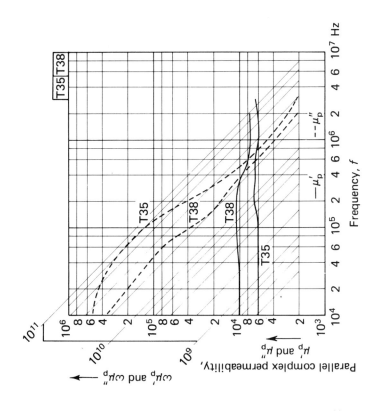

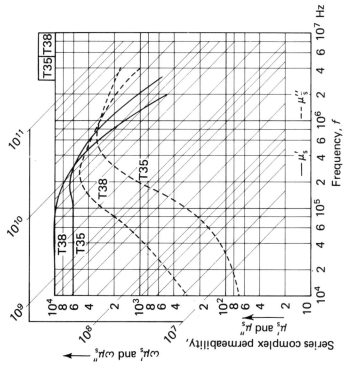

Figure 3.11.4

	Code No.	Manufr	Class
MnZn ferrite:	T35, T38	6	III

96 Properties of some manganese zinc and nickel zinc ferrites

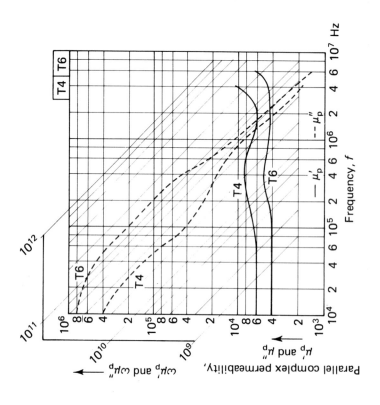

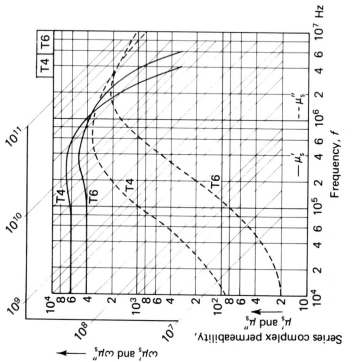

Figure 3.11.5

MnZn ferrite: Code No. T4, T6 Manufr 7 Class III

Properties of some manganese zinc and nickel zinc ferrites   97

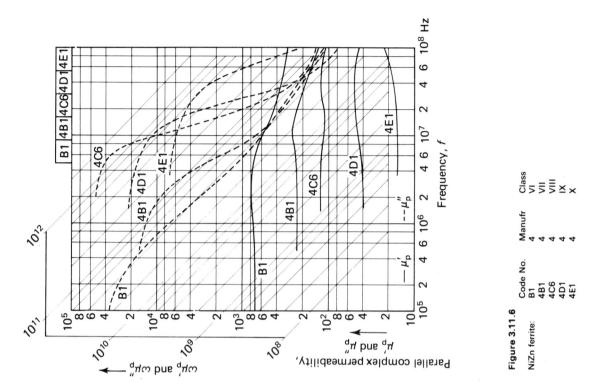

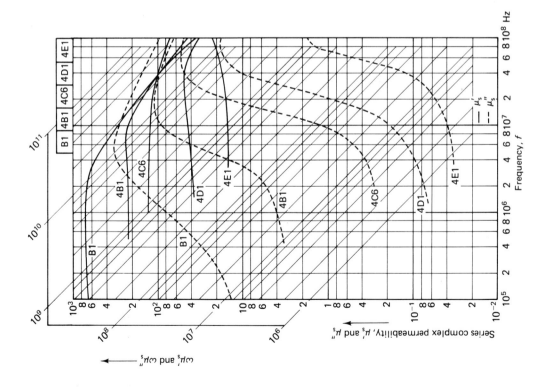

Figure 3.11.6

NiZn ferrite:

Code No.	Manufr	Class
B1	4	VI
4B1	4	VII
4C6	4	VIII
4D1	4	IX
4E1	4	X

Properties of some manganese zinc and nickel zinc ferrites

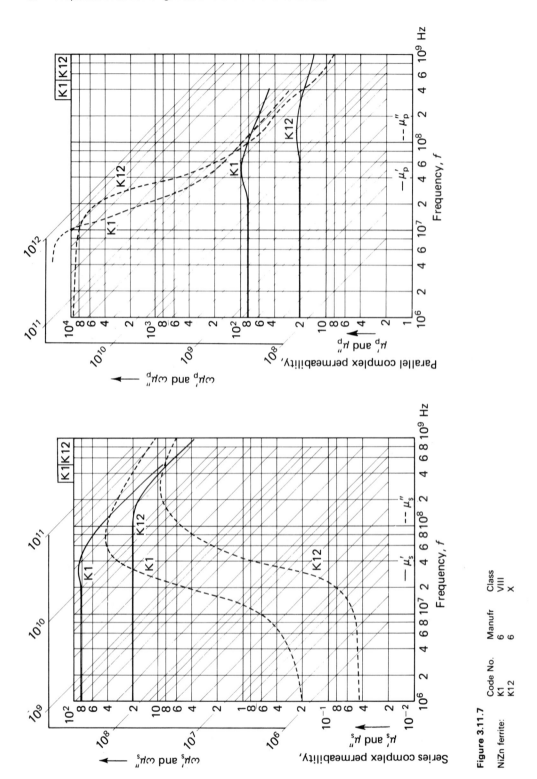

Figure 3.11.7

NiZn ferrite:	Code No.	Manufr	Class
	K1	6	VIII
	K12	6	X

Residual loss factor spectrum, Figure 3.12

$B < 0.1$ mT (1 Gs)
$\theta \simeq 20°C$

The residual loss factor $(\tan \delta_r)/\mu_i$ is the principal loss parameter at low flux densities. For an inductor with a given effective permeability, a ferrite with the lowest residual loss factor will give the lowest core loss.

The residual loss factor is strongly influenced by the permeability in the denominator so that at low frequencies the higher permeability materials tend to have the lower loss factors. However, as described in the introduction to Figure 3.11, the higher the permeability the lower is the frequency of the ferromagnetic resonance dispersion. This may be seen clearly in the graphs of the present figure where the ferrites having the lower loss factors tend to cut off at the lower frequencies.

Using the residual loss factor as the criterion and referring to Figure 3.12.2 as an example, the MnZn ferrite grade 3H3 ($\mu_i \simeq 2000$) is best up to a frequency of about 400 kHz, where grade 3D3, due to its lower permeability ($\mu_i \simeq 500$) and consequent higher cut-off frequency, emerges as the better material. Above about 2 MHz the NiZn ferrites, having a wide range of lower permeabilities, generally take over from the MnZn ferrites. Other things being equal, the lower the permeability the higher is the frequency band in which the residual loss factor is superior, see for example Figure 3.12.12 where the initial permeabilities of the grades 3F, 2F, 1E and 2E are 125, 40, 25 and 13 respectively.

So for each ferrite there is a frequency range within which it may potentially exhibit the lowest loss factor and therefore be the best choice for a low loss application. Of course, some materials may not show superior residual loss performance at any frequency. Such materials may be of early development or they may have special characteristics such as higher B_{sat} or lower cost which gives them a particular field of application.

At frequencies well below the ferromagnetic resonance the residual loss is believed to arise at least in part from thermally activated domain wall movements.

See also: Sections 2.2.1, 2.2.5, 4.2.3, 5.7.5, Figure 3.11, and References 38, 41–45 and 52.

$$\frac{\tan \delta_r}{\mu_i} = \omega L_o G \qquad \text{(see Eqn (4.46))}$$

where $L_o = \mu_o N^2 / C_1$

G = residual loss conductance

(thus loss factor is proportional to the residual loss conductance measured across a winding of N turns on a core of given core factor C_1)

$$\frac{(\tan \delta_r)_{gapped}}{\mu_e} \simeq \frac{\tan \delta_r}{\mu_i}$$

$$(\tan \delta_r)_{gapped} \simeq \mu_e \left(\frac{\tan \delta_r}{\mu_i} \right) \qquad \text{(see Eqns (2.56), (4.50))}$$

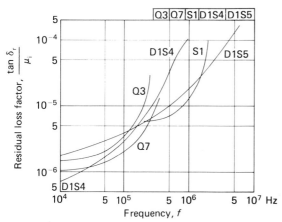

Figure 3.12.1

	Code No.	Manufr	Class
MnZn ferrite:	Q3, Q7	5	I
	S1	5	II
	D1S4	2	I
	D1S5	2	II

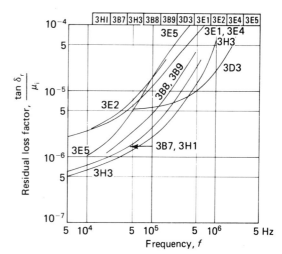

Figure 3.12.2

	Code No.	Manufr	Class
MnZn ferrite:	3H1, 3B7, 3H3	4	I
	3B8	4	IV
	3B9	19	I
	3D3	4, 19	II
	3E1, 3E2, 3E4	4	III
	3E5	4, 19	III

100 Properties of some manganese zinc and nickel zinc ferrites

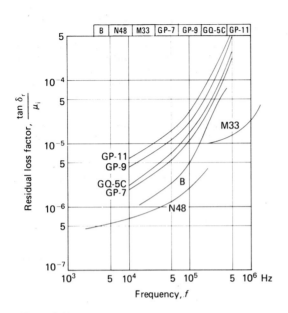

Figure 3.12.3

	Code No.	Manufr	Class
MnZn ferrite:	B	18	III
	N48	6	I
	M33	6	II
	GP-7, GP-9, GQ-5C, GP-11	24	III

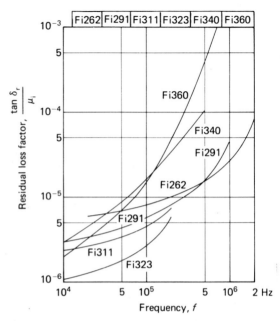

Figure 3.12.5

	Code No.	Manufr	Class
MnZn ferrite:	Fi262, Fi291	8	II
	Fi311, Fi323	8	I, IV
	Fi340, Fi360	8	III

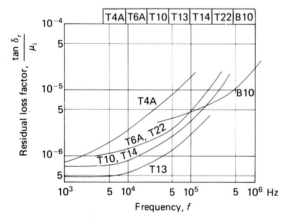

Figure 3.12.4

	Code No.	Manufr	Class
MnZn ferrite:	T4A, T6A	7	III, XII
	T10, T13, T14	7	I
	T22	7	III, IV
	B10	7	II

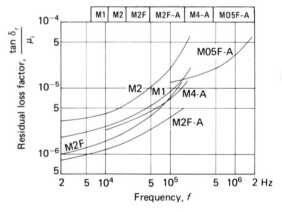

Figure 3.12.6

	Code No.	Manufr	Class
MnZn ferrite:	M1, M2F, M2F-A	10	I
	M2, M4-A	10	I, III
	M05F-A	10	II

Properties of some manganese zinc and nickel zinc ferrites 101

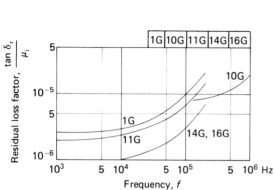

Figure 3.12.7

	Code No.	Manufr	Class
MnZn ferrite:	1G, 11G, 14G, 16G	11	I
	10G	11	II

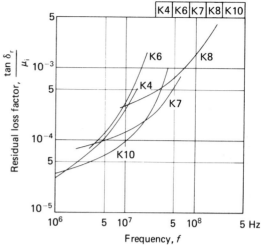

Figure 3.12.9

	Code No.	Manufr	Class
NiZn ferrite:	K4	5	VII
	K6, K10	5	VIII
	K7	5	IX
	K8	5	X

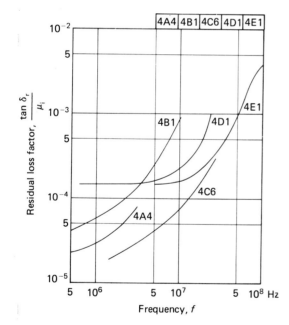

Figure 3.12.8

	Code No.	Manufr	Class
NiZn ferrite:	4A4	4	VI
	4B1	4	VII
	4C6	4	VIII
	4D1	4	IX
	4E1	4	X

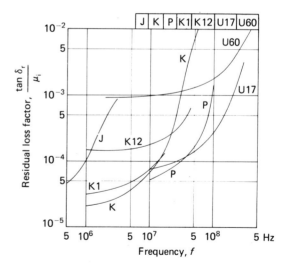

Figure 3.12.10

	Code No.	Manufr	Class
NiZn ferrite:	J	18	VI
	K	18	VIII
	P	18	IX
	K1	6	VIII
	K12	6	X
	U17, U60	6	XI

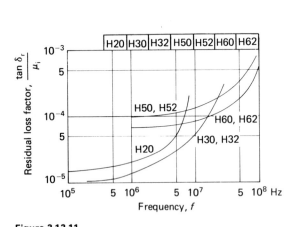

Figure 3.12.11

	Code No.	Manufr	Class
NiZn ferrite:	H20	7	VII
	H30, H32	7	VIII
	H50, H52	7	IX
	H60, H62	7	X

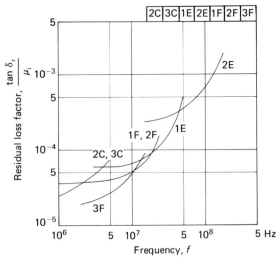

Figure 3.12.12

	Code No.	Manufr	Class
NiZn ferrite:	2C, 3C	11	VII
	1E, 2E	11	X
	1F, 3F	11	VIII
	2F	11	IX

Residual loss factor as a function of temperature, Figure 3.13

$\hat{B} < 0.1$ mT (1 Gs)

The temperature dependence of $(\tan \delta_r)/\mu_i$ at various frequencies is shown in these graphs for a number of MnZn ferrites. In every case there is a minimum in the vicinity of 0°C. This is the result of the anisotropy having been arranged to go through zero at such a temperature, see Sections 1.3.2 and 1.3.7. By proper control of the composition and sintering conditions the amount of ferrous iron can be adjusted to place the anisotropy compensation at an appropriate temperature. As seen in Figure 3.8, the compensation temperature is usually placed at the lower end of the operating temperature range in order to either raise the permeability or to control the temperature coefficient of permeability over the operating range. This provision is incompatible with achieving the minimum residual loss in the centre of the operating range, so the manufacturing process control must be adjusted to give the best compromise.

The temperature of minimum loss decreases as the frequency increases. A study of the graphs suggests that this is due to the more rapid increase of loss with frequency at the higher temperatures; the increase in permeability with temperature would tend to lower the ferromagnetic resonance frequency.

See also: Sections 1.3.2, 1.3.7, and Figures 3.8, 3.12, 3.17, 3.20.

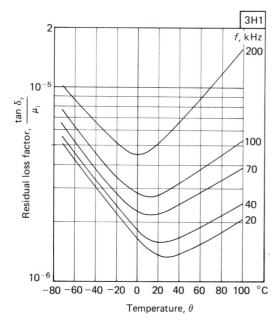

Figure 3.13.1

	Code No.	Manufr	Class
MnZn ferrite:	3H1	4	I

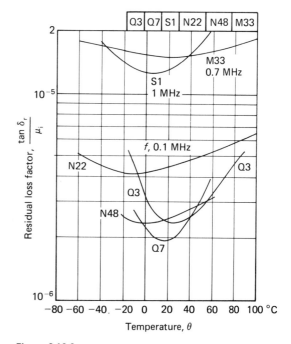

Figure 3.13.2

	Code No.	Manufr	Class
MnZn ferrite:	Q3, Q7	5	I
	S1	5	II
	N22	6	I, XII
	N48	6	I
	M33	6	II

104 Properties of some manganese zinc and nickel zinc ferrites

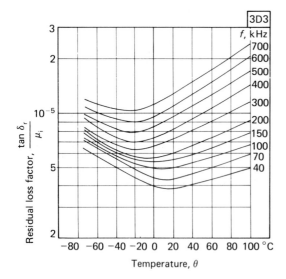

Figure 3.13.3

	Code No.	Manufr	Class
MnZn ferrite:	3D3	4, 19	II

Residual loss tangent as a function of frequency with superimposed static field as a parameter, Figure 3.14

$\hat{B} < 0.1$ mT (1 Gs)
$\theta \simeq 20°C$

These graphs show the influence of a static magnetic field on the residual loss tangent, the static field and the measuring field being parallel. At high frequencies the loss is associated, at least in part, with the ferromagnetic resonance. The frequency of this resonance, i.e. the frequency at which μ_s'' reaches a maximum value, is increased when a polarizing field is applied. This reduces the loss at any given frequency immediately below the cut-off region. The basic residual loss, extending over the lower part of the spectrum, is believed to arise largely from thermally activated domain wall movements. Since a polarizing field reduces the number of domain walls the residual loss is decreased. Therefore the overall effect of a static field is to move the curve towards the r.h. bottom corner of the graph.

Since the permeability also decreases when a steady field is applied the variation of residual loss factor $(\tan \delta_r)/\mu$ depends on whether $\tan \delta_r$ or μ decreases at the greater rate. In fact $\tan \delta_r$ usually decreases at the greater rate at low values of the static field so that $(\tan \delta_r)/\mu$ initially falls; at higher fields the situation is reversed.

See also Section 6.2.3, Figures 3.6, 3.12, and References 49, 50.

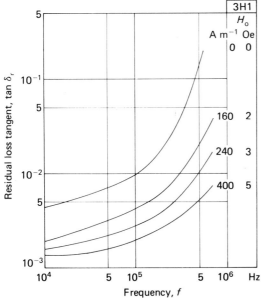

Figure 3.14.1

	Code No.	Manufr	Class
MnZn ferrite:	3H1	4	I

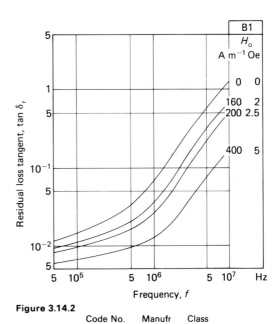

Figure 3.14.2

	Code No.	Manufr	Class
NiZn ferrite:	B1	4	VI

106 Properties of some manganese zinc and nickel zinc ferrites

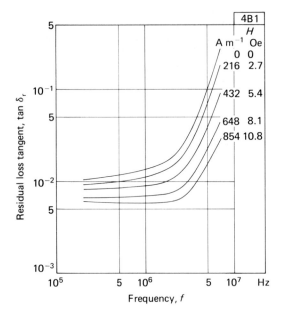

Figure 3.14.3

	Code No.	Manufr	Class
NiZn ferrite:	4B1	4	VII

Hysteresis loss factor as a function of flux density, Figure 3.15

$f = 10 \text{ kHz}$

$\theta \simeq 20°C$

In the discussion of loss expressions in Section 2.2.6 it was pointed out that in practice the higher powers of f and B are usually ignored in considering the low-amplitude loss terms and as a result simple relations such as the Legg expression (Eqn (2.61)) are obtained. At low frequencies and low flux densities the coefficients may be regarded as material constants, but at higher frequencies or flux densities the higher power terms become significant and are usually allowed for by making the coefficients themselves functions of frequency and flux density. In this figure the relation between the hysteresis loss factor and the flux density is shown for the same manganese zinc ferrite specimens as used for the data in Figure 3.5. From Eqn (2.63),

$$\frac{\tan \delta_h}{\mu} = \eta_B \hat{B}$$

Thus if the hysteresis constant η_B were indeed constant the hysteresis loss factor would be proportional to $\hat{B}$. In practice the relation is not linear so, if the above equation is used in an experimental determination of η_B, the range of flux densities over which the slope of the curve is measured must be specified. The relevant IEC Publication[51] states that the peak flux density should not exceed 5 mT; a typical peak flux density range found in manufacturers' specifications is 1.5–3.0 mT (15–30 Gs). In laboratory measurements the tangent at the origin may be used.

The Rayleigh and Peterson coefficients may also be obtained from these graphs. Strictly both of these coefficients relate to the hysteresis loss with field strength as the independent variable. However at very low flux densities the permeability closely approaches the initial permeability and Eqns (2.33) and (2.37) may be used to obtain these coefficients. Using the tangent of the curves at the origin, the values of the coefficients have been obtained for each of the specimens referred to in this figure, (Table 3.7).

These results may be compared with those derived from the variation in permeability with amplitude in Figure 3.5. It is seen that there is a rough correspondence between the respective values of v. In the Rayleigh relation the permeability rise and the hysteresis loss are both dependent on the value of a single constant v.

The significance of the hysteresis coefficients in the calculation of hysteresis loss, third harmonic distortion and third order intermodulation products is discussed in detail in Chapter 2.

Table 3.7

Specimen	μ_i	Rayleigh coefficient v (SI units)	Peterson coefficient a_{02} (SI units)	(CGS)
3H1	2090	15.4	7.7	616
3H3	2000	18.5	9.3	735
3D3	690	1.6	0.8	64
3E2	5160	530	265	21200

From the loss point of view the hysteresis loss tangent is a particularly useful parameter when determining the extra dissipation, or the degradation of Q-factor, occurring in an inductor when it is being operated at a flux density that is not vanishingly small. It is shown in Section 4.2.3 that if a core has a material loss factor $(\tan \delta)/\mu$ at a given frequency and flux density, arising from any origin then, when it is gapped to an effective permeability μ_e and operated at the same frequency and flux density, the effective loss tangent will be $\mu_e \times (\tan \delta)/\mu$. Thus the curves of this figure may be used directly for any gapped core provided the effective peak flux density $\hat{B}_e$ is known and the frequency is not too high. The dependence of $(\tan \delta_h)/\mu$ on frequency is shown in the next figure.

See also Sections 2.2.3, 2.2.6, 4.2.3, Figures 3.5, 3.16, 3.17, 3.18, and Reference 51.

$$\frac{\tan \delta_h}{\mu_a} = \frac{4v\hat{B}}{3\pi\mu_o\mu_a^3} \qquad \text{(see Eqn (2.33))}$$

$$= \frac{8a_{02}\hat{B}}{3\pi\mu_o\mu_a^3} \qquad \text{(see Eqn (2.37))}$$

as $\hat{B} \to 0$, $\mu_a \to \mu_i$

$$\frac{\tan \delta_h}{\mu} = \frac{a\hat{B}}{2\pi} = \eta_B \hat{B} \qquad \text{(see Eqns (2.62), (2.63) and (4.52))}$$

$$(\tan \delta_h)_{\text{gapped}} = \mu_e \left(\frac{\tan \delta_h}{\mu}\right) \qquad \text{(see Eqn (4.50))}$$

$$\frac{E_{3a}}{E_a} = 0.6 \tan \delta_h \qquad \text{(see Eqn (2.68))}$$

for a material operating in the Rayleigh region.

108 Properties of some manganese zinc and nickel zinc ferrites

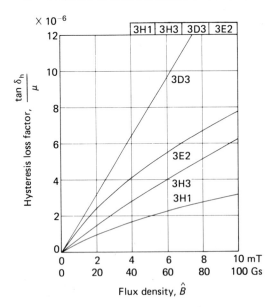

Figure 3.15.1

	Code No.	Manufr	Class
MnZn ferrite:	3H1, 3H3	4	I
	3D3	4, 19	II
	3E2	4	III

Hysteresis loss factor as a function of frequency, Figure 3.16

$\hat{B} = 0.5 \, \text{mT} \, (5 \, \text{Gs})$

$\theta \simeq 20°\text{C}$

In considering the shape of the curves in this figure it is necessary to have a clear definition of the dependent variable. At low frequencies and at low flux densities it may be seen from Section 2.2.6 that:

$$\frac{\tan \delta_h}{\mu} = \eta_B \hat{B}$$

where η_B is the hysteresis material constant.

Eqns (4.53) and (4.54) are expressions that are relevant to measuring η_B using an admittance bridge. It follows that:

$$\frac{\tan \delta_h}{\mu} = \frac{\hat{B}\omega^2 N^3 \mu_o}{\sqrt{2}C_2} \times \frac{\Delta G}{\Delta U}$$

Thus at a given frequency and flux density the hysteresis loss factor is proportional to the quotient of the increase in conductance and the increase of voltage producing it. At low frequencies this increase of conductance is undoubtedly the result of simple hysteresis loss, i.e. due to irreversible domain wall movements, but at higher frequencies it may be caused by other magnetic loss processes. In particular, as the ferromagnetic resonance is approached a part of the total loss is due to the damping of this resonance and it is possible that level dependent components of this damping contribute to the change in conductance. It is interesting to speculate at this stage whether such contributions increase the magnetic distortion and intermodulation. Some light is thrown on this matter by the data given in Figure 3.18.

Summing up, although it is convenient to refer to this variable as the hysteresis loss factor it must be recognized that this will be a misnomer at the higher frequencies.

Some evidence of a contribution from a ferromagnetic resonance process is apparent in the figure. It is seen that the loss factor of the highest permeability material, 3E2, rises at a relatively low frequency while that of the lowest permeability material shown, 3H1, rises at a higher frequency. The specimens used for these measurements were identical with those for which data are given in Figures 3.5, 3.15 and 3.18.

See also Sections 2.2.3, 2.2.6, 4.2.3, and Figures 3.5, 3.12, 3.15, 3.18.

$\dfrac{\tan \delta_h}{\mu} = \dfrac{a\hat{B}}{2\pi} = \dfrac{4v\hat{B}}{3\pi\mu_o\mu^3}$ (see Eqns (2.63), (2.33))

$\eta_B B = \dfrac{\mu_o \omega^2 N^3}{\sqrt{2}C_2} \dfrac{\Delta G}{\Delta U}$ (see Eqn (4.53))

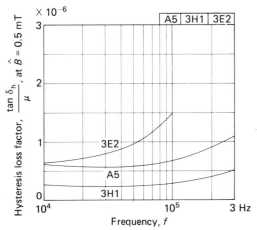

Figure 3.16.1

	Code No.	Manufr	Class
MnZn ferrite:	A5, 3H1	4	I
	3E2	4	III

Hysteresis loss factor as a function of temperature, Figure 3.17

$f = 10$ kHz

$\hat{B} = 1$ mT (10 Gs)

This figure shows results obtained on three MnZn ferrites. It is interesting to note that each curve has a minimum coinciding approximately with the temperature at which the secondary peak occurs in the permeability/temperature relation and also the minimum in the residual loss factor/temperature curves. It is at this temperature that the magnetocrystalline anisotropy goes through zero, so some connection between the hysteresis process and the anisotropy is apparent. This lends support to the theory that the residual loss is at least in part due to thermally activated hysteresis loss as proposed by Néel.[38, 52]

See also Figures 3.8, 3.13, and References 38, 52.

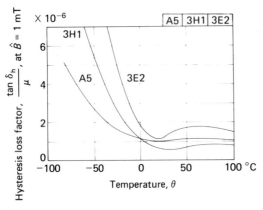

Figure 3.17.1

	Code No.	Manufr	Class
MnZn ferrite:	A5, 3H1	4	I
	3E2	4	III

Magnetic distortion, Figure 3.18

$\hat{B} = 0.5$ mT (5 Gs)

$\theta \simeq 20°$C

The graphs in this figure show the variation of the magnetic distortion with frequency for a number of typical ferrites. With the exception of those for 4C6, the measured results, shown by the full lines, were obtained on the actual toroidal specimens used to obtain the data for Figures 3.5, 3.15 and 3.16. The results refer to the distortion e.m.f. (open-circuit voltage) due to a sinusoidal current, and are expressed in decibels relative to the voltage of the applied signals. The abscissa represents the fundamental frequency or the mean of the applied frequencies.

The third harmonic was measured by passing the signal through a low pass filter and applying it to a winding on the specimen through a relatively high inductance coil. The e.m.f. across the winding on the specimen was applied through another high inductance coil to a high pass filter and thence to a spectrum analyser. The filters and inductances were composed of air-cored coils to prevent spurious distortion. The low pass filter was designed to suppress the distortion from the source and the high pass filter prevented the fundamental from reaching the detector. A number of pairs of filters and inductors were needed.

For the intermodulation measurements,[53] two independent signals were applied to the winding on the specimen via a bridge network which was adjusted to prevent coupling between the sources. Each amplitude was equivalent to 0.5 mT (5 Gs). The frequencies, f_a and f_b, were chosen to be on either side of a nominal frequency and removed from it by about 20%. The nominal frequency is represented along the abscissa. Again the specimen was separated from the source and detector impedances by high inductance air-cored coils and a simple filter was used to prevent the applied signals reaching the spectrum analyser.

It has been seen in Chapter 2 that the distortion products depend on one or both of the Peterson (or similar) coefficients, e.g. a_{11} and a_{02}. Figure 3.5 shows $\Delta\mu/\mu_i$ as a function of $\hat{B}$ from which a_{11}, at a particular frequency, may be calculated using Eqn (2.35). The value of a_{11} calculated from this and similar graphs falls with frequency; for the MnZn specimens it reaches about half of its low-frequency value by about 300 kHz. Figure 3.16 shows the variation of hysteresis loss factor with frequency; from these data the coefficient a_{02} may be calculated using Eqn (2.37). Since a_{02} is proportional to $(\tan\delta_h)/\mu$ it follows that a_{02} is a rising function of frequency for these specimens.

From Eqn (2.68) the relative amplitude of the third harmonic may be calculated and similarly from Eqns (2.70) and (2.71) the relative amplitudes of the sum and difference products may be obtained. The starting data are the Peterson coefficients obtained from Figures 3.5 and 3.15 and similar graphs obtained at other frequencies. These calculated results are shown as broken curves in the following graphs. Whereas the measured difference products show reasonable agreement with theory, the sum products show a discrepancy of about 5 to 10 dB at the lower frequencies. However the separation of the measured third order summation product curve from the third harmonic curve is in reasonable agreement with theoretical prediction; from the equations quoted below, the third harmonic amplitude should be 4.6 dB below that of the summation product.

As the frequency rises the calculated third harmonic and sum products rise following the rise of $(\tan\delta_h)/\mu$ with frequency (Figure 3.16) but the corresponding measured curves fall. This tends to confirm the suggestion made in the introduction to Figure 3.16, that at the higher frequencies the measured hysteresis loss factor contains components not having their origins in the simple hysteresis phenomena that arise from the irreversible components of domain wall movement. Further, the falling characteristics measured for the third harmonic and third order summation products tend to confirm the theory that in high permeability ferrites the domain wall movement relaxes at frequencies in the order of 100 kHz. In the case of the lower permeability ferrites, e.g. 3D3 and 4C6, the droop is not established at frequencies much below 1 MHz. No clear dispersion of the third order difference products is apparent within the range of measurements.

In view of the above observations, it would appear more reliable, when comparing materials for magnetic distortion, to use directly measured distortion data rather than measured hysteresis factors or coefficients; the theoretical relations between these parameters appear to be unreliable, particularly at the higher frequencies.

See also Sections 2.2.3, 2.2.6, 2.2.7, 4.2.3, Figures 3.5, 3.15, 3.16, 3.17, and References 53, 54.

Peterson coefficients:

$$a_{11} = d\mu_a/dH = \mu_o\mu_a d\mu_a/dB \qquad \text{(see Eqn (2.35))}$$

$$a_{02} = \frac{3\pi}{8}\frac{\mu_o\mu_a^2}{\hat{B}}\tan\delta_h \qquad \text{(see Eqn (2.37))}$$

as $\hat{B} \to 0$, $\mu_a \to \mu_i$

Product amplitudes:

$$\frac{E_{3a}}{E_a} = \frac{8}{5\pi}\frac{a_{02}\hat{B}_a}{\mu_o\mu^2} = 0.6\tan\delta_h \qquad \text{(see Eqn (2.68))}$$

$$\frac{E_{2a+b}}{E_a} = \frac{128}{15\pi^2}\frac{a_{02}\hat{B}_a}{\mu_o\mu^2} = 1.019\tan\delta_h \qquad \text{(see Eqn (2.71))}$$

$$\frac{E_{2a-b}}{E_a} = \frac{16}{15\pi}\frac{a_{02}\hat{B}_a}{\mu_o\mu^2}\sqrt{\left(\frac{a_{11}}{a_{02}}\right)^2 + \left(\frac{8}{3\pi}\right)^2}$$

$$= 0.869\tan\delta_h \text{ if } \frac{a_{11}}{a_{02}} = 2 \qquad \text{(see Eqn (2.70))}$$

$\hat{B}_a$ is the flux density of the applied signal of frequency f_a.

112 Properties of some manganese zinc and nickel zinc ferrites

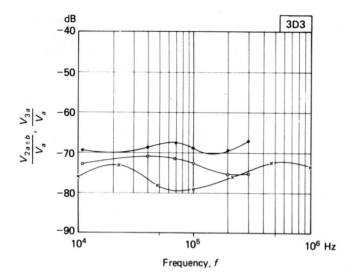

Figure 3.18.1

	Code No.	Manufr	Class
MnZn ferrite:	3H1	4	I

	Sum	Difference	3rd Harmonic
Measured	o—o	•—•	x—x
Calculated	o- -o	•- -•	x- -x

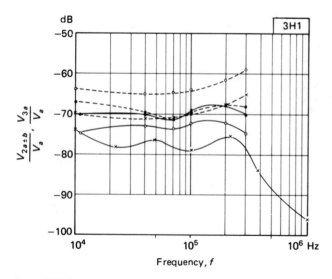

Figure 3.18.2

	Code No.	Manufr	Class
MnZn ferrite:	3D3	4, 19	II

Properties of some manganese zinc and nickel zinc ferrites 113

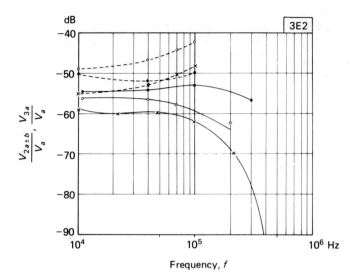

Figure 3.18.3

	Code No.	Manufr	Class
MnZn ferrite:	3E2	4	III

	Sum	Difference	3rd Harmonic
Measured	o—o	•—•	x—x
Calculated	o- -o	•- -•	x- -x

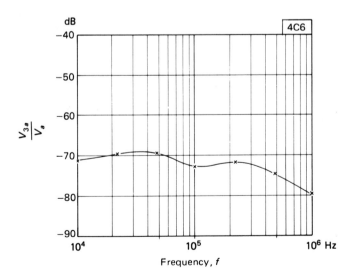

Figure 3.18.4

	Code No.	Manufr	Class
NiZn ferrite:	4C6	4	VIII

Magnetic power loss as a function of flux density and frequency, Figure 3.19

The graphs in this figure show the variation of the magnetic power loss (volume) density, P_M, with peak flux density and at a number of frequencies in the range 10–500 kHz. Magnetic power loss has been defined in Section 2.3 as the power loss of a ferrite material measured at high amplitudes when eddy current loss has been excluded; this is loosely called hysteresis loss but will usually include losses of non-hysteresis origin. Most of the materials represented are MnZn power ferrites.

In quoting data from manufacturers' catalogues it is not always possible to be sure of the measuring conditions and, in particular, whether eddy current loss has been excluded. It is to be expected that most measurements have been made on test toroids, typically 30 mm in diameter, and with symmetrical sine wave flux density. Assuming that the eddy current component is negligible, then the total loss in a given core may be predicted by multiplying P_M by the effective core volume and adding the eddy current loss estimated using Eqn (2.46) or, more conveniently, Eqn (9.19).

The data in Figure 3.19.3 represents the results of extensive measurements carried out under the supervision of the author. The measurements were made on a wide range of commercial E and U cores, the eddy current loss being deducted as described above. Three distinct modes of excitation were used; symmetrical sine and square wave voltage, and unidirectional square wave voltage. It was found that at a given frequency the magnetic power loss was not strongly dependent on the mode of excitation provided the total flux density excursion, B_{P-P}, remained the same, see also Section 9.3. Indeed, in one experiment, an initially symmetrical sine wave excitation, B_{P-P}, was progressively biased until it became a unidirectional excitation (from remanence) of amplitude B_{P-P}; the magnetic power loss remained within about 5% of the mean value.[55, 56] It was found that at frequencies up to about 100 kHz the power loss at a temperature of 100°C could be expressed by the single formula:

$$P_M = kf^{1.3}B^{2.5} \qquad (3.4)$$

where B is the peak or the peak-to-peak value and for a practical core would be the effective flux density.

The exponents of f and B are typical of a number of MnZn power ferrites. Specifically, for the ferrite grade 3C8 represented in Figure 3.19.3,

$$P_M = 4.23 \times 10^{-6} f^{1.3} \hat{B}^{2.5}$$

or $\quad P_M = 0.748 \times 10^{-6} f^{1.3}(B_{P-P})^{2.5} \quad \mu W\,mm^{-3}$

where f is in kHz and B is in mT.

The range of application of Figure 3.19.3 is emphasized by the inclusion of a B_{P-P} scale at the top of the graph.

It is of interest to note that if the magnetic power loss were due entirely to low-frequency hysteresis (irreversible) processes then it would be expected that the loss/cycle would be constant and therefore the loss density would be proportional to frequency. In fact, for the above ferrite, the slope of the log P_M vs. log f graph varies from about 1.1 at 10 kHz to 1.5 at 100 kHz, the value of 1.3 being a mean slope over that range. The upward curvature suggests that as the frequency approaches the onset of ferromagnetic resonance, additional loss mechanisms associated with this resonance occur.

Finally, it should be noted that for a typical ferrite power core the heating effect of the losses generally restricts the power loss density to the low hundreds of $\mu W\,mm^{-3}$, depending on core size. Extrapolation of the curves to 1000 $\mu W\,mm^{-3}$ can be irrelevant.

See also Sections 2.3, 4.2.3, 9.3, 9.5.2, and References 55, 56.

For a given core:

$$\text{magnetic loss} \simeq P_M V_{\hat{e}} = kf^m \hat{B}^n V_{\hat{e}} \qquad \text{(see Eqn (4.57))}$$

$$\text{eddy current loss} \simeq \frac{\pi f^2 \hat{B}_{\hat{e}}^2 A_{\hat{e}} V_{\hat{e}}}{4\rho} \times 10^{-15} \quad W$$

$$\text{(see Eqn (9.19))}$$

where f is in kHz, $\hat{B}_{\hat{e}}$ is in mT, $A_{\hat{e}}$ and $V_{\hat{e}}$ are in mm units, and ρ is in $\Omega\,m$

Properties of some manganese zinc and nickel zinc ferrites   115

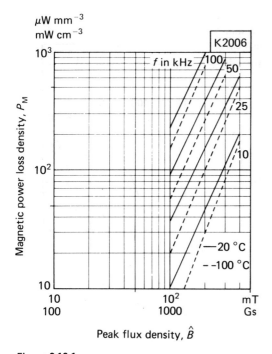

Figure 3.19.1

	Code No.	Manufr	Class
MnZn ferrite:	K2006	1	IV

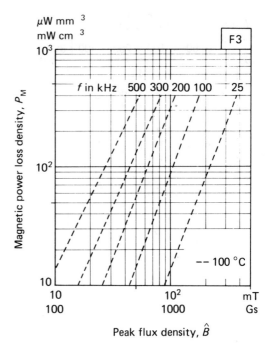

Figure 3.19.2

	Code No.	Manufr	Class
MnZn ferrite:	F3	3	IV

116   Properties of some manganese zinc and nickel zinc ferrites

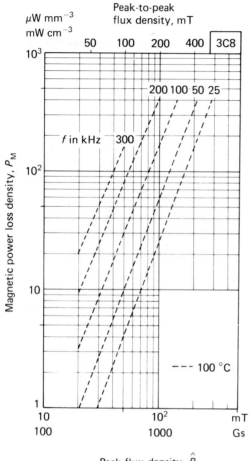

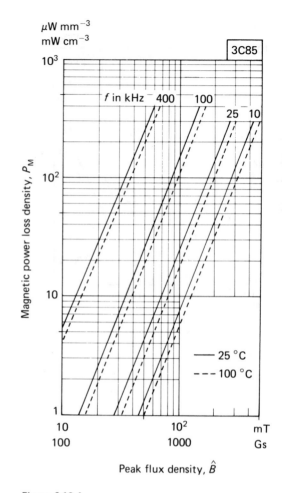

Figure 3.19.3
MnZn ferrite: Code No. 3C8   Manufr 4, 19   Class IV

Figure 3.19.4
MnZn ferrite: Code No. 3C85   Manufr 4, 19   Class IV

Properties of some manganese zinc and nickel zinc ferrites 117

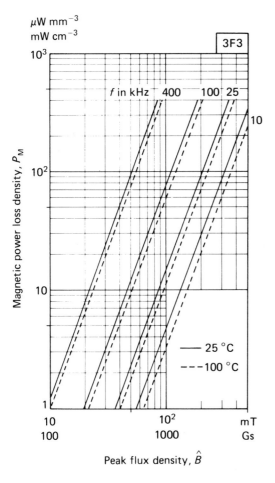

Figure 3.19.5

	Code No.	Manufr	Class
MnZn ferrite:	3F3	4, 19	IV

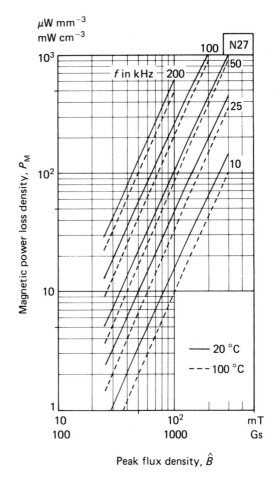

Figure 3.19.6

	Code No.	Manufr	Class
MnZn ferrite:	N27	6	IV

118 Properties of some manganese zinc and nickel zinc ferrites

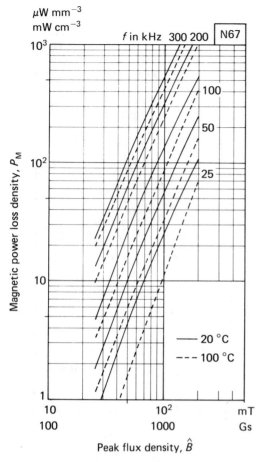

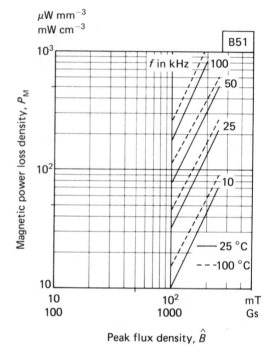

Figure 3.19.7

	Code No.	Manufr	Class
MnZn ferrite:	N67	6	IV

Figure 3.19.8

	Code No.	Manufr	Class
MnZn ferrite:	B51	7	IV

Properties of some manganese zinc and nickel zinc ferrites   119

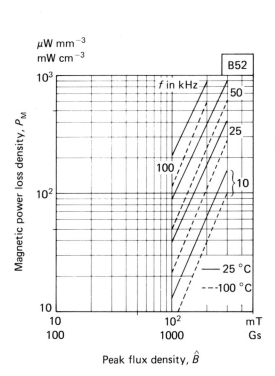

Figure 3.19.9

	Code No.	Manufr	Class
MnZn ferrite:	B52	7	IV

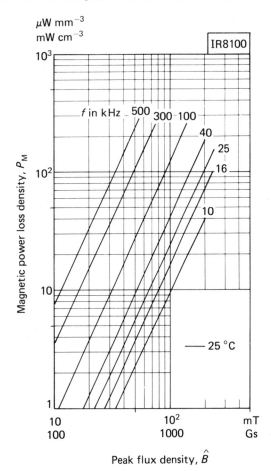

Figure 3.19.10

	Code No.	Manufr	Class
MnZn ferrite:	IR8100	15	IV

120   Properties of some manganese zinc and nickel zinc ferrites

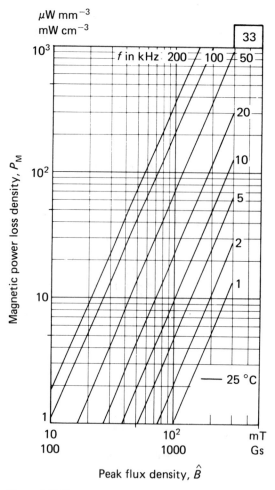

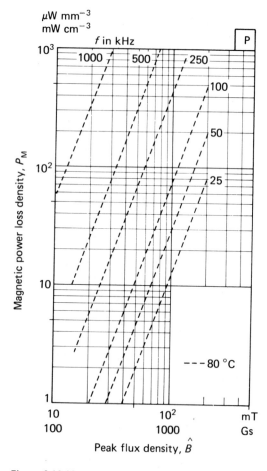

Figure 3.19.11
MnZn ferrite: Code No. 33   Manufr 23   Class IV

Figure 3.19.12
MnZn ferrite: Code No. P   Manufr 22   Class IV

Properties of some manganese zinc and nickel zinc ferrites 121

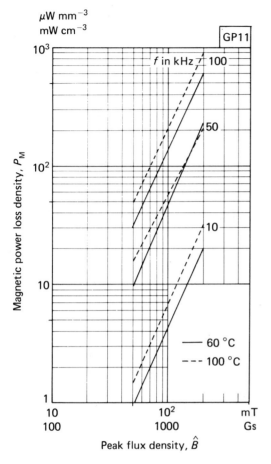

Figure 3.19.13
MnZn ferrite: Code No. GP11   Manufr 24   Class III

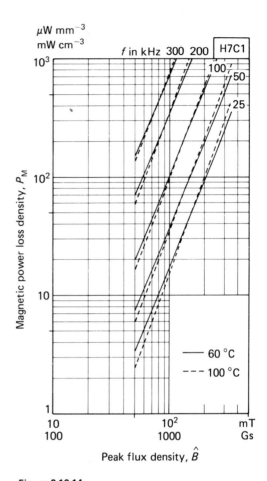

Figure 3.19.14
MnZn ferrite: Code No. H7C1   Manufr 26   Class IV

122 Properties of some manganese zinc and nickel zinc ferrites

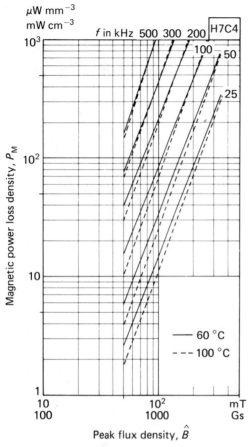

Figure 3.19.15

	Code No.	Manufr	Class
MnZn ferrite:	H7C4	26	IV

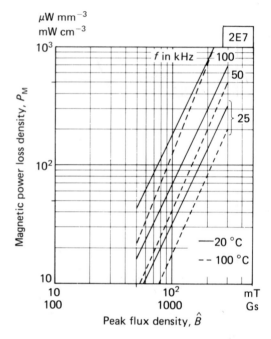

Figure 3.19.16

	Code No.	Manufr	Class
MnZn ferrite:	2E7	28	IV

Magnetic power loss as a function of temperature, Figure 3.20.

It has been seen, in Figures 3.13 and 3.17, that the low-amplitude losses in ferrites show a minimum at the anisotropy compensation temperature. The high-amplitude losses show the same effect.

Originally, power ferrites had minimum loss densities in the region of room temperature so under normal working conditions the loss increased with temperature. In modern power ferrites the minimum is usually placed at or near the expected working temperature, some manufacturers giving a choice of several minimum-loss temperatures by offering grade variants.

This figure shows the temperature dependence of the magnetic loss density for a number of MnZn power ferrites.

See also Sections 1.3.2, 1.3.7, Figures 3.13, 3.17, 3.19, and Reference 57.

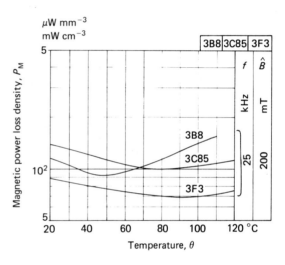

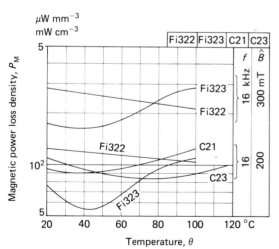

Figure 3.20.1

	Code No.	Manufr	Class
MnZn ferrite:	Fi322, Fi323	8	IV
	C21, C23	2	IV

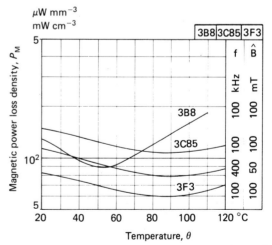

Figure 3.20.2

	Code No.	Manufr	Class
MnZn ferrite:	3B8	4	IV
	3C85, 3F3	4, 19	IV

124 Properties of some manganese zinc and nickel zinc ferrites

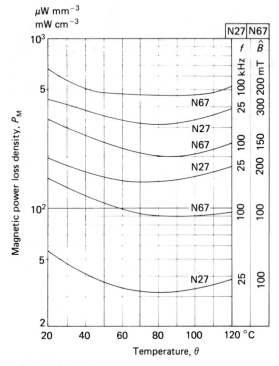

Figure 3.20.3

	Code No.	Manufr	Class
MnZn ferrite:	N27, N67	6	IV

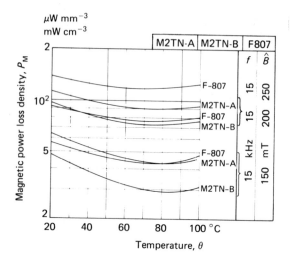

Figure 3.20.5

	Code No.	Manufr	Class
MnZn ferrite:	M2TN-A, M2TN-B	10	IV
	F-807	13	IV

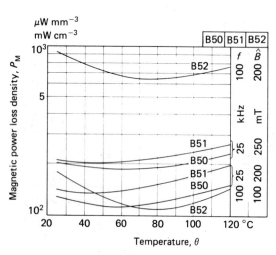

Figure 3.20.4

	Code No.	Manufr	Class
MnZn ferrite:	B50, B51, B52	7	IV

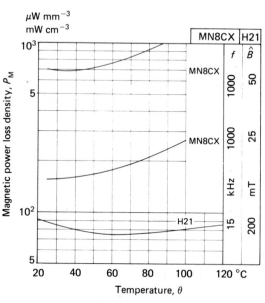

Figure 3.20.6

	Code No.	Manufr	Class
MnZn ferrite:	MN8CX	16	IV
	H21	14	IV

Properties of some manganese zinc and nickel zinc ferrites 125

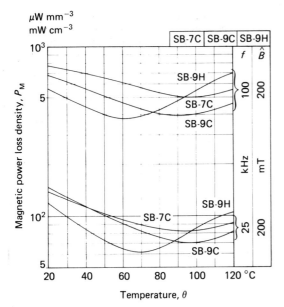

Figure 3.20.7

	Code No.	Manufr	Class
MnZn ferrite:	SB-7C, SB-9C, SB-9H	24	IV

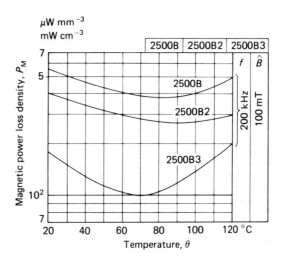

Figure 3.20.9

	Code No.	Manufr	Class
MnZn ferrite:	2500B, 2500B2, 2500B3	27	IV

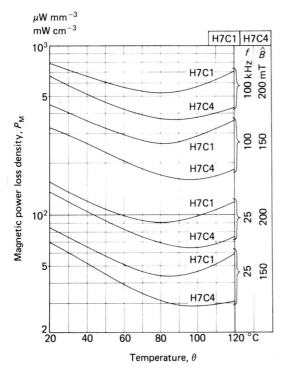

Figure 3.20.8

	Code No.	Manufr	Class
MnZn ferrite:	H7C1, H7C4	26	IV

Resistivity as a function of temperature, Figure 3.21

$f \to 0$

At room temperature most manganese zinc ferrites have resistivities between about 0.01 and 10 Ω m (1 and 1000 Ω cm); nickel zinc ferrites normally have much higher values, e.g. $> 10^3$ Ω m (10^5 Ω cm). As ferrites are semiconductors the resistivity falls with rising temperature.

The graphs in this figure show the d.c. resistivity as a function of temperature for a number of grades of ferrite. It is seen that over the temperature range -70 to $+100°C$ the resistivity of manganese zinc ferrites falls by a ratio that is between 30 and 100; the corresponding figures for nickel zinc ferrites are 10^3–10^4.

Only polycrystalline ferrites are considered, so the bulk resistivity arises from a combination of the crystallite resistivity and the resistivity of the crystallite boundaries. The boundary resistivity is much greater than that of the crystallite so the boundaries have the greatest influence on the d.c. resistivity.

The resistivity ρ at an absolute temperature T is given by:[58]

$$\rho = \rho_\infty \exp(E_\rho / kT) \quad (3.5)$$

where ρ_∞ is the resistivity extrapolated to $T = \infty$

E_ρ is the activation energy

k is Boltzmann's constant

If E_ρ is to be expressed in electron-volts

$k = 8.62 \times 10^{-5}$ eV K^{-1}.

From this equation, E_ρ has been derived for each curve, or in some cases a region of a curve, and the value has been placed against the curve to which it applies.

See also Section 2.2.4, Figures 3.22, 3.23, and References 58–63.

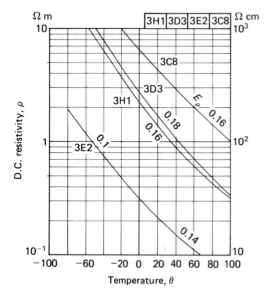

Figure 3.21.1

	Code No.	Manufr	Class
MnZn ferrite:	3H1	4	I
	3D3	4, 19	II
	3E2	4	III
	3C8	4, 19	IV

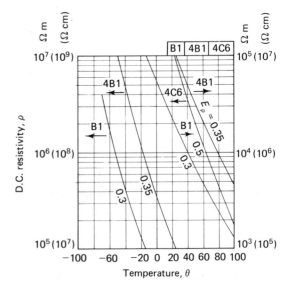

Figure 3.21.2

	Code No.	Manufr	Class
NiZn ferrite:	B1	4	VI
	4B1	4	VII
	4C6	4	VIII

Resistivity and permittivity as functions of frequency, Figure 3.22

$\theta \approx 25°C$

As the frequency rises from a low value the bulk resistivity, ρ, and permittivity, ε, of a polycrystalline ferrite stay constant at first and then fall to become asymptotic to lower values at high frequencies. This variation has the characteristic of a relaxation and is attributable to the granular structure of ferrites, in which crystallites (grains) are separated by boundaries having much higher resistivity than the crystallites. Thus the structure behaves as a compound dielectric and in Section 2.2.4 the theoretical characteristics of such a structure have been considered.

For manganese zinc ferrites typical values are:

relative boundary thickness, a	10^{-4}
crystallite resistivity, ρ_1	$10^{-3}\,\Omega\,m$
	$(0.1\,\Omega\,cm)$
boundary resistivity, ρ_2	$10^4\,\Omega\,m$
	$(10^6\,\Omega\,cm)$
crystallite permittivity, ε_1	
boundary permittivity, ε_2	10

Therefore, from Eqn (2.44), the dispersion of ρ and ε in a ferrite for which the simple model applies would be:

at low frequency $\quad \rho \rightarrow 1.0\,\Omega\,m\ (100\,\Omega\,cm)$
$\quad\quad\quad\quad\quad\quad \varepsilon \rightarrow 10^5$
at high frequency $\quad \rho \rightarrow 10^{-3}\,\Omega\,m\ (0.1\,\Omega\,cm)$
$\quad\quad\quad\quad\quad\quad\quad \varepsilon \rightarrow 10$

The relaxation time, calculated from Eqn (2.45) is $8.85 \times 10^{-10}\,s$ giving a relaxation frequency of 180 MHz.

For nickel zinc ferrites typical values are:

relative boundary thickness, a	0.3×10^{-2}
crystallite resistivity, ρ_1	$30\,\Omega\,m$
	$(3 \times 10^3\,\Omega\,cm)$
boundary resistivity, ρ_2	3×10^6 to
	$3 \times 10^7\,\Omega\,m$
	$(3 \times 10^8$ to
	$3 \times 10^9\,\Omega\,cm)$
crystallite permittivity, ε_1	
boundary permittivity, ε_2	10

Therefore using the same model:

at low frequency $\quad \rho \rightarrow 10^4$ to $10^5\,\Omega\,m\ (10^6$ to $10^7\,\Omega\,cm)$
$\quad\quad\quad\quad\quad\quad \varepsilon \rightarrow 3 \times 10^3$
at high frequency $\quad \rho \rightarrow 30\,\Omega\,m\ (3 \times 10^3\,\Omega\,cm)$
$\quad\quad\quad\quad\quad\quad\quad \varepsilon \rightarrow 10$

The corresponding relaxation time is $8.8 \times 10^{-7}\,s$ and the relaxation frequency is 180 kHz.

The experimental results shown in this figure show dispersions broadly in accordance with these boundary values but the nature of the relaxation suggests that the dielectrics cannot be adequately represented by the simple structure shown in the diagram of Figure 2.5.

See also Section 2.2.4, Figures 2.5, 3.21, and References 64, 65.

The shunt capacitance, C_p, and resistance, R_p, of a block of material having resistivity ρ and permittivity ε are given by:

$$C_p = \varepsilon_0 \varepsilon A / l \quad F$$
$$R_p = \rho l / A \quad \Omega \quad \text{(see Eqn (2.41))}$$

where $\varepsilon_0 = 8.854 \times 10^{-12}\ F\,m^{-1}$

For two-layer dielectric, or crystallite and boundary:

$$\left. \begin{array}{ll} \text{when } f \rightarrow 0 & \rho \rightarrow a\rho_2 \\ & \varepsilon \rightarrow \varepsilon_2/a \\ \text{when } f \rightarrow \infty & \rho \rightarrow \rho_1 \\ & \varepsilon \rightarrow \varepsilon_1 \end{array} \right\} \quad \text{(see Eqn (2.44))}$$

$$\tau = \varepsilon_0 \frac{\varepsilon_1 + \varepsilon_2/a}{1/\rho_1 + 1/a\rho_2} \quad s \quad \text{(see Eqn (2.45))}$$

where the suffix 1 refers to the crystallite and 2 refers to the boundary.

128 Properties of some manganese zinc and nickel zinc ferrites

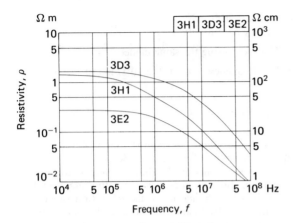

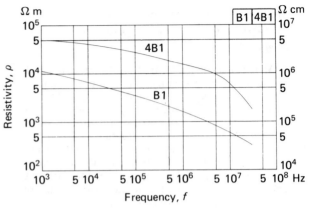

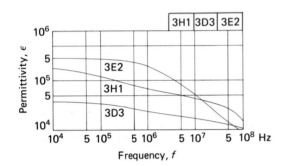

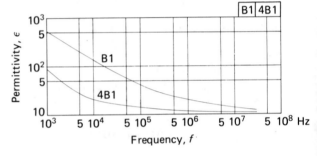

Figure 3.22.1

	Code No.	Manufr	Class
MnZn ferrite:	3H1	4	I
	3D3	4, 19	II
	3E2	4	III

Figure 3.22.2

	Code No.	Manufr	Class
NiZn ferrite:	B1	4	VI
	4B1	4	VII

High-frequency resistivity and permittivity as functions of temperature, Figure 3.23

$f = 1$ MHz

This figure shows the resistivity and permittivity of a number of typical manganese zinc and nickel zinc ferrites measured against temperature. As at low frequencies (see Figure 3.21) the resistivity falls with temperature. The permittivity generally increases slowly with temperature over the range considered.

See also Section 2.2.4, Figures 2.5, 3.21, 3.22, and References 66, 67.

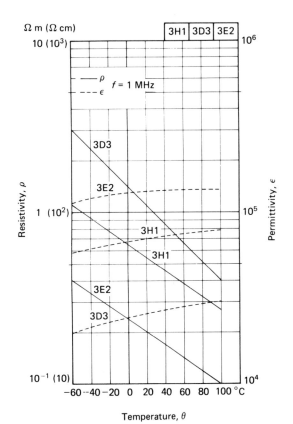

Figure 3.23.1

	Code No.	Manufr	Class
MnZn ferrite:	3H1	4	I
	3D3	4, 19	II
	3E2	4	III

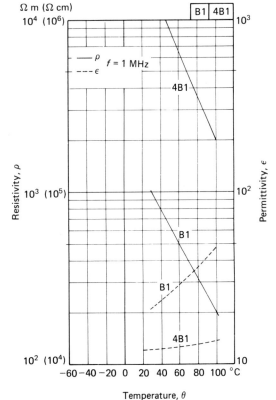

Figure 3.23.2

	Code No	Manufr	Class
NiZn ferrite:	B1	4	VI
	4B1	4	VII

Static magnetostriction, Figure 3.24

$\theta \simeq 20°C$

This figure shows the magnetostriction, i.e. the fractional change of length, as a function of steady field strength. Toroidal specimens were used and they were demagnetized before each measurement was made.

The manganese zinc ferrites, 3E2, A9 and 3H1 have very low magnetostriction; the saturation values are less than 10^{-6} and may be positive or negative. The curves show that the saturation value may be much smaller than the value at lower field strengths and indeed sign reversal has been observed on some specimens.

The range of nickel zinc ferrites is represented by the high permeability grade B1 and, at the other extreme, the low permeability grade 4E1 (nickel ferrite). For these ferrites the saturation magnetostriction is relatively large and negative; the saturation magnetostrictions for the other nickel zinc grades may be expected to lie in intermediate positions depending on the nickel/zinc ratio (e.g. the curve for 4C6).

These static magnetostriction curves are somewhat analogous to the initial $B-H$ curve; they may be useful for predicting the steady-state situation but, for alternating (vibratory) conditions, data on the dynamic magnetostriction are required.[69] Pursuing the analogy, the dynamic magnetostriction corresponds to the incremental permeability and cannot be deduced very well from the static curves.

See also References 68–71.

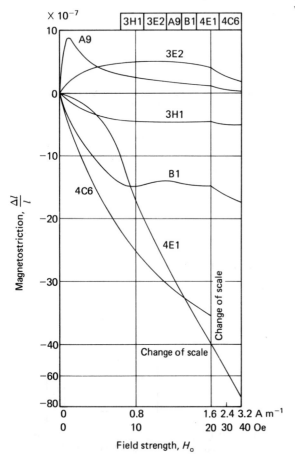

Figure 3.24.1

	Code No.	Manufr	Class
MnZn ferrite:	3H1	4	I
	3E2	4	III
	A9	4	IV
NiZn ferrite:	B1	4	VI
	4E1	4	X
	4C6	4	VIII

Initial permeability as a function of stress, Figure 3.25

$f \simeq 1 \text{ kHz}$

$\theta \simeq 20°C$

Components using ferrite cores are often assembled with clips or clamps that set up a stress in the ferrite. It is therefore useful to have data showing the influence of mechanical stress on the initial permeability. This figure shows such data for a number of typical ferrites.

It is notable that a large change of permeability may be produced by quite a moderate stress. In the case of the manganese zinc ferrites this may at first sight seem inconsistent with the very low magnetostriction observed in the previous figure. High permeability depends on the total anisotropy being small and one of the components of this total is the stress anisotropy which equals the product of the stress and the magnetostriction. If the other components, e.g. magnetocrystalline anisotropy, are very small, the stress anisotropy will predominate and it will have a controlling influence on the permeability. Changes in externally applied stress will then produce large changes in permeability. It follows that low loss, high permeability ferrites are likely to be particularly stress-sensitive. It has been shown[72] that if the magnetostriction is negative a small externally applied compressive stress will usually raise the permeability and vice versa. Large stresses, compressive or tensile invariably lower the permeability.

Stress-sensitivity applies also to losses, stress of either sign generally increasing the losses.[57, 74]

See also References 57 and 72–79.

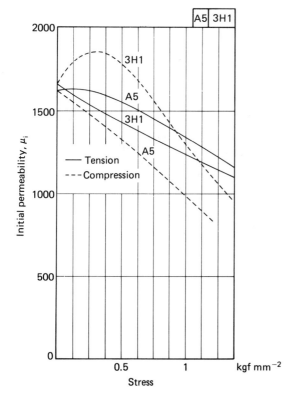

Figure 3.25.1

	Code No.	Manufr	Class
MnZn ferrite:	A5, 3H1	4	I

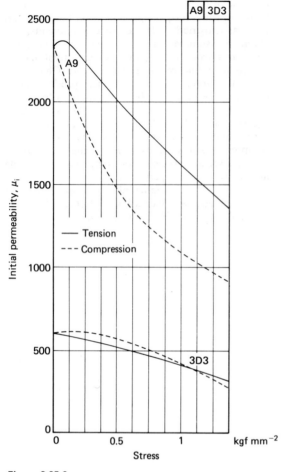

Figure 3.25.2

	Code No.	Manufr	Class
MnZn ferrite:	A9	4	IV
	3D3	4, 19	II

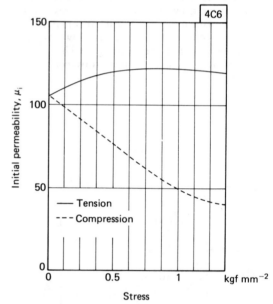

Figure 3.25.3

	Code No.	Manufr	Class
NiZn ferrite:	4C6	4	VIII

3.5 References and bibliography

1. Information on ferrite materials appearing in manufacturers' catalogues of transformer and inductor cores, *Int. Electrotechnical Commission, Publication 401*, Geneva (1972)
2. STUIJTS, A. L., VERWEEL, J. and PELOSCHEK, H. P. Dense ferrites and their applications, *Trans. IEEE, Commun. Electron.*, No. 75, 726 (1964)
3. SELLWOOD, D. Kingston Technical College, private communication (1966)
4. NISHIKAWA, T. *et al*. Mechanical properties of ferrites, *Natn. tech. Rep.*, **10**, Part 1, 305, Part 2, 477 (1964)
5. TANAKA, T. Young's and shear moduli, hardness and bending strength of polycrystalline Mn-Zn ferrites, *Jap. J. appl. Phys.*, **14**, 1897 (1975)
6. VAN DER MEER, A. B. D. and SLIJKERMAN, N. P. Mechanical strength of magnesium zinc ferrites for yoke rings, *Ferrites: Proc. of Int. Conf. on Ferrites, 1980*, edited by H. Watanabe, S. Iida and M. Sugimoto. 301, Centre for Academic Publications, Japan (1981)
   BERDIKOV, V. F., BOGOMOLOV, N. I., PUSHKAREV, O. I. and GAVRICHENKO, V. V. Brittle and strength properties of hot pressed manganese-zinc ferrites by the microidentation method, *All-union Sci.—Res. Inst. of Abrasives and Grinding, USSR Poroshk. Metall. (USSR)*, **22**, 87 (1983). Trans in: *Sov. Powder Metall. and Met. Ceram. (USA)*, **22**, 672 (1983)
7. VERHAEGHE, J. L., ROBRECHT, G. G. and BRUYNOOGHE, W. M. On the specific heat of Mn-Zn and Ni-Zn ferrite between 20°C and 350°C, *Appl. scient. Res., B*, **8**, 128 (1960)
8. HESS, J. and ZENGER, M. The influence of material properties of ferrite cores on the temperature rise of transformers, *Fourth Int. Conference on Ferrites, Part II*, edited by F. F. Y. Wang; *Advances in Ceramics*, **16**, 501 (1985)
   BEKKER, YA. M. Some features of the thermal resistance of ferrites, *Soviet Phys. solid St.*, **7**, 1240 (1965)

Figure 3.1

9. KORNETZKI, M. Die Hystereseverluste von Ferriten mit anamaler Magnetisierungsschliefe, *Z. angew. Phys.*, **10**, 368 (1958)
10. ECKERT, O. Ferrite with constricted loops and thermal magnetic treatment, *Proc. Inst. elect. Engrs*, **104**, B, 428 (1957)
11. MICHALOWSKI, L. The influence of the anisotropy constant on the perminvar effect in ferrites, *Physica Stat. Sol.*, **8**, 543 (1965)
    EDWARDS, G. W. and GLAISTER, R. M. Cobalt-ferrous ferrite with re-entrant hysteresis loop, *Trans. IEEE*, **MAG-2**, 696 (1966)
    KOSHKIN, L. I. and STRYGIN, YU. F. Magnetic anisotropy in polycrystalline ferrites, induced by electrothermal treatment, *Soviet Phys. solid St.*, **8**, 380 (1966)

Figure 3.3

12. SMIT, J. and WIJN, H. P. J. *Ferrites*, Sec. 13.2, Philips Technical Library, Eindhoven (1959)

Figure 3.7

13. HANNA, C. R. Design of reactances and transformers which carry direct current. *J. Am. Inst. elect. Engrs*, **46**, 128 (1927)
    LEGG, V. E. Optimum air gap for various magnetic materials in cores of coils subject to superposed direct current, *Trans. Am. Inst. elect. Engrs*, **64**, 709 (1945)

Figure 3.8

14. SMIT, J. and WIJN, H. P. J. *Ferrites*, Sec. 48.1, Philips Technical Library, Eindhoven (1959)
15. ENZ, U. Permeability, crystalline anisotropy and magnetostriction of polycrystalline manganese-zinc-ferrous ferrite, *Proc. I.E.E.*, **109**, Part B Suppl. No. 21, 246 (1962)
16. PELOSCHEK, H. P. and PERDUIJN, D. J. High-permeability MnZn ferrites with flat µ-T curves, *IEEE Trans. Magn.*, **MAG-4**, 453 (1968)
17. KNOWLES, J. E. Permeability mechanisms in manganese zinc ferrites, *J. de Physique, Colloque C1*, Suppl. au No. 4, **38**, C1-27 (1977)
18. HOEKSTRA, B., GYORGY, E. M., GALLAGHER, P. K., JOHNSON, Jr., D. W., ZYDZIK, G. and VAN UITERT, L. G. Initial permeability and intrinsic magnetic properties of polycrystalline MnZn-ferrites, *J. Appl. Phys.*, **49**, 4902 (1978)
19. STOPPELS, D. Relationship between magnetocrystalline anisotropy, including second-order contribution, and initial magnetic permeability for monocrystalline MnZn ferrous ferrite, *J. Appl. Phys.*, **51**, 2789 (1980)
20. ROESS, E. Magnetic properties and microstructure of high-permeability Mn-Zn ferrites, in *Ferrites: Proc. of Int. Conf. on Ferrites, 1970*, edited by Y. Hoshino, S. Iida, and M. Sugimoto, 203, University Park Press, Tokyo (1971)
21. ARAI, T. and IDO, T. Ni-Zn-Co type ferrites with very small temperature coefficient of initial permeability, *ibid*, 225 (1971)
22. GILES, A. D. and WESTENDORP, F. F. Simultaneous substitution of cobalt and titanium in linear manganese zinc ferrites, *J. de Physique, Colloque C1*, Suppl. au No. 4, **38**, C1-47 (1977)
23. NISHIYAMA, T. and KOTANI, T. Studies on Ni-Co ferrites with negative temperature coefficient of permeability, *Ferrites: Proc of Int. Conf. on Ferrites, 1980*, edited by H. Watanabe, S. Iida, and M. Sugimoto, 317, Centre for Academic Publications, Japan (1981)

Figure 3.10

24. Cores for inductors and transformers for telecommunication, Part 1: Measuring methods, *Int. Electrotechnical Commission, Publication 367-1*, Clauses 6 and 8, Geneva (1982)
25. SNELLING, E. C. Disaccommodation and its relation to the stability of inductors having manganese zinc ferrite cores, *Mullard tech. Commun.*, **6**, 207 (1962)
26. The International Electrotechnical Vocabulary, *Int. Electrotechnical Commission, Publication 50*, Chap. 221, items 221-02-54 to 56, Geneva (1988)
27. KRUPICKA, S. and GERBER, R. A contribution to studying the mechanism of permeability

disaccommodation in ferrites, *Czech J. Phys.*, **10**, 158 (1960)
28. KAMPEZYK, W. and ROESPEL, G. Some mechanisms governing the permeability versus time of ferrite cores, *Siemens Rev.*, **31**, 312 (1964)
29. BRAGINSKI, A. Magnetic after effects in iron-rich ferrites containing vacancies, *Physica Stat. Sol.*, **11**, 603 (1965)
30. OKADA, T. and AKASHI, T. Richter type after effect in manganese zinc ferrite, *J. Phys. Soc., Japan*, **20**, 639 (1965)
31. KOTSCHY, J and ROESPEL, G. Disaccommodation of permeability of ferrite cores after partial shocks, *Siemens Rev.*, **37**, 263 (1970)
32. MATSUBARA, T., KAWAI, J., SUGANO, I. and AKASHI, T. Disaccommodation in Mn-Zn ferrites, *Ferrites: Proc of Int. Conf. on Ferrites, 1970*, edited by Y. Hoshino, S. Iida, and M. Sugimoto, 214, University Park Press, Tokyo (1971)
33. KNOWLES, J. E. and RANKIN, P. Disaccommodation of permeability in manganese-zinc-titanium ferrites, *J. de Physique, Colloque C1*, Suppl. au No. 2-3, **32**, C1-845 (1971)
34. KNOWLES, J. E. Magnetic after-effects in ferrites substituted with titanium or tin, *Philips Res. Repts*, **29**, 93 (1974)
35. POSTUPOLSKI, T. and PIOTRKIEWICZ, M. Influence of the direction of demagnetization on the parameters of diffusion after-effect, *Phys. Stat. Sol.*, **32**, 247 (1975)
36. GYORGY, E. M., JOHNSON, D. W. and VOGEL, E. M. The prediction of long term disaccommodation in ferrites, *Ferrites: Proc of Int. Conf. on Ferrites, 1980*, edited by H. Watanabe, S. Iida, and M. Sugimoto, 196, Centre for Academic Publications, Japan (1981)
37. WILLEY, R. J. and MULLIN, J. T. Temperature and time stability of MnZn ferrites, *J. Magn. and Magn. Mater.*, **26**, 315 (1982)
38. GILES, A. D. and WESTENDORP, F. F. Some loss relationships in Mn-Zn ferro ferrites and their response to magnetic disturbance, *IEEE Trans. Magn.*, **MAG-18**, 944 (1982)
39. POSTUPOLSKI, T. and OKONIEWSKA-PSZCZOLKOWSKA, E. Magnetic conditioning—does it always bring on the neutral state?, *Acta Physica Polonica*, **A68**, 35 (1985)
40. POSTUPOLSKI, T. and WISNIEWSKA, A. Combined domain and migrational after-effect, *Acta Physica Polonica*, **A68**, 41 (1985)

Figures 3.11 and 3.12
41. LAX, B. and BUTTON, K. J. *Microwave ferrites and ferrimagnetics*, McGraw-Hill Book Co. (1962)
42. RADO, G. T. Magnetic spectra of ferrites, *Rev. mod. Phys.*, **25**, 81 (1953)
43. SNOEK, J. L. Dispersion and absorption in magnetic ferrites at frequencies above 1 Mc/s, *Physica Amsterdam*, **14**, 207 (1948)
44. POLDER, D. On the theory of ferromagnetic resonance, *Phil. Mag.*, **40**, 99 (1949)
45. BROESE VAN GROENOU, A., BONGERS, P. F. and STUIJTS, A. L. Magnetism, microstructure and crystal chemistry of spinel ferrites, *Mater. Sci. and Eng.*, **3**, 317 (1969)

46. NAITO, Y. On the permeability dispersion of a spinel ferrite, *Electron. and Commun. in Japan*, **56-C**, 118 (1973)
47. NAITO, Y. Formulation of frequency dispersion of ferrite permeability, *Electron. and Commun. in Japan*, **59-C**, 100 (1976)
48. DIXON, M., STAKELON, T. S. and SUNDAHL, R. C. Complex permeability variations of iron excess CoNiZn ferrites with annealing temperature. *IEEE Trans. Magn.*, **MAG-13**, 1351 (1977)

Figure 3.14
49. ANGEL, Y. Comportement des ferrites dans des champs magnétiques croisés, *Acta electron.*, **7**, Part 1, 7, Part 2, 119 (1963)
50. KRAMAR, G. P. and PANOVA, YA. I. The permeability dispersion in ferrites at different levels of direct current magnetic biasing, *Phys. Stat. Sol. (a)*, **86**, 283 (1984)

Figure 3.15
51. Cores for inductors and transformers for telecommunication, Part 1: Measuring methods, *Int. Electrotechnical Commission, Publication 367-1*, Cl. 11.1.5, Geneva (1982)

Figure 3.17
52. NEEL, L. Some theoretical aspects of rock magnetism, *Adv. Phys. (Phil. Mag. Suppl.)*, **4**, 191 (1955)

Figure 3.18
53. SNELLING, E. C. Magnetic intermodulation in manganese zinc ferrites, *IEEE Conf. Pub. No. 12*, 34-1 (1965)
54. KOEHLER, J. W. L. Non-linear distortion phenomena of magnetic origin, *Philips tech. Rev.*, **2**, 193 (1937)

Figure 3.19
55. ANNIS, A. D. unpublished report
56. SNELLING, E. C. and GILES, A. D. *Ferrites for inductors and transformers*, pp 136–137, Research Studies Press, Letchworth, Hertfordshire, England; distrib. by John Wiley & Sons, New York (1983)

Figure 3.20
57. CHAMBERLAYNE, J. W. and GILES, A. D. Some aspects of ferrite cores for power transformers, *2nd Conf. on Adv. in Magn. Mater.*, 17, IEE, London (1976)

Figure 3.21
58. SMIT, J. and WIJN, H. P. J. *Ferrites*, p 233, Philips Technical Library, Eindhoven (1959)
59. KOMAR, A. P. and KLIVSHIN, V. V. Temperature dependence of the electrical resistivity of ferrites, *Izv. Akad. Nauk SSSR, Ser. fiz. (Bull. Acad. Sci. USSR, Phys.)*, **18**, 400 (1954)
60. BROZ, J. A study of the electrical conductivity of MnZn ferrite, *Czech. J. Phys.*, **6**, 321 (1965)
61. VAN UITERT, L. G. Dielectric properties of and conductivity in ferrites, *Proc. Inst. Radio Engrs*, **44**, 1294 (1956)
62. PARKER, R., GRIFFITHS, B. A. and ELWELL, D. The effect of cobalt substitution on electrical conduction in nickel ferrite, *Br. J. appl. Phys.*, **17**, 1269 (1966)
63. STIJNTJES, TH. G. W., KLERK, J. and BROESE VAN GROENOU, A. Permeability and conductivity of

Ti-substituted MnZn ferrites, *Philips Res. Reports*, **25**, 95 (1970)

Figure 3.22
64. KROETZSCH, M. Über die Niederfrequenzdispersion der Dielektrizitätskonstanten und der elektrischen Leitfähigkeit polykristalliner Ferrite, *Physica Stat. Sol.*, **6**, 479 (1964)
65. SATO, H. and WATANABE, M. Dielectric properties of ferrites, *Electron. Communs. Japan*, **48**, 92 (1965)

Figure 3.23
66. MURTHY, V. R. K. and SOBHANADRI, J. Dielectric properties of some nickel-zinc ferrites at radio frequency, *Phys. stat. sol. (a)*, **36**, K133 (1976)
67. MIROSHKIN, V. P., PANOVA, YA. I. and PASSYNKOV, V. V. Dielectric relaxation in polycrystalline ferrites, *Phys. stat. sol. (a)*, **66**, 779 (1981)

Figure 3.24
68. BOZORTH, R. M., TILDEN, E. F. and WILLIAMS, A. J. Anisotropy and magnetostriction of some ferrites, *Phys. Rev.*, **99**, 1788 (1955)
69. VAN DER BURGT, C. M. Piezomagnetic ferrites, *Electron. Tech.*, **37**, 330 (1960)
70. OHTA, K. and KOBAYASHI, N. Magnetostriction constants of Mn-Zn-Fe ferrites, *Japan. J. Appl. Phys.*, **3**, 576 (1964)
71. KULIKOWSKI, J. and BIENKOWSKI, A. Magnetostriction of Ni-Zn ferrites containing cobalt, *J. Magn. and Magn. Mater.*, **26**, 297 (1982)

Figure 3.25
72. SMIT, J. and WIJN, H. P. J. *Ferrites*, Sec. 49.2. Philips Technical Library, Eindhoven (1959)
73. KNOWLES, J. E. The effect of surface grinding upon the permeability of manganese-zinc ferrites, *J. Phys. D: Appl Phys.*, **3**, 1346 (1970)
74. SNELLING, E. C. The effects of stress on some properties of MnZn ferrites, *IEEE Trans. Magn.*, **MAG-10**, 616 (1974)
75. LOAEC, J., GLOBUS, A. LE FLOC'H, M. and JOHANNIN, P. Effect of hydrostatic pressure on the magnetization mechanisms in Ni-Zn ferrites, *IEEE Trans. Magn.*, **MAG-11**, 1320 (1975)
76. LOAEC, J., LE FLOC'H, M. and JOHANNIN, P. Effect of hydrostatic pressure on the susceptibility frequency spectrum of polycrystalline Mn-Zn and Ni-Zn ferrites, *IEEE Trans. Magn.*, **MAG-14**, 915 (1978)
77. LE FLOC'H, M., LOAEC, J., PASCARD, H. and GLOBUS, A. Effect of pressure on soft magnetic materials, *IEEE Trans. Magn.*, **MAG-17**, 3129 (1981)
78. LOAEC, J., LE FLOC'H, M. and KONN-MARTIN, A. M. Effect of the hydrostatic pressure on both sides of the secondary peak of the thermal spectra of Mn-Zn ferrites, *J. Magn. and Magn. Mater.*, **26**, 309 (1982)
79. VISSER, E. G. Effect of uniaxial tensile stress on the permeability of monocrystalline MnZnFe^{II} ferrite, *J. Appl. Phys.*, **55** (6), 2251 (1984)

4 Magnetic circuit theory

4.1 Introduction

The foregoing chapters have been concerned principally with the intrinsic properties of ferrite materials. These properties have been expressed mainly in terms of the electrical impedance of an ideal winding on an ideal core shape, i.e. a uniform toroid having small radial thickness. For such a core it may be assumed that the field strength is uniform, and the magnetic path length and cross-sectional area equal the mean circumference and physical cross-sectional area respectively. It is now necessary to consider the magnetic properties of more practical core shapes and, as far as possible, to relate these to the properties of the material.

Before these relations are considered the underlying assumptions must be emphasized. In all the expressions relating core properties to material properties it is assumed that the ferrite, on a macroscopic scale, is homogeneous and isotropic. It is also assumed that any arbitrary core made of a particular grade of ferrite will have the same intrinsic properties as an ideal toroid made of the same material. These assumptions are seldom completely valid; the more complicated the form of the practical core the greater is the discrepancy likely to be. The reason is that in the manufacture of a simple toroid it is possible to make the pressed density fairly uniform throughout the core volume and to control accurately the sintering conditions in the immediate vicinity of the core. The properties of the ferrite are sensitive to both pressed density and firing conditions. When the same powder is pressed into a more complicated shape, e.g. a half pot core, it is more difficult to ensure uniformity either in density or exposure to kiln conditions. Differences will also result if the cores have very different cross-sectional areas. For these reasons the performance of a particular core shape may differ from that predicted from the relations that are derived in this chapter.[1] In addition some of these relations involve approximations which may add to the discrepancy.

To avoid these difficulties manufacturers now usually state the performance of a given core in terms of the core properties; data on materials are given mainly as a guide. For example, the performance of a transformer core is better specified in terms of the minimum inductance that will be obtained for a given number of turns (i.e. a property of the particular core) than in terms of minimum initial permeability (a material property). On the other hand a designer mainly interested in predicting the performance of a new or proposed ferrite core must have access to the intrinsic properties of the ferrite, such as those given in Chapter 3. The designer is not, however, concerned at this stage with the performance limits of the core and therefore some discrepancies between predicted and realized performance are usually not too serious.

Subject to the above limitations, the relations set out in the following sections provide a useful and reasonably accurate approach to the design of components using magnetically soft ferrite cores.

4.2 Closed magnetic cores

4.2.1 The effective dimensions of a core

When considering a uniform toroid having a very small radial thickness it is possible to speak of its magnetic length, l, cross-sectional area, A, and volume, V, without ambiguity. If however the cross-section is not small or uniform, the effect of the core geometry on the core properties is more complicated. The problem is to find the effective dimensions, l_e, A_e and V_e, which would define a hypothetical toroid having the same properties as the non-uniform core. Once this has been done, these effective dimensions may be used to calculate the performance of the non-uniform core just as though it were an ideal toroid. Strictly, this concept should be confined to cases where the flux densities are low so that the material may be assumed to obey approximately the Rayleigh or Peterson relations. However, it is often in practice extended to higher flux densities where the merit of convenience outweighs the loss of theoretical validity.

The first approach is rigorous but is limited in its practical application. The line integral of the field strength along an elementary path linking a winding of N turns carrying a current I is

$$\oint H ds = NI \quad \text{A} \tag{4.1}$$

where s is distance along the elementary path.

For an elementary path of length l the line integral equals Hl where H is the effective or average field strength round the magnetic path. A practical closed magnetic path may be regarded as a bundle of elementary paths in parallel, each elementary path being a tube of magnetic flux, i.e. containing a uniform flux $d\Phi$ but varying in cross-sectional area dA in accordance with the contour of the main body of the core.

Then

$$d\Phi = \mu_o \mu H dA = (\mu_o \mu NI dA)/l$$

and $\quad \Phi = \mu_o NI \int \dfrac{\mu dA}{l} \quad$ Wb $\tag{4.2}$

the integral being the sum of all the elementary areas across any section, each element being multiplied by the permeability of the elementary circuit and divided by its length. Thus it is in the nature of a parallel integration.

From the Peterson relation (Eqn (2.35)) the permeability depends on the field strength. Ignoring the higher powers of H this expression for μ may be substituted in Eqn (4.2).

$$\Phi = \mu_o NI \left(a_{10} \int \dfrac{dA}{l} + a_{11} NI \int \dfrac{dA}{l^2} \right) \quad \text{Wb}$$

The flux in an equivalent ideal toroid having dimensions l_e and A_e is

$$\Phi = \mu_o NI \mu \dfrac{A_e}{l_e}$$

$$= \mu_o NI \left(a_{10} \dfrac{A_e}{l_e} + a_{11} NI \dfrac{A_e}{l_e^2} \right) \quad \text{Wb}$$

For equivalence these two fluxes will be equal so, by equating coefficients, the following identities are obtained

$$\left. \begin{array}{l} \dfrac{A_e}{l_e} = \int \dfrac{dA}{l} \\[2mm] \dfrac{A_e}{l_e^2} = \int \dfrac{dA}{l^2} \end{array} \right\} \tag{4.3}$$

(4.1) $\quad \oint H ds = \dfrac{4\pi NI}{10} \quad$ A

(4.2) $\quad \Phi = \dfrac{4\pi NI}{10} \int \dfrac{\mu dA}{l} \quad$ Mx

These results may be corroborated by use of Peterson's loss expression (Eqn (2.36)). The hysteresis loss per cycle in an elementary magnetic path is given by

$$W_h = \dfrac{8}{3} \mu_o a_{02} \hat{H}^3 l dA$$

$$= \dfrac{8}{3} \mu_o a_{02} (N\hat{I})^3 \dfrac{dA}{l^2}$$

$\therefore$ Total loss $= \dfrac{8}{3} \mu_o a_{02} (N\hat{I})^3 \int \dfrac{dA}{l^2} \quad$ J cycle^{-1}

$$= \dfrac{8}{3} \mu_o a_{02} (N\hat{I})^3 \dfrac{A_e}{l_e^2}$$

for the equivalent toroid.

The integrations expressed in Eqns (4.3) are practicable only when a manageable expression for l may be obtained in terms of the position of the element of area, dA. Such a case is a radially thick toroid of rectangular cross-section having inner and outer radii r_1 and r_2 respectively and an axial thickness h, see Figure 4.1.

Then $\quad dA = h dr \quad$ and $\quad l = 2\pi r$

$$\therefore \quad \dfrac{A_e}{l_e} = \int_{r_1}^{r_2} \dfrac{h dr}{2\pi r} = \dfrac{h \log_e (r_2/r_1)}{2\pi}$$

and $\quad \dfrac{A_e}{l_e^2} = \int_{r_1}^{r_2} \dfrac{h dr}{4\pi^2 r^2} = \dfrac{h}{4\pi^2} \left(\dfrac{1}{r_1} - \dfrac{1}{r_2} \right)$

Hence $\quad A_e = \dfrac{h \log_e^2 (r_2/r_1)}{(1/r_1) - (1/r_2)} \quad$ m^2 $\tag{4.4}$

$$l_e = \dfrac{2\pi \log_e (r_2/r_1)}{(1/r_1) - (1/r_2)} \quad \text{m} \tag{4.5}$$

The core factor, C_1, see Eqn (4.9), is given by:

$$C_1 = \dfrac{l_e}{A_e} = \dfrac{2\pi}{h \log_e (r_2/r_1)} \tag{4.6}$$

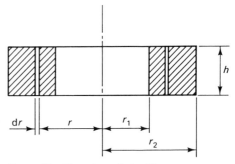

Figure 4.1 Dimensions of a toroid

138 Magentic circuit theory

and the effective volume

$$V_e = A_e l_e = \frac{2\pi h \log_e^3(r_2/r_1)}{[(1/r_1) - (1/r_2)]^2} \quad m^3 \quad (4.7)$$

In practice an alternative approach[2] depending on integration along the magnetic path is often more useful, if less exact. It is necessary to assume that, within the region of low flux densities being considered, the variation of permeability with field strength may be ignored.

As before

$$NI = \oint H \mathrm{d}s$$

$$= \frac{\Phi}{\mu_o} \oint \frac{\mathrm{d}s}{\mu A}$$

$$\therefore \Phi = \frac{\mu_o NI}{\oint(\mathrm{d}s/\mu A)} \quad \text{Wb} \quad (4.8)$$

The line integral is, of course, the reluctance of the magnetic path. If μ is uniform the reluctance is usually written

$$\frac{1}{\mu} \Sigma(l/A)$$

The reluctance of the equivalent ideal toroid is $l_e/\mu A_e$.

Therefore $\dfrac{l_e}{A_e} = \Sigma(l/A)$

i.e. a series summation of elements of path length measured along the mean magnetic path, divided by the corresponding areas. This summation is referred to as the core factor C_1.

$$C_1 = \Sigma(l/A) = l_e/A_e \quad (4.9)$$

To obtain separate expressions for l_e and A_e use may be made of another parameter, namely l/A^2. This arises naturally when the hysteresis loss is integrated along the mean magnetic path. The hysteresis loss per cycle is, from Eqn (2.36),

$$W_h = \frac{8}{3} \mu_o a_{02} \oint \hat{H}^3 \, A \mathrm{d}s$$

$$= \frac{8}{3} \frac{a_{02}}{\mu_o^2 \mu^3} \oint \hat{B}^3 \, A \mathrm{d}s$$

$$= \frac{8}{3} \frac{a_{02} \hat{\Phi}^3}{\mu_o^2 \mu^3} \oint \frac{\mathrm{d}s}{A^2} \quad \text{J cycle}^{-1}$$

$$= \frac{8}{3} \frac{a_{02} \hat{\Phi}^3}{\mu_o^2 \mu^3} \frac{l_e}{A_e^2} \quad \text{for the equivalent ideal toroid.}$$

$$(4.10)$$

(4.8) $\Phi = \dfrac{4\pi NI}{10 \oint(\mathrm{d}s/\mu A)} \quad \text{Mx}$

(4.10) $W_h = \dfrac{2}{3\pi} \dfrac{a_{02} \hat{\Phi}^3 10^{-7}}{\mu^3} \dfrac{l_e}{A_e^2} \quad \text{J cycle}^{-1}$

The line integral is usually written $\Sigma(l/A^2)$ and referred to as the core factor C_2. Thus

$$C_2 = \Sigma(l/A^2) = l_e/A_e^2 \quad (4.11)$$

Combining this with Eqn (4.9),

$$A_e = \frac{\Sigma(l/A)}{\Sigma(l/A^2)} = \frac{C_1}{C_2} \quad (4.12)$$

$$l_e = \frac{(\Sigma(l/A))^2}{\Sigma(l/A^2)} = \frac{(C_1)^2}{C_2} \quad (4.13)$$

and $V_e = A_e l_e = \dfrac{(\Sigma(l/A))^3}{(\Sigma(l/A^2))^2} = \dfrac{(C_1)^3}{(C_2)^2} \quad (4.14)$

These expressions are in common use for calculating the effective dimensions of an arbitrary core. However, the results must be rather approximate because the calculation assumes that the mean magnetic path is known and that it is possible to assign an effective area to each element of it. In practice the mean path is usually taken to coincide with a surface, perpendicular to the plane of the flux path, which divides the cross-section of the core into two equal areas. However, it is clear from the more exact treatment of the radially thick toroid that the actual mean magnetic path lies somewhat inside this approximate path.

In fact, in practical cores there are usually sharp corners and in the region of these the flux tends to concentrate round the inside of the bend, further shortening the actual mean path. Figure 4.2 shows the computed flux distribution round the magnetic path of one of a pair of E cores. The mid-flux line has been clearly marked.

IEC Publication 205 lists standard formulae for calculating core factors and effective dimensions for a number of widely used shapes.[3] These standard formulae, whilst not being theoretically accurate for the reasons just explained, do ensure that a common yardstick is used when calculating the properties of a core from electrical measurements on an associated winding. Figure 4.3 sets out the calculation of the core factors for a simple core shape. This example illustrates typical approximations that have to be made to deal with difficult sections, e.g. corners (see also Appendix A and Reference 4).

Within the limits of the assumptions, the use of the effective dimensions will simplify calculations involving reluctance and hysteresis loss of a non-uniform magnetic core. In particular they give rise to an effective flux density, B_e. From Eqn (4.10) the hysteresis loss per cycle in a core of non-uniform cross-section is given by

Magnetic circuit theory   139

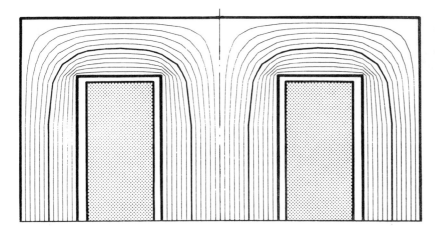

Figure 4.2 Computed plot of flux in one of a pair of ungapped E cores; the mid-flux line is emphasized

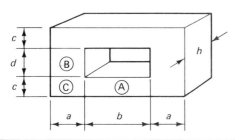

Section	l	A
A	b	ch
B	d	ah
C	1/4 circumference of mean circle $= \frac{2\pi}{4} \times \frac{(a+c)}{4} = \frac{\pi}{8}(a+c)$	Mean of terminating areas $\frac{h(a+c)}{2}$
Total $\Sigma l/A$	$\frac{2b}{ch} + \frac{2d}{ah} + \frac{\pi}{h}$	
Total $\Sigma l/A^2$	$\frac{2b}{c^2h^2} + \frac{2d}{a^2h^2} + \frac{2\pi}{h^2(a+c)}$	

Figure 4.3 Example of core factor calculation showing typical approximations

$$W_h = \frac{8}{3} \frac{a_{02}}{\mu_o^2 \mu^3} \frac{\hat{\Phi}^3}{A_e^3} V_e$$

$$= \frac{8}{3} \frac{a_{02} \hat{B}_e^3 V_e}{\mu_o^2 \mu^3} \quad \text{J cycle}^{-1} \quad (4.15)$$

where $\hat{B}_e = \hat{\Phi}/A_e$ \quad (4.16)

(4.15) \quad $W_h = \frac{2}{3\pi} \frac{a_{02} \hat{B}_e^3 V_e}{\mu^3} \times 10^{-7}$ \quad J cycle^{-1}

It follows from Eqn (2.13) that

$$E = \frac{\omega \hat{B}_e A_e N}{\sqrt{2}} \quad \text{V} \quad (4.17)$$

From Eqn (4.8), the inductance of a winding on such a core is

$$L = \frac{N\hat{\Phi}}{\hat{I}} = \frac{\mu_o N^2}{\Sigma(l/\mu A)} \quad \text{H} \quad (4.18)$$

$$\therefore L = \mu_o \mu N^2 A_e/l_e = \mu_o \mu N^2/C_1 \quad \text{H} \quad (4.19)$$

Thus the inductance is calculated for the equivalent ideal toroid having the same material permeability.

4.2.2 The effect of an air gap on core reluctance and effective permeability

Air gaps are introduced into magnetic cores for a variety of reasons. In the design of a permanent magnet, for instance, an air gap is essential to make the stored magnetic energy accessible. An inductor usually has an air gap to dilute unwanted effects of the core material and, by proper choice of gap length, to improve the overall performance. In a choke or transformer carrying d.c. an optimum air gap will ensure that the maximum inductance is obtained for a given number of turns. In this section the effect of the air gap will be studied in relation to the core reluct-

(4.17) \quad $E = \frac{\omega \hat{B}_e A_e N}{\sqrt{2}} \times 10^{-8}$ \quad V

(4.18) \quad $L = \frac{N\hat{\Phi}}{\hat{I}} \times 10^{-8} = \frac{4\pi N^2}{\Sigma(l/\mu A)} \times 10^{-9}$ \quad H

(4.19) \quad $L = \frac{4\pi \mu N^2 A_e}{l_e} \times 10^{-9} = \frac{4\pi \mu N^2}{C_1} \times 10^{-9}$ \quad H

ance and effective permeability. The effect on losses will be considered in the following section.

The air gap or gaps may be regarded as a part of the magnetic path having length l_g, cross-sectional area A_g and unity relative permeability in a generalized core having a reluctance as expressed in Eqn (4.8). This equation may be written as

$$\Phi = \frac{\mu_o NI}{\frac{l_g}{A_g} + \Sigma \frac{l_m}{\mu A_m}} \quad \text{Wb} \quad (4.20)$$

where the subscript m refers to the magnetic core.

Although this equation is general it is usually restricted in its use to cores having air gap lengths that are small compared with the dimensions of the core cross-section adjacent to the gap. Only under these conditions is the fringing flux a small proportion of the total flux and is it possible to estimate a value for the effective area of the gap. Eqn (4.20) may be written

$$\Phi = \frac{\mu_o NI}{\frac{l_g}{A_g} + \frac{l_e - l_g}{\mu A_e}} \quad \text{Wb} \quad (4.21)$$

where l_e is the total effective magnetic path length.

By analogy with the derivation of Eqn (4.18)

$$L = \frac{\mu_o N^2}{\frac{l_g}{A_g} + \frac{l_e - l_g}{\mu A_e}} \quad \text{H} \quad (4.22)$$

If the gap length is small compared with the total magnetic path length this equation reduces to

$$L = \frac{\mu_o N^2}{\frac{l_g}{A_g} + \frac{C_1}{\mu}} \quad \text{H} \quad (4.23)$$

Assuming that the material permeability is not affected by the air gap (as it would be if the core were subjected to a static field, see Figure 3.6) it is seen that the effect of an air gap is to reduce the inductance. Thus the core behaves as though it had a reduced permeability, referred to as the effective permeability, μ_e:

$$L = \frac{\mu_o \mu_e N^2 A_e}{l_e} \quad \text{H} \quad (4.24)$$

Combining Eqns (4.22) and (4.24)

$$\mu_e = \frac{l_e/A_e}{\frac{l_g}{A_g} + \frac{l_e - l_g}{\mu A_e}} \quad (4.25)$$

Again if $l_g \ll l_e$ this simplifies to

$$\mu_e = \frac{C_1}{\frac{l_g}{A_g} + \frac{C_1}{\mu}} \quad (4.26)$$

If the permeability of the material is high and the gap length is not too small then l_g/A_g may be large compared with C_1/μ. Then, if A_g may be taken approximately equal to A_e, μ_e approaches l_e/l_g, i.e. it becomes largely dependent on the gap ratio and relatively independent of the material permeability.

Developing the more exact expression, Eqn (4.25), and putting $A_g = A_e$,

$$\frac{l_e}{\mu_e} = \frac{l_e - l_g}{\mu} + l_g$$

$$\therefore \mu_e = \frac{l_e}{\frac{l_e - l_g}{\mu} + l_g} \quad (4.27)$$

from which

$$\frac{\mu}{\mu_e} = \frac{\mu}{l_e}\left(\frac{l_e - l_g}{\mu} + l_g\right)$$

$$= 1 + \frac{l_g(\mu - 1)}{l_e}$$

$$(4.20) \quad \Phi = \frac{4\pi NI}{10\left(\frac{l_g}{A_g} + \Sigma \frac{l_m}{\mu A_m}\right)} \quad \text{Mx}$$

$$(4.21) \quad \Phi = \frac{4\pi NI}{10\left(\frac{l_g}{A_g} + \frac{l_e - l_g}{\mu A_e}\right)} \quad \text{Mx}$$

$$(4.22) \quad L = \frac{4\pi N^2 10^{-9}}{\frac{l_g}{A_g} + \frac{l_e - l_g}{\mu A_e}} \quad \text{H}$$

$$(4.23) \quad L = \frac{4\pi N^2 10^{-9}}{\frac{l_g}{A_g} + \frac{C_1}{\mu}} \quad \text{H}$$

$$(4.24) \quad L = \frac{4\pi \mu_e N^2 A_e 10^{-9}}{l_e} \quad \text{H}$$

$$\therefore \quad \frac{l_g}{l_e} = \left(\frac{\mu}{\mu_e} - 1\right) \bigg/ \left(\mu - 1\right)$$

which may be rearranged to give

$$1 - \frac{l_g}{l_e} = \frac{\mu(\mu_e - 1)}{\mu_e(\mu - 1)} \qquad (4.28)$$

This relation is useful in the analysis of the effect of the gap on the stability and the loss of cores. If the gap is small compared with the dimensions of the cross-section then A_g may be taken as equal to the area of the core face forming the gap. However, as the gap length increases the value of A_g becomes larger due to the fringing flux.

This can be seen in Figure 4.4, which shows the computed flux distribution in and around one of a gapped pair of E cores. The fringing flux reduces the reluctance of the gap and results in a higher value of μ_e than that predicted by Eqns (4.25) or (4.26). Any attempt to analyse the problem on a general basis is complicated by the fact that, as the gap becomes larger, the fringing or leakage flux becomes dependent not only on the gap geometry but also on that of the winding and the rest of the magnetic circuit.

The method of conformal transformations may be used to obtain an expression for the reluctance of the air gap. Astle[5] has shown that the effect of the fringing flux is to increase the effective semi-width of the gap (e.g. the radius in the case of a circular pole face) by an amount:

$$\left(0.241 + \frac{1}{\pi} \log_e b_a/l_g\right) l_g$$

where b_a is the total inside length of the limb containing the air gap (e.g. for a pot core b_a equals the width of the winding space). Eqn (4.26) may now be modified to allow for fringing flux:

$$\mu_e = \frac{C_1}{\dfrac{1}{\dfrac{A}{l_g} + \left(0.241 + \dfrac{1}{\pi} \log_e \dfrac{b_a}{l_g}\right) P} + \dfrac{C_1}{\mu}} \qquad (4.29)$$

where A is the actual area of the pole face and P is the length of the pole face perimeter, e.g. the circumference for a circular pole face. Eqn (4.29) is accurate provided l_g is much less than the width of the air gap, and provided b_a/l_g is larger than about 5. Calculations based upon this formula have given values for the effective permeability of pot cores accurate to within 2% over wide ranges of gap length.

One of the principal benefits of an air gap is that any changes in the value of the material permeability are reduced in their effect on the inductance of a winding on the core. From Eqn (4.25)

$$\frac{d\mu_e}{d\mu} = \frac{\dfrac{l_e}{A_e} \dfrac{l_e - l_g}{\mu^2 A_e}}{\left(\dfrac{l_g}{A_g} + \dfrac{l_e - l_g}{\mu A_e}\right)^2} = \frac{\mu_e^2}{\mu^2}(1 - l_g/l_e)$$

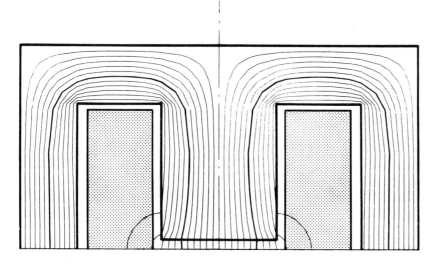

Figure 4.4 Computed plot of flux in one of a pair of gapped E cores; the mid-flux line is emphasized

142 Magentic circuit theory

$$= \frac{\mu_e^2}{\mu^2} \cdot \frac{\mu(\mu_e - 1)}{\mu_e(\mu - 1)} \quad \text{from Eqn (4.28)}$$

$$\therefore \frac{d\mu_e}{d\mu} = \frac{\mu_e(\mu_e - 1)}{\mu(\mu - 1)} \tag{4.30}$$

or $\dfrac{d\mu_e}{\mu_e} = \dfrac{d\mu}{\mu} \cdot \dfrac{(\mu_e - 1)}{(\mu - 1)}$ (4.31)

Thus the fractional change in the effective permeability is smaller by a factor $(\mu_e - 1)/(\mu - 1)$ than the change $d\mu/\mu$ that produces it. The factor $(\mu_e - 1)/(\mu - 1)$ is called the dilution ratio. In practice μ_e and μ are usually much greater than unity so the dilution ratio approximates to μ_e/μ. Then

$$\frac{d\mu_e}{\mu_e} = \frac{d\mu}{\mu} \cdot \frac{\mu_e}{\mu} \tag{4.32}$$

The cause of the change of material permeability may be a temperature change, a time effect, mechanical pressure, magnetic polarization, etc. The first two causes give rise to the factors mentioned in Section 2.2.2.

(1) Temperature factor

If the temperature coefficient of permeability is $\Delta\mu/\mu\Delta\theta$ then the temperature coefficient of effective permeability is, from Eqn (4.32),

$$\frac{\Delta\mu_e}{\mu_e\Delta\theta} = \frac{\Delta\mu}{\mu\Delta\theta} \frac{\mu_e}{\mu} = \frac{\Delta\mu}{\mu^2\Delta\theta}\mu_e \tag{4.33}$$

$\Delta\mu/\mu^2\Delta\theta$ is called the temperature factor, designated a_F. It is a material property which when multiplied by μ_e gives the temperature coefficient of the gapped core.

(2) Disaccommodation factor

By the same approach, the disaccommodation coefficient (of permeability) d defined by Eqn (2.26) gives rise to a disaccommodation factor D_F ($= d/\mu$, see Eqn (2.27)) which when multiplied by μ_e gives a measure of the disaccommodation of the gapped core.

Considering now the B–H relation of a given core, the effect of an air gap is to change the horizontal scale so that the loop is less inclined relative to the horizontal, and this is consistent with a reduction of the effective permeability. This effect is referred to as the shearing of the B–H relation and applies to both the initial magnetization and the hysteresis loop. It is illustrated in Figure 4.5 by reference to an idealized loop.

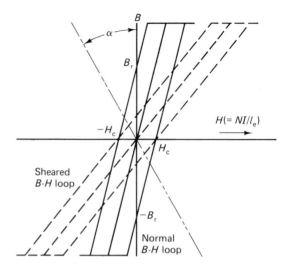

Figure 4.5 The shearing of an idealized B–H loop due to an air gap

In the analysis that follows it will be assumed that the core is of uniform cross section. The B–H loop, to be representative of a material, must refer to such a core otherwise the flux density would depend on the particular cross section chosen for the measurement. For a continuous uniform core, having the idealized B–H loop shown in Figure 4.5, the initial magnetization curve is given, below saturation, by

$$(B)_0 = \frac{\mu_0\mu(NI)_0}{l_e} \quad \text{T} \tag{4.34}$$

where the subscript denotes the zero gap condition. The corresponding two arms of the B–H loop may be expressed by

$$(B)_0 = \mu_0\mu\left(\frac{(NI)_0}{l_e} \pm H_c\right) \quad \text{T} \tag{4.35}$$

where H_c is the coercivity, and the left-hand and right-hand limbs of the loop are represented by taking the positive and negative sign respectively.

If now an air gap of length l_g ($\ll l_e$) and effective area A_g is introduced it follows from Eqn (4.21) that the magnetization curve is given by

(4.34) $\quad (B)_0 = \dfrac{4\pi\mu(NI)_0}{10 l_e} \quad$ Gs

(4.35) $\quad (B)_0 = \mu\left(\dfrac{4\pi(NI)_0}{10 l_e} \pm H_c\right) \quad$ Gs

$$B = \frac{\mu_o NI}{A_e \left(\frac{l_g}{A_g} + \frac{l_e}{\mu A_e}\right)} = \frac{\mu_o \mu NI}{l_e \left(\frac{\mu A_e l_g}{A_g l_e} + 1\right)} \quad T \quad (4.36)$$

and for the corresponding loop:

$$B = \frac{\mu_o \mu}{\frac{\mu A_e l_g}{A_g l_e} + 1} \left(\frac{NI}{l_e} \pm H_c\right) \quad T \quad (4.37)$$

(when $B = 0$ the total $NI/l_e = \pm H_c$ since the m.m.f across the gap must be zero).

The field strengths for equal flux densities, with and without air gap, may be obtained from these equations.

For the initial magnetization curve the sheared value of the field strength is

$$H = \frac{NI}{l_e} = \frac{(NI)_0}{l_e}\left(1 + \frac{\mu A_e l_g}{A_g l_e}\right) \quad A\,m^{-1} \quad (4.38)$$

and the corresponding value on the loop is

$$H = \frac{NI}{l_e} = \left(\frac{(NI)_0}{l_e} \pm H_c\right)\left(1 + \frac{\mu A_e l_g}{A_g l_e}\right) \mp H_c \quad A\,m^{-1} \quad (4.39)$$

Thus if the flux density is to remain unchanged when an air gap is introduced, the field strength on the initial magnetization curve must be increased by a factor $(1 + \mu A_e l_g / A_g l_e)$; the extra field strength being required to overcome the gap reluctance. This result applies equally well to a practical B–H loop, in which case μ is replaced by the amplitude permeability. A B–H loop may be graphically sheared by calculating new values of NI/l_e from Eqn (4.39) for a number of values of B. A simpler method is to draw an inclined

---

(4.36) $\quad B = \dfrac{4\pi NI}{10 A_e \left(\dfrac{l_g}{A_g} + \dfrac{l_e}{\mu A_e}\right)} = \dfrac{4\pi \mu NI}{10 l_e \left(\dfrac{\mu A_e l_g}{A_g l_e} + 1\right)} \quad$ Gs

(4.37) $\quad B = \dfrac{\mu}{\dfrac{\mu A_e l_g}{A_g l_e} + 1}\left(\dfrac{4\pi NI}{10 l_e} \pm H_c\right) \quad$ Gs

(4.38) $\quad H = \dfrac{4\pi NI}{10 l_e} = \dfrac{4\pi (NI)_0}{10 l_e}\left(1 + \dfrac{\mu A_e l_g}{A_g l_e}\right) \quad$ Oe

(4.39) $\quad H = \dfrac{4\pi NI}{10 l_e} = \left(\dfrac{4\pi (NI)_0}{10 l_e} \pm H_c\right)\left(1 + \dfrac{\mu A_e l_g}{A_g l_e}\right) \mp H_c \quad$ Oe

---

Magnetic circuit theory 143

B-axis as shown by the chain dotted line in Figure 4.5. The tangent of the angle a is clearly the ratio of the extra field strength to the corresponding flux density i.e.

$$\tan a = \frac{(NI)_0}{l_e} \frac{\mu A_e l_g}{A_g l_e} \frac{1}{B} = \frac{A_e l_g}{\mu_o A_g l_e} \quad (4.40)$$

This new B-axis having been constructed, the values for the sheared loop may be read off from the original loop, the B scale remaining unchanged in the vertical direction and the H values being measured off horizontally from the new B-axis in terms of the unchanged H scale.

Apart from reduction of effective permeability it is clear from Figure 4.5 that the presence of an air gap reduces the remanence.

4.2.3 Magnetic losses

In this section the core is assumed to have a general shape characterized by core factors C_1 and C_2 or effective dimensions l_e, A_e and V_e (see Eqn 4.9 to 4.14). The core is also assumed to be operated at low amplitude such that the Rayleigh relations apply, to have uniform permeability and to have losses that may be expressed by a relation having the form of Eqn (2.60).

If secondary effects such as changes of temperature may be ignored then the total loss in watts per unit volume of material is a function only of the frequency and flux density (or the field strength applying to a particular element). If these are constant then the dissipation of the core, i.e. the power that it draws from the circuit, is constant. It is very useful to keep this principle in mind when considering the loss expressions for a general core.

There are two fundamental ways of expressing the losses of a core in terms of circuit elements. They are illustrated in Figure 4.6. The first is in terms of current and a series impedance and the second is in terms of voltage and parallel impedance or admittance.

In the first case the magnetizing current and therefore the total applied field strength are the reference parameters and the loss is expressed in terms of series resistance, i.e. the core loss is $I^2 R_s$. While this is convenient in certain instances, difficulties arise when considering the influence of an air gap because the relation between the flux density and the current involves the effective permeability, and thus the current and the frequency do not alone define the con-

---

(4.40) $\quad \tan a = \dfrac{4\pi (NI)_0}{10 l_e} \dfrac{\mu A_e l_g}{A_g l_e} \dfrac{1}{B} = \dfrac{A_e l_g}{A_g l_e}$

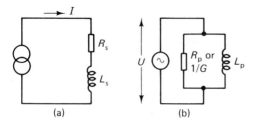

Figure 4.6 Series and parallel representations of core loss

ditions in the core. For example, consider the Legg hysteresis term from Eqn 2.61

$$\frac{R_h}{\mu f L} = a\hat{B}_e$$

It follows from Eqn (2.56) that for a gapped core of general shape:

$$\frac{R_h}{\mu_e f L} = a\hat{B}_e$$

$$\therefore R_h = a\mu_e f L \hat{B}_e = \sqrt{2}a\mu_0\mu_e^2 f L N I / l_e$$

Substituting for N from Eqn (4.24)

$$R_h = aIf\sqrt{\left(\frac{2\mu_0\mu_e^3 L^3}{V_e}\right)} \ \Omega \qquad (4.41)$$

Thus when the series loss resistance is expressed in terms of a material hysteresis loss coefficient the relation is somewhat unwieldy and, due to uncertainty in practice of the value of μ_e, the relation may be unreliable. For this reason the measurement of R_h on a gapped core may not be an accurate indication of the material hysteresis coefficient. However, it is by definition a direct measure of the hysteresis loss contribution of the gapped core (i.e. the component property) and it is sometimes used for this purpose. More generally it gave rise to a hysteresis factor F_h which is defined as

$$F_h = \frac{R_h}{I f L^{3/2}} \qquad (4.42)$$

Inasmuch as the hysteresis coefficient a is a constant, this factor is constant for a given core having a given effective permeability. From it, the hysteresis resistance may be calculated for particular values of current, frequency and inductance. So for a general core

$$\tan\delta_h = \frac{R_h}{2\pi f L} = \frac{F_h I \sqrt{L}}{2\pi} \qquad (4.43)$$

Although this series expression of hysteresis loss has been widely used in specifying the performance of loading coils for telephone lines, it is not now the hysteresis parameter usually quoted in inductor core data. It has been superseded by the IEC hysteresis core constant,[6] η_i, which is recommended for expressing the hysteresis properties of gapped cores; it is defined by

$$\tan\delta_h = \eta_i \hat{I}\sqrt{L} \qquad (4.44)$$

Thus the series impedance, or current, approach is suitable for expressing the hysteresis properties of gapped cores.

Turning to the alternative method, the reference parameter is voltage or the corresponding flux density. It is convenient now to repeat the derivation of Eqn (2.54) in terms of a general core. If a loss-free winding having N turns is placed on a general magnetic core and an alternating voltage U is applied, a power loss $P = U^2 G$ will be observed, where G is the conductance appearing across the winding due to the magnetic loss in the core. The induced e.m.f., the frequency and the flux density are related by Eqn (4.17). If the loss tangent is much less than unity the induced e.m.f. will approximately equal the applied voltage, so

$$U = \frac{\omega \hat{B}_e A_e N}{\sqrt{2}} \ \text{V} \qquad (4.45)$$

Thus for a given effective core cross-sectional area and number of turns, the voltage and frequency determine the effective flux density.

The loss tangent due to the magnetic core loss is

$$\tan\delta_m = \frac{\omega L_p}{R_p} = \omega L_p G$$

Since $\tan\delta_m \ll 1$, the distinction between series and parallel inductance may be dropped (see Figure 2.2).

Substituting $L_p = \mu_0 \mu N^2 A_e / l_e$

$$\tan\delta_m = \omega\mu_0\mu N^2 A_e G / l_e \qquad (4.46)$$

Putting $G = P_m A_e l_e / U^2$,

where P_m is the power loss density corresponding to a flux density $\hat{B}_e$,

---

(4.45) $\quad U = \dfrac{\omega \hat{B}_e A_e N}{\sqrt{2}} \times 10^{-8}$ V

(4.46) $\quad \tan\delta_m = \dfrac{4\pi\omega\mu N^2 A_e G}{l_e} \times 10^{-9}$

(4.41) $\quad R_h = aIf\sqrt{\left(\dfrac{8\pi\mu_e^3 L^3 10^7}{V_e}\right)} \ \Omega$

$$\tan \delta_m = \frac{\mu_o \mu}{\pi f B_e^2} P_m \quad (4.47)$$

The effect of introducing an air gap may now be considered. An air gap of length l_g introduced into a core having a total magnetic length l_e reduces the volume by a factor $(1 - l_g/l_e)$. If the frequency and flux density (applied voltage) are unchanged so that the power loss (volume) density is constant, the total power loss will also be reduced in the ratio $(1 - l_g/l_e)$. Therefore the conductance due to the gapped core

$$= (\text{power loss})/U^2 = \frac{P}{U^2}(1 - l_g/l_e)$$

$$= G \frac{\mu(\mu_e - 1)}{\mu_e(\mu - 1)} \quad \Omega^{-1} \quad (4.48)$$

by substitution from Eqn (4.28). Thus the conductance is reduced from the ungapped value by the ratio of the gapped to the ungapped volumes. The new loss tangent is therefore given by

$$(\tan \delta_m)_{\text{gapped}} = \omega L \frac{\mu_e}{\mu} \times G \frac{\mu(\mu_e - 1)}{\mu_e(\mu - 1)} = \omega L G \frac{(\mu_e - 1)}{(\mu - 1)}$$

or

$$\frac{(\tan \delta_m)_{\text{gapped}}}{(\mu_e - 1)} = \frac{\tan \delta_m}{(\mu - 1)} \quad (4.49)$$

These expressions are referred to as loss factors. Because μ_e and μ are usually much greater than unity, the above equation is usually approximated to

$$\frac{(\tan \delta)_{\text{gapped}}}{\mu_e} = \frac{\tan \delta}{\mu} \quad (4.50)$$

The subscript m has been dropped in this general equation because clearly the loss tangent need not be due to the total loss; it may be due to any specific form of loss. From Eqns (2.62) and (4.50) the loss tangent for a core, gapped or ungapped, may be expressed in terms of the Legg coefficients

$$\tan \delta_m = \frac{\mu_e}{2\pi} (a\hat{B}_e + ef + c) \quad (4.51)$$

The effective flux density is now used because the core shape is no longer restricted to the ideal. If the core is ungapped, $\mu_e = \mu$.

Considering first the hysteresis component

$$\tan \delta_h = \frac{\mu_e a \hat{B}_e}{2\pi} = \mu_e \eta_B \hat{B}_e \quad (4.52)$$

This relation gives the hysteresis loss tangent for a perfectly general core in terms of the Legg hysteresis coefficient or the IEC hysteresis material constant η_B (see Eqn (2.63)), assuming that the hysteresis power loss is proportional to B^3 and the gap is physically small.

It follows that a and η_B may be related to the hysteresis loss conductance of a general core. Putting $\tan \delta_h = \omega L G_h$ and substituting for L and B_e from Eqns (4.24) and (4.45) the above expression gives

$$a = \frac{\omega \mu_o N^2 A_e G_h}{l_e} \cdot \frac{2\pi A_e N \omega}{\sqrt{2} U} = \sqrt{2\pi} \frac{\mu_o \omega^2 N^3}{C_2} \frac{G_h}{U}$$

$$= 55.8 \frac{\omega^2 N^3}{C_2} \frac{G_h}{U} 10^{-7}$$

and similarly

$$\eta_B = \frac{a}{2\pi} = 8.89 \frac{\omega^2 N^3}{C_2} \times \frac{G_h}{U} \times 10^{-7}$$

$\quad$ T^{-1} $\quad$ (4.53)

where U is the applied voltage. This relation may be used to measure the hysteresis coefficient of a given gapped core. In practice the voltage applied to the winding is changed by an amount ΔU and the corresponding change ΔG in the total measured conductance is noted. Then

$$\frac{G_h}{U} = \frac{\Delta G}{\Delta U} \quad (4.54)$$

This differential method eliminates those parts of the loss conductance which are due to other causes, e.g. residual loss, and which do not depend on flux density. Since magnetic materials do not, in general, obey the simple relation of Eqn (2.31), G_h is not exactly proportional to U so it is usual to make the measurement between two specified low flux densities.

There is one difficulty with this measurement. When U increases, the permeability increases slightly in accordance with the Rayleigh relation. The corresponding increase in inductance changes the value of the loss conductance due to the other losses, e.g. the copper loss. The copper loss conductance equals $R_{dc}/\omega^2 L^2$ so the measured change in conductance will be partly due to an increase of hysteresis loss and partly due to a decrease of copper loss. The residual loss and the eddy current copper loss may also have some effect. If a Rayleigh material is assumed, it may be shown that the resultant error will be less than 5% if the overall Q-factor is greater than 100. This is usually the case.

Summarizing, where the operating level is defined in terms of current, the series hysteresis resistance or the hysteresis factor, F_h, is an appropriate means of

---

(4.47) $\tan \delta_m = \dfrac{4\mu P_m 10^7}{f \hat{B}^2}$

(4.53) $a = 55.8 \dfrac{\omega^2 N^3}{C_2} \dfrac{G_h}{U} \times 10^{-17}$ Gs^{-1}

expressing the hysteresis performance of specific cores. However, for the purpose of relating the hysteresis of the material to the hysteresis performance of, or measured on, a given core, the hysteresis loss tangent, $\tan \delta_h$, the IEC hysteresis material constant, η_B, or the Legg coefficient, a, expressed in terms of $\hat{B}_e$ are more convenient parameters.

The remaining two loss terms in Eqn (4.51) do not depend on the value of the flux density (assuming the change of μ_e with amplitude is negligible) so the question of the current or voltage approach does not arise. From Eqn (2.59) the loss factor due to eddy currents in a core of simple geometry is given by

$$\frac{\tan \delta_F}{\mu} = \frac{\pi \mu_o d^2 f}{\rho \beta} = \frac{(\tan \delta_F)_{gapped}}{\mu_e}$$

where d and β are defined following Eqn (2.46). So, in general,

$$\tan \delta_F = \frac{\pi \mu_o \mu_e d^2 f}{\rho \beta} \qquad (4.55)$$

For cores of more complicated shape than those covered by Eqn (2.46) the factor corresponding to d^2/β must be calculated from first principles.

At high frequencies or with large cross sections, dimensional resonance or skin effect may occur. The basic phenomena have been described in Chapter 2 but as they are essentially properties of a specific core the more practical aspects may be considered here.

For most practical purposes these phenomena are confined to the high permeability, fairly low resistivity ferrites such as manganese zinc ferrite. In nickel zinc ferrites the permittivity is relatively low so dimensional resonance can only occur at frequencies beyond the normal range of application.

From Eqn (2.49), the half-wavelength corresponding to dimensional resonance has been calculated as a function of frequency for three MnZn ferrites of different permeabilities and the results are shown in Figure 4.7. Strictly the curves apply only to an infinite plate of ferrite but they may, in fact, be used also to estimate the performance of practical cores. If it is proposed to use a MnZn core in the frequency range where dimensional resonance may occur then the least cross-sectional dimension should be much less than $\lambda/2$ if dimensional resonance is to be avoided. It is seen that the higher the initial permeability the lower is the frequency of dimensional resonance in a core of a given size (or the smaller must the core be to avoid dimensional resonance at a given frequency).

Figure 2.6 shows the complex permeability and

(4.55) $\tan \delta_F = \frac{4\pi^2 \mu_e d^2 f}{\rho \beta} \times 10^{-9}$

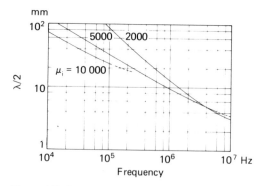

Figure 4.7 Dimensional resonance in MnZn ferrite cores, calculated from typical properties of three ferrites with different permeabilities. To avoid dimensional resonance at a given frequency, the least cross-sectional dimension of the core must be much less than $\lambda/2$

permittivity components as functions of frequency in the region of dimensional resonance calculated by Brockman et al.[7] for a typical manganese zinc ferrite. It may be observed that while μ'_s goes to zero, the impedance of a winding on such a core (proportional to $\sqrt{[\mu'^2_s + \mu''^2_s]}$) does not. In fact the impedance stays fairly constant in the vicinity of resonance, but the phase angle rotates.

For most manganese zinc ferrites, μ_s falls rapidly with frequency between about 1 and 10 MHz due to the ferromagnetic resonance. So any dimensional resonance that might occur at frequencies above a few MHz merges with this ferromagnetic resonance. Although dimensional resonance could have a serious effect on transformer and inductor performance, in practice it rarely does so. At higher frequencies smaller cores are normally used and at frequencies above a few MHz it may be preferable to change over to nickel zinc ferrite cores to avoid the ferromagnetic resonance dispersion that occurs in manganese zinc ferrites at these frequencies. These two reasons tend to make the avoidance of dimensional resonance automatic. Therefore it is in the design of devices requiring large manganese zinc ferrite cores at frequencies between about 100 kHz and 5 MHz that the possibility of dimensional resonance should be checked (see also Section 7.3.1). Skin effect in ferrites is even rarer; it would only occur in manganese zinc ferrite cores at frequencies greater than about 50 MHz.

An air gap which reduces the permeability to an effective value, μ_e, will increase the frequency of dimensional resonance. For a loss-free medium the frequency is increased by the ratio $(\mu/\mu_e)^{1/2}$ since in this case μ_e must be substituted for μ in Eqn (2.48).

Dimensional resonance is troublesome mainly in the measurement of the complex permeability or permittivity of high permeability ferrites at frequencies higher than 1 MHz. The most practical way of avoiding error due to this phenomenon is to ensure that the least cross-sectional dimension of the specimen, measured perpendicularly to the field, is made much smaller than the half-wavelength given by Figure 4.7.

Moving on from eddy current effects, the last of the loss components in Eqn (4.51) is the residual loss. The nature of this loss has been discussed briefly in the introduction to Figure (3.12). As a material property it is usually expressed by the residual loss factor, $(\tan \delta_r)/\mu$, as a function of frequency. For a gapped core, it follows from Eqn (4.50) that

$$(\tan \delta_r)_{\text{gapped}} = \left(\frac{\tan \delta_r}{\mu}\right) \mu_e \qquad (4.56)$$

This is the simplest of the losses to calculate. Provided the material properties are constant, the residual loss tangent is independent of the size and shape of the core and depends only on the material parameter, i.e. the residual loss factor, and the effective permeability.

Since, for a given core, $\tan \delta_F$ and $\tan \delta_r$ are functions only of frequency they are often lumped together and designated $\tan \delta_{r+F}$ in specifications for particular ferrite cores.

The core losses in a ferrite resulting from high-amplitude flux density excursions have been discussed briefly in Section 2.3. There are two distinguishable components of this loss; the eddy current loss, which is almost invariably the minor component, and the loss that, in the present context, can be called the magnetic loss. The latter is the loss measured on a magnetic material as a function of frequency and flux density when the eddy current loss is eliminated. As indicated in Section 2.3, it is often called hysteresis loss. However, since hysteresis is by definition a property of the static B–H loop,[8] it is preferable, in the case of higher measuring frequencies, to refer to it as high-amplitude magnetic loss. At frequencies in excess of a few kilohertz this magnetic loss includes contributions from origins other than hysteresis, for example, ferromagnetic resonance loss may become significant. So the loss is no longer necessarily proportional to frequency as it would be if the Rayleigh relations applied, see Eqns (2.31) and (2.32).

It is seen from Figure 3.19 that the high-amplitude magnetic loss of a ferrite material is expressed as a power loss (volume) density which is independent of the size and shape of the core and has the form of Eqn (2.74), repeated here for convenience

$$P_M = k f^m \hat{B}^n \qquad (4.57)$$

where, over the frequency range 10 to 100 kHz, m has a value of about 1.3 for a typical manganese zinc ferrite intended for power applications and n lies between 2 and 3.

Ferrite-cored power transformers have been used for many years to handle a variety of waveforms, notably the line scan waveform in television receivers. However, it has been usual for manufacturers to provide the power loss data only in relation to a symmetrical sinusoidal flux density. As indicated in Section 2.3, this is convenient both for specification and measurement and, of course, such data are directly applicable to designs involving only sinusoidal excitation. In practice, the use of the sinusoidal data has been extended with reasonable success to applications where the flux waveform departs substantially from sinusoidal, for example, where the excitation is by voltage having rectangular waveform resulting in an approximately triangular flux waveform. The resulting estimates of core performance have generally been sufficiently accurate.

In Chapter 9 it is seen that a major application of power ferrites is in cores for transformers used in electronically controlled Switched Mode Power Supplies. In such equipment the voltage waveform that excites the core is often rectangular and may be symmetrical about the origin or not. It is observed in some of the data presented in Figure 3.19 that for a typical power ferrite the power loss (volume) density depends primarily on the total peak-to-peak excursion of the flux density and only to a minor extent on the waveform or its symmetry.[9] Where this is the case it is preferable to express the loss as a function of B_{p-p} rather than $\hat{B}$ as this avoids the ambiguity when asymmetrical waveforms are being considered.

For a general core, having a non-uniform cross-section, the total flux, Φ must first be calculated. From this the flux density, Φ/A, may be obtained for each section or limb of the core having cross-sectional area A and then the power loss in that section or limb may be obtained from Eqn (4.57) by multiplying P_M by the corresponding volume. The total magnetic loss may thus be obtained by summation.

For sinusoidal excitation the flux is given by

$$\hat{\Phi} = \sqrt{2} E/\omega N \quad \text{Wb} \qquad (4.58)$$

$$\Phi_{p-p} = \sqrt{2} E/\pi f N$$

where E is the r.m.s. voltage.

For symmetrical rectangular waves, from Eqn (2.12),

$$\hat{\Phi} = E/4fN \quad \text{Wb} \qquad (4.59)$$

$$\Phi_{p-p} = E/2fN$$

where E is the voltage pulse amplitude.

(4.58) $\hat{\Phi} = \sqrt{2E \times 10^8}/\omega N$ Mx

(4.59) $\hat{\Phi} = E \times 10^8/4fN$ Mx

For the general case, again from Eqn (2.12),

$$\Delta\Phi = \bar{E}\Delta t/N \quad \text{Wb} \qquad (4.60)$$

where $\bar{E}$ is the average e.m.f. during time Δt.

If eddy current loss is likely to be significant this must also be calculated. In the case of a symmetrical square wave voltage, the flux wave is triangular so the modulus of the rate of change of flux is constant. Giles[10] has shown that the eddy current loss for a sinusoidal excitation is 23% greater than for square wave excitation, having the same flux density excursion, so calculations based on Eqn (2.46) will be somewhat pessimistic. Such calculation may be applied to each section or limb and the results, having been summed, may be added to the magnetic loss to obtain the total core loss. Since the eddy current loss contribution is usually a small fraction of the total core loss, an approximate value will generally suffice. For a hypothetical core of cylindrical cross-sectional area $A(=\pi d^2/4)$, Eqn (2.46) may be written in the form

$$P_F = \pi \hat{B}^2 f^2 A / 4\rho \quad \text{W m}^{-3}$$

It is usually sufficiently accurate, in the case of a general core, to substitute the effective area, A_e, in place of A:

$$P_F = \pi \hat{B}^2 f^2 A_e / 4\rho \quad \text{W m}^{-3} \qquad (4.61)$$

Where a standard core shape is being considered it is, of course, possible, and indeed preferable, to quote the total core loss directly as a core property. This can, at least for sinusoidal excitation, be easily measured on the core and used as the basis of the core performance specification and quality control. In this case, the parameters may be chosen to be more convenient to the circuit designer, who generally works in terms of voltages, currents and frequencies (or times) rather than in terms of B and H. This leads to core data in the form of graphs of total core loss (which includes eddy current loss) as a function of the peak-to-peak flux expressed in Webers or volt seconds per turn or volts per kilohertz turn, with frequency as a parameter. This approach is discussed in more detail in Chapter 9.

4.2.4 Distortion and intermodulation

In Chapter 2 the generation of wave form distortion and intermodulation products in a magnetic material is considered. The analysis assumes that the material has a Rayleigh or Peterson parabolic loop and that the core is of an ideal shape. The results are derived by calculating the amplitude distortion products in the flux density when a sinusoidal field strength is applied. In this section the way in which these results may be applied to a general core will be briefly considered.

The effect of an air gap may be inferred from Eqn (4.50). Applying this, for example, to Eqn (2.68) the e.m.f. distortion ratio for a gapped core is

$$\frac{E_{3a}}{E_a} = 0.6 \, (\tan \delta_h)_{\text{gapped}} = 0.6 \, (\tan \delta_h) \frac{\mu_e}{\mu} \qquad (4.62)$$

This may be derived more convincingly by considering the derivation of Eqn (2.68) and replacing the Peterson coefficient, which depends on μ, by the Legg coefficient, which does not. From Eqns (2.61) and (2.37), the amplitude of the 3rd harmonic given by Eqn (2.66) becomes

$$\hat{B}_{3a} = \frac{a}{2\pi} \frac{\mu_0^2 \mu^3 \hat{H}_a^2}{5}$$

If the magnetic circuit has an effective permeability, μ_e, then this may be substituted for μ. It then follows that

$$\frac{\hat{B}_{3a}}{\hat{B}_a} = \frac{a}{2\pi} \frac{\mu_e \hat{B}_a}{5}$$

or

$$\frac{\hat{E}_{3a}}{\hat{E}_a} = \frac{3}{5} \frac{a\mu \hat{B}_a}{2\pi} \frac{\mu_e}{\mu} = 0.6 (\tan \delta_h)\frac{\mu_e}{\mu}$$

from Eqn (2.62)

The other intermodulation amplitude ratios expressed in terms of $\tan \delta_h$ in Eqns (2.70) to (2.72) may be similarly converted.

The derivation of the relative magnitudes of the distortion and intermodulation e.m.fs in Section 2.2.7 was based on the assumption that the current wave is sinusoidal. However, the magnitude of the distortion e.m.f. will be substantially the same even if the applied voltage waveform is sinusoidal, for although the resulting current waveform will be non-sinusoidal, the amount of distortion considered in an analysis confined to the Rayleigh region is so small that the error in neglecting it will be of second order.

Referring to Figure 4.8(a), Z_m represents the impedance due to the magnetic material and it has in series with it a distortion generator, E_{na}, where n represents the order of the distortion product.[11] The impedance Z_A represents the source impedance, the winding resistance being assumed negligible. If $Z_A \to \infty$, the situation is in accordance with the analysis in Section 2.2.7. The sinusoidal source generator, E_a, will drive a sinusoidal current at frequency f_a

---

(4.60) $\Delta\Phi = \bar{E}\Delta t \times 10^8/N$ Mx

(4.61) $P_F = \pi \hat{B}^2 f^2 A_e \times 10^{-16}/4\rho$ W m^{-3}

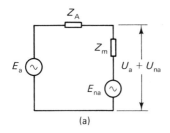

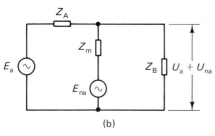

Figure 4.8 Distortion voltage across the terminals of a loaded inductor or transformer

through the inductor and the full distortion e.m.f. will appear across the inductor terminals together with the fundamental voltage, U_a. Therefore the distortion ratio $E_{na}/U_a = U_{na}/U_a$ may be measured across the terminals of the inductor.

If Z_A is not infinite, the distortion generator will not have an open-circuit and a distortion current, I_{na}, will flow.

$$I_{na} = \frac{E_{na}}{Z_A + Z_m}$$

where Z_A and Z_m are the impedances observed at the distortion frequency.

Assuming that the amplitude of the fundamental e.m.f. across the inductor is unchanged and the winding resistance is negligible, the distortion voltage ratio across the terminals will be

$$\frac{U_{na}}{U_a} = \frac{E_{na} - I_{na}Z_m}{U_a}$$

$$= \frac{E_{na}}{U_a} \frac{Z_A}{Z_A + Z_m} \qquad (4.63)$$

If, as in the low-frequency equivalent circuit of a transformer (see Figure 7.3), there is also a load impedance, the circuit appears as in Figure 4.8(b). In this case

$$\frac{U_{na}}{U_a} = \frac{E_{na}}{U_a} \frac{Z}{Z + Z_m} \qquad (4.64)$$

where $Z = Z_A Z_B/(Z_A + Z_B)$, all impedances corresponding to the distortion frequency.

In a well designed transformer Z_m is usually much greater than Z at the distortion frequencies so the distortion voltage ratio will be much less than the distortion e.m.f ratio.

4.3 Open magnetic cores

4.3.1 General

The magnetic cores considered so far have either had no air gaps or air gaps so small that the flux could be assumed approximately constant round the magnetic path, i.e. the lines of flux leave the magnetic material mainly at the surfaces forming the air gap. When the gap becomes an appreciable fraction of the total magnetic path length, the greater reluctance of the gap causes the flux to leave the magnetic material before crossing the pole faces which form the gap and the flux is not constant within the core. This invalidates the foregoing treatment and calls for an approach which takes this fringing or leakage flux into account.

The most common form of ferrite core having an air gap large enough to cause appreciable leakage flux is a simple cylindrical or rod core and this section will be mainly concerned with this shape.

Figure 4.9(a) represents an infinitely long solenoid containing no magnetic material. The solenoid carries a current I so the internal field strength is NI/l. The flux density inside the solenoid, represented by the long arrows, is $B = \mu_0 H$ and the flux density at all points outside is zero.

The next diagram (b) shows the same solenoid containing a very long magnetic core. In addition to the applied field H there is now a field due to the alignment of the magnetic moments of the atoms (electron spins). In a ferromagnetic or ferrimagnetic material the net effect of the atomic moments is to augment the applied field giving rise to an increased flux density represented by the short arrows. The total flux density in the solenoid is now

$$B = \mu_0 H + J \quad T \qquad (4.65)$$

where J is the magnetic polarization or intrinsic magnetic flux density in tesla, see Section 2.1. The exterior flux density is still zero because the fields due to the atomic currents cancel at all points outside the core.

(4.65) $B = H + 4\pi M$ Gs

where M is the intensity of magnetization or magnetic moment per cm³.

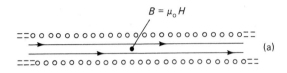

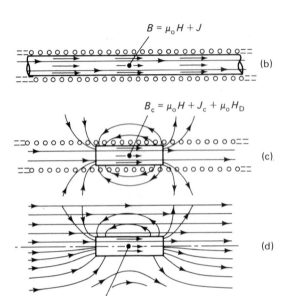

Figure 4.9 Magnetic flux associated with (a) a very long air-cored solenoid, (b) a very long solenoid enclosing a very long ferromagnetic cylinder, (c) a very long solenoid enclosing a short ferromagnetic cylinder (d) a short ferromagnetic cylinder immersed in a uniform magnetic field (upper half of diagram shows component fields; lower half shows resultant field)

If all but a short centre section of the core is removed it is clear that the total field acting in the remaining section will be diminished, since the atomic fields of the removed portions no longer contribute (see Figure 4.9(c)). This reduction may be considered to be due to a reverse or demagnetizing field which has neutralized the atomic fields that were previously there. The moments of the atomic currents no longer cancel in their effect outside the solenoid and so an exterior or leakage field exists. Within the remaining portion of the cylindrical core the resultant field will, in general, vary from a maximum in the centre to a minimum at the ends. This may be regarded as due to a non-uniform distribution of the demagnetizing field. In the special case of the remaining portion being an ellipsoid the internal field and flux density are constant, and the demagnetizing field is constant.

The demagnetizing field at the centre of the core may be expressed as

$$H_D = -NJ_c/\mu_0 = -NM_c \quad \text{A m}^{-1} \quad (4.66)$$

where J_c is the magnetic polarization at the centre of the core and N is called the demagnetization factor. This factor is considered in more detail later. The flux density in the centre of the core is now given by

$$B_c = \mu_0 H + J_c + \mu_0 H_D \quad (4.67)$$

$$= J_c + \mu_0 \left(H - \frac{NJ_c}{\mu_0} \right) \quad \text{from (4.66)} \quad (4.68)$$

$$\therefore \frac{B_c - J_c}{J_c} = \frac{\mu_0 H_c}{B_c - \mu_0 H_c} = \frac{\mu_0 H}{B_c - \mu_0 H_c} - N$$

since the magnetic polarization at the centre, $J_c = B_c - \mu_0 H_c$ from Eqn (4.65); H_c ($= H - NJ_c/\mu_0$) is the actual field strength at the centre of the rod. The permeability of the material is $\mu = B_c/\mu_0 H_c$, while the overall permeability is denoted by μ_{rod} and is defined by $\mu_{rod} = B_c/\mu_0 H$ (H being the applied field).

Simplifying the above equation

$$\frac{1}{\mu - 1} = \frac{1}{\mu_{rod}} \left(\frac{1}{1 - 1/\mu} \right) - N \quad (4.69)$$

Since μ is nearly always much greater than unity, this equation simplifies to

$$\frac{1}{\mu} \approx \frac{1}{\mu_{rod}} - N \quad (4.70)$$

So far μ_{rod} has been defined as the ratio of the flux density at the centre of the cylindrical core in Figure 4.9(c) to the flux density in the centre of the solenoid in Figure 4.9(a). If instead of being enclosed by a long close-fitting solenoid, the short cylindrical core were introduced into a relatively large region in which there existed, before the presence of the core, a uniform field H, then the situation would be similar to that depicted in Figure 4.9(c) except that the field, H, would no longer be confined to the interior of the solenoid but would occupy the whole region. It would combine with the leakage field to give a resultant field distribution as shown in Figure 4.9(d). The upper part of this figure shows the component fields while the

---

(4.66) $H_D = -NM_c$ Oe

Note: $(N)_{CGS} = 4\pi(N)_{SI}$

(4.67) $B_c = H + 4\pi M_c + H_D$ Gs

(4.68) $B_c = 4\pi M_c + H - NM_c$ Gs

(4.69) $\dfrac{1}{\mu - 1} = \dfrac{1}{\mu_{rod}} \left(\dfrac{1}{1 - 1/\mu} \right) - \dfrac{N}{4\pi}$

(4.70) $\dfrac{1}{\mu} \approx \dfrac{1}{\mu_{rod}} - \dfrac{N}{4\pi}$

lower half shows the resultant field. The value of μ_{rod} derived above may now be given an additional definition; it is the ratio of the flux density at the centre of a cylindrical core aligned in a uniform field, to the flux density existing there in the absence of the core. μ_{rod} thus differs from μ_e, the latter referring to a gapped core in which the total flux does not vary greatly along the magnetic path.

The value of the demagnetization factor N depends on the geometry of the core and to a lesser extent on its permeability. Figure 4.10 gives values for cylinders and ellipsoids of revolution. These have been calculated from formulae derived in the literature.[12–14] It will be noted that the demagnetization factors of the ellipsoids do not depend on the material permeability.

For any body, the sum of the demagnetization factors, relating to three orthogonal axes of that body, is unity. Thus all the ellipsoid demagnetizing factors approach 1/3 as the ellipsoid shape approaches that of a sphere. The demagnetization factors of the cylinders depend on both the dimensional ratio, m (which in this case equals length/diameter), and also on the material permeability.

Using some of this data in Eqn (4.69), μ_{rod} has been calculated as a function of length/diameter ratio for cylinders, with the material permeability as a parameter. The results are shown in Figure 4.11. This graph shows that when the material permeability is low, the value of μ_{rod} is asymptotic to the material permeability as the rod becomes more slender. This is

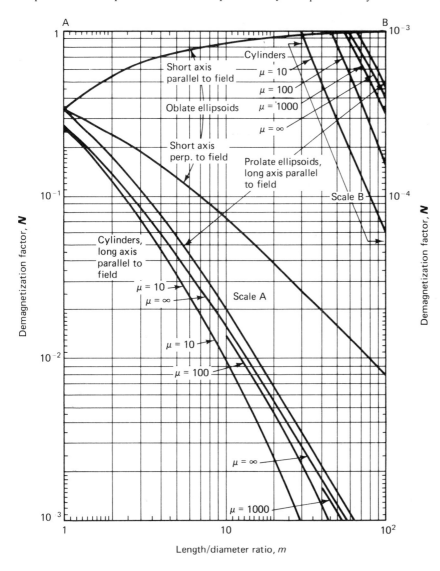

Figure 4.10 Demagnetization factors for ellipsoids and cylinders as functions of m (=long axis/short axis for ellipsoids and length/diameter for cylinders) (after Bozorth et al.[12]). Note that the values of N are appropriate to SI units; $(N)_{SI} = (1/4\pi)(N)_{CGS}$

152 Magnetic circuit theory

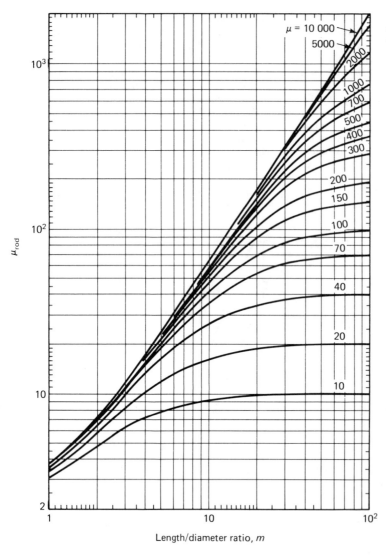

Figure 4.11 μ_{rod} as a function of m with material permeability as a parameter

because the demagnetizing factor becomes very small; from another point of view it could be said that the effective air gap becomes very small. When the material permeability is high the demagnetizing factor or the effective air gap does not become negligible within the practical range of slenderness considered. Even so, the graph shows that rod permeabilities up to 200 may easily be obtained with practical ferrite rods.

If the ferrite core is a tube having the same material permeability and the same outside dimensions as a rod, then μ_{rod} will be the same, i.e. the flux density in the ferrite half way between the ends of the tube will be $\mu_o\mu_{rod}$ times the field strength which would exist there in the absence of core. However, the total flux passing through the centre portion of the core would be less than that for a solid rod by the ratio of the cross-sectional areas.

4.3.2 Flux distribution along a cylinder immersed in a magnetic field

It has been seen that when a cylindrical core is immersed in a uniform field, parallel to the field direction, the flux density varies along the length of the core. The variation depends on the dimensional ratio of the core and on the permeability. It has been calculated by Warmuth[14] for cores of infinite permeability.

Figure 4.12(a) shows the measured distribution for a number of cylinders representing typical combinations of permeability and dimensional ratio.

Three types of distribution may be distinguished:
(1) Magnetically long cylinders, i.e. m large enough to make $\mu_{rod} \to \mu$. This gives a rather flattened curve which falls to a low value at the ends of the rod. The lower the material permeability or the higher the value of m the flatter the distribution.
(2) Intermediate cylinders, i.e. m such that μ_{rod} is less than, say, 0.8μ. An approximately parabolic distribution is obtained. This distribution is similar to that calculated by Warmuth for a cylinder of infinite permeability. If the rod is geometrically short, i.e. such that $m \to 3$ or less, then the next result applies even if $\mu_{rod} \to \mu$.
(3) Geometrically short cylinders, i.e. where $m \to 1$. This distribution is approximately parabolic but is shallower than in (2), i.e. it gives a relatively high value of flux density at the ends of the rod. In the limit as $m \to 0$ clearly the flux density will become uniform along the axis.

Any actual example may be identified by its value of μ and m as corresponding to, or lying between these types. When the field becomes very large the cylinder may approach saturation at the centre and the permeability may vary from a low value at the centre to a high value at the ends. This tends to make the flux density more uniform over the centre region and the

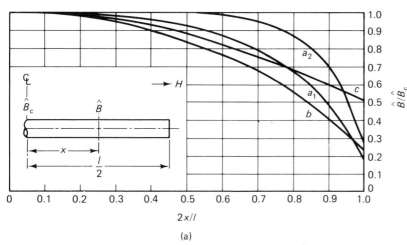

Figure 4.12 Distribution of flux density measured along various types of ferrite cylinders immersed in a uniform field. (a) Flux density distribution as a function of fractional distance from centre; (b) e.m.f. averaging factor F_A as a function of the averaging length, centrally located

	m	μ_i	μ_{rod}	Remarks
a_1	80	50	49.5	Magnetically long, $\mu_{rod} \to \mu_i$
a_2	80	180	172	
b	10 / 15	1000 / 1000	62 / 114	Intermediate
c	3	50–1000	10.5–12	Physically short, i.e. $m < 5$

154 Magentic circuit theory

distribution approaches that of (1) above.

A short coil placed in the centre of a rod, will have an e.m.f, E_c, induced in it corresponding to the central flux density $\hat{B}_c$. If the length of the coil is now increased without altering the number of turns the e.m.f will fall since it will correspond to the average flux density in the part of the rod covered by the coil. The ratio of this e.m.f. to the centre e.m.f., E/E_c equals $\hat{B}$ averaged over the length of coil divided by $\hat{B}_c$. This ratio is called the e.m.f. averaging factor, F_A. This factor is given in Figure 4.12(b) as a function of the fraction of the rod covered by the coil, for the three distributions distinguished above. From this graph the approximate value of F_A may be found for any type of ferrite cylinder immersed in a uniform field.

4.3.3 Flux distribution along a cylinder energized by a winding

Another type of flux distribution of importance is that resulting from an energized winding embracing all or part of the cylinder. This distribution is not very dependent on the permeability of the ferrite or on the dimensional ratio m. It is however strongly dependent on the fraction of the rod covered by the winding. Some measured distributions are shown in Figure 4.13(a). It is seen that if the winding covers the whole of the cylinder the flux density distribution is similar to that for a cylinder of high permeability material immersed in a uniform magnetic field (see Figure 4.12(a) curve b). As the winding becomes shorter the flux falls away more steeply from its central value.

Figure 4.13(b) shows the averaging factors $F_A = \hat{B}_{av}/\hat{B}_c$ for each distribution; they correspond to the curves of Figure 4.12(b) for field-immersed cylinders. The curves of Figure 4.13 enable the flux linkage with a winding or between windings to be estimated and also provide a basis for estimating core losses due to a current in an energizing winding.

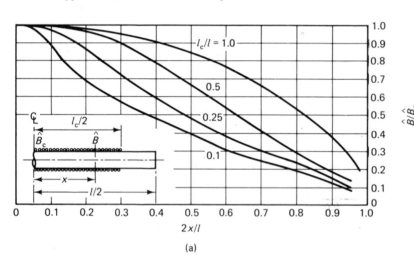

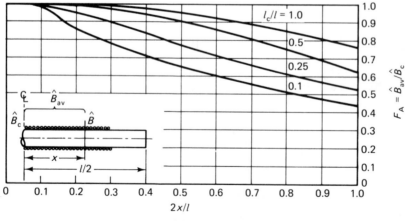

Figure 4.13 Distribution of flux density measured along a ferrite cylinder energized by a central solenoid, the parameter being the fraction of the cylinder covered by the solenoid. The result is almost independent of μ_{rod}. (a) Flux distribution as a function of fractional distance from the centre; (b) e.m.f. averaging factor F_A as a function of the averaging length, centrally located

4.3.4 The inductance of a winding having a cylindrical core

If L_a denotes the inductance of a winding having no magnetic core and L is the inductance of the same winding when embracing a cylindrical magnetic core then, by definition,

$$\frac{L}{L_a} = \mu_{coil} \qquad (4.71)$$

Sometimes this is called the inductance ratio or the apparent permeability. Both L and μ_{coil} are important parameters in the design of inductors having cylindrical cores.

Direct calculation of μ_{coil} is difficult since it depends on the geometry of the winding and the core and to some extent on the core permeability. A simpler approach has been adopted here; it applies only to centrally placed windings.

If a magnetically long cylinder of cross sectional area A, is covered along its whole length, l, with a winding of N turns, it approximates to a magnetic circuit having a small gap and an effective permeability, μ_{rod}. Therefore the inductance is given by

$$L = \mu_0 \mu_{rod} \frac{N^2 A}{l} \quad \text{H}$$

If the winding does not equal the length of the cylinder but is nevertheless centrally placed, the inductance will be modified because of the changed flux distribution along the axis. It has been seen in Figure 4.13 that the flux distribution is very nearly independent of μ_{rod}; it depends almost solely on the ratio of winding length to rod length l_c/l. It follows that in the special case of $l_c/l = 1$ the flux distribution is also independent of μ_{rod}. Therefore the above equation holds for fully wound cylinders of almost any dimensional ratio. If the cylinder is not fully wound the inductance is altered by an amount depending only on l_c/l. Therefore

$$L \propto \mu_{rod} \frac{N^2 A}{l}$$

the constant of proportionality depending only on l_c/l. To test this theory, the constant of proportionality was measured as a function of l_c/l for a wide variety of ferrite cylinders; the initial permeabilities ranged from 15 to 1000 and the dimensional ratios, m, ranged from 3 to 18. The results are enclosed by the shaded area of Figure 4.14. Further measurements were made to show that, within reason, these results do not depend on the radial thickness of the winding. This graph may be used to estimate the inductance of a wide variety of rod-cored inductors; if a mean curve is used the error will be less than about 12%. The values of N, A and l are given and the value of μ_{rod} may be obtained from Figure 4.11.

To obtain μ_{coil} it only remains to calculate the air cored inductance, L_a, of the winding. This may be done readily by reference to Section 11.5 where general design data is given. Eqn (4.71) then gives μ_{coil}.

4.3.5 Magnetic losses

The low-level loss in a cylindrical core[15] may in principle be calculated from the general Eqn (2.54)

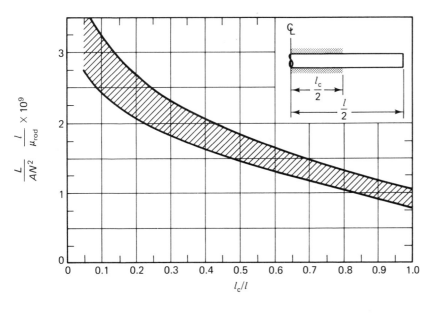

Figure 4.14 $Ll/AN^2\mu_{rod}$ as a function of l_c/l. The results for a wide variety of ferrite rods are contained within the shaded area (dimensions in mm, L in H)

Magentic circuit theory

$$P_m = \frac{\pi f \hat{B}^2}{\mu_o} \frac{\tan \delta_m}{\mu} \quad \text{W m}^{-3} \qquad (4.72)$$

This expression may be applied to each element of length and the total loss is

$$P = \frac{2\pi f A}{\mu_o} \int_0^{l/2} \hat{B}^2 \frac{\tan \delta_m}{\mu} dx \quad \text{W} \qquad (4.73)$$

where x is the distance from the centre (see Figure 4.12)

If the hysteresis loss predominates and the Rayleigh relations apply then $(\tan \delta_m)/\mu$ is proportional to B (see Eqn (2.33)). If eddy current loss is the only loss considered then $(\tan \delta_m)/\mu$ is a constant depending on the diameter of the rod and its resistivity (see Eqn (2.47)). If only residual loss is considered then $(\tan \delta_m)/\mu$ is a material parameter which is a function of frequency. Where more than one loss must be taken into account the corresponding loss factors must be summed for each element.

As in the case of any other magnetic circuit, the losses in a cylinder of magnetic material may be expressed in terms of the tangent of the resulting loss angle, δ, measured at the terminals of a winding embracing the core. This parameter will be designated $(\tan \delta_m)_{rod}$ to distinguish it from the corresponding material parameter. If a voltage, U, is applied to the winding, the resultant core loss may be represented as a conductance, G, in parallel with the winding. Then $P = U^2 G$. If the inductance of the winding is L the effective core loss tangent is given by

$$(\tan \delta_m)_{rod} = \omega L G = \omega L \frac{P}{U^2} \qquad (4.74)$$

From the e.m.f equation, assuming that the overall loss angle is small,

$$U^2 = \frac{(\omega \hat{B}_{av} A N)^2}{2}$$

where $\hat{B}_{av}$ is the peak flux density averaged over the length of the winding. Combining this with Eqns (4.73) and (4.74)

$$(\tan \delta_m)_{rod} = \frac{2L}{\mu_o A N^2 \hat{B}_{av}^2} \int_0^{l/2} \hat{B}^2 \frac{\tan \delta_m}{\mu} dx \qquad (4.75)$$

(4.72) $\quad P_m = \frac{f \hat{B}^2 10^{-7}}{4} \frac{\tan \delta_m}{\mu} \quad \text{W cm}^{-3}$

(4.73) $\quad P = \frac{f A 10^{-7}}{2} \int_0^{l/2} \hat{B}^2 \frac{\tan \delta_m}{\mu} dx \quad \text{W}$

(4.75) $\quad (\tan \delta_m)_{rod} = \frac{L 10^9}{2\pi A N^2 \hat{B}_{av}^2} \int_0^{l/2} \hat{B}^2 \frac{\tan \delta_m}{\mu} dx$

This equation may be applied to the general case of a cylinder of magnetic material surrounded over all or part of its length by a winding carrying an alternating current. The distribution of flux density along the cylinder depends on the fraction of the cylinder covered by the winding. This has been discussed in Section 4.3.3. If the distribution is known, together with the relation between $(\tan \delta)/\mu$ and $\hat{B}$, then the integral may be evaluated. At low flux densities $(\tan \delta_m)/\mu$ is virtually independent of $\hat{B}$ and may be removed from the integral. Given the distribution of $\hat{B}$ and the fraction of the cylinder covered by the winding, the value $\hat{B}_{av}$ may be readily deduced. The remaining factor L/AN^2 is a function of the cylinder and winding geometry and empirical data have been given in the previous section.

Two limiting cases will now be considered. The first is a long cylinder covered by a winding of equal length. If it is assumed that the distribution of $\hat{B}$ is parabolic and that $\hat{B}$ reaches zero at the ends of the rod (see Figure 4.12(a)), then Eqn (4.75) may be evaluated. The flux density at distance x from the centre is

$$\hat{B} = \hat{B}_c \left(1 - \frac{4x^2}{l^2}\right)$$

If this expression is integrated over the half rod length it is found that the average value of the flux density is

$$\hat{B}_{av} = \frac{2}{3} \hat{B}_c$$

$$\therefore \hat{B}_{av}^2 = \frac{4}{9} \hat{B}_c^2 \qquad (4.76)$$

If $\hat{B}^2$ is similarly integrated the result is

$$\int_0^{l/2} \hat{B}^2 dx = \frac{8}{15} \hat{B}_c^2 \frac{l}{2} \qquad (4.77)$$

Using these results in Eqn (4.75) and assuming that $(\tan \delta_m)/\mu$ is constant

$$(\tan \delta_m)_{rod} = \frac{lL}{\mu_o A N^2} \frac{6}{5} \frac{\tan \delta_m}{\mu} \qquad (4.78)$$

This is probably the most practical form of expression for the effective loss tangent of a fully wound cylinder. The approximate value of L/AN^2 may be obtained from Figure 4.14 (note: the dimensions in the present discussion are in m, in Figure 4.14 they are in mm). If the inductance is put in terms of μ_{coil}, see Eqn (4.71), an interesting analogy results. Assuming the winding is a long close-fitting solenoid.

$$L = \mu_o \mu_{coil} \frac{A N^2}{l}$$

(4.78) $\quad (\tan \delta_m)_{rod} = \frac{lL 10^9}{4\pi A N^2} \frac{6}{5} \frac{\tan \delta_m}{\mu}$

Therefore, Eqn (4.78) becomes

$$(\tan \delta_m)_{rod} = \frac{6}{5} \frac{\tan \delta_m}{\mu} \mu_{coil} \simeq \frac{\tan \delta_m}{\mu} \mu_{coil} \quad (4.79)$$

This is similar in form to the result for a toroid having a small air gap, see Eqn (4.50).

If the winding is not close-fitting or if the cylinder has a central hole a simple modification is required. It is assumed that the actual cross-sectional area of the ferrite cylinder or tube is A and the effective aperture of the winding is A_N. If A_N is not very much greater than A it will be seen that the general Eqn (4.75) will still apply; for the same value of A the flux linking the winding will to a close approximation be unchanged and the flux distribution will be as before. However, in deriving Eqn (4.79) the expression for L must now use A_N instead of A. Therefore this equation for the effective loss tangent must become

$$(\tan \delta_m)_{rod} \simeq \frac{A_N}{A} \frac{\tan \delta_m}{\mu} \mu_{coil} \quad (4.80)$$

The other limiting case is that of a very short winding on a long cylinder. Figure 4.13(a) shows a typical distribution of flux density. By graphical integration it has been found that the average value of $\hat{B}^2$ is 0.22 $\hat{B}_c^2$. Because the winding is short the flux density may be taken as constant over its length, therefore $\hat{B}_{av} = \hat{B}_c$. Putting these results in Eqn (4.75)

$$(\tan \delta_m)_{rod} \simeq \frac{lL}{\mu_0 A N^2} 0.22 \frac{\tan \delta_m}{\mu} \quad (4.81)$$

In this case there is no simple approximation for L/AN^2; for any given cylindrical core a value may be obtained from Figure 4.14.

Finally attention may be drawn to the evaluation of losses in a cylinder operating at higher levels. At the higher flux densities the core loss is usually expressed in terms of the power loss in watts per unit volume as a function of flux density and frequency. If the flux distribution is known or may be estimated from the foregoing data then the total loss in watts may be obtained by graphical integration of the loss density along the length of the cylinder.

4.3.6 Temperature coefficient

The temperature coefficient of inductance of a gapped core having an effective permeability μ_e follows from Eqn (4.33)

$$\frac{\Delta L}{L \Delta \theta} = \frac{\Delta \mu}{\mu^2 \Delta \theta} \times \mu_e$$

(4.81) $\quad (\tan \delta_m)_{rod} = \frac{lL 10^9}{4\pi A N^2} 0.22 \frac{\tan \delta_m}{\mu}$

The factor $\Delta\mu/\mu^2\Delta\theta$ is called the temperature factor, a_F, and is a material parameter. In the particular case of a cylindrical core[15] μ_e may, without great error, be replaced by μ_{coil} (see Section 4.3.4).
So

$$\text{Temp. coeff.} = \text{Temp. factor} \times \mu_{coil} \quad (4.82)$$

Thus if a winding on a cylindrical ferrite core has $\mu_{coil} = 20$ and the t.f. of the ferrite is $10 \times 10^{-6} \,°C^{-1}$ then the t.c. will be 200 ppm/°C.

4.4 References and bibliography

Section 4.1
1. Information on magnetic materials appearing in manufacturers' catalogues of transformer and inductor cores, *Int. Electrotechnical Commission, Publication 401*, Geneva (1972)

Section 4.2.1
2. OLSEN, E. *Applied magnetism, a study in quantities*, Philips Technical Library, Eindhoven (1966)
3. Calculation of the effective parameters of magnetic piece parts, *Int. Electrotechnical Commission, Publication 205*, Geneva (1966); see also Amendments No. 1 (1976) and No. 2 (1981), First supplement, 205A (1968) and Second supplement, 205B (1974)
4. DISTEFANO, M. I. Adjust ferrite-core constants, to suit your coil design needs, *Electron. Des.*, **24**, 154 (1977)

Section 4.2.2
5. ASTLE, B., private communication

Section 4.2.3
6. The International Electrotechnical Vocabulary, *Int. Electrotechnical Commission, Publication 50*, Chap. 221, item 221-03-34, Geneva (1988)
7. BROCKMAN, F. G., DOWLING, P. H. and STENECK, W. G. Dimensional effects resulting from a high dielectric constant found in a ferromagnetic ferrite, *Phys. Rev.*, **77**, 85 (1950)
8. The International Electrotechnical Vocabulary, *Int. Electrotechnical Commission, Publication 50*, Chap. 221, item 221-01-19, Geneva (1988)
9. SNELLING, E. C. and GILES, A. D. *Ferrites for inductors and transformers*, Section 5.3.3.3, Research Studies Press, Letchworth, Hertfordshire, England; distrib. by John Wiley & Sons, New York (1983)
10. *Ibid.*, Section 3.5.2

Section 4.2.4
11. PARTRIDGE, N. Harmonic distortion in audio-frequency transformers, *Wireless Engr.*, **19**, Part 1, 394, Part 2, 451 (1942)
12. BOZORTH, R. M. and CHAPIN, D. M., Demagnetizing factors of rods, *J. appl. Phys.*, **13**, 320 (1942)
13. STABLEIN, F. and SCHLECHTWEG, H. Uber den Entmagnetisierungsfaktor zylindrischer Stäbe, *Z. Phys.*, **95**, 630 (1935)
14. WARMUTH, K. Uber den ballistischen Entmagnetisierungsfaktor zylindrischer Stäbe, *Arch. Elektrotech.*, **33**, 747 (1939) and **41**, 242 (1954)

Section 4.3.5 and 4.3.6
15. VAN SUCHTELEN, H. Ferroxcube aerial rods, *Electron. appl. Bull.*, **13**, 88 (1952)

5 Inductors

5.1 Introduction

Inductance has been one of the principal passive elements of frequency-selective electrical networks since the beginning of telecommunications engineering and it is an important constituent in a wide variety of electronic equipment, ranging from telephony[1,2] to domestic radio[3] and television. The performance requirements have varied widely but it was in the design and construction of wave filters for Frequency Division Multiplex (FDM) telephony that the most stringent requirements emerged and it was this application that stimulated the development of the modern high-performance inductor.

With the advent of integrated circuits, the constraints on network design changed, there being a strong incentive to use only those circuit elements that can be realized on silicon. Clearly the physical inductor is not eligible as an integrated circuit element. If the network requires inductive reactance then either this must be provided by a separate, off-chip, inductor or simulated on-chip by circuits such as the gyrator. The realization of resonant elements within the micro-electronic concept is possible, in some circumstances, by the use of acoustic bulk wave or surface wave devices. More usually the integrated circuit designer, implementing an analogue network design, tends to avoid the need for inductive reactance by employing active filter networks based on selectivity provided by resistance/capacitance combinations.

A more powerful trend against the inductor has been the widespread supersession of analogue systems by digital circuit techniques which virtually eliminate the need for inductance.

The modern inductor is, however, the product of a long and successful evolution. It can provide very high performance, it is very flexible in its design and application and it can be inexpensive. It is still used in large numbers in a variety of applications. In particular, millions of high performance inductors have been used annually to provide filtering functions in FDM telephony transmission equipment and most of this is still in service. The introduction of digital transmission has recently led to a decline but not a cessation in the demand for such inductors. Because of the very high cost of development of integrated circuits, particularly those embodying analogue functions, their use is usually viable only when a very large demand for the specific application is assured. Thus the application needs to be well standardized, with a good expectation that it will retain a fixed performance requirement for a reasonable product life. For applications where the demand is likely to be relatively small or specialized, or where there is a need for the design to remain flexible, the inductor remains an economic alternative; its design can often be completed in a few minutes and its manufacture does not require large capital investment.

The modern high-performance inductor invariably uses a ferrite core. Standard ranges of such cores were developed initially to meet the stringent performance requirements of analogue telephony and it is with this class of inductor that this chapter is primarily concerned. Other classes of inductor, generally having less severe requirements, are embraced by the general treatment. For example, inductors used in radio and television receivers are often air-cored or have simple cylindrical cores; although they are outside the present scope, much of this chapter, together with Chapter 10 and Chapter 4, Section 4.3, is relevant to their design.

Before considering the theory and practice of inductor design in detail it may be useful to survey the requirements of a high-quality inductor. The basic requirement of the network designer is usually an inductance-capacitance combination which, ideally, resonates at the correct frequency irrespective of time or environmental conditions, which has negligible energy loss and which involves the smallest possible volume and cost. The capacitor is by no means a perfect partner in this alliance and its properties must be taken into account when considering the design of the inductor.

The requirements of the network may be translated into the following general inductor requirements:

(1) an inductance value which, once adjusted, is substantially constant at a given temperature during the service life of the equipment,
(2) a temperature coefficient of inductance that is

within close limits about an appropriate nominal value,
(3) very low electrical and magnetic losses,
(4) small cost and volume.

Before considering the problems of satisfying these requirements it will be useful to amplify them and make them more specific.

5.1.1 Inductance

The network design usually yields precise inductance and capacitance values, the latter often being constrained to standard values for reasons of economy. The capacitors are usually supplied having a tolerance range within ± 2% of the required nominal value. The inductor is designed to have a nominal inductance that is equal to the required value but it must, in accordance with current practice, be adjustable so that the correct LC product may be obtained. The adjustment range must be adequate i.e. it must at least equal the sum of the tolerances for which it must compensate. These are typically ± 2% for the capacitor, ± 1% for stray capacitance, an allowance for the effect of winding geometry and the integral increments in the number of turns, and the tolerance on the effective permeability of the inductor. An adjustment range of ± 7% is typical. Within this range the resolution of adjustment must be such that the resonant frequency may be set to an accuracy of about ± 100 ppm.

The value of the inductor and capacitor will in general change as functions of temperature, time, etc. These unwanted changes are called variability. To some extent the variability of the inductor may be reduced by increasing the air gap to lower the effective permeability (see Section 4.2.2). Perhaps the least difficult aspect of variability is the temperature coefficient. From the circuit point of view it is the total change of the LC product over a given temperature range that is important; the temperature coefficient of the LC product may typically be limited to 0 ± 100 ppm/°C. Practical limits for the temperature coefficient of inductance might be $+50$ to $+150$ ppm/°C.

A typical requirement for the long-term constancy of the LC product is that it should not change by more than ± 0.1% during its service life. The corresponding limit for inductance drift might be ± 500 ppm.

5.1.2 Q-factor

A high Q-factor for a given volume is nearly always desirable because this enables a better network (e.g. a filter with a lower pass-band insertion loss or a sharper cut-off characteristic) to be made from a given number of inductors or alternatively a given attenuation/frequency characteristic to be obtained with fewer inductors. At the higher frequencies, i.e. greater than a few tens of kilohertz, the inductor is usually designed to have the maximum Q-factor at a given frequency and in a given volume. At lower frequencies it is not usually possible to achieve maximum Q-factor without an unreasonable increase of variability. This is because a maximum Q-factor would require a high value of effective permeability in order to achieve a balance between the winding and core losses. Thus high Q-factor and low variability are often incompatible requirements; they are also to some extent interchangeable. If the Q-factor is made high at the expense of stability then a sharp cut-off or resonance may be obtained but a sufficiently large margin or guard-band must be provided to allow for frequency drift. In such a case the high Q-factor may to some extent be wasted, and a similar overall performance might be obtained using a larger air gap which, although it would result in a lower Q-factor, would give increased stability; the cut-off frequency or resonance will be less sharp but a smaller guard-band will be needed.

Q-factors currently specified for inductors range from 50 at frequencies below 300 Hz, through values of 500 to 1000 at 100 kHz, falling to the region of 200 at 20 MHz.

5.1.3 Hysteresis effects

The effects of hysteresis are, of course, required to be as small as possible. Usually filter inductors are operated at very low amplitudes and the loss contribution due to hysteresis is negligible (however this is not always true when making measurements of inductor Q-factor and precautions must be taken to ensure that the measuring voltage corresponds to a sufficiently low flux density).

Other effects of hysteresis are the increase of inductance with amplitude and the introduction of waveform distortion. The former effect is not usually serious at the flux densities normally employed, but again care must be taken that the amplitude used during inductance adjustment is not too high or an erroneous inductance value will be obtained. Usually the most important effect of hysteresis is waveform distortion which gives rise to the generation of harmonics and intermodulation products. These may be particularly troublesome where a large difference in power levels exists between two separate signals in the same circuit, e.g. at the common connections of two-wire line filters.

160 Inductors

5.1.4 Miscellaneous

It is usually required that the magnetic coupling between adjacent inductors shall be very small. Precise requirements depend on the circumstances but a voltage ratio of between 50 and 70 dB measured between equal windings on adjacent inductors normally is acceptable.

Among the features that facilitate the design and construction of inductors are the following: an adequate range of sizes and effective permeabilities, simple provision for windings and lead-out wires, and simple but effective means of assembly, terminating and mounting.

5.2 Form of core

Until the introduction of ferrites, the inductors used in telecomunication equipment used almost exclusively the various grades of powdered iron, or nickel iron, cores. Because of the low bulk permeability, adequate magnetic isolation between adjacent inductors could be achieved only by the use of toroidal cores. Such cores present some winding and adjustment problems. When ferrites, which have relatively high permeabilities, were introduced on a commercial scale it became possible to make magnetic circuits of more convenient shape whilst maintaining a high degree of magnetic screening. One of the most successful of the forms of ferrite core developed for inductors is shown, in its basic form, in Figure 5.1(a). It is called a pot core and is essentially a cylinder of ferrite having an enclosed annular space for the winding. In practice this basic shape is modified to provide functional design features; a typical pot core is shown in Figure 5.1(b). It usually consists of two similar half pot cores having the mating surfaces finely ground to ensure a stable magnetic joint of low reluctance. An air gap is usually provided by grinding back one or both of the centre core faces. Slots or holes are provided in the outer wall so that the terminal leads of the winding may be brought out and it is usual to provide a central hole to facilitate adjustment of the inductance.

A number of related core configurations have evolved from the basic pot core. The most successful has been the RM core, shown in Figure 5.2. This occupies a square plan area and so improves the packing density of cores on a printed circuit board. It has two relatively large sections cut away to provide access for the winding terminations. These use pins which are integral with the coil former and are spaced on the standard p.c.b. grid. Chamfers and recesses are provided at the remaining corners of the square to accommodate spring clips for assembly of the core halves.

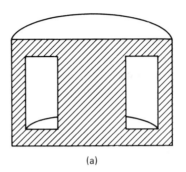

(a)

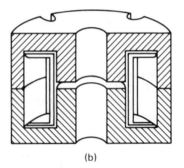

(b)

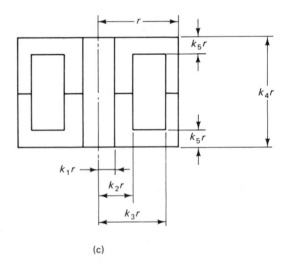

(c)

Figure 5.1 The geometry of a ferrite pot core: (a) basic shape; (b) a practical design; (c) symbols used in analysis

There is a range of core sizes, each size designated RMx where x is the nominal dimension, in tenths of an inch, of the side of the square. At present this range, which is the subject of an IEC standard,[4] extends from $x = 4$ to $x = 14$, and has largely superseded the cylindric pot core for both inductors and

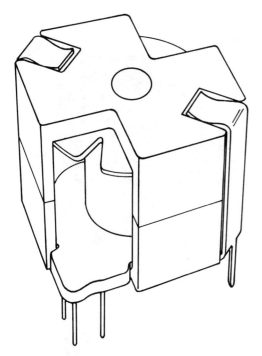

Figure 5.2 A typical RM core

transformers. Figure 5.3 shows a range of commercially available RM cores.

Before considering the best design of an inductor on a given core the main considerations in the design of the core itself will be discussed briefly in the light of the requirements noted at the beginning of this chapter.

A high Q-factor implies a low magnetic loss in the ferrite and also that the shape of the core utilizes the core material to the best advantage. Assuming a given ferrite material, the problem is to find the proportions of the core that will give the highest performance. The criterion of performance may vary from one application to another but over a wide range of inductances and frequencies it is true that the core which can contain a winding giving the lowest ratio of d.c. resistance to inductance has the best overall performance (see Section 5.7.8). In the following analysis the simple pot core shape is assumed and the optimum proportions of the core are derived according to the above criterion. Similar analysis was performed on the more complicated RM core shape during its development in order to optimize its proportions.

Using the symbols defined in Figure 5.1(c) the ratio of the d.c. winding resistance to the inductance, R_{dc}/L, may be expressed in terms of the core proportions k_1 to k_5. The inductance of N turns on a pot core with an effective permeability μ_e is (see Eqn (4.24))

$$L = \frac{\mu_0 \mu_e N^2}{C_1} \quad \text{H} \tag{5.1}$$

where C_1 is the core factor and is defined in Eqn (4.9). The resistance of the same winding, assumed to be wound with copper conductor having resistivity ρ_c, is given (see Eqn (11.11)) by

$$R_{dc} = \frac{\rho_c N^2 l_w}{A_a F_a} \quad \Omega \tag{5.2}$$

where l_w is the mean turn length $= \pi r(k_2 + k_3)$
A_a is the window area $= r^2(k_3 - k_2)(k_4 - 2k_5)$
(see Figure 11.1)
F_a is the overall copper factor
$$= \frac{\text{total copper cross section}}{A_a}$$

The winding is assumed to fill a hypothetical coil former so that the overall copper factor may be considered to be independent of the core proportions. It is also assumed that the spaces between the core and the inner and outer surfaces of the winding are equal.

Then
$$\frac{R_{dc}}{L} = \frac{\pi \rho_c (k_2 + k_3)}{F_a r(k_3 - k_2)(k_4 - 2k_5)} \frac{C_1}{\mu_0 \mu_e} \quad \Omega \text{H}^{-1} \tag{5.3}$$

where the core factor C_1 is a function of k_1 to k_5 and may be calculated by analogy with Figure 4.3. An example of such a calculation is given in Appendix A. If $(C_1)_1$ denotes the core factor for a core of the same proportions but of unit radius then since the dimensions of C_1 are [length]$^{-1}$

$$C_1 = \frac{(C_1)_1}{r}$$

∴ Eqn (5.3) becomes

$$\frac{R_{dc}}{L} = \frac{\pi \rho_c}{F_a \mu_0 \mu_e} \frac{k_2 + k_3}{r^2(k_3 - k_2)(k_4 - 2k_5)} (C_1)_1 \quad \Omega \text{H}^{-1} \tag{5.4}$$

For a given set of core proportions, i.e. a given set of values for k_1 to k_5, R_{dc}/L will thus vary inversely as the square of the linear dimensions. The effect of size

---

(5.1) $L = \dfrac{4\pi \mu_e N^2}{C_1} \times 10^{-9}$ H

(5.3) $\dfrac{R_{dc}}{L} = \dfrac{\rho_c(k_2 + k_3)}{F_a r(k_3 - k_2)(k_4 - 2k_5)} \dfrac{C_1 10^9}{4\mu_e} \quad \Omega \text{H}^{-1}$

(5.4) $\dfrac{R_{dc}}{L} = \dfrac{\rho_c 10^9}{4 F_a \mu_e} \dfrac{k_2 + k_3}{r^2(k_3 - k_2)(k_4 - 2k_5)} (C_1)_1 \quad \Omega \text{H}^{-1}$

162 Inductors

Figure 5.3 A range of RM cores (courtesy Philips Components)

may be removed by multiplying through by $V_T^{2/3}$ where V_T is the overall volume i.e. $\pi r^3 k_4$:

$$\frac{R_{dc} V_T^{2/3}}{L} = \frac{\pi \rho_c}{F_a \mu_o \mu_e} \frac{(\pi k_4)^{2/3}(k_2 + k_3)}{(k_3 - k_2)(k_2 - 2k_5)} (C_1)_1 \quad \Omega\, H^{-1} \tag{5.5}$$

$$(5.5) \quad \frac{R_{dc} V_T^{2/3}}{L} = \frac{\rho_c 10^9}{4 F_a \mu_c} \frac{(\pi k_4)^{2/3}(k_2 + k_3)}{(k_3 - k_2)(k_4 - 2k_5)} (C_1)_1 \quad \Omega\, H^{-1}$$

This expression may be minimized with respect to the proportionality factors, $k_2 \ldots k_5$, remembering that $(C_1)_1$ is a function of these factors. Thus the proportions for optimum utilization of a given volume may be found. An alternative approach would be to fix the height and diameter arbitrarily and to calculate the remaining proportions. In practice there may be other constraints. As mentioned in Section 1.3.6, the length to thickness ratio of walls should not exceed about 5 if the core is to be successfully pressed in production, so this may in fact mean that k_3 is determined only by k_4 and k_5. It should also be noted that k_1, the normalized radius of the central hole, is not a variable in this problem. Its value should ideally be zero but in practice the smallest value consistent with an adequate means of adjustment must be chosen.

Table 5.1 shows the results obtained using two arbitrary values of k_1.

Table 5.1 Pot core proportions for minimum R_{dc}/L (see figure 5.1.(c))

k_1	k_2	k_3	k_4	k_5
0	0.390	0.840	1.30	0.226
0.210	0.442	0.854	1.17	0.206

These proportions may be modified slightly in practice. The provision of slots for the lead-out wires will result in a small correction. Again, as the minimum value of R_{dc}/L obtained by the variation of any one proportion is rather shallow, it is possible to shift the proportions marginally away from optimum to obtain some other advantage. For example, if the centre core radius is increased by 5% this might result in only 1% degradation in R_{dc}/L, but the centre core cross-sectional area increases by at least 10% and this will produce a significant improvement in the hysteresis factor. Alternatively, for a small pot core used mainly at the higher frequencies, it might be decided to reduce the centre core diameter slightly to obtain a better spacing of the winding from the core. Whatever approach is adopted Eqn (5.5) provides a means of assessing the effect of changes of core proportions on R_{dc}/L for a fully wound coil.

The actual dimensions as distinct from the proportions depend on the sizes chosen. Table 5.2 gives the approximate diameters and heights of two standard ranges of pot cores each having approximately logarithmic progression of diameters, volumes and values of R_{dc}/L. The derived parameters (Section 4.2.1) are also given. The corresponding values for the standard range of RM cores are also listed.

5.3 Air gap and the calculation of inductance

The effect of an air gap on core properties in general is discussed in Section 4.2.2. In practice the air gap in an inductor core is provided during manufacture. Originally it was common for a given size of pot core to be available with a number of standard air gap lengths. This was inappropriate because the designer is not directly interested in gap length but rather in the effect of the air gap. It is now common practice for the gap to be adjusted to give one of a series of standard values of effective permeability or, alternatively, standard values of some factor relating inductance and number of turns.

The effective permeability, μ_e, is defined in the derivation of Eqn (4.24). It is subsequently shown that this is a property on which much of the core performance depends, e.g. the temperature coefficient and the residual loss tangent are proportional to μ_e. If therefore a range of cores of various sizes are available with effective permeabilities taken from a series of standard values, the inductor design is simplified because the designer may readily translate a trial design from one size of core to another in the range and expect to find the same values of μ_e available.

In practice, the standardization of effective permeabilities has, in many cases, been superseded by the standardization of inductance factor, A_L, values generally being chosen from the logarithmic series 63, 100, 160, 250, 400, 630, etc. The inductance factor is defined as the inductance of a coil on a given core divided by the square of the number of turns; it is generally expressed in nanohenries:

$$L = A_L N^2 \qquad (5.6)$$

where L is in henries $\times 10^{-9}$

Thus, from Eqn (5.1),

$$L = A_L N^2 10^{-9} \quad \text{H}$$
$$= \frac{\mu_o \mu_e N^2}{C_1}$$

If C_1 is in mm^{-1} then

$$A_L = \frac{\mu_o \mu_e 10^6}{C_1} \quad \text{nH for 1 turn} \qquad (5.7)$$

It should be noted that whereas A_L is susceptible to precise measurement and may therefore be stated within given limits, the value of μ_e may only be deduced, in practice, from a knowledge of A_L and a somewhat arbitrary calculation of the core factor C_1. Therefore the precise value of μ_e is not generally known for a pot core or RM core where the geometry of the flux path is complicated. However, in addition to the specified A_L value, the core data always includes a nominal value of μ_e calculated from a value of C_1 obtained according to a standard formula such as that shown in Appendix A. In practice, the nominal value of μ_e is sufficiently reliable to enable the

(5.7) $\quad A_L = \dfrac{4\pi\mu_e}{C_1} \quad$ nH for 1 turn

where C_1 is in cm^{-1}

Table 5.2 Standard inductor core ranges
This table gives the principal physical dimensions together with the core factors and effective dimensions; for the two IEC ranges the values are taken from the appropriate IEC standard. Notes: (1) values appearing in manufacturers' catalogues may differ slightly from the standard values due to differing dimensional limits and rounding. (2) RM6R and RM6S are variations of the basic RM6 geometry.

MULLARD 'VINKOR' RANGE — In accordance with BS 4061[5], range 1

Type designation	Nominal overall dimensions Dia. (mm)	Nominal overall dimensions Height (mm)	Overall volume (mm³)	Core factors $C_1 = \Sigma l/A$ (mm⁻¹)	Core factors $C_2 = \Sigma l/A^2$ (mm⁻³)	l_e (mm)	A_e (mm²)	V_e (mm³)	R_{dc}/L^* (Ω/H)
10	10.0	6.8	0.534 × 10³	1.02	77.3 × 10⁻³	13.4	13.2	0.177 × 10³	1029
12	12.0	7.8	0.882 × 10³	0.800	40.0 × 10⁻³	16.0	20.0	0.321 × 10³	696
14	14.0	9.0	1.38 × 10³	0.720	27.8 × 10⁻³	18.7	25.9	0.484 × 10³	564
18	18.0	11.2	2.85 × 10³	0.558	12.6 × 10⁻³	24.7	44.3	1.09 × 10³	346
21	21.5	13.6	4.94 × 10³	0.425	5.88 × 10⁻³	30.7	72.3	2.22 × 10³	228
25	25.4	16.0	8.11 × 10³	0.364	3.64 × 10⁻³	36.4	99.9	3.63 × 10³	162
30	29.5	18.8	12.8 × 10³	0.283	1.88 × 10⁻³	43.2	153	6.59 × 10³	119
35	35.5	22.8	22.6 × 10³	0.236	1.06 × 10⁻³	52.5	223	11.7 × 10³	79.4
45	45.0	29.2	46.4 × 10³	0.187	0.518 × 10⁻³	67.5	361	24.3 × 10³	47.2

IEC RANGE OF POT CORES — In accordance with IEC Publication 133[6] and BS 4061[5], range 2

Type designation	Dia. (mm)	Height (mm)	Overall volume (mm³)	$C_1 = \Sigma l/A$ (mm⁻¹)	$C_2 = \Sigma l/A^2$ (mm⁻³)	l_e (mm)	A_e (mm²)	V_e (mm³)	R_{dc}/L^* (Ω/H)
9 × 5	9.15	5.25	0.345 × 10³	1.25	125 × 10⁻³	12.5	10.0	0.125 × 10³	1406
11 × 7	11.2	6.45	0.635 × 10³	1.00	63.0 × 10⁻³	15.9	15.9	0.252 × 10³	931
14 × 8	14.1	8.35	1.30 × 10³	0.80	32.0 × 10⁻³	20.0	25.0	0.50 × 10³	584
18 × 11	18.0	10.55	2.68 × 10³	0.60	13.9 × 10⁻³	25.9	43	1.12 × 10³	336
22 × 13	21.6	13.4	4.91 × 10³	0.50	7.9 × 10⁻³	31.6	63	2.00 × 10³	250
26 × 16	25.5	16.1	8.22 × 10³	0.40	4.3 × 10⁻³	37.2	93	3.46 × 10³	161
30 × 19	30.0	18.8	13.3 × 10³	0.33	2.43 × 10⁻³	45	136	6.1 × 10³	113
36 × 22	35.6	21.7	21.6 × 10³	0.26	1.29 × 10⁻³	52	202	10.6 × 10³	79.2
42 × 29	42.4	29.6	41.8 × 10³	0.26	0.98 × 10⁻³	69	265	18.3 × 10³	50.9

IEC RANGE OF RM CORES — In accordance with IEC Publication 431[4]

Type designation	Side (mm)	Height (mm)	Overall volume (mm³)	$C_1 = \Sigma l/A$ (mm⁻¹)	$C_2 = \Sigma l/A^2$ (mm⁻³)	l_e (mm)	A_e (mm²)	V_e (mm³)	R_{dc}/L^* (Ω/H)
RM4	9.7	10.4	0.98 × 10³	1.90	172 × 10⁻³	21.0	11.0	0.232 × 10³	1024
RM5	12.1	10.4	1.52 × 10³	1.00	48.0 × 10⁻³	20.8	20.8	0.430 × 10³	590
RM6R	14.5	12.4	2.61 × 10³	0.80	25.0 × 10⁻³	25.6	32.0	0.820 × 10³	407
RM6S	14.4	12.4	2.57 × 10³	0.86	27.5 × 10⁻³	26.9	31.3	0.840 × 10³	437
RM7	16.9	13.4	3.83 × 10³	0.74	18.4 × 10⁻³	29.8	40	1.20 × 10³	336
RM8	19.3	16.4	6.11 × 10³	0.67	12.8 × 10⁻³	35.1	52	1.84 × 10³	246
RM10	24.2	18.6	10.9 × 10³	0.50	6.0 × 10⁻³	42	83	3.47 × 10³	164
RM14	34.3	28.9	34.0 × 10³	0.40	2.25 × 10⁻³	71	178	12.6 × 10³	79.6

*Calculated according to Eqn (5.3.) putting $\rho_c = 1.709 \times 10^{-8}$ Ωm, $\mu_c = 100$ and $F_a = 0.3$

performance of a gapped core to be estimated from the material parameters with reasonable accuracy. Table 5.3 lists the effective permeabilities of the RM range.

Where a means of inductance adjustment is to be provided, the air gap length is usually chosen such that the specified value of A_L refers to the core without the adjuster. The tolerance on A_L is typically $\pm 1\%$ for the lower values rising to $\pm 3\%$ when $A_L \geqslant 630$.

The value of A_L measured on a given core depends to some extent on the disposition of the winding in the winding space. Reasonably reproducible results may be obtained if the coil former is wound full. Figure 5.4 illustrates typical variations of A_L values with winding height. The specified values of A_L usually relate to a winding specially defined for the purpose of testing.

Another winding configuration of practical interest is the single layer. The A_L is somewhat dependent on the position of the layer as shown in Figure 5.5. This can be used to advantage when, for a winding with few turns, it is desirable to bias the A_L value away from its specified value, e.g. to overcome the integral turns problem (see the next section).

5.4 Inductance adjustment

Although a variety of methods have in the past been proposed for the adjustment of the inductance of ferrite cored inductors only two basic methods have gained practical acceptance. One method, applicable mainly to cylindrical pot cores is to shape the centre faces asymmetrically; a possible geometry is shown in Figure 5.6(a). If one half of the core is rotated relative to the other a change of effective permeability occurs. This method depends on the redistribution of the fringing field; if the flux lines were all parallel to the axis no change would be obtained as the mean length and geometric cross-section of the air gap are unchanged. The disadvantage of this method is that the inductance must be set before the assembly of the pot core is finally secured and mounted with the rest of the circuit, and subsequent adjustment is difficult. The other method is far more widely used so it will now be described in some detail with reference to Figure 5.6(b) and (c).

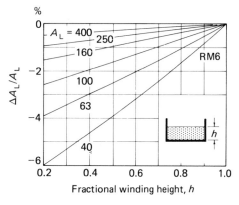

Figure 5.4 Dependence of A_L on winding height of an RM6 core. (courtesy Philips Components)

Table 5.3 Effective permeabilities of the RM range of inductor cores, calculated from $\mu_e = \dfrac{A_L C_1}{0.4\pi}$ where C_1 is taken from Table 5.2.

The effective permeability values given in this table may differ slightly from those appearing in manufacturers' catalogues due to small divergencies in C_1 values (see Table 5.2).

Core type	A_L (nH/turn²)									
	40	63	100	160	250	315	400	630	1000	1600
RM4	60.5	95.3	151	242	378	476	605			
RM5	31.8	50.1	79.6	127.3	199	251	318			
RM6R	25.5	40.1	63.7	101.9	159	201	255	401		
RM6S	27.4	43.1	68.4	109.5	171	216	274	431		
RM7	23.6	37.1	58.9	94.2	147.2	185	236	371		
RM8	21.3	33.6	53.3	85.3	133.3	168	213	336	533	
RM10		25.1	39.8	63.7	99.5	125.3	159	251	398	637
RM14				50.9	79.6	100.3	127.3	201	318	509

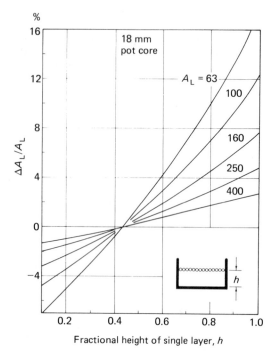

Figure 5.5 Dependence of A_L on the position of a single layer winding in an 18 mm pot core. (Courtesy Philips Components)

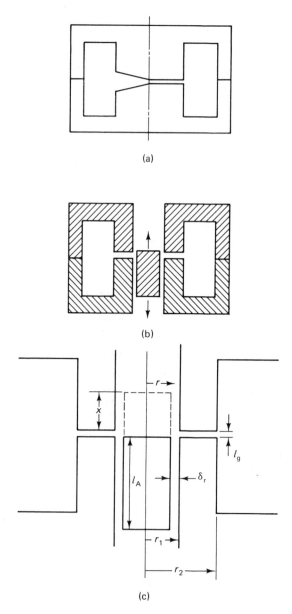

Figure 5.6 Ferrite pot core adjustment: (a) adjustment by chamfered centre pole; (b) adjustment by magnetic shunt; (c) symbols used in analysis of magnetic shunt

A cylindrical magnetic core is introduced into a central hole so that it partially shunts the air gap. If the axial position is varied the effective reluctance of the air gap is changed and with it the inductance of a winding on the core.

A simplified analysis will indicate the salient features of the method. It is assumed that the reluctance of the cylinder is negligible compared with that of the air gap surrounding it (in practice this is true if its permeability is greater than about 100). It is also assumed that the main air gap length l_g is so small that it is negligible compared with length l_A of the shunt, and that fringing in the absence of the shunt is zero so that the magnetic area of the air gap is $A_g = \pi(r_2^2 - r_1^2)$.

∴ Reluctance of the air gap without the shunt

$$= \frac{l_g}{A_g} \quad (5.8)$$

The reluctance in parallel with this when the shunt is in the position shown by the broken line, i.e. a distance x beyond the main air gap is:

$$\frac{\delta r}{2\pi r x} + \frac{\delta r}{2\pi r(l_A - x)}$$

(where δr is the path length of the radial gap between the shunt and the hole, and r is the mean radius of this gap).

Therefore this shunt reluctance $= \dfrac{\delta r}{2\pi r} \dfrac{l_A}{x(l_A - x)}$ (5.9)

It should be noted that this varies from ∞ when $x = 0$, through a minimum when $x = l_A/2$ (as may be shown by differentiation) to ∞ again when $x = l_A$

The minimum value is

$$\dfrac{\delta r}{2\pi r} \dfrac{4}{l_A} \qquad (5.10)$$

The smaller this value is, the larger will be the total adjustment range assuming a given value for l_g and assuming that the shunt starts from the all-out position. Thus to increase the adjustment range δr should be small and r and l_A should be large. The total air gap reluctance is obtained from the parallel combination of the reluctances given by Eqns (5.8) and (5.9), i.e.

$$\dfrac{\dfrac{\delta r l_A}{2\pi r x(l_A - x)} \times \dfrac{l_g}{A_g}}{\dfrac{\delta r l_A}{2\pi r x(l_A - x)} + \dfrac{l_g}{A_g}} = \dfrac{\delta r l_A l_g}{\delta r l_A A_g + 2\pi r x l_g(l_A - x)}$$

From Eqn (4.26) the resultant effective permeability is given by

$$\left. \begin{aligned} \mu_e &= C_1 \times \left[C_1/\mu + \dfrac{\delta r l_A l_g}{\delta r l_A A_g + 2\pi r x l_g(l_A - x)} \right]^{-1} \\ &\text{and the extreme values are} \\ (\mu_e)_{min} &= \dfrac{C_1}{C_1/\mu + l_g/A_g} \quad \text{and} \\ (\mu_e)_{max} &= \dfrac{C_1}{C_1/\mu + l_g/(A_g + \pi r l_g l_A/2\delta r)} \end{aligned} \right\} \quad (5.11)$$

Figure 5.7 shows the change of effective permeability as a function of x calculated from the Eqn (5.11) assuming a typical pot core. The corresponding curve measured on the same pot core is also given. The divergence at the lower end is due to the fact that the shunt begins to reduce the effective air gap before it has crossed the lower boundary of the actual air gap. This fringing effect is ignored in the analysis. If it is desired to allow for the effect of a low permeability shunt then its effective reluctance may be added to Eqn (5.9). This reluctance will be approximately of the form $kl_A/\mu A_A$ where l_A, μ and A_A refer to the shunt and k is a factor, less than 1, to allow for the non-uniform flux distribution in the shunt. Reference 7 gives a more detailed treatment.

The undulations appearing towards the top of the measured curve shown in Figure 5.7 illustrate the

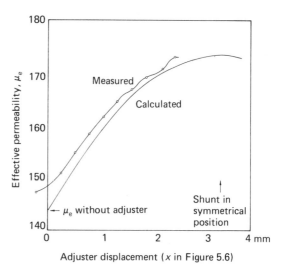

Figure 5.7 Calculated and measured adjustment curves for a 25 mm pot core having nominal $\mu_e = 160$

effect of eccentricities in the adjuster mechanism; this is an undesirable effect and is briefly considered at the end of this section.

The adjustment range obtained with a given shunt in a given core depends on the length of the main air gap; the larger the air gap the greater the adjustment range (see Eqn (5.11)). When the effective permeability is high, i.e. the main air gap is small, it is sometimes difficult to achieve a sufficient adjustment range. For this reason it is desirable to make the radial clearance between the shunt and the centre core as small as possible and the length of the shunt as long as possible. Increasing the permeability of the shunt does not help because, as stated earlier, when it exceeds about 100 the reluctance of the shunt itself becomes negligible. At the other extreme, a low effective permeability may result in a larger adjustment range than the minimum required. This in itself is not an undesirable feature. However, it is accompanied by an increase of the maximum slope for the adjustment curve and if this is allowed to become too large the stability of inductance with temperature and time will suffer. Accordingly it is usual to limit the maximum slope. This may conveniently be done by using a shunt of very low permeability or by increasing the radial clearance.

In the introduction to this chapter, reference was made to the minimum range of adjustment needed to ensure that the required LC product may be obtained, allowing reasonable manufacturing tolerances on parameters affecting the LC product. A typical minimum range is 14% (total) made up as follows: $\pm 3\%$ for the tolerance of A_L, $\pm 2\%$ for the tolerance of the resonating capacitor and ±2% margin for miscellaneous contributions such as stray capacitance. It was noted in Section 5.3 that the tolerance of A_L, and therefore μ_e, depends on its value, the higher the value of A_L the larger the tolerance. This effect aggravates the difficulty of obtaining sufficient adjustment range with the higher μ_e values; even where variability is not a limiting factor lack of adjustment range will normally limit the use of high effective permeabilities, e.g. 400, for inductor cores.

The number of turns required on a winding to give any nominal inductance will, in general, be non-integral when calculated from the nominal A_L value. The nearest integral number of turns will give an inductance that differs from the required value and this difference must also be taken up by the adjuster. Normally this is a negligibly small amount, but for low inductances requiring relatively few turns the possible difference will be appreciable and extra adjustment range will be necessary if gaps in the range of attainable inductances are to be avoided; e.g. if the required inductance lies midway between values given by 9 and 10 turns respectively, an extra margin of about ±10% is necessary. Low inductances are required mainly at high frequencies and at these frequencies low effective permeabilities are usual to obtain optimum Q-factors. The low μ_e values are usually accompanied by large adjustment ranges, so to some extent the extra margin occurs naturally.

For very low inductances, the gap between integral turn values may be bridged by having available several versions of the required core with closely spaced effective permeabilities. These permeabilities would be so chosen that the corresponding net adjustment ranges just overlap, thus providing, in effect, a very wide range of adjustable inductance with a given number of turns. The use of fractional turns, utilizing the availability of multiple lead-out slots, is to be deprecated because this generally leads to one turn coupling into a magnetic circuit consisting of two adjacent limbs, by-passing the air gap. This ungapped circuit will contribute high loss and temperature coefficient. The subject is discussed more fully in Reference 8. A useful alternative, when the winding consists of only one or very few layers, is to set the position of the winding so that the nominal A_L value is suitably modified,[9] see Figure 5.5.

Figure 5.8 shows three typical adjusting arrangements. In the first the shunt is in the form of a tube carried on a resilient plastic carrier. This carrier is an easy sliding fit in the hole. When rotated by a screwdriver engaged with the slot at the top of the carrier the shunt moves axially along the stud. The latter has a fine thread which gives good adjustment accuracy. There is no prepared thread in the plastic; the thread form is impressed by that of the stud. This arrangement gives a shake-proof adjustment that is free of backlash. In manufacture the plastic carrier is injection moulded with the tube located in the mould so that the surface of the tube is held concentric with the axis of the carrier. This results in concentric rotation in use and enables δr to be made very small.[10]

The next arrangement illustrated consists of a tube carried by a plastic screw. Usually the screw has a prepared thread but sometimes it makes its own thread when it is screwed into the threaded bush. In some variations the tube is merely cemented into position on the screw. This is simple but if the tube has an eccentric bore then the rotation becomes eccentric. Alternatively the tube may have the screw injection moulded through it in a similar way to the first design. A disadvantage of this example is that the position of the tube relative to the gap depends on a length of unsupported plastic and if this is dimensionally unstable inductance instability will result.

In the third example a threaded cylindrical shunt is used and this cuts its own thread into a plastic sleeve which lines the hole. Here again the position of the shunt depends on a length of unsupported plastic. The design is simple and the action is concentric but

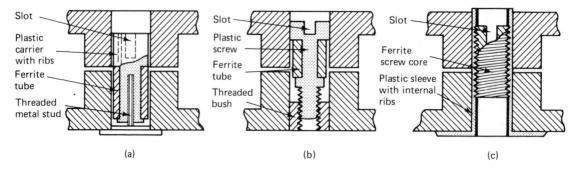

Figure 5.8 Typical adjusters using the principle of the magnetic shunt

the interposition of the sleeve between the threaded shunt and the pot core prevents the value of δr from being made very small.

In practice all three methods, and other similar ones, can give satisfactory results.

To facilitate rapid and accurate adjustment it is very desirable that the adjustment curve should be monotonic. If the magnetic shunt in the adjusting mechanism is not concentric and either its axis is slightly inclined to that of the hole or the holes in the two half-cores are not accurately aligned, then undulations may appear in the adjustment curve, see Figure 5.7. It is required by the IEC performance specification[11] that, within the adjustment range quoted by the manufacturer, there shall be no reversal of slope on the adjustment curve.

5.5 Constancy of inductance

5.5.1 General

In the introduction to this chapter it was observed that constancy of the *LC* product is an important requirement of a resonant circuit. The change of inductance (and capacitance) with time should be made as small as possible and the change of inductance with temperature should, ideally, balance exactly any corresponding change in capacitance.

These changes in inductance are sometimes referred to as variability. The variability may originate in the ferrite or it may be due to the non-ferrite parts of the inductor. In either case it may be reversible e.g. a function of temperature or irreversible, e.g. drift with time.

It is not too difficult to make a ferrite-cored inductor with a fairly small variability, i.e. having a temperature coefficient of inductance which is within 50 ppm/°C of the required value and a long-term drift of less than 0.1%. However if a performance substantially better than this is required great care must be taken in the design and construction of the inductor assembly. There are many factors which contribute small changes of inductance and they are difficult to diagnose and remedy. Much work has been done on this subject and as a result appreciable reductions in inductance variability have been obtained.

In this section the main factors affecting inductance variability will be considered and some typical experimental results will be shown.

5.5.2 Variability due to the ferrite

The contribution of the ferrite material to the variability of inductance depends on the constancy of the material permeability. The methods of expressing the variability of a magnetic core have been considered in Sections 2.2.2 and 4.2.2. In general the permeability may vary reversibly or irreversibly with temperature and it may vary with time, flux density, frequency, polarizing field strength and mechanical stress. The characteristics of some of these variations have been described in Chapter 3. In ferrites intended for high quality inductors the unwanted variations of permeability are kept to a minimum by manufacturing control.

In a ferrite-cored inductor, any change in the value of the material permeability will be reduced in its effect on the inductance by the dilution ratio, μ_e/μ_i, as shown by Eqn (4.32). Although several aspects of inductor design are influenced by the choice of μ_e it often happens, particularly at low frequency, that the value is governed mainly by the inductance constancy requirements.

The permeability is a function of temperature and provided the rate of temperature change is not too great, and other disturbances are avoided, this function may be regarded as reversible. In Figure 3.8 there are a variety of μ–θ graphs. In general these are not linear but they are usually taken to be so over a restricted temperature range. As described in Section 4.2.2 the temperature dependence of permeability is normally expressed as a temperature factor, T.F. This is the average material temperature coefficient, T.C., obtained from measurements at each end of the temperature range divided by the material permeability, see Eqn (4.33). The temperature coefficient of an ideal gapped core made of the same material is (T.F.) × μ_e. The T.F. will have a nominal value and limits, and the corresponding nominal value and limits of the effective temperature coefficient may in principle be reduced or adjusted by the appropriate choice of μ_e.

Ferrite cored inductors are often resonated with polystyrene foil capacitors and the temperature coefficients of the two components can be made to compensate partially. The T.C. of a polystyrene capacitor may range from -90 to -210 ppm/°C although it is possible to obtain closer limits, e.g. -115 to -145 ppm/°C. A ferrite core intended to be used with such a capacitor might have a temperature factor of $(1 \pm 0.5) \times 10^{-6}/°C$. If it also has $\mu_e = 100$, the resulting limits of T.C. would be $+50$ and $+150$ ppm/°C. The corresponding limits of T.C. for the *LC* product using the smaller tolerance capacitor would be -95 and $+35$ ppm/°C and the limits on the temperature coefficient of resonant frequency would be -47.5 and $+17.5$ ppm/°C. If this tolerance is not acceptable then the nominal and limit values of the material T.F. must be specified more closely to improve the compensation.

Often it is more difficult to obtain a suitable nominal T.F. value than to reduce the tolerance. If the nominal compensation cannot then be improved by adjusting the effective permeability (perhaps due to loss considerations) then special ceramic capacitors designed for T.C. compensation may be used. These have high values of positive or negative T.C. and constitute a small but appropriate fraction of the total resonating capacitance, the rest being made up, for example, with low-loss polystyrene capacitors. Even so, the ferrite T.F. must have a close tolerance if the overall T.C. of resonant frequency is to be within close limits.

The above discussion assumes that the permeability is an approximately linear function of temperature. As the slope of the function is made to approach zero at a particular temperature the curvature becomes pronounced (e.g. see Figure 3.8.3). When there is considerable departure from linearity the concept of temperature coefficient is more difficult to apply. One approach is to divide the temperature range into zones within which the slope of the μ-temperature curve is within stated limits.

For more specific requirements it is perhaps better to consider graphical areas, bounded on each side by the temperature limits and at the top and bottom by lines defining the permissible limits of the temperature-dependent parameter. The basic parameter is resonant frequency. From such a representation the boundaries of the corresponding LC product area may be obtained and this may be split into areas defining the permissible limits of change of the inductance and capacitance values. In this way an unambiguous specification of temperature dependence of permeability may be derived. Figure 5.9 illustrates this approach.

The next most important contribution of the material permeability to inductance variability is the time variation or disaccommodation.[12] This phenomenon has already been considered in some detail (see Sections 2.2.2 and 4.2.2 and Figure 3.10). Although time variation may be caused by any magnetic, thermal or mechanical disturbance, in the normal applications of telephony inductors the most common disturbance during assembly and subsequent service is temperature change. The permeability will rise above the stable value during the temperature disturbance and then when the temperature change ceases, it will return to the stable value appropriate to that temperature approximately in proportion to the logarithm of time. The effect is that a small overshoot on the permeability–time curve occurs when the temperature is raised from one steady value to another. The extent of this overshoot depends on the disaccommodation value of the material and the rate and magnitude of the temperature change. A temperature change of 50°C might produce a permeability overshoot of 1–2% in an ungapped core and, at constant temperature, a stable permeability would be substantially restored in about 24 h. The overshoot in a gapped core would be reduced by the dilution ratio.

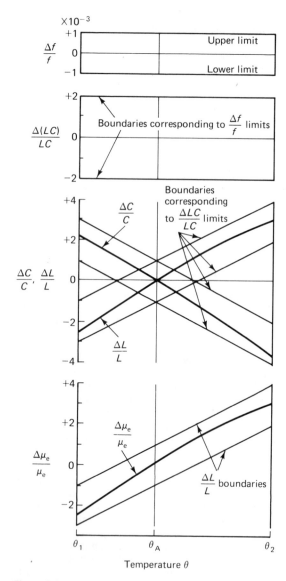

Figure 5.9 Typical boundaries of temperature dependence of an LC circuit and its components. θ_A is the temperature at which the LC product is adjusted to the correct value

Under conditions of small irregular temperature excursions, such as might occur during service the overall effect would be to raise the average permeability by an amount which would be quite small compared with the variations due to the temperature coefficient. Indeed it is an advantage that the natural temperature fluctuations prevent the permeability from ever going too far along the asymptotic approach to the ultimate quiescent value, a state that can only be achieved in practice by prolonged exposure to very constant temperature conditions. An obvious precaution during the manufacture of inductors is to ensure that an adequate interval (e.g. between one and three days) is left between any process involving a large temperature excursion or mechanical stress (e.g. impregnation, cementing or encapsulation) and the final process of inductance adjustment.

Other possible contributions to the inductance variability associated with permeability changes arise from changes of amplitude and the superposition of a steady magnetic field. Provided the effective flux density is not too high its influence on the inductance may be deduced from the graphs in Figure 3.5, using the dilution ratio μ_e/μ (see Eqn (4.32)). If the effective flux density is low, e.g. less than 1 mT (10 Gs), the variations in amplitude will normally have negligible effect. Figure 5.10 shows the variation of inductance as a function of flux density for a typical inductor core. The effect of amplitude is more often encountered during test measurements than during operation. A filter may work at very low amplitudes where

variation of inductance due to this cause is negligible but during measurement of inductance or of the insertion-loss characteristic the signal level may be quite high and may result in erroneous measurements or adjustments.

The variation of μ_e due to a static magnetic field (polarizing ampere turns) is not so easily deduced from the material properties and it is better for it to be measured on the core being considered. Figure 5.11 gives some typical results. As far as possible d.c. polarization in high quality inductors should be avoided if stability is important.

Mechanical stress is another factor that must sometimes be taken into account. The ferrite core is often stressed during assembly either by design or accident and if this stress is excessive a substantial change in initial permeability will result. Figure 3.25 gives typical results. Stress also changes the slope of the μ–θ curve and so influences the T.C. of inductance. It may be concluded that mechanical stress should be kept to a minimum and where it is unavoidable it should be constant.

The initial permeabilities of inductor ferrites do not vary significantly with frequency within the frequency range for which the material may be used in low loss applications. This may be seen in Figure 3.11 where the dispersion in μ_s' is shown to coincide with the large increase in residual loss. For this reason, and because the inductance is finally adjusted at the operating frequency, the influence of frequency may be ignored.

5.5.3 Contribution of the non-ferrite parts—construction and assembly

The variability of inductance arising from causes other than the variation of the material permeability is now considered. Many contributory causes are possible. Their main effect is to produce a drift or irreversible change of inductance; in most cases the reversible change, i.e. contribution to the temperature

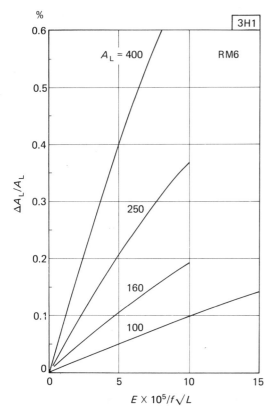

Figure 5.10 Dependence of A_L of an RM6 core on signal amplitude. E is in volts r.m.s., f is in Hz and L is in mH. Note that, for a given core, $B \propto E/f\sqrt{L}$. (Courtesy Philips Components)

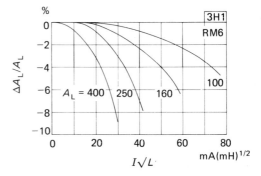

Figure 5.11 Dependence of A_L of an RM6 core on static field expressed in terms of current and inductance. (Courtesy Philips Components)

coefficient, is small but it will be seen that there are one or two notable exceptions to this.

The most important general category is the change of inductance due to changes in the effective reluctance of the total air gap, and again there are many possible mechanisms. The central air gap may change in length due to strain in the ferrite. Clearly pressure near the centre of the core faces will produce strains tending to close the central air gap. To avoid this it is essential to apply any clamping or fastening pressure over the centre of area of the ungapped limbs so that only simple compression is produced. Another possible source of strain is the differential linear expansion between the ferrite and other members that might be cemented or otherwise attached to the core. A marked reduction of temperature coefficient of inductance may result from a tag board which is firmly cemented to a core face and therefore causes convex distortion of the face as the temperature rises. The direct effect of stress on the initial permeability has been mentioned in the previous section.

If the halves of the ferrite core are clamped together or fastened to a tag or pin board by a force distributed around the periphery, the value of this force must be carefully chosen. If it is very small, the effective air gap at the mating surfaces will be appreciable in spite of the fact that these surfaces are given a finely ground finish during manufacture. As the force increases the surfaces are brought into more intimate contact and the resultant air gap becomes small and stable. If the force is further increased a point is reached at which the reduction of initial permeability due to stress becomes significant and the inductance begins to decrease. It is probably best to limit the force to a value well below this point. Changes of force due to thermal expansion or creepage in the clamping parts should be avoided and of course the mating surfaces should be perfectly clean.

An alternative method of fastening the core halves together is to cement them at the mating surfaces. The general method and the precautions to be taken are discussed in Section 1.4.2. If the thickness of the adhesive is too great its thermal expansion will affect the T.C. of the inductance. More serious is the effect of creep, which may be observed on cemented joints in general. If the joint is clamped during curing and then released it will tend to expand with time; on the other hand an increase of pressure will cause the joint to contract. However, if the cementing process is done with care and the above-mentioned precautions are observed, a stable and inexpensive assembly may be obtained.

In the RM range of cores the spring clips are designed to apply near-optimum clamping pressure throughout the range of dimensional tolerances and the point of contact is approximately at the centre of area of the outer limbs.

It has been seen that a typical adjuster is designed to adjust the inductance over a range of about ± 10%. From the moment the adjustment is completed, the adjuster should not contribute to further inductance change during the service life of the inductor. The adjuster should therefore be carefully designed to make it insensitive to shock, vibration and temperature variations.[13, 14] In particular, steps must be taken to avoid changes of adjuster position due to thermal expansion and creep of parts, particularly those made of organic materials.

Unwanted inductance change may also result from movement of the winding. If the whole coil former is free to move about inside the core, the resulting variation of inductance is easily demonstrated. It is therefore good practice to prevent this movement either by cementing the coil former to one half of the core with a resilient adhesive or by inserting some soft packing. There is some evidence that the thermal expansion of the winding can influence the T.C. of inductance and this effect depends on the winding geometry.

Finally, processes such as encapsulation could obviously give rise to stresses producing variability of inductance. If such protection is essential the precautions given in Section 1.4.3 should be observed.

5.5.4 Overall inductance variability

Of the processes of inductance variability considered in the foregoing sections, those which are reversible with temperature will combine to give an overall temperature coefficient of inductance. This will in general be somewhat different from $(T.F.) \times \mu_e$ due mainly to the contribution of the non-ferrite parts, and it must be one of the objects of a good inductor construction to make this discrepancy as small as possible. However, only in the most exacting applications is this aspect of variability serious.

Similarly all the irreversible processes will combine to give an inductance drift with time. Most irreversible processes may be accelerated by temperature cycling. Therefore, when a high order of inductance constancy is required it is advisable to subject the inductor, before final adjustment, to a number of temperature cycles beyond the range of temperatures to which it will be exposed in service. This will have the effect of relieving stresses and making the residual changes reversible with temperature. If the inductor is to be a part of a network it would perhaps be better to temperature cycle the whole assembly as this would have a similarly advantageous effect on other network components. There is of course one irreversible effect which is not reduced but rather induced by thermal disturbance, i.e. disaccommodation. The previous advice still applies; after thermal disturbance there should be an adequate time allowed for the disaccom-

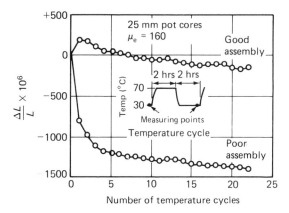

Figure 5.12 Inductance drifts measured on typical ferrite pot cores (without adjusters). The half cores were cemented together with epoxy adhesive and the assemblies were subjected to a series of dry temperature cycles. The results illustrate good and poor examples

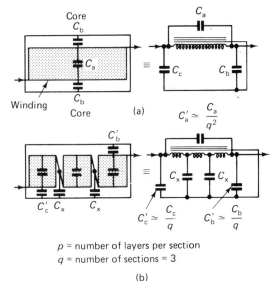

p = number of layers per section
q = number of sections = 3

(b)

Figure 5.13 Representation of winding self capacitances for: (a) a single section winding; (b) a three-section winding

modation to become negligible before the final inductance adjustment is made.

Inductance variability, both the reversible part and the long-term drift, may be studied experimentally by subjecting an inductor to a series of temperature cycles.[15] After allowing adequate time for acclimatization at the upper and lower temperature of each cycle, the inductance is accurately measured and the results are presented graphically. Figure 5.12 shows the results of actual measurements on typical ferrite pot cores cemented together with epoxy resin; it illustrates good and poor examples. In the latter example most of the drift occurs during the first few cycles and is due to creep of the cemented joint. Had this specimen been given five stabilizing cycles before the test its subsequent drift would have been comparable with the good example.

5.6 Self capacitance

The analysis of the self capacitance of a winding is given in Section 11.6. It is sufficient here to discuss only the practical aspects of the subject as they affect inductor design.

The self capacitance of a single section, multi-layer, winding consists of the three components[16] illustrated in Figure 5.13(a). If the number of layers is large, C_c is the direct capacitance between the first layer and the core (assuming the core may be regarded as an electrode) and C_b is the direct capacitance between the top surface of the winding and the core. C_a is the value of a hypothetical capacitance appearing across the winding and is equivalent to the distributed capacitances between turns and between layers. In practice the contribution of the inter-turn capacitance is usually negligible, so from item 5 of Table 11.3

$$C_a = \frac{4C_l(p-1)}{3p^2} \tag{5.12}$$

where C_l is the direct capacitance between adjacent layers when they are considered as forming a series of parallel plate capacitors, and p is the number of layers, assumed identical.

In Section 11.6 methods for calculating the three components of the self capacitance of a practical winding are given. Table 5.4 gives typical values of C_a and C_c for full, unimpregnated single-section windings on ferrite RM cores. The values of C_b depend strongly on the fullness of the winding and are therefore too uncertain to quote. Wax impregnation of a winding will increase C_a by 15 to 30%; the increase in C_c will be smaller unless the whole inductor is impregnated.

If the core is not, in effect, connected to either end of the winding, then the total self capacitance is

$$C_s = C_a + \frac{C_b C_c}{C_b + C_c} \tag{5.13}$$

Inductors are often used in this way, with the cores floating with respect to the circuit, if the minimum self capacitance is required. However it is often convenient, or necessary, to connect the core to one end of the winding. This, in effect, short circuits one of the end capacitances and puts the other directly in parallel with C_a. Experience suggests that the outer end

Table 5.4 Typical values of component self capacitances for full, unimpregnated, single-section windings on the ferrite RM cores quoted in Table 5.2. (Courtesy of Philips Components).

Core type	Typical values of C_a (pF)		Typical values of C_c (pF)	
	Insulated copper wire IEC grade 1	Silk covered bunched conductors	IEC grade 1 copper wire diameters 0.025 mm to 0.170 mm	IEC grade 1 copper wire >0.170 mm and bunched conductors
RM5	8	6	5	4
RM6	11	8	9	7
RM7	12	9	12	10
RM8	16	11	17	15
RM10	18	14	23	21

capacitance C_b is generally the larger and is certainly the less predictable of the end capacitances so it is usually recommended that if a connection must be made, the core should be connected to the finish of the winding.

The loss tangent of the self capacitance depends, in practice, mainly on the conductor insulation, although the material of the coil former may make a small contribution. For a dry unimpregnated winding of enamelled copper wire the loss tangent of the self capacitance would be typically 0.01 to 0.02. For a winding of bunched conductors the loss tangent of the self capacitance depends very much on the textile covering. The following loss tangents are typical: natural silk, 0.015; polypropylene fibre, 0.003; rayon, 0.035.

The effect of impregnation depends on the material used; good quality wax will not increase the loss tangent very much, whereas some varnishes have very high loss factors and would certainly reduce the overall Q-factor of the inductor at the high frequencies. Impregnation is not normally used as a means of protecting inductors intended to have high Q-factors at high frequencies. However some protection is required, for moisture in the winding will certainly cause the loss tangent of the self capacitance to be excessively high. The preferred practice has been to seal hermetically the whole network into a container from which all water vapour has been removed.

The most effective way of reducing the self capacitance of a winding is to divide it into sections wound side-by-side. Figure 5.13(b) illustrates a three section winding in comparison with the single section winding. Item 6 in Table 11.3 gives an expression for C_a' for various multi-section windings.

Sectioning reduces C_a in two ways; C_l is divided by the number of sections, q (ignoring the inevitable wastage of winding space due to the separation of the sections), and the number of layers, p, is multiplied by q. Thus from Eqn (5.12), C_a is divided approximately by q^2.

The end capacitances are clearly divided approximately by q. There are in addition capacitances from the intersection positions, denoted by C_x. These are each equal to $C_b' + C_c'$. If one end of the winding is connected to the core then these capacitances are transformed across the terminals of the winding in accordance with the usual transformation ratio, i.e. in the arrangement illustrated in Figure 5.13(b) the additional shunt self capacitance is $C_x(1/9 + 4/9)$. If the core is floating the situation is more complicated. The star mesh consisting of one of the capacitances, C_x, and the two parts into which it divides the winding may be transposed into the equivalent delta mesh. The result of this transposition is that each of the capacitances, C_x, is equivalent to additional end capacitances, adding to C_b' and C_c', and a negative capacitance in parallel with C_a'.

At frequencies above a few megahertz very low values of self capacitance are essential. Usually the inductance values are quite small so relatively few turns are required. Under these circumstances a single layer winding gives the best results. The self capacitance of a single layer winding depends almost only on the overall dimensions; Figure 11.19 provides a means of estimating this self capacitance assuming that the capacitance to core may be neglected.

Self capacitance introduces loss and possibly variability in proportion to its magnitude relative to the total resonating capacitance. Thus at low frequencies it will normally have negligible effect but at high frequencies where resonating capacitances are small, self capacitance is a major design consideration and should be made as small as possible.

5.7 Q-factor

5.7.1 General

The Q-factor requirements of an inductor and their relation to the need for inductance constancy are briefly discussed in Section 5.1. In the design of an inductor the following features are, in principle, independent variables; choice of ferrite material, size of core, effective permeability and form of winding. Usually some of these features are determined by considerations other than Q-factor, e.g. availability may dictate the grade of ferrite, the equipment construction may influence the choice of core size, and variability or hysteresis specifications will impose a maximum value of μ_e. Within such boundaries an inductor must usually be designed to have the maximum possible Q-factor; a lower limit will often be specified.

The Q-factor is the reciprocal of the total loss tangent, $\tan \delta_{tot}$. Provided $\tan \delta_{tot} \ll 1$ it may be taken as equal to the sum of all the partial or contributory loss tangents corresponding to the various origins of loss. For this reason it is convenient to work in terms of loss tangents

$$\frac{1}{Q} = \tan \delta_{tot} = \Sigma \tan \delta \qquad (5.14)$$

The design problem therefore is to calculate the contributory loss tangents and to minimize the total. The following sections first consider these loss factors separately and then in combination. Finally the overall Q-factor is considered in relation to frequency, inductance, type of winding conductor, and the effective permeability of the core.

5.7.2 Loss due to the d.c. winding resistance

If R_{dc} is the d.c. resistance of a winding and L is the inductance then the corresponding loss tangent is

$$\tan \delta_{dc} = R_{dc}/\omega L \qquad (5.15)$$

At a given frequency this factor depends primarily on the geometry of the core. The influence of the core shape on the ratio R_{dc}/L has been analysed in Section 5.2. If the core size and shape are fixed then this ratio depends mainly on the effective permeability and the fullness of the winding. When the loss due to d.c. winding resistance is important the winding should occupy as much of the available space as possible. From Eqns (5.2) and (5.6)

$$R_{dc} = \frac{\rho_c N^2 l_w}{A_a F_a} \quad \Omega$$

and $\qquad L = N^2 A_L 10^{-9} \quad H$

$$\therefore \quad \tan \delta_{dc} = \frac{\rho_c l_w 10^9}{\omega A_a F_a A_L} = \frac{4 \rho_c l_w 10^9}{\omega A_L N \pi d^2} \qquad (5.16)$$

from Eqn (11.5); A_L is in nH/turn^2.

Section 11.2 considers the relation between the window area A_a, the actual available winding space, the type of conductor and the overall copper factor F_a. Calculation of $\tan \delta_{dc}$ from Eqn (5.16) inevitably involves some judgement of copper factor, since this will depend on the type of winding and the margin that must be allowed if over-full windings are to be avoided. A more direct calculation is possible if, for the coil formers being used, there are reliable tables giving the number of turns for a reasonably full winding, and the corresponding resistance, for a range of conductor sizes and types. Such tables are usually based on experimentally determined copper factors. From the required inductance value the number of turns may be calculated and, from the winding tables, the most suitable conductor diameter may be found. The d.c. resistance may be obtained directly from the tables or, if the actual number of turns required differs appreciably from the number corresponding to the most suitable conductor diameter, the d.c. resistance may be found by interpolation.

The reciprocal of $\tan \delta_{dc}$ is the Q-factor due to the d.c. winding resistance. It is usually impracticable to obtain high Q-factors at low frequencies because (1) in this region $\tan \delta_{dc}$ ($= R_{dc}/\omega L$) is almost the only loss contribution so the Q-factor is proportional to f, and (2) R_{dc}/L can only be decreased significantly by using larger cores or larger effective permeabilities. Economics limit core size, and the need for inductance constancy or adequate adjustment range limits the value of μ_e.

5.7.3 Loss due to eddy currents in the winding conductors

As the frequency increases additional losses occur in the windings due to eddy currents induced in the conductors by the magnetic fields within the winding. These eddy current phenomena are considered in detail in Section 11.4 where graphs may be found which express the functions necessary in the calculation of the eddy current losses.

In the design of inductors both skin effect and proximity effect need to be considered. Both effects depend on the ratio of conductor diameter, d, to penetration depth, Δ.

Skin effect is the tendency for the alternating current to flow near the surface of the conductor. It is due to eddy currents in the conductor which arise from the magnetic field associated with the current in the conductor itself (see Figure 11.5). It causes the resistance to increase by an amount, R_{se}, above the d.c. value. Therefore the a.c. resistance due to skin effect may be expressed by

$$R_{ac} = R_{dc} + R_{se} = R_{dc}(1 + F) \quad \Omega$$

where F is the skin effect factor. It is given as a function of d/Δ in Figure 11.6; Δ may be obtained from Table 11.2 or Figure 11.7.

The combined loss tangent, $\tan \delta_{ac}$, may be evaluated by putting R_{ac} in place of R_{dc} in Eqn (5.15). Alternatively the skin effect loss tangent, $\tan \delta_{se}$, may be calculated separately from

$$\tan \delta_{se} = \frac{R_{dc}F}{\omega L} \quad (5.17)$$

As stated in Section 11.4.2, F is proportional to f^2 at low values of d/Δ, so at low frequencies $\tan \delta_{se}$ is proportional to f. It reaches a maximum when $d/\Delta \simeq 6$. At higher values of d/Δ, F becomes approximately proportional to $f^{1/2}$ so that $\tan \delta_{se}$ approaches proportionality to $f^{-1/2}$ at higher frequencies.

In most inductor designs the skin effect loss is small compared with other losses. If solid conductors are used, then at low frequencies $\tan \delta_{se}$ is very small while at frequencies for which $\tan \delta_{se}$ is near maximum, proximity effect will usually be very much larger than the skin effect. With bunched conductors, i.e. conductors formed as a rope of insulated strands, the skin effect is usually very much reduced. A simply twisted bunch of insulated strands would have the same skin effect as a solid conductor having the same copper cross-section. However, measurements have shown that most bunched conductors behave as though the strands weave between the outer surface and the centre thus preventing skin effect by ensuring a uniform current distribution. This arises because bunched conductors are usually made by twisting together groups of strands which have, in turn, been twisted together. Thus most bunched conductors behave as though the strands are almost perfectly transposed, so the skin effect is negligible.

The major cause of eddy current loss in inductor windings is proximity effect. This is the effect of the magnetic field of the winding as a whole. In most cases this field is substantially perpendicular to the axis of the conductor at any point. Eddy currents flow and return along the length of each conductor in such a way as to oppose the field (see Figure 11.5). This effect is examined in more detail in Section 11.4.3 and it is shown that the associated loss tangent is given by

$$\tan \delta_{pe} = \frac{k_e f N n d^4 G_r}{\mu_e} = \frac{k_E f N n d^4 G_r}{A_L} \quad (5.18)$$

where k_e and k_E are constants for a given core and winding geometry (see Eqn (11.26))
N is the number of turns,

n is the number of strands in the conductor if bunched conductor is used,
d is the diameter of the solid conductor or strand
G_r is a factor that arises from the tendency of the eddy currents to prevent the field from penetrating the conductor; it is a function of d/Δ, see Figure 11.6

The proximity effect constant, k_e or k_E, must usually be determined experimentally for each core type by the method described in Section 11.4.3. It depends on the value of μ_e and also on the cross section of the winding.

Table 5.5 gives values for the proximity effect constant for ferrite cores in the three standard ranges listed in Table 5.2.

At the lower frequencies solid conductor gives obvious advantages; it has a better copper factor and is very much cheaper than equivalent bunched conductor. It is of interest to consider what is the optimum diameter of solid conductor and at what point the change to bunched conductor should be made.[17]

At frequencies where d/Δ is less than about 1.5, the factor $G_r \simeq 1$. Assuming that the skin effect is negligible, the total copper loss tangent is then given by Eqns (5.16) and (5.18)

$$\tan \delta_{dc} + \tan \delta_{pe} = \frac{4\rho_c l_w 10^9}{\omega A_L N \pi d^2} + \frac{k_E f N d^4}{A_L} \quad (5.19)$$

Differentiating with respect to d and equating to zero, the combined loss is minimum when

$$\frac{2\rho_c l_w 10^9}{\omega A_L N \pi d^2} = \frac{k_E f N d^4}{A_L}$$

i.e. when $\tan \delta_{pe} = \tfrac{1}{2} \tan \delta_{dc}$

or the combined
loss tangent $= 1.5 \tan \delta_{dc}$ $\quad (5.20)$

or $R_{AC} = 1.5 R_{dc}$

Under these conditions, the optimum conductor diameter, d_{opt}, is given by

$$d_{opt} = \left(\frac{\rho_c l_w A_L}{\pi^2 f^2 L k_E}\right)^{1/6} = \left(\frac{\rho_c l_w \mu_e}{\pi^2 f^2 L k_e}\right)^{1/6} \quad (5.21)$$

The second version corresponds to the alternative expression in Eqn (5.18). d_{opt} is in mm when l_w is in mm, ρ_c is in Ω mm ($\rho_c \simeq 1.71 \times 10^{-6}\,\Omega$ mm at 20°C), A_L is in nH/turn^2, L is in H, and the proximity effect constants have the units used in Table 5.5.

At the lower end of the frequency range under consideration, the optimum diameter will be larger than that which the core can accommodate. In this case the winding should employ the largest conductor diameter that will reasonably fill the coil former, as is

Table 5.5 The proximity effect constants for ferrite inductor cores. These figures are applicable to Eqn (5.18) when d is in mm and A_L is in nH/(turn)^2

MULLARD 'VINKOR' RANGE					Core dia./ht.				
	10.0/6.8	12.0/7.8	14.0/9.0	18.0/11.2	21.5/13.6	25.4/16.0	29.5/18.8	35.5/22.8	45.0/29.2
μ_c					$k_c \times 10^6$				
40	15	11.3	6.5	5.7	3.7	2.5	1.9	1.4	0.9
63	14	10.9	6.3	5.4	3.7	2.5	1.9	1.4	0.9
100	12.7	10.2	6.0	5.1	3.5	2.4	1.8	1.3	0.9
160	11.2	9.4	5.6	4.6	3.2	2.2	1.6	1.2	0.8
250	9.4	8.3	5.1	3.5	2.7	1.9	1.4	1.1	0.7
400	7.1	6.7	4.2	2.6	2.2	1.6	1.2	0.9	0.6

IEC RANGE OF POT CORES				Core dia./ht.			
	14.1/8.35	18.0/10.55	21.6/13.4	25.5/16.1	30.0/18.8	35.5/21.7	
A_L				$k_E \times 10^6$			
40	11.6	12.1					
63	11.2	11.7	8.2	8.5			
100	10.9	11.3	8.0	8.5	7.7	6.4	
160	10.4	10.3	7.8	8.2	7.7	6.4	
250	10.0	9.5	7.5	7.9	7.3	6.4	
400	9.0	8.3	6.8	7.6	7.0	5.9	
630		6.7	6.1	7.0	6.2	5.5	
1000			4.8	6.0	5.5	5.0	
1600				4.4		4.3	

IEC RANGE OF RM CORES						
	RM5	RM6R	RM6S	RM7	RM8	RM10
A_L				$k_E \times 10^6$		
40	12.1	8.9	8.3	9.0		
63	11.7	9.0	8.3	9.0	7.9	6.8
100	11.2	8.8	8.2	8.8	7.7	6.8
160	10.4	8.6	8.0	8.7	7.5	6.8
250	9.5	8.3	7.7	8.2	7.3	6.5
315	9.2	8.1	7.6			
400	8.7	7.8	7.3	7.5	6.9	6.3
630		6.9	6.4		6.3	5.5
1000						4.5

usual in low frequency inductor designs. At higher frequencies, as the optimum diameter decreases, a point is reached when the optimum diameter is less than the maximum that could be accommodated by the core. It is then advantageous to use the optimum diameter and only partially fill the coil former, preferably spacing the winding away from the centre pole and the fringing flux around the air gap.

If the design frequency is such that the proximity effect reduces d_{opt} to the point where the overall copper loss of solid conductor is too high, then bunched conductors should be used, the simple twist of the strands cancelling the e.m.f.s induced by the flux passing between the strands, see Section 11.4.3.

Eqns (5.19) and (5.21) can be rewritten for bunched conductors having n strands of diameter d:

$$\tan \delta_{dc} + \tan \delta_{pe} = \frac{4\rho_c l_w 10^9}{\omega A_L N n \pi d^2} + \frac{k_E f N n d^4}{A_L}$$

and the optimum strand diameter is

$$d_{opt} = \left(\frac{\rho_c l_w A_L}{\pi^2 f^2 n^2 L k_E} \right)^{1/6} = \left(\frac{\rho_c l_w \mu_e}{\pi^2 f^2 n^2 L k_e} \right)^{1/6} \quad (5.22)$$

With bunched conductor it is almost invariably beneficial to fill the coil former, so for a given core size and a given number of turns there is a relation between d and n which can usually be obtained from manufacturers' winding data. Given this relation, Eqn (5.22) may be used (iteratively) to select the largest strand diameter to give the minimum overall copper loss; the proximity factor G_r can be taken as unity for most bunched conductors operating below about 1 MHz. Tables B4 in Appendix B give data on a range of bunched conductors.

At much higher frequencies, a winding of solid conductor may have a lower proximity effect loss than a similar winding of bunched conductor. When d/Δ is large the eddy currents tend to reduce the field strength inside the conductor, the current distribution becomes more uniform and the eddy current losses are reduced. This is expressed by the rapid decrease of G_r shown in Figure 11.6. The proximity effect loss tangent, $\tan \delta_{pe}$, which is proportional to f when $G_r \simeq 1$, reaches a maximum when $d/\Delta \simeq 3.5$ and subsequently approaches proportionality to $f^{1/2}$ (see Section 11.4.3) and may become very small. A change to stranded (bunched) conductor would then decrease d/Δ and therefore increase G_r. If it happens that this change is accommodated in the straight region of the log G_r/log (d/Δ) graph (see Figure 11.6), i.e. d/Δ for the strand is greater than about 5, then at a given frequency, $G_r \propto d^{-3}$. For a given copper cross section the number of strands, n, is proportional to d^{-2}, so referring to Eqn (5.18) it may be readily deduced that, within the limits of these assumptions, $\tan \delta_{pe} \propto d^{-1}$, i.e. stranding will increase the proximity effect. Thus for high frequency inductors, e.g. operating at frequencies of about 10 MHz or higher it may be advantageous to use solid wire unless very finely stranded conductors are available.

This effect is illustrated in Figure 5.14. The winding resistance and eddy current loss tangents have been calculated for three particular windings on a given typical pot core. The windings have equal numbers of turns and the copper cross-sectional areas are equal so that the inductance and d.c. resistance are the same for each. Winding A has a solid conductor while windings B and C have bunched conductors of different strand diameter. Figure 5.14(a) shows the component loss tangents. It has been assumed that the bunched conductors are perfectly transposed, i.e. the only skin effect and proximity effect is that occurring in the individual strands. It is seen that the skin effect causes the loss tangent to rise above the $R_{dc}/\omega L$ line at high frequencies but even for the solid conductor the contribution is small compared with that of the proximity effect. In practice, the skin effect will be smaller than the value estimated from Eqn (5.17) because the factor F defined in Section 11.4.2 applies to an isolated conductor. In a closely-packed winding the circulating fields within the conductor will be reduced by those in the adjacent conductors. For bunched conductors the skin effect within the

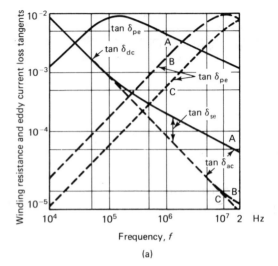

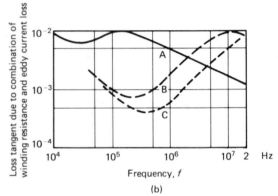

Figure 5.14 Winding resistance and eddy current loss tangents as functions of frequency for three particular windings: (a) component loss tangents; (b) combined loss tangents.
Inductor details:
  18 mm pot core, $\mu_e = 100$, $k_e = 5.1 \times 10^{-6}$
  Number of turns = 19, inductance = 82 μH
Winding A: 0.6 mm solid conductor
  B: 74×0.07 mm bunched conductor
  C: 225×0.04 mm bunched conductor
  Equal conductor
  cross-sectional areas:
  $R_{dc} = 41.6$ mΩ
Note: bunched conductor assumed to be perfectly transposed

strands is quite negligible. The proximity effect loss tangent has a maximum at the frequency which corresponds to $d/\Delta \simeq 3.5$. It is clear that for the solid conductor at the higher frequencies $\tan \delta_{pe} \propto f^{-1/2}$ and in consequence it ultimately falls below the value of $\tan \delta_{pe}$ for the bunched conductors. Figure 5.14(b) shows the total loss tangent due to winding resistance and eddy current losses. The advantage of bunched conductors for the middle frequency range is illustrated and the superiority of solid conductor at the higher frequencies may be seen.

5.7.4 Loss due to stray capacitance

The nature of the stray capacitances associated with an inductor winding has been discussed in Section 5.6 and a table of typical values is given. The effect of the stray capacitance on the loss tangent of an inductor will now be considered. It is assumed that the stray capacitances may be regarded as a single self capacitance, C_s, connected in parallel with the winding. This self capacitance may contribute to the total loss tangent in two ways. The first arises simply from the loss angle associated with C_s and is referred to as dielectric loss and the second is called shunt capacitance loss and arises in certain circumstances due to circulating currents in the LC_s circuit. Each loss will now briefly be considered.

5.7.4.1 Dielectric loss

If the loss angle of the self capacitance is δ_d then the corresponding loss conductance appearing in parallel with the inductance is

$$G_s = \omega C_s \tan \delta_d \qquad (5.23)$$

(note: for practical purposes $\tan \delta_d$ equals the power-factor, $\cos \varphi$, of the self capacitance.)

This loss contributes to a loss tangent, $\tan \delta_{cp}$, to the overall loss tangent of the inductor:

$$\left. \begin{array}{l} \tan \delta_{cp} = \omega L G_s = \omega^2 L C_s \tan \delta_d \\ \qquad = \dfrac{C_s}{C_{res}} \tan \delta_d \end{array} \right\} \qquad (5.24)$$

where C_{res} is the total capacitance required to resonate the inductance L at the frequency $\omega/2\pi$, i.e. $C_{res} = 1/\omega^2 L$.

Normally $\omega^2 L C_s \ll 1$, i.e. the inductor will not be near self resonance at its design frequency. If, however, a specification calls for a high inductance at a high frequency then self resonance may be approached and the degradation of Q-factor due to the dielectric loss would become appreciable. In the limit $\tan \delta_{cp}$ would equal $\tan \delta_d$.

The value of $\tan \delta_d$ depends on the dielectrics associated with the self capacitance; typical values are quoted in Section 5.6. As an example of the effect of dielectric loss, an inductor with the following properties may be considered

Q-factor without dielectric loss $= 500$

$\tan \delta_d = 0.01$

$C_s/C_{res} = 0.04$

By Eqn (5.24)

$\tan \delta_{cp} = 0.04 \times 0.01 = 0.0004$

∴ the Q-factor including this dielectric loss is given by

$$\frac{1}{1/500 + 0.0004} = 417$$

5.7.4.2 Loss due to circulating currents in self capacitance

This occurs in a series resonant circuit when the inductor is shunted by a self capacitance. The equivalent circuit is shown in Figure 5.15; R represents the loss associated with the inductor and C_s is at this stage considered loss-free. The impedance in series with the resonating capacitance, C, is*

$$\frac{R + j\omega L}{j\omega C_s \{R + j(\omega L - 1/\omega C_s)\}} \simeq R(1 + 2\omega^2 L C_s) +$$

$$+ j\omega L(1 + \omega^2 L C_s) \qquad (5.25)$$

This assumes that $R/\omega L$ and $(\omega^2 L C_s)^2$ are negligible compared with unity, i.e. that the inductor Q-factor is reasonably high and that the inductance is far from self-resonance. Both of these conditions are normally

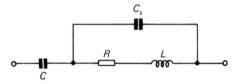

Figure 5.15 Equivalent circuit to illustrate the loss due to circulating currents in the self capacitance

---

Proof:
The impedance of the parallel circuit in Figure 5.15 is

$$\begin{aligned} Z &= (R + j\omega L)/j\omega C_s \{R + j(\omega L - 1/\omega C_s)\} \\ &= \frac{R + j\omega L(1 - \omega^2 L C_s - C_s R^2/L)}{(1 - \omega^2 L C_s)^2 + \omega^2 C_s^2 R^2} \end{aligned}$$

Let $\omega^2 L C_s \ll 1$ so that $(\omega^2 L C_s)^2$ is negligible compared with 1 and let $R/\omega L$ be negligible compared with 1. Now $\omega^2 C_s^2 R^2 = (\omega^2 L C_s)^2 R^2/\omega^2 L^2$ and this is negligible compared with $\omega^2 L C_s$.
Then

$$Z \simeq \{R + j\omega L(1 - \omega^2 L C_s - C_s R^2/L)\}(1 + 2\omega^2 L C_s)$$
$$\simeq R(1 + 2\omega^2 L C_s) + j\omega L(1 - \omega^2 L C_s - C_s R^2/L + 2\omega^2 L C_s$$
$$- 2\omega^4 L^2 C_s^2 - 2\omega^2 C_s^2 R^2)$$

The last two terms are negligible and so is $C_s R^2/L$ since this equals $\omega^2 L C_s \times R^2/\omega^2 L^2$

$$\therefore Z \simeq R(1 + 2\omega^2 L C_s) + j\omega L(1 + \omega^2 L C_s)$$

satisfied in practice. The real part of this expression represents the series loss resistance and this is greater than R by an amount $2R\omega^2 LC_s$. Since the Q-factor of the inductor is more readily estimated than the value of R, this increase in loss resistance is more conveniently expressed as $2\omega^3 L^2 C_s/Q$, i.e. the loss tangent of the inductor is increased by $2\omega^2 LC_s/Q$ or $2C_s/C_{res}Q$.

If the dielectric loss of the self capacitance, $\tan \delta_d$, is also taken into account, by including a resistance in series with the capacitance, an expression corresponding to Eqn (5.25) may be obtained in which the real part is

$$R + (2/Q + \tan \delta_d)\omega^2 L^2 C_s \tag{5.26}$$

assuming $\tan \delta_d \ll 1$

Thus the total loss factor, $\tan \delta_{cs}$, due to self capacitance in a series circuit is given by*

$$\begin{aligned}\tan \delta_{cs} &= (2/Q + \tan \delta_d)\omega^2 LC_s \\ &= (2\tan \delta_L + \tan \delta_d)C_s/C_{res}\end{aligned} \tag{5.27}$$

where Q and $\tan \delta_L$ refer to the inductor.

Using again the example illustrating dielectric loss which was quoted above but this time assuming a series-connected circuit:

$$\tan \delta_{cs} = (2 \times 0.002 + 0.01) \times 0.04$$

$$= 0.00056$$

$$\therefore \quad Q_{tot} = 390$$

Here the major degradation is from the dielectric loss; however had the Q-factor of the inductor been 200 then the two contributions would have been equal.

5.7.5 Core losses

In Chapter 4 the relations between the intrinsic properties of the ferrite material and the properties of a given core of general shape were considered. The concepts of core factors, the effective dimensions of a core and the effective flux density were introduced. Using these principles the contribution of the core loss to the total loss tangent of an inductor may be calculated from the material properties (e.g. see Eqns (4.50) and (4.51)).

It was, however, observed that this approach, while useful for analysis, is limited in practical usefulness because the actual material properties in a given core of intricate shape cannot be accurately known or controlled except by destructive testing. It is therefore common practice to specify the core properties explicitly in terms of parameters measured on the core and controlled during manufacture. This makes the calculation of the core loss contribution very straightforward.

The residual loss tangent of a gapped core at a given frequency is given approximately by

$$\tan \delta_r = \mu_e \left(\frac{\tan \delta_r}{\mu}\right) = \mu_e(\omega L_o G) \tag{5.28}$$

where $\tan \delta_r/\mu$ is the residual loss factor measured on the ungapped core (see Eqns (4.50), (2.18) and (2.14)).

Similarly the eddy current core loss tangent is given by

$$\tan \delta_F = \text{const} \times f \times \mu_e \tag{5.29}$$

(see Eqn (4.55)).

Since both of these loss tangents are functions of frequency and are proportional to μ_e they are almost inseparable in a given core. It is usual to specify, at a given frequency or frequencies, the sum of these loss tangents, i.e. $\tan \delta_{r+F}$. Thus the value of the core loss contribution at very low amplitudes is available directly for the particular core type and effective permeability being considered.

The hysteresis loss may be expressed in a variety of ways and these have been considered in Sections 2.2.6 and 4.2.3. Three commonly used methods will be quoted here by way of illustration. Some manufacturers specify a hysteresis factor such as F_h (see Eqn (4.42)) or core constant η_i (see Eqn (4.44)). These parameters are normalized representations of the series loss resistance and are therefore essentially core parameters. They are related to the hysteresis loss tangent as follows:

$$\tan \delta_h = \frac{F_h \hat{I}\sqrt{L}}{2\pi} \quad (= \eta_i \hat{I}\sqrt{L}) \tag{5.30}$$

Alternatively a hysteresis coefficient (material parameter) may be specified, e.g. the Legg hysteresis coefficient, a, or the hysteresis material constant η_B (see Eqns (4.52) and (2.63)). Then

$$\tan \delta_h = \frac{\mu_e a \hat{B}_e}{2\pi} \quad (= \mu_e \eta_B \hat{B}_e) \tag{5.31}$$

Since these parameters are not constant with ampli-

---

*This result also follows from Eqns (5.25) and (5.24). The increase in resistance due to C_s when this is loss-free is $2R\omega^2 LC_s$. Therefore the loss factor due to this is $\tan \delta_{cs} = 2\omega C_s R$. If now C_s has a loss factor $\tan \delta_d$ then from Eqn (5.24) the combined loss factor due to C_s is

$$\tan \delta_{cs} = 2\omega C_s R + \frac{C_s}{C_{res}} \tan \delta_d$$
$$= (2/Q + \tan \delta_d) C_s/C_{res}$$

since

$$C_{res} = 1/\omega^2 L$$

tude the flux density must also be specified. Another approach is to specify the value of $\tan \delta_h$ at a given flux density for a particular core having a stated effective permeability.

Whichever method is used the required value of hysteresis loss tangent is readily obtained, whether it is needed for its contribution to the total loss tangent or, in another context, to enable the waveform distortion to be estimated.

5.7.6 The combined loss tangents

The total loss tangent is the sum of all the contributory loss tangents set out in the foregoing sections. Table 5.6 summarizes the expressions.

The loss tangent due to the d.c. winding resistance is inversely proportional to f while other loss tangents are approximately proportional to f or f^2. It follows that, other parameters being constant, there will be a frequency at which the total loss tangent is a minimum. Similarly some loss tangents are proportional to μ_e while others are inversely proportional. Thus at a given frequency, other parameters being constant, there will be a particular value of μ_e for which $\tan \delta_{tot}$ will be a minimum.

Figure 5.16 shows a typical variation of the contributory loss tangents with frequency for an arbitrary inductor design. The upper curve is the total loss tangent. At low frequencies the total loss tangent is due almost entirely to the d.c. resistance of the winding. As the frequency rises this contribution falls and, in the vicinity of minimum total loss, it is exceeded by one of the rising losses. The relative magnitude of the contributory loss tangents at any frequency depends on the inductor design. At the higher frequencies the total loss tangent may be dominated by the loss due to the proximity effect, the self capacitance or the core.

In this and the following sections the hysteresis loss contribution has been ignored. This is because most filter inductors normally operate at very low amplitudes. Moreover, the general inclusion of hysteresis loss complicates the presentation because it adds another independent variable; if the amplitude is not

Table 5.6 Summary of loss expressions for inductors
Since the loss tangents are dimensionless the expressions apply equally to SI units and CGS units: the constants k_e, k_E and a have the dimensions [time].[length]$^{-4}$, [time].[length]$^{-4}$.[henries] and [flux density]$^{-1}$ respectively and must be expressed in appropriate units.

Loss tangent due to:	Symbol	Expression	Eqn No.	Page No.
DC resistance	$\tan \delta_{dc}$	$\dfrac{R_{dc}}{\omega L} =$	5.15	175
		$\dfrac{\rho_c l_w 10^9}{\omega A_a F_a A_L}$	5.16	175
Skin effect	$\tan \delta_{sc}$	$\dfrac{R_{dc} F}{\omega L}$	5.17	176
Proximity effect	$\tan \delta_{pe}$	$\dfrac{k_e f N n d^4 G_r}{\mu_e} = \dfrac{k_E f N n d^4 G_r}{A_L}$	5.18	176
Dielectric loss in self capacitance (parallel resonant circuit only)	$\tan \delta_{cp}$	$\omega^2 L C_s \tan \delta_d = \dfrac{C_s}{C_{res}} \tan \delta_d$	5.24	179
Circulating currents in self capacitance (series resonant circuit only)	$\tan \delta_{cs}$	$(2/Q + \tan \delta_d)\omega^2 L C_s$ $= (2 \tan \delta_L + \tan \delta_d) \dfrac{C_s}{C_{res}}$	5.27	180
Residual loss in core	$\tan \delta_r$	$\mu_e \left(\dfrac{\tan \delta_r}{\mu} \right)$ $\Big\} \tan \delta_{r+F}$	5.28	180
Eddy current loss in core	$\tan \delta_F$	const $f \mu_e$	5.29	180
Hysteresis loss in core	$\tan \delta_h$	$\dfrac{F_h \hat{I} \sqrt{L}}{2\pi} (= \eta_I \hat{I} \sqrt{L})$	5.30	180
		$= \dfrac{\mu_e a \hat{B}_e}{2\pi} (= \mu_e \eta_B \hat{B}_e)$	5.31	180

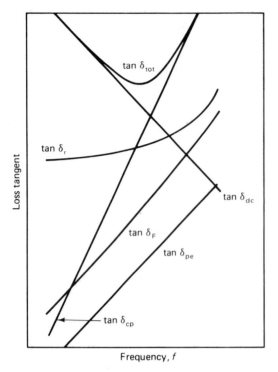

Figure 5.16 Typical variation of contributory loss tangents with frequency

negligible it might have any value. Allowance for hysteresis loss is easily made; the value of $\tan \delta_h$ appropriate to the particular design has only to be added to the otherwise total loss tangent or $1/Q$.

5.7.7 The presentation of Q-factors

It is common practice to present the Q-factor of an inductor as a function of frequency. Such a curve will give some information about the performance attainable with a given core type but since it will depend very much on the inductance, the type of wire and the winding geometry, many Q-curves would be required if the possible performance is to be adequately described. Figure 5.17 shows some typical Q-curves. At the lower frequencies where the winding resistance loss predominates, high Q-factors may be achieved only with large cores and/or high effective permeabilities. Since variability requirements usually set an upper limit to the effective permeability, an inductor core can seldom be used to attain its maximum possible Q-factor at low frequencies. At frequencies between about 40 kHz and 200 kHz the optimum effective permeability with respect to Q-factor matches the variability requirements quite well and the maximum Q-factors may be exploited. At higher frequencies the residual core loss factors are rising (see Figure 3.12) and this causes a reduction in the maximum Q-factors.

Q-curves are often obtained by measurements made on representative inductors. A glance at Table 5.6 will show that the results of such measurements depend on many parameters, e.g. copper factor, conductor diameter, core loss, and these will have manufacturing tolerances. In order to ensure that the measured results are typical, average values of these parameters must be used or one must take the average of measurements made on many samples representing the spread of all the parameters. The alternative is calculation.

Calculation based on the expressions given in Table 5.6 is quite straightforward provided reliable values of certain parameters are available. Such parameters are F_a, k_e, C_s and $\tan \delta_d$; they must usually be determined experimentally and it may be necessary to express the results as functions of conductor diameter, winding geometry or effective permeability. As these parameters are basic to many possible calculations great care in their determination is justifiable. Other parameters, such as conductor diameter and core loss, have known limits and their values may be carefully chosen to be nominal or typical. Given accurate data, calculations based on the loss tangent expressions will provide reliable Q-curves. In general they will represent the typical performance of an inductor more accurately than a few measurements on a random sample.

Individual Q-curves, however obtained, are of limited value as a guide to an inductor designer. To be useful, a Q-curve must represent a core type, effective permeability, inductance and type of conductor corresponding to the designer's requirements. There are a large number of combinations. The solution is to compute the Q-curves for a wide range of possibilities and express the results in the form of Q-contours on inductance-frequency coordinates.[18,19] This form was described by Welsby[20] as a means of representing measured results; it is particularly valuable for the display of a large set of computed Q-factors.

A typical Q-map showing a set of Q-contours is illustrated in Figure 5.18. Such a chart is computed for a particular core with a certain effective permeability and some assumption must be made about the winding, e.g. the type of conductor, the type of coil former and how full it is wound. If the calculation is based on a full winding, with an appropriate copper factor to allow for practical considerations, then a particular conductor diameter will determine the number of turns and the corresponding inductance. The l.h. and r.h. scales represent these parameters in the correct relation. A horizontal line at any inductance will intersect the Q-contours to give the Q-factor as a function of frequency for that inductance.

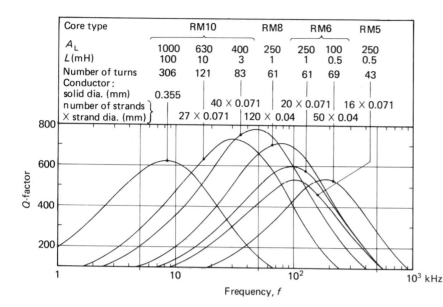

Figure 5.17 Typical Q-curves for some RM cores. Note: A_L is a nominal value without adjuster; L is inductance with adjuster.

For each core type and effective permeability, several Q-maps may be prepared representing different types of conductor, solid and bunched. In this way a set of Q-maps may be built up representing the performance of all inductors falling within the boundaries of the computation. Such graphs make the design of an inductor a relatively simple matter of inspection and choice, and they also provide a valuable indication of the influence of the design parameters on the Q-factor.

5.7.8 The effect of design parameters on Q-factor

Figure 5.18 is a computed Q-map for an RM6 core having an effective permeability of 161 ($A_L = 250$) and a full winding of solid, enamel covered conductor. In this and the following Q-map (Figure 5.20) the hysteresis loss is assumed to be negligible, and the loss due to the circulating currents in the self-capacitance has been omitted as these charts are intended to represent only the performance of inductors in parallel resonant circuits.

Figure 5.19 gives the contributory loss tangents as functions of frequency corresponding to three inductance values from Figure 5.18. In these and the following graphs the residual and eddy current losses have been combined into a single loss tangent, $\tan \delta_{r+F}$.

Figure 5.19(a) is for a high inductance value. The winding has many turns of fine wire. Only two of the loss tangents are of importance, that due to d.c. winding resistance at low frequencies and that due to dielectric loss in the self capacitance at high frequencies. The emphasis of the latter loss is not the result of high self capacitance but due to the high inductance leading to the approach of self-resonance at a fairly low frequency (see Eqn (5.24)).

Figure 5.19(b) is for a medium inductance value approximately corresponding to the peak of the Q-surface. The maximum Q-factor is determined mainly (50%) by the d.c. winding loss with the core loss, proximity effect loss and the dielectric loss of the self capacitance accounting for the remainder in approximately equal proportions. Again the high-frequency Q-factor is due mainly to the dielectric loss.

Figure 5.19(c) is for a low inductance. The maximum Q-factor depends principally on the d.c. winding resistance and the proximity effect. The curvature of the latter function, due to G_r in Eqn (5.18) falling below unity, is very marked. It causes a distortion of the $\tan \delta_{tot}$ curve and results in the dielectric and core losses rising to predominance at high frequency.

As a result of this analysis the reason for the shape of this particular Q-surface becomes apparent. The l.h. side is determined by the d.c. winding resistance. Because the copper factor is not very sensitive to the number of turns, R_{dc}/L and $\tan \delta_{dc}$ are also insensitive. This is apparent in the graphs of Figure 5.19, and is the reason why the l.h. contours of Figure 5.18 are almost parallel with the L axis. The top r.h. slope of the surface is due to the high inductance values emphasizing the dielectric loss of the self capacitance while the lower r.h. slopes depend on the proximity effect. Skin effect has negligible effect over the range of this analysis. The peak of the surface depends mainly on d.c. winding resistance with the core loss, dielectric loss, and proximity loss each making a relatively small contribution. It is clear, therefore,

184 Inductors

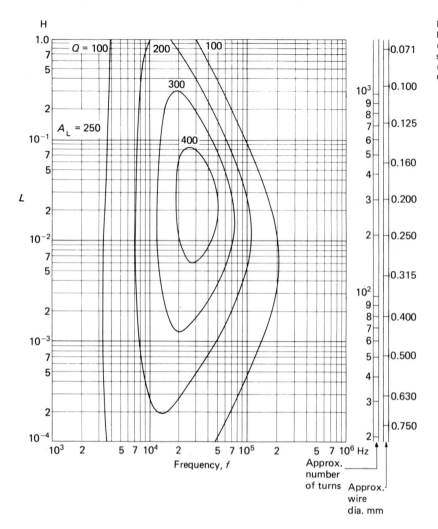

Figure 5.18 Q-map for an RM6R core, $A_L = 250$ ($\mu_e = 161$), fully wound with solid enamelled conductor. (Courtesy Philips Components)

that with $\mu_e = 161$ the core loss is not greatly contributing to the total loss, and higher Q-factors could be attained with higher values of effective permeability. Variability or the need for sufficient adjustment range usually sets a limit to the increase in μ_e under these circumstances.

Figure 5.20 shows another computed Q-map. It is for the same RM6 core and the same effective permeability (161) but this time it is for a full winding of bunched conductors consisting of insulated strands of 0.04 mm diameter. The number of strands determines the overall diameter and therefore determines the number of turns and the inductance. Compared with the previous Q-map, it is seen that the use of bunched conductors has resulted in the region of maximum Q-factor moving to a higher frequency and lower inductance; the maximum value of Q-factor is much higher.

Figure 5.21 gives the contributory loss tangents as functions of frequency, again corresponding to three inductance values from the Q-map. The following general observations may be made. The poorer copper factor of bunched conductors has appreciably raised the position of the $\tan \delta_{dc}$ line but stranding has made proximity effect negligible over the whole range.

Figure 5.21(a) is for an inductance corresponding approximately to that of Figure 5.19(b). The total loss depends mainly on d.c. winding resistance and dielectric loss in the self capacitance while the residual loss in the core significantly affects the minimum value.

Figure 5.21(b) is for an inductance that is typical of the values used at about 100 kHz. The total loss curve is almost exclusively composed from the d.c. winding loss and core loss curves with the dielectric loss making a small contribution at the higher frequencies.

Figure 5.21(c) is for a low inductance. Here the

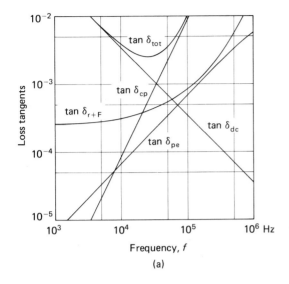

(a)

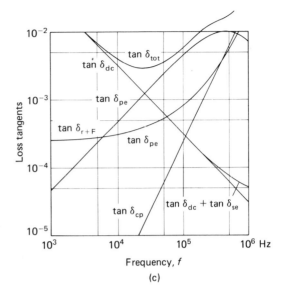

(c)

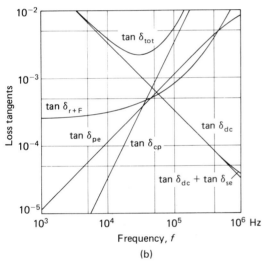

(b)

Figure 5.19 Contributory loss tangents at three sections of Figure 5.18: (a) at $L=91$ mH, 583 turns of 0.140 mm diameter solid enamelled copper wire; (b) at $L=36$ mH, 367 turns of 0.180 mm diameter solid enamelled copper wire; (c) at $L=2.7$ mH. 100 turns of 0.355 mm diameter solid enamelled copper wire

dielectric loss has become negligible over the entire frequency range and the total loss curve depends only on the d.c. winding loss and the core loss.

Returning to the Q-map in Figure 5.20, the l.h. side is again dependent mainly on d.c. winding losses. In this region the Q-factors are less than the corresponding ones in the Q-map for solid conductors because windings of bunched conductors have smaller copper factors. Again the slope at the top r.h. side of the surface is due to dielectric loss in the self capacitance. The higher contours and the slope at the bottom r.h. side depend mainly on the core loss.

An increase in effective permeability will lower the winding loss and increase the core loss so that the peak of the surface will move to lower frequencies. A decrease in effective permeability will tend to have the reverse effect but because the core loss rises rapidly at frequencies above about 100 kHz the maximum Q-factors will be decreased. To get the best Q-factors at higher frequencies it would be necessary to use a lower permeability grade of ferrite; the residual loss factor would be greater at low frequencies but the frequency at which it would begin to rise rapidly would be higher, see Figure 3.12.

A decrease in core size would increase the d.c. winding loss tangent for the same effective permeability. This would tend to increase the frequency at which the peak occurs but rising core loss would depress its height.

This type of analysis could be extended to different winding configurations, e.g. sectioned windings, and could provide Q-maps which allow for the hysteresis loss corresponding to a particular amplitude or Q-maps representing lower limits of performance.

Reference to Figure 5.21 suggests that, for the majority of practical inductor designs using bunched conductors, the minimum loss tangent is determined mainly by the sum of the winding d.c. loss tangent and the core loss tangent.[21] When this is so,

$$\tan \delta_{tot} = \frac{R_{dc}}{\omega \mu_e L_o} + \left(\frac{\tan \delta_{r+F}}{\mu}\right)\mu_e \quad (5.32)$$

from Eqn (5.15) in which $L_o = L/\mu_e = \mu_o N^2/C_1$ and is a

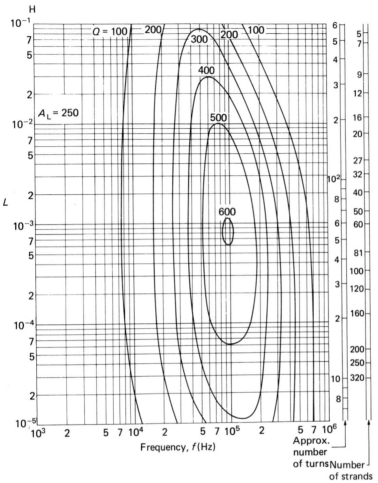

Figure 5.20 Q-map for an RM6R core, $A_L=250$ ($\mu_e=161$), fully wound with bunched conductor having strand diameter=0.04 mm. (Courtesy Philips Components)

constant for a given core and winding, and from Eqn (5.28) in which $\tan \delta_r$ is replaced by $\tan \delta_{r+F}$.

By differentiation with respect to μ_e, the total loss tangent will be a minimum when

$$\frac{R_{dc}}{\omega \mu_e L_o} = \left(\frac{\tan \delta_{r+F}}{\mu}\right) \mu_e \qquad (5.33)$$

and this occurs when

$$\mu_e^2 = \frac{R_{dc}}{\omega L_o (\tan \delta_{r+F})/\mu} \qquad (5.34)$$

$$\therefore Q_{max} = \frac{\omega \mu_e L_o}{2 R_{dc}} = \frac{\omega L_o}{2 R_{dc}} \left[\frac{R_{dc}}{\omega L_o (\tan \delta_{r+F})/\mu}\right]^{1/2}$$

i.e. $Q_{max} = \left(\frac{\omega L_o}{R_{dc}}\right)^{1/2} \times \frac{1}{2\left(\frac{\tan \delta_{r+F}}{\mu}\right)^{1/2}} \qquad (5.35)$

At a given frequency, the term involving core loss is a constant for a given ferrite material and the first term depends only on the core and winding geometry. From Eqn (5.4), L_o/R_{dc} is proportional to (length)^2, so Q_{max} at a given frequency is proportional to the linear dimensions of the core and winding assembly.

Alternatively, for a given core shape and size, Q_{max} at a given frequency is inversely proportional to the square root of the core material loss factor.

In practice, the Q-factor requirements for a particular class of application tend to remain constant so from Eqn (5.35) it follows that for a specified value of Q_{max} at a given frequency the linear dimensions of a ferrite inductor are proportional to $(\tan \delta_{r+F}/\mu)^{1/2}$.

Developments in ferrite materials have progressively reduced the core loss factor and this has led to a corresponding reduction in core size for a given class of application.

At high frequencies, e.g. 2 to 20 MHz, the highest

Inductors 187

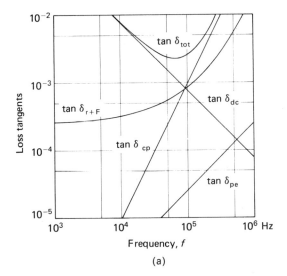

(a)

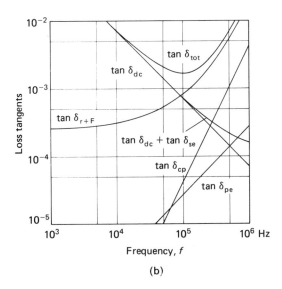

(b)

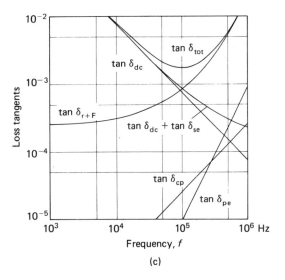

(c)

Figure 5.21 Contributory loss tangents at three sections of Figure 5.20: (a) at $L=20$ mH, 273 turns of 12×0.04 mm diameter bunched conductor; (b) at $L=0.93$ mH, 59 turns of 60×0.04 mm diameter bunched conductor; (c) at $L=0.21$ mH, 28 turns of 120×0.04 mm diameter bunched conductor

5.8 Waveform distortion and intermodulation

The way in which waveform distortion and intermodulation arise in a magnetic material is considered in Section 2.2.7 and the results of some analyses are given in terms $\tan \delta_h$ (see Eqns (2.70)–(2.72)). For gapped cores the e.m.f distortion is reduced by the ratio μ_e/μ (see Eqn (4.62)). For some ferrites directly measured e.m.f. distortion data are available (see Figure 3.18) and in such cases it is usually preferable to use this data rather than infer the distortion from the hysteresis parameters.

In a network such as a filter it may not be easy to calculate the overall effect of distortion generated by a particular inductor operating at a voltage determined by its resonance condition. However, with modern circuit analysis CAD it is, in principle, possible to input the hysteresis or distortion data discussed above and obtain the signal distortion levels as a part of the overall network performance.

Intermodulation is most troublesome when large differences in amplitude exist between two separate signals in one circuit, e.g. at the junction of a pair of two-wire line filters where the send signal may be at a level of $+17$ dB and the receive signal may be at a level 60 dB lower than this. The intermodulation products would typically be required to be 60 dB below the receive signal, i.e. -120 dB with respect to

Q-factors are obtained with windings that occupy only a fraction of the available space, thus the relation between conductor diameter and inductance is lost. In this case it is possible to compute a Q-map in which each point represents the highest Q-factor obtainable over a range of conductor sizes and types, i.e. the surface is the envelope of many possible surfaces. The optimum conductor size and type may be indicated by an overlaid set of curves.[22]

the send signal that generates them. A typical specification for the inductors at the common ends of the filter would require hysteresis factors (F_h) of about 0.1 compared with 4 to 30 obtained on normal ferrite inductor cores. Low values of F_h are obtained by using very large air gaps.

5.9 References and bibliography

Section 5.1
1. TALLEY, D. *Basic carrier telephony*, Hayden Book Co., Rochelle Park, New Jersey (1977)
2. ANDREWS, M. J. Frequency-division multiplex line and terminal equipment, *Post Office elec. Engrs. J.*, **74**, 260 (1981)
3. Ferroxcube in I.F. transformers for F.M., *Matronics*, No. 3, 72 (1953)

Section 5.2
4. Dimensions of square cores (RM-cores) made of magnetic oxides and associated parts, *Int. Electrotechnical Commission, Publication 431*, Geneva (1983)
5. Dimensions of pot cores made of ferromagnetic oxides for use in telecommunication and allied electronic equipment, *British Standards Inst., BS4061*, London (1966)
6. Dimensions of pot-cores made of magnetic oxides and associated parts, *Int. Electrotechnical Commission, Publication 133*, Geneva (1985)

Section 5.4
7. POSTUPOLSKI, T. and OKONIEWSKA, E. Calculation of adjustment of cored inductors, *Electron Technology*, **9**, 133, Institute of Electron Technology, Warsaw (1976)
8. LONGHURST, C. E. Performance of coils with fractional turns when used with Vinkor assemblies, *Mullard tech. Commun.*, **6**, 183 (1962)
9. BANZI, F. J. Higher Q from pot core inductors, *IEEE Trans.*, **PHP-13**, 371 (1977)
10. NEWCOMB, C. V. A new adjuster for Mullard ferrite inductor cores, *Mullard tech. Commun.*, **11**, No. 107, 142 (1970)
11. Inductor and transformer cores for telecommunications, Part 1: Generic specification, *Int. Electrotechnical Commission, Publication 723-1*, Clause 13.6.2, Geneva (1982)

Section 5.5.2
12. SNELLING, E.C. Disaccommodation and its relation to the stability of inductors having ferrite cores, *Mullard tech. Commun.*, **6**, 207 (1962)

Section 5.5.3
13. NEWCOMB, C. V. The influence of adjuster mechanisms on the temperature coefficient of ferrite-core assemblies, *Mullard tech. Commun.*, **12**, No. 117, 202 (1973)
14. POSTUPOLSKI, T. and OKONIEWSKA, E. Thermal instability of ferrite inductors equipped with adjustment system, *Electron Technology*, **9**, 121, Institute of Electron Technology, Warsaw (1976)

Section 5.5.4
15. NEWCOMB, C. V. and SNELLING, E. C. Analysis of variability in cored inductors, *Philips tech. Rev.*, **28**, 184 (1967)

Section 5.6
16. LONGHURST, C. E. Stray capacitances of coils wound on Mullard Vinkor pot-core assemblies, *Mullard tech. Commun.*, **5**, 218 (1961)

Section 5.7.3
17. SNELLING, E. C. Ferrites for power applications, Paper No. 6 in Colloq. on Trends in forced commutation components, *IEE, Digest No. 1978/3*, London (1978)

Section 5.7.7
18. MEYER, R. and SIBILLE, R. Essais de présentation des charactéristiques principales d'un pot en ferrite, *Onde élect.*, **42**, 638 (1962)
19. SNELLING, E. C. Q contour charts, *Philips tech. Rev.*, **28**, 186 (1967)
20. WELSBY, V. G. *The theory and design of inductance coils*, Macdonald and Co., London, 2nd Edn (1960)

Section 5.7.8
21. SNELLING, E. C. and GILES, A. D. *Ferrites for inductors and transformers*, p. 94, Research Studies Press, Letchworth, Hertfordshire, England; distrib. by John Wiley & Sons, New York (1983)
22. Mullard Technical Handbook, Book 3, Part 4, Section G (1979)

General
MACFADYEN, K. A. *Small transformers and inductors*, Chapman and Hall, London (1954)
VAN SUCHTELEN, H. Tolerances and temperature coefficient of coils with ferroxcube slugs, *Electron. Applic. Bull.*, **14**, 27 (1953)

6 High-frequency transductors

6.1 Introduction

Because the relation between the flux density and the field strength in a magnetic material is essentially non-linear it follows that the permeability depends on the magnitude of the field strength. In Section 2.1 the main features of the B–H loop are described and it is seen that a number of different permeabilities may be distinguished. Of these it is the incremental or reversible permeability that is of interest in transductor design. Figure 3.6 shows the incremental permeability as a function of field strength for simple magnetic circuits composed of typical ferrites. It is seen that in the absence of an air gap the incremental permeability varies from a value equal to the initial permeability when the superimposed static field is small, to a value approaching unity when the static field is large enough to saturate the material.

This phenomenon has long been used as a means of controlling current at power frequencies. In its simplest form the current is controlled by a choke, the reactance of which may be varied by passing a direct current through a separate control winding. More elaborate forms of control by this method are described in the literature on magnetic amplifiers.

The extension of the method to high-frequency circuits was described in 1938[1] but the limitations of the available high-frequency core materials prevented its practical use. Laminated cores gave a large range of control but the losses were too high, whereas powdered iron cores which had relatively low h.f. losses were unsatisfactory because of the small range over which it was possible to vary the permeability.

When ferrites became available these difficulties were greatly reduced. Ferrites have low losses and relatively high permeabilities which can be readily varied with a polarizing field. With these materials came the possibility of controlling the inductance in h.f. circuits electrically, remotely and automatically[2-4]. A wide variety of applications have been described. They include:

(1) Remote tuning of receivers, e.g. manually in communication receivers or electronically in panoramic receivers[5],
(2) Remote or automatic antenna matching (or tuning)[6],
(3) Automatic frequency control,
(4) Frequency modulation in f.m. generators, frequency sweep generators and telemetering systems[7],
(5) H.F. switching and attenuation[8],
(6) Variable coupling transformers,
(7) Variable frequency filters[9],
(8) Control of frequency in particle accelerators such as synchrotrons[10-11],
(9) Television raster correction[12].

6.2 General mode of operation

6.2.1 The magnetic circuits

The basic magnetic and electrical circuits are shown schematically in Figure 6.1. The h.f. core is situated between the poles of a magnetic yoke. Usually the h.f. core and the yoke are separate items but in some designs they may be formed from the same piece or pieces of ferrite. A control winding on the yoke applies a field to the h.f. core. The associated flux divides equally between the two sides of the core. The signal winding, the inductance of which it is proposed

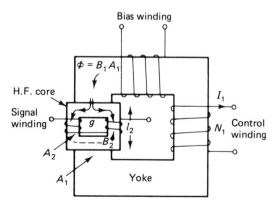

Figure 6.1 Schematic diagram of a transductor

to control, is wound on the h.f. core in two series-aiding halves, one on each side. This arrangement is used to avoid magnetic coupling between the h.f. winding and the windings on the yoke; the flux linkages between the yoke and the two halves of the h.f. winding ideally cancel. A high degree of isolation is essential because otherwise the impedance associated with the control winding would be reflected into the h.f. circuit giving rise to additional losses and self capacitance. The bias winding is not always necessary; it permits the h.f. core to be set to a selected working point on the B–H curve, but sometimes the mean current in the control winding has the appropriate value. In some designs bias is achieved by means of a permanent magnet; this reduces the operating power. Practical arrangements will be discussed later in this section.

The operation may be explained qualitatively with reference to Figure 6.2. In (a) a typical relation between the ampere-turns applied to the yoke and flux in the yoke is shown. If the steady or bias m.m.f. is represented by $(NI)_b$ and the control m.m.f. is varied over a total excursion $\Delta(N_1 I_1)$ then the flux, Φ, in the yoke varies over the range $\Delta\Phi$. Assuming no leakage flux, the sum of the control fluxes in the two arms of the h.f. core equals the flux in the yoke. If the subscript 1 refers to the yoke and 2 refers to either of the side limbs, the following simplified analysis of the control conditions may be written:

$$B_2 = \frac{B_1 A_1}{2 A_2} \qquad (6.1)$$

$$N_1 I_1 = \frac{\Phi}{\mu_o}\left(\frac{l_1}{\mu_1 A_1} + \frac{l_2}{2\mu_2 A_2}\right) \quad \text{A} \qquad (6.2)$$

where N_1 is the number of turns on the control winding

$$N_1 I_1 = \frac{B_1 A_1}{\mu_o}\left(\frac{l_1}{\mu_1 A_1} + \frac{l_2}{2\mu_2 A_2}\right)$$

$$\therefore \quad B_2 = \frac{\mu_o N_1 I_1}{\frac{2 l_1 A_2}{\mu_1 A_1} + \frac{l_2}{\mu_2}} \quad \text{T} \qquad (6.3)$$

This relation is of limited usefulness because (1) it ignores the reluctance of the horizontal branches of the h.f. core, (2) μ_2 is a function of B_2 and (3) the permeability of the h.f. core varies along the local magnetic path length so there is no simple relation between B_2 and the effective incremental permeability of the h.f. core. However in practice it is often sufficient to estimate the limiting conditions. When the control current is small the h.f. core has a permeability that is near the initial (un-polarized) permeability. As the control current increases and the h.f. core progresses towards saturation, the yoke reluctance, in an efficient device, should become negligible. Eqn (6.3) then reduces to

$$B_2 = \frac{\mu_o \mu_2 N_1 I_1}{l_2} \quad \text{T}$$

or $\quad H_2 = \dfrac{B_2}{\mu_o \mu_2} = \dfrac{N_1 I_1}{l_2} \quad \text{Am}^{-1} \qquad (6.4)$

So far only the control magnetic circuit has been considered. Figure 6.2(b) shows the corresponding conditions in the h.f. core. Such a diagram requires careful interpretation. An h.f. signal corresponding to a field strength excursion ΔH_s acting around the h.f. magnetic circuit will produce a corresponding flux density ΔB_s. The effective incremental permeability will be complicated by the varying degree of polarization round the h.f. magnetic circuit. Further, ΔH_s and ΔB_s refer to a different magnetic circuit to that which determines the steady or control conditions. Indeed, the major loop drawn in this diagram cannot refer directly to the h.f. magnetic circuit as such but only to those parts of it influenced by the control field. And the conditions in these depend to some extent on the conditions in the yoke.

To obtain a qualitative picture of the operation these difficulties may be ignored. The steady field strength may be regarded as being controlled over the extent of the major loop, and the minor h.f. loop will change its slope successively as shown. Thus the effective incremental permeability ($=\Delta B_s/\mu_o \Delta H_s$) of the h.f. core may be varied and the inductance of the signal winding will vary accordingly. Figure 6.2(c) shows a typical relation between μ_Δ and $N_1 I_1 / l$. Within the limitations mentioned above the quantitative relation for a particular core may be estimated from the graphs showing μ_Δ as a function of NI/l for the (ungapped) material. Such graphs are shown in Figure 3.6. When the control current (plus the bias current, if any) approaches zero the effective incremental permeability of the h.f. core will approach the initial permeability and the inductance of a winding

---

(6.2) $\quad N_1 I_1 = \dfrac{10\Phi}{4\pi}\left(\dfrac{l_1}{\mu_1 A_1} + \dfrac{l_2}{2\mu_2 A_2}\right) \quad$ A

(6.3) $\quad B_2 = \dfrac{0.4\pi N_1 I_1}{\frac{2 l_1 A_2}{\mu_1 A_1} + \frac{l_2}{\mu_2}} \quad$ Gs

(6.4) $\quad H_2 = \dfrac{B_2}{\mu_2} = \dfrac{0.4\pi N_1 I_1}{l_2} \quad$ Oe

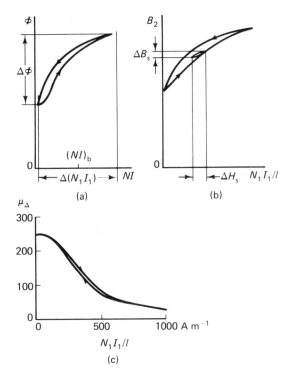

Figure 6.2 Curves illustrating the operation of a transductor. (a) The relation between the control flux and the control m.m.f. (b) The relation between the flux density and field strength in the h.f. core due to currents in the signal winding, as a function of the operating point of the h.f. core. (c) The incremental permeability of a typical h.f. core as a function of the control field strength

on that core may be calculated. When the control current is sufficient to saturate the vertical limb of the h.f. core the effective incremental permeability will approach unity. However, the shape of the curve in Figure 6.2(c) and the curves in Figure 3.6 suggests that the extra inductance range obtainable from complete saturation is not worth the large increase in control current that is necessary.

So far the hysteresis effects illustrated in Figure 6.2 have not been introduced into the discussion. The hysteresis shown by the composite control magnetic circuit, Figure 6.2(a), is partly due to the yoke and party due to the h.f. core. Often the h.f. core is made of nickel zinc ferrite which may have a large amount of magnetic hysteresis. The overall hysteresis may be minimized by making the yoke of low hysteresis material such as manganese zinc ferrite. Other methods depend on the electronic regulation of the control current and will be briefly mentioned later. The control magnetic hysteresis is reflected in the effective B–H curve of the h.f. magnetic circuit, Figure 6.2(b), and the incremental permeability becomes a two-valued function of N_1I_1/l, Figure 6.2(c); where l is the effective path length of the control magnetic circuit.

6.2.2 Temperature dependence

An important consideration in the design of a transducer is the variation of temperature coefficient of inductance of the signal inductor as the control range is traversed. Figure 6.3 shows the B–H relation at two temperatures. Assuming the temperature coefficient of the initial permeability is positive the curve for the higher temperature will have the greater slope at the origin. However, at the higher temperature the material will saturate at a lower flux density so the two curves will cross. If the slope of the curve is taken as a rough indication of the magnitude of the incremental permeability then it is clear that at H_1 the temperature coefficient is positive, at H_3 it is negative and at H_2 it has an average value of zero.

In Figure 3.9 the incremental permeability of a number of ferrites is shown as a function of temperature for increasing values of polarizing field. The polarizing field depresses the permeability and also lowers the temperature at which it peaks. Thus if it is required to have a zero temperature dependence at a particular temperature the appropriate static field strength will be the value that places the peak at that temperature.

Sometimes magnetic polarization has been used for the express purpose of reducing the temperature coefficient of an inductor but usually this phenomenon is merely a by-product of operation of a transductor. If the required range of control is small, e.g. where a substantially linear inductance variation is required, then it is usually possible to arrange that the average temperature coefficient approaches zero over the control range. More often a large range of control

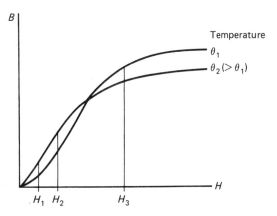

Figure 6.3 Typical B–H curves at two different temperatures, illustrating the way in which the temperature dependence varies with H

6.2.3 Core loss

The application of an external magnetic field increases the internal magnetic anistropy and this moves the ferromagnetic resonance to higher frequencies. In addition, the static field reduces the number of domain walls in the material and this tends to reduce phenomena associated with domain wall movement, e.g. hysteresis loss and to some extent residual loss (see the caption to Figure 3.12). Thus there will be a general lowering of the tan δ curve as the steady field is increased. Experimental results are shown in Figure 3.14.

In estimating the total loss of an inductor having an ungapped core the loss contribution from the core is represented by the core loss tangent. It is seen that this progressively falls as the steady field is increased. Therefore it is to be expected that the Q-factor of the h.f. inductor will rise as the field is increased, and in practice this is found to be true. However, at the same time as the core loss is decreasing the permeability and inductance are decreasing so that the winding loss tangent ($=R_{AC}/\omega L$) increases. This causes the Q-factor to peak at an intermediate value of polarizing field and then to fall as the field is further increased. If however the inductor is resonated with a fixed capacitance, C, the resonant frequency will increase as the inductance falls; the winding loss tangent is then given by $R_{AC}\sqrt{(C/L)}$. Often this loss is negligible in h.f. transductors.

The behaviour of the loss factor, $(\tan \delta)/\mu$, which is the usual criterion of core quality, may also be mentioned here although it is not very relevant to the design unless the h.f. core is gapped. Since both numerator and denominator of this factor decrease with increasing static field the overall effect depends on which decreases at the greater rate. At the higher frequencies tan δ initially has the greater rate of decrease and the loss factor decreases as the field rises. When it reaches an intermediate value the rate of decrease of μ becomes larger; the loss factor then shows a minimum. (See also Figure 3.14.)

6.2.4 General

The overall performance of a h.f. transductor is often expressed in graphical form, e.g. as in Figure 6.4. Such graphs show the frequency at which the signal inductor resonates with a given capacitor, as a function of control current; the Q-factors are indicated at intervals along the curve.

In choosing the best ferrite for the signal core account must be taken of the required range of inductance control, the operating frequency range and the required Q-factors. Temperature coefficient and hysteresis may also be important. Factors affecting the choice of ferrite are discussed at the end of Section 6.4.

Throughout the foregoing discussion it has been assumed that the polarizing field is as far as possible directed along the same magnetic axis as the h.f. field and all the data relate to this mode of operation. It is quite possible to make a transductor in which the static field is everywhere perpendicular to the h.f. field. The behaviour, while similar to the parallel case, is generally inferior; in particular the range of inductance control is usually smaller. However, this mode of operation is useful at very high frequencies where the core may be operated at a static field strengths that take it above the ferromagnetic resonance. Using planar hexaferrite, see Section 1.1, good performance at frequencies as high as 500 MHz has been reported[13].

6.3 Applications and performance

The basic characteristics described in the previous section may now be put into their proper context by reference to practical examples drawn from the literature and elsewhere. The examples have been chosen to illustrate the wide range of applications and to give data on the performance that may be expected.

6.3.1 Experimental models

The author and some colleagues[14] have investigated a number of ferrites and configurations to assess their suitability as cores for transductors to operate at frequencies between 3 and 25 MHz. Figure 6.4 shows some of the results obtained. The most interesting configuration was the 'picture frame' core made of nickel zinc ferrite placed between the central poles of an otherwise standard ferrite pot core. The control winding is in the position of the normal pot core winding and because the pot core is designed to make efficient use of the winding space a good sensitivity was achieved. In the example illustrated the signal inductor was resonated with a capacitance of 100 pF and the resonant frequency was variable from 7 to 17 MHz with a Q-factor of about 100. The maximum control winding power was 0.4 W and the overall diameter was 21 mm.

6.3.2 Antenna tuning

The most obvious application is the remote or automatic tuning of r.f. circuits in communication systems. Newall, Gomard and Ainlay[6] have described

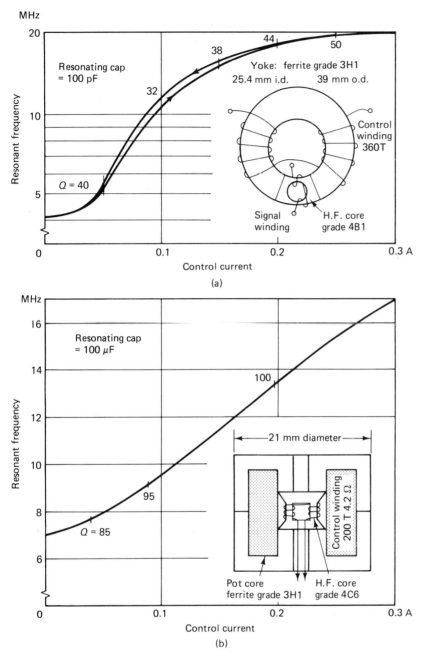

Figure 6.4 The construction and performance of two experimental transductors

the use of a transductor for the remote adjustment of the antenna matching circuit in a 12 W marine transmitter operating in the 2 to 4 MHz band. Figure 6.5 shows the practical construction, the performance and the basic circuit in which it was used.

The signal inductor uses a ferrite pot core and is mounted between the poles of a silicon iron yoke which carries the control windings. Copper foil is placed between the ferrite and the poles to prevent the h.f. flux from entering the iron and thereby increasing the loss in the signal inductor. The coaxial cable from the antenna is connected to the tank circuit (T)

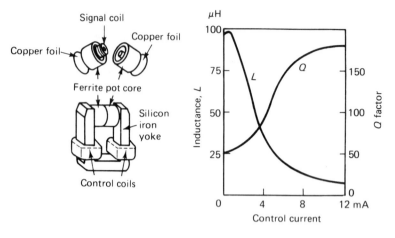

Figure 6.5 The construction, performance and circuit of a remotely controlled antenna matching system. (After Newhall et al.[6])

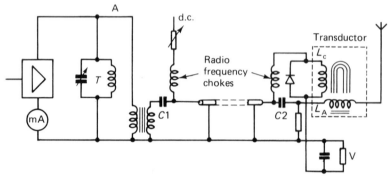

through a wide-band ferrite-cored transformer.

The tank circuit is tuned with the circuit broken at A, the milliammeter indicating the resonance by a minimum in the current. The circuit is then restored, and if the secondary circuit is reactive the transmitter will be detuned. The direct current to the transductor control winding, L_C, is then varied by means of a rheostat until the h.f. inductor, L_A, has the correct value to tune the antenna and the secondary circuit becomes resistive. The correct adjustment is indicated by the restoration of tank circuit resonance, the tank capacitance not entering into this adjustment. The radio frequency chokes isolate the control circuit from the radio frequency while the capacitors C_1 and C_2 block the d.c. from the r.f. circuits. The diode limits the back-e.m.f. developing across the control winding when the d.c. is switched off. The thermistor, V, is arranged to compensate the temperature coefficient of the signal inductor L_A through the medium of the control current.

It is obviously possible to make this adjustment automatic by linking the control current adjustment to the tank circuit tuning.

6.3.3 Control of a synchrotron field

This application concerns the control of the frequency of the accelerating field in a proton synchrotron. As the protons accelerate they are held in an orbital track by the increasing intensity of the transverse magnetic field. Because the track radius is constant the accelerating field must increase in frequency to keep step with the speed of revolution. For a particular device described by Pressman and Blewett[10] the range of control required is 343 to 4181 kHz (1:12.2) with 0.3% stability.

The construction chosen was a ferrite toroid with a slot cut in the side to receive the h.f. winding. Figure 6.6 shows the arrangement and the performance achieved. By winding the h.f. inductor in two halves, one on each limb formed by the slot, coupling between the h.f. and control winding is avoided. The core and h.f. winding were enclosed in an annular box which carried the toroidal control winding. It was necessary to pump oil at constant temperature through the box in order to achieve the necessary temperature stability.

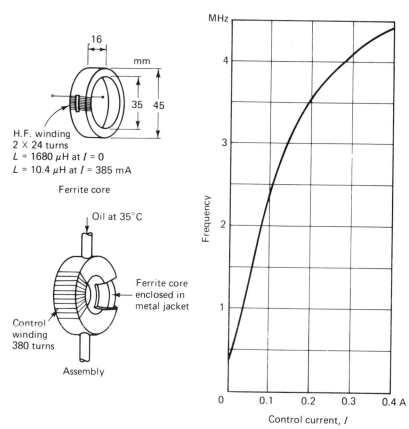

Figure 6.6 The construction and performance of a transductor controlling the frequency of the accelerating field in a proton synchroton. (Courtesy Pressman et al.[10])

6.3.4 Dynamic raster correction in colour television

In colour television it is necessary to design the deflection coils to give optimum spot quality and such a design gives rise to some pin cushion distortion of the raster, see Figure 6.7(a). This distortion may be corrected by modulating the line-scanning current with a parabolic waveform at the same frequency as the field (or frame) scan and modulating the field-scanning current with a parabolic waveform at the line-scanning frequency, see Figure 6.7(b).

One method of achieving this raster correction has been described by Wölber[12], it uses a transductor between the line and field deflection circuits. Figure 6.7(c) shows a simplifed circuit which illustrates the basic principles. The transductor uses a double aperture ferrite core, Figure 6.7(d). The centre limb carries a winding which is connected in series with the field deflection coils so that the core is taken towards saturation at each end of the field scan. The two outer limbs carry windings series-connected so that, when the core is unsaturated, there is no coupling with the winding on the centre limb. Thus the impedance of the series-connected windings decreases towards each end of the line scan and, because they shunt the line deflection coils, they modulate the line deflection current in the required manner.

As the field current increases towards the saturation value, the balance of flux in the core is disturbed because in one outer limb the fluxes due to the line and field current add together while in the other limb they are in opposition. Thus one of the outer limbs is driven further towards saturation than the other. It follows that there is a coupling between the line and field circuits which varies from zero at the centre of the field scan to equal and opposite maxima at each end of the field scan. If the voltage due to this coupling were injected directly in series with the field deflection circuit the correction would be in the wrong sense. This is overcome by connecting a capacitor across the centre-limb winding such that it is resonated below the line frequency. The inductor L_t is introduced to control the correction required. The resultant current at line frequency provides the necessary modulation of the field scanning current.

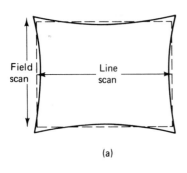
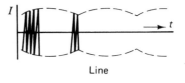
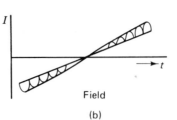
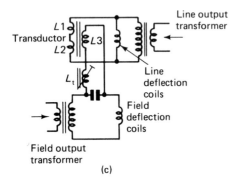
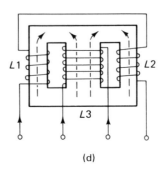

Figure 6.7 Dynamic raster correction in a colour television receiver: (a) pin cushion raster distortion, (b) corrective modulation of the line and field waveforms, (c) basic correction circuit, (d) transductor arrangement. (Courtesy Wöbler[12])

6.3.5 Commercially available transductors

Several ranges of transductors are commercially available. An example taken from the literature of the Wiltek Inc[15] will serve as an illustration. Figure 6.8 shows the details of the construction and performance. The core is formed from two rings of ferrite, the control and bias winding being applied to the two rings in parallel while the signal winding is in two series halves, each embracing one ring and wound in opposite directions so that coupling with the control winding is avoided. In such a construction the inductance of the control winding varies with control current in proportion to the inductance of the signal winding. The control winding has a nominal inductance of 6 H and a d.c. resistance of 1000 Ω. The peak control current is limited to 50 mA and the r.m.s. control current is limited to 30 mA. Thus the peak control power required to achieve the maximum frequency is 2.5 W and the maximum continuous control power is 0.9 W.

This saturable reactor covers the a.m. broadcast band and would be applicable to swept-frequency signal generators operating in this band. It could also be used for remote tuning of a radio receiver but due to the steady control current being limited to 30 mA the range of tuning would be inadequate to cover the whole a.m. band. In either of these applications it may be desirable to use several tuned circuits operating in unison, i.e. tracking. Specially matched groups are available for this purpose.

6.4 Design and operating techniques

Because a transductor is current operated it is preferable that the impedance of the control winding should be low; resistance will absorb power from the control circuit, and inductance may lead to long time constants and resonance effects. The control winding should be supplied from a constant current (high internal impedance) source. This will ensure that the control ampere-turns are determined only by the

High-frequency transductors   197

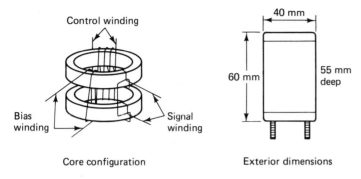
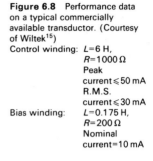
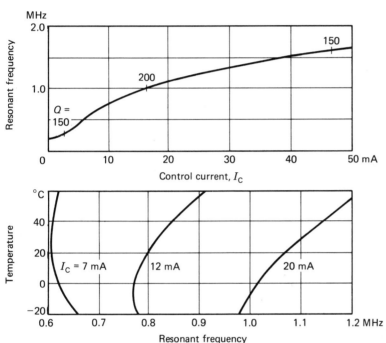

Figure 6.8 Performance data on a typical commercially available transductor. (Courtesy of Wiltek[15])
Control winding: $L=6$ H, $R=1000\,\Omega$
Peak current $\leqslant 50$ mA
R.M.S. current $\leqslant 30$ mA
Bias winding: $L=0.175$ H, $R=200\,\Omega$
Nominal current $=10$ mA

control circuit and are not affected by any variations of control winding resistance or inductance. The resonant frequency of the control winding should be well above the maximum control current frequency.

The operating point, as previously mentioned, may be set by means of a magnet, a bias winding or by the average control current. Assuming that there is no feedback in the control circuit, the choice of operating point will depend on the relative importance of control range, linearity and temperature coefficient.

If the maximum control range is required then the linearity requirement reduces merely to accommodating the control range ($\Delta(N_1I_1)$ in Figure 6.2(a)) between saturation and the point of maximum μ_Δ. This latter point may, due to hysteresis, lie a little to the left of the Φ axis. The greatest inductance linearity will be obtained with a small control range centred about the point of inflection of the $\mu_\Delta - N_1I_1/l$ curve. A possible alternative requirement is for a linear variation of resonant frequency with control current. An approximation to this is obtainable over a limited range by choosing a working point such that μ_Δ is proportional to $(N_1I_1)^{-2}$. In practice linear frequency control is more conveniently obtained by a feedback system as described later.

It was mentioned in Section 6.2.2 that the temperature coefficient of inductance depends on the working point. In practice zero temperature coefficient may be approached only if the control range is very small. If the control range is not small then the best compromise must be sought for the temperature coefficients at each end of the range, see Figure 6.8. When no other attempt is made to reduce the temperature coefficient the fractional change of inductance may typically range from 0.01 to 2% per °C. These relatively high values arise from the use of an ungapped h.f. core. They may be reduced, as in ferrite cored inductors, by introducing an air gap (see Sections 2.2.2, 4.2.2 and 5.5.2). This should be done in such a way that the gap does not intercept the control flux, e.g. by placing it at g in Figure 6.1. The control of the incremental permeability of the side limbs of the h.f. core will be unaffected but, of course, the gap will reduce the value, and the variation, of the effective permeability so the effective control range will be reduced. The loss tangent will also be reduced. Therefore the gapping technique may be reserved for stable, high Q-factor inductors which do not require a high sensitivity.

Another obvious technique is to enclose the device in a temperature controlled box such as an oven of the type used for crystal resonators. By this means the resonant frequency may be held to better than 500 ppm over a wide ambient temperature range. Yet another method that has been used successfully is to shunt or series connect the control or bias winding with a network comprising thermistors and resistors arranged to compensate the temperature coefficient of the h.f. core by the automatic adjustment of the working point. Since the required compensation varies over the control range (see Figure 6.8) this method is of limited application.

There is in practice considerable difficulty in isolating the control and signal circuits. Coupling can result in a very large decrease in Q-factor. If the control and signal windings share a common magnetic core, minor unbalances in magnetic symmetry result in appreciable coupling. If the signal core is separate as in Figure 6.1, the h.f. field may leak into the yoke particularly if the h.f. core has air gaps near the pole faces of the yoke. The yoke material may have a very large magnetic loss at the signal frequency and so contribute to a decrease in Q-factor. A suitable copper foil screen between the signal core and the yoke will prevent this form of coupling but it will introduce some eddy current loss into the signal inductor and in addition it introduces a gap in the control magnetic circuit. The minimum thickness for such a screen may be judged from Figure 11.7.

So far, the discussion has been concerned with the performance of ferrite transductors as isolated devices. The stability and hysteresis problems may be overcome without sacrifice of control range by using a closed loop control circuit. Figure 6.9 shows a very simplified block schematic diagram intended to illustrate the basic principle. The transductor controls an oscillator which has a feedback path to a frequency discriminator. This circuit is arranged to give a d.c. signal proportional to the deviation of the frequency of the oscillator from a given value. This may conveniently be done by taking a part of the available oscillator signal, passing it through an RC circuit, R_1C_1, so that its amplitude varies with frequency, and then applying it to a detector. The magnitude of the d.c. voltage obtained is dependent on the oscillator frequency and amplitude. The other part of the oscillator output is simply rectified and applied, through a variable potentiometer, R_2, in a series opposing connection with the frequency-sensitive d.c. voltage. The resultant voltage depends on the frequency of the oscillator, the setting of the potentiometer and the amplitude of the oscillator output. This error signal is amplified and applied to the control circuit of the transductor in such a way as to reduce the error signal. The oscillator will be automatically set to a frequency such that the error signal is zero; since this means that the two components of the error signal are equal and opposite, the setting is immune to changes of oscillator amplitude. The frequency depends only on the setting of the potentiometer so the effects of temperature and hysteresis are virtually eliminated. The circuit could easily be used to control a number of matched transductors connected in cascade. Clearly the potentiometer could be replaced by some electronic means of attenuation or gain control so that automatic control of sweep may be used.

Finally it is useful to consider the relative merits and limitations of the various grades of ferrite; Figures 3.6, 3.9 and 3.14 show relevant properties of materials that may find application in transductor design. Planar hexaferrites, which may find application at very high frequencies[13], are not considered here.

The high permeability manganese zinc ferrites have the greatest sensitivities; inductance ratios of 300:1 are obtainable without excessive control fields, e.g. 200 A m^{-1}, providing the core may be ungapped. If excessive core loss is to be avoided the lowest frequency of the control range should not exceed about 400 kHz (e.g. see Figure 3.14.1); it is assumed that as the control current is increased and the operating frequency rises the polarizing field will prevent any appreciable rise in loss factor. As previously stated, an air gap will raise the Q-factor, lower the temperature coefficient and restrict the control range.

At higher frequencies the choice must be confined to nickel zinc ferrites. These give rather small inductance ratios, e.g. 6:1 with a static field excursion of 1000 Am^{-1} for ferrite grade 4C6. The operating frequency range for such a ferrite would be situated between about 5 and 50 MHz. Ferrite transductors operating up to 400 MHz have been reported.

When using nickel zinc ferrites for transductors it is

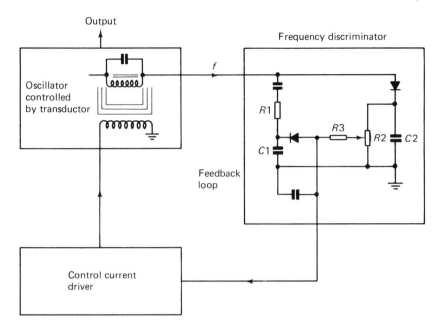

Figure 6.9 Simplified schematic diagram of closed loop control of transductors

necessary to take simple precautions to avoid mechanical resonance of the core. Nickel zinc ferrite has a relatively large magnetostriction so that under the influence of combined a.c. and d.c. fields mechanical resonances are likely to occur. As these resonances have many overtones the electrical performance of the transductor may be impaired by the spectrum of the motional impedance. The application of heavy mechanical damping such as varnish, immersion in oil or encapsulation in an elastomer is usually completely effective in preventing this trouble.

6.5 References and bibliography

Section 6.1
1. DE KRAMOLIN, L. Magnetic tuning devices, *Wireless Wld.*, **42**, 160 (1938)
2. STIBER, S. Electronically tunable circuit elements, *Trans. Inst. Radio Engrs.*, **MIL-4**, 527 (1960).
3. KISER, J. L. The electrically variable inductor, *Electron. Inds.*, **20**, no. 6, 144 (1961)
4. VERWEEL, J. On the HF permeability of dense ferrites in polarizing fields, *IEEE Trans. Magn.*, **MAG-5**, 622 (1969)
5. STIBER, S. Remote tuning receiver has no moving parts, *Electronics*, **26**, no. 7, 186 (1953)
6. NEWHALL, E., GOMARD, P. and AINLAY, A. Saturable reactors as R.F. tuning elements, *Electronics*, **25**, no. 9, 112 (1952)
7. SLATER, F. A ferrite frequency modulator, *Marconi Instrum.*, **4**, 186 (1954)
8. JACKSON, R. C. and SIMPSON, A. W. A high frequency ferrimagnetic attenuator, *Electron. Engng*, **34**, 246 (1962)
9. BESKOROVAINYI, B. M., VOL'F, V. M., GORBENKO, V. S., KAROVSKII, M. L., SHOTSKIL, B. I and IUR'EV, A. A. Variable ferrite filters, *Radio Eng. USSR.*, **15**, no. 9, 79 (1960)
10. PRESSMAN, A. I. and BLEWETT, J. P. A 300- to 4000-kilocycle electrically tuned oscillator, *Proc. Inst. Radio Engrs*, **39**, 74 (1951)
11. LOMBARDINI, P. P., SCHWARTZ, R. F. and DOVIAK, R. J. Measurements of the properties of various ferrites used in magnetically tuned resonant circuits in the 2·5 to 45 Mc/s region, *J. appl. Phys.*, **29**, 395 (1958)
12. WÖLBER, J. A saturable core reactor for raster correction in colour television receivers, *Electron. Applic.*, **26**, 43 (1965)

Section 6.2
13. REMIS, I. and BADY, I. Ferrites for permeability tuning in the VHF region using the above resonance mode, *IEEE Trans.*, **MAGN-6**, 46 (1970)

Section 6.3.1
14. WHILLIER, R. T. and WHITBOURN, E. A. Performance of ferrite saturable reactors in the frequency range 1–40 Mc/s, unpublished report (1962)

Section 6.3.2
6. NEWHALL, E., *et al.*, see Section 6.1, Reference 6

Section 6.3.3
10. PRESSMAN, A. I., *et al.*, see Section 6.1, Reference 10

Section 6.3.4
12. WÖLBER, J., see Section 6.1, Reference 12

Section 6.3.5
15. *Track Increductors*, technical literature by Wiltek, Inc., Wilton, Conn. USA

General

HARVEY, R. L., GORDON, I. and BRADEN, R. A. The effect of a d-c magnetic field on the U.H.F. permeability and losses of some hexagonal magnetic compounds, *R.C.A. Rev.*, **22**, 648 (1961)

BARKER, R. C. Nonlinear magnetics, *Electro-Technology*, **71**, no. 3, 95 (1963)

GEYGER, W. A. *Nonlinear-magnetic control devices*, McGraw-Hill Book Co. (1964)

STUIJTS, A. L., VERWEEL, J. and PELOSCHEK, H. P. Dense ferrites and their applications, *Trans. I.E.E.E.*, **Commun. Electron.**, no. 75, 726 (1964)

7 Wide-band transformers

7.1 Introduction

Wide-band transformers are used extensively in communication systems. They may be used to match different impedances, provide accurate current or voltage ratios, provide interconnection between balanced and unbalanced circuits and to provide d.c. isolation. Because the signal power levels in a communication system are low except at the output stages most wide-band transformers are not required to transmit appreciable amounts of power. On the other hand most high-power transformers operate at a single frequency or over a narrow frequency band. The design procedure for such transformers is considered in Chapter 9; it differs fundamentally from the approach to wide-band transformers. However, it is sometimes required to design a wide-band high-power transformer and then it is necessary to combine both procedures, considering the power limitations of the core and winding together with the limitations due to leakage inductance and self capacitance. In this chapter the emphasis will be on low power wide-band transformers but most of the transmission relations apply to transformers of any rating.

The voltage applied to a wide band transformer is usually complex, containing energy distributed over a wide spectrum and ideally it is required that the voltage waveform appearing across the load shall have lost nothing by the transformation. The fractional bandwidth may be defined as the ratio f_2/f_1 where f_2 is the highest frequency to be transmitted satisfactorily and f_1 is the corresponding lowest frequency. Figure 7.1 shows a typical transmission characteristic of a wide-band transformer, indicating the main features. In practice, bandwidths are rarely less than 10 and may in some applications be required to exceed 1000. Such bandwidths are possible only when the primary and secondary windings are tightly coupled; narrow bandwidth transformers such as those used in intermediate frequency amplifiers are very loosely coupled.

Pulse transformers, which are considered in the next chapter, may be regarded as a special case of a wide-band transformer. The information characterizing the pulse is usually distributed over a wide band and if an effective bandwidth may be assigned to it then the design procedures discussed in this chapter may be used. However since the avoidance of pulse distortion involves both the attenuation and phase shift characteristics it is usually better to design a pulse transformer using the pulse characteristic directly, as described in the next chapter.

7.2 The transmission and reflection characteristics in terms of the equivalent circuit elements

7.2.1 Some basic concepts

The relations between the equivalent circuit of a transformer and its transmission and reflection characteristics are a special case of the more general passive network theory. Since, in the present context, the emphasis is on the application of ferrite in practical transformer design, much of the fundamental theory must be assumed. However, a summary of the basic network concepts is given in this section and, in the following sections, the essential transformer design equations and functions are presented. Derivations and a fuller treatment may be found in the standard text books.

Figure 7.2 illustrates some basic properties of passive networks. If a source having resistance R_a is connected to a load of resistance R_b, the power in the load is given by $E_a^2 R_b/(R_a + R_b)^2$. This is a maximum when $R_a = R_b$. The maximum power in the load is called the applied power and is designated P_a

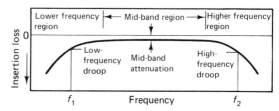

Figure 7.1 Idealized transmission characteristic of a wide-band transformer showing the main features

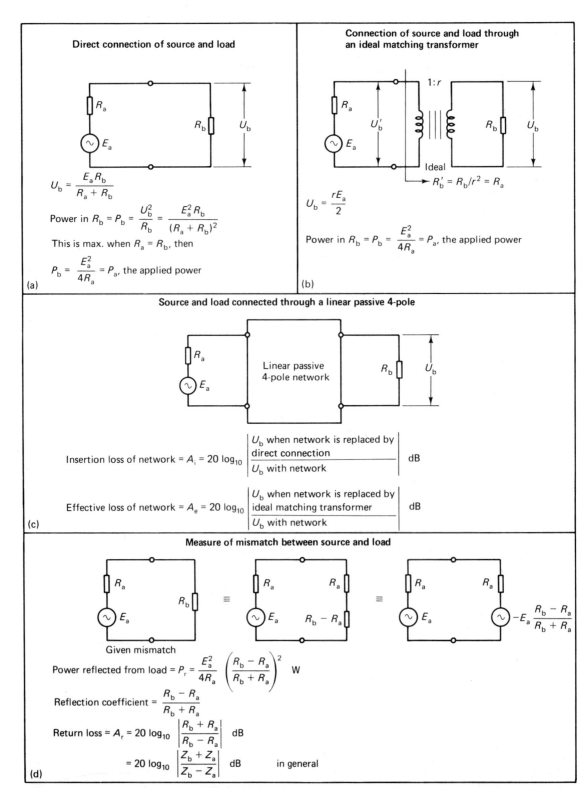

Figure 7.2 A summary of useful expressions and definitions relating to passive networks

$$P_a = \frac{E_a^2}{4R_a} \text{ W} \qquad (7.1)$$

If $R_a \neq R_b$ the power in R_b may be made equal to the applied power by interposing an ideal transformer as shown in Figure 7.2(b). Since an ideal transformer has no energy losses or series reactances, the terminal voltages will equal the winding e.m.f.s. Considering the magnetic induction in the core

$$\frac{d\Phi}{dt} = \frac{U_b'}{N_1} = \frac{U_b}{N_2} \text{ V} \qquad (7.2)$$

where Φ is the flux in the core and N_1, N_2 are the primary and secondary turns respectively

$$\therefore \quad U_b = U_b' \frac{N_2}{N_1} = U_b' r \text{ V} \qquad (7.3)$$

where r is the turns ratio N_2/N_1

If I_1 and I_2 are the primary and secondary currents respectively, then:

input power = $I_1 U_b'$ = output power = $I_2 U_b$

$$\therefore \quad I_2 = I_1/r \text{ A} \qquad (7.4)$$

The input resistance

$$\frac{U_b'}{I_1} = \frac{U_b}{r} \times \frac{1}{rI_2} = \frac{R_b}{r^2} = R_b' \text{ }\Omega \qquad (7.5)$$

where R_b' is the load resistance as it appears at the primary terminals. When the transformation ratio is such that $R_a = R_b'$, i.e. when $r = (R_b R_a)^{1/2}$, the transformer is said to match the source and load, and the power in the load equals the applied power.

If the circuit shown in (a) is broken and a linear passive 4-pole network is inserted as shown in (c), then in general the power in the load will change. The change may be expressed as the insertion loss A_i. This is the ratio, in dB, of the power in the load without the network to the power in the load when the network is inserted, i.e.

$$A_i = 20 \log_{10} \left| \frac{\text{Load voltage when network is replaced by direct connection}}{\text{Load voltage with network}} \right| \text{ dB} \qquad (7.6)$$

The insertion loss may be negative, i.e. a gain; this could be the case when the 4-pole is a matching transformer, the improvement in matching more than

---

(7.2) $\quad \frac{d\Phi}{dt} \times 10^{-8} = \frac{U_b'}{N_1} = \frac{U_b}{N_2} \text{ V}$

compensating for the loss introduced by the transformer. To avoid this difficulty when expressing the attenuation of a transforming network, the following definition is adopted. The effective loss, A_e, of a network is the ratio, in dB, of the applied power to the power in the load when the network is inserted, i.e.

$$A_e = 20 \log_{10} \left| \frac{\text{Load voltage when network is replaced by ideal matching transformer}}{\text{Load voltage with network}} \right| \text{ dB} \qquad (7.7)$$

The effective loss is normally used to describe the attenuation of a transformer since it compares it to an ideal matching transformer. If $R_a = R_b$ then the effective loss equals the insertion loss.

The insertion and effective losses include not only the dissipative components of the attenuation occurring within the network but also the mismatch losses occurring at its terminals. For example, mismatch between the source impedance and the input impedance of the network has the effect of reflecting a part of the applied power. This is illustrated in Figure 7.2(d) and the following expressions are derived:

$$\text{Reflection coefficient} = \left| \frac{R_b - R_a}{R_b + R_a} \right| \qquad (7.8)$$

$$\text{Return loss} = A_r = 10 \log_{10} \left| \frac{\text{Applied power}}{\text{Reflected power}} \right|$$

$$= 10 \log_{10} \left| \frac{E_a^2}{4R_a} \times \frac{4R_a}{E_a^2} \left\{ \frac{R_b + R_a}{R_b - R_a} \right\}^2 \right|$$

$$= 20 \log_{10} \left| \frac{R_b + R_a}{R_b - R_a} \right| \text{ dB} \qquad (7.9)$$

These expressions have been derived in terms of resistances but it may be shown that they are true also for general impedances Z_a and Z_b in which case the expression within the modulus brackets will, in general, be complex. Return loss can, and often does, occur when there is simply a direct connection between source and load; in this case R_b is the load resistance. Generally, it is common practice to express any circuit impedance in terms of its return loss relative to a nominal or specified value.

Reflection or mismatch is often more troublesome than the attenuation and it is common to find that the return loss requirement in a wide-band transformer specification is more severe than that for insertion loss.

If the inserted network has no dissipation, i.e. it

contains only pure reactances, then the attenuation must be due entirely to reflection and there will be a relation between the return loss and the effective loss. Of the applied power P_a let a proportion, P_r, be reflected, the remainder $(P_a - P_r)$ being transmitted into the load. The power in the load is

$$P_b = P_a - P_r$$

but

$$\frac{P_a}{P_b} = \text{antilog}_{10} \frac{A_e}{10} \quad \text{and} \quad \frac{P_a}{P_r} = \text{antilog}_{10} \frac{A_r}{10}$$

$$\therefore \quad \frac{1}{\text{antilog}_{10}(A_e/10)} = 1 - \frac{1}{\text{antilog}_{10}(A_r/10)} \quad (7.10)$$

This relation is true for any passive dissipationless network connected between any source and load. If the source and load impedances are equal then the relation concerns only the network. On the other hand the network may be a direct connection of unequal source and load impedances in which case the relation refers only to the effective loss and return loss between source and load. In general both forms of mismatch may be present. The alternative ordinate scales of Figures 7.7 and 7.8 are nomograms for the above relation.

7.2.2 The equivalent circuit of a transformer

The foregoing summary refers to the general linear passive 4-pole network. The present section is concerned with a particular form of such a network, namely, the equivalent circuit of a transformer. The usual form of this circuit, inasmuch as it may be represented by lumped components, is shown in Figure 7.3 in which the physical significance of each of the components is indicated. All the components that normally belong to the secondary side of the transformer have been transferred to the primary side and carry a prime to indicate this. Direct capacitive coupling between the primary and secondary circuits would cause the higher frequency signals to by-pass the transformer and this should be avoided; usually this capacitive coupling is eliminated by inserting an electrostatic screen between the windings. The actual transformation is provided by the ideal transformer shown at the r.h.s. of the diagram. In most of the discussion that follows, this ideal transformer will be ignored and attention will be confined to the equivalent circuit referred to the primary side. This is inserted between the source resistance R_a and the reflected load resistance R'_b. The attenuation of this equivalent circuit or of part of it may, in this context, be expressed as insertion loss. In most of the following discussion it is this insertion loss that will be

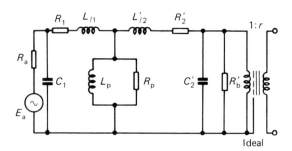

Figure 7.3 Lumped-element equivalent circuit of a transformer.
E_a is the source e.m.f.
R_a is the source resistance
R_1 is the resistance of the primary winding
R'_2 is the resistance of the secondary winding referred to the primary side
R'_b is the load resistance referred to the primary side
R_p is the shunt resistance representing the total loss in the core
C_1 is the effective capacitance shunting the terminals of the primary winding. This is usually the self capacitance associated with the primary winding plus any other capacitance shunting the input circuit
C'_2 is the corresponding secondary capacitance referred to the primary side
L_{l1} is the primary leakage inductance
L'_{l2} is the secondary leakage inductance referred to the primary side. Since leakage inductances are much smaller than the shunt inductance, L_p, the two leakage inductances are usually treated as one, i.e. $L_l = L_{l1} + L_{l2}$
L_p is the open circuit inductance of the primary
The direct capacitance between primary and secondary is usually eliminated by insertion of a screen as shown

referred to. This will avoid any ambiguity which might arise when the transformer does not provide nominal impedance matching.

Referring to Figure 7.1, the transmission characteristic of a wide-band transformer may be divided into three parts, namely, the lower frequency region, the mid-band region, and the higher frequency region. These are usually far enough apart on a frequency basis to be considered separately, e.g. the factors which affect the lower frequency region usually have negligible effect in the higher frequency region and vice versa.

A typical transformer specification would state the maximum permitted effective loss at mid-band and the maximum permitted additional effective loss (or droop) at f_1 and f_2 respectively. The mid-band effective loss might be limited to about 0.5 dB and the specified maximum droop would be typically between 0.1 and 3 dB depending on the required flatness of the characteristic. If the specification included return loss requirements, these, as previously observed, are usually more stringent than the attenuation requirements. A typical specification might set the minimum return loss to a value lying somewhere between 14 and 25 dB over all or part of the frequency band. For a dissipationless network these figures correspond to effective losses of 0.18 and 0.015 dB respectively, thus illustrating relative severity of the return loss requirements.

In certain special applications the phase shift of the load voltage due to the transformer may be specified.

The three main regions of the transmission characteristic will now be considered. For each region the insertion loss, return loss and, in some cases, the phase shift will be related quantitatively to the elements of the equivalent circuit. The resulting expressions will facilitate the first stage of transformer design, i.e. to translate the specification into limit values of the equivalent circuit elements.

7.2.3 The mid-band region

For the majority of low and medium frequency transformers, and in particular those using ferrite cores, the only elements of the equivalent circuit which are likely to influence the transmission at mid-band are the winding resistances, $R_1 + R_2' (= R_s)$, and the shunt resistance, R_p, representing the core loss. In certain circumstances the mid-band transmission of high-frequency transformers may also be influenced by the shunt inductance; this possibility is considered in Section 7.4.1 and is disregarded in the present discussion.

It will be seen in Section 7.3.1 that, for low-frequency transformers having ferrite cores, the shunt loss resistance, R_p, is usually so large that it does not have a significant effect on the mid-band loss. On the other hand the effect of the winding resistance is usually negligible in the design of medium and high frequency transformers.

From Eqn (7.6) it is readily derived that the insertion loss due to the series resistance is

$$A_i = 20 \log_{10} \left(1 + \frac{R_s}{R_a + R_b'} \right) \quad \text{dB} \qquad (7.11)$$

while for the shunt resistance

$$A_i = 20 \log_{10} \left(1 + \frac{R}{R_p} \right) \quad \text{dB} \qquad (7.12)$$

where $R_s = R_1 + R_2'$

$R = R_a R_b'/(R_a + R_b')$

Figure 7.4 represents these equations in graphical form. If both R_s and R_p are present and their individual contributions are small then the total insertion loss may be taken as the sum of the separate losses. It should be noted that if R_s significantly increases with frequency due to eddy current effects the mid-band characteristic will droop towards the upper end.

Eqn (7.9) gives the return loss of one resistance relative to another. Using this equation and referring to the diagram in Figure 7.4 the input resistance R_i may be expressed in terms of its return loss relative to R_b':

$$A_r = 20 \log_{10} \left\{ \frac{2 + \dfrac{R_b'}{R_p} + \dfrac{R_s}{R_b'}}{\dfrac{R_b'}{R_p} - \dfrac{R_s}{R_b'}} \right\} \quad \text{dB} \qquad (7.13)$$

This assumes that $R_p \gg R_1$ or R_2'.

This function is represented graphically in Figure 7.5. If both R_s and R_p are present they compensate when $R_s/R_b' = R_b'/R_p$ and then the return loss is infinite (perfect matching). If $R_p = \infty$, the monotonic curve gives the return loss due to R_s, while if $R_s = 0$ the intercepts with the l.h. ordinate give the return loss due to R_p. An inspection of this graph suggests that it is not difficult to achieve a good mid-band return loss.

7.2.4 The lower frequency region

This is the region in the vicinity of f_1 (see Figure 7.1). The term is relative; for a high frequency transformer f_1 might for example be in the neighbourhood of 10 MHz. In general the droop in the transmission characteristic in the lower frequency region is due to the shunt impedance. This impedance diminishes as the frequency decreases and causes a progressive increase in attenuation which is additional to the mid-band loss. In most cases the contribution of R_p to this droop is negligible; an exception arises when the flux density ($\propto f^{-1}$) is high enough to cause significant hysteresis loss and then R_p will decrease as the frequency falls. Ignoring this effect, the droop may be expressed in terms of the shunt inductance:

$$A_i = 10 \log_{10} \left\{ 1 + \left(\frac{R}{\omega L_p} \right)^2 \right\} \quad \text{dB} \qquad (7.14)$$

The corresponding return loss of the input impedance relative to R_b' is

$$A_r = 10 \log_{10} \left\{ 1 + \left(\frac{2\omega L_p}{R_b'} \right)^2 \right\} \quad \text{dB} \qquad (7.15)$$

and the phase shift, φ, is given by

$$\tan \varphi = \frac{R}{\omega L_p} \qquad (7.16)$$

These three functions are expressed in Figure 7.6, where the independent variable is $\omega L_p/R$ for insertion loss and phase shift, and $\omega L_p/R_b'$ for return loss. A limit on any of these three parameters at a given frequency will set a minimum value for L_p. This is usually the starting point for wide-band transformer design.

It is possible under certain circumstances to improve the characteristic in the lower frequency region

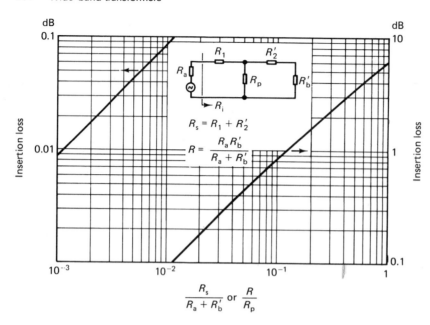

Figure 7.4 Insertion loss due to series or shunt resistance

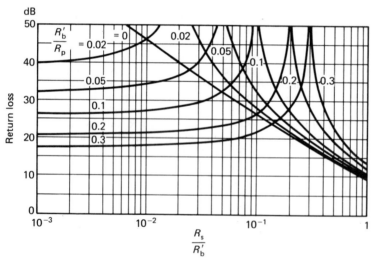

Figure 7.5 Return loss (R_i with respect to R_b') due to any combination of series and shunt resistance

by inserting capacitance in series with the primary and/or secondary winding. This may be regarded as an analogue of the more usual practice for the higher frequency region which is considered in the next section. The technique for the lower frequency region is described in Section 7.2.6.

7.2.5 The higher frequency region—filter techniques

The transmission characteristic in the higher frequency region (i.e. in the region of f_2 in Figure 7.1) is influenced mainly by the leakage inductance and the shunt capacitances, and their relation to R_a and R_b'. The winding resistance, if comparable to the leakage reactance or capacitive reactance, may also influence the shape of the characteristic but it is usual to assume that the winding resistance can be neglected in this respect.

It often happens that the shape of the higher frequency characteristic is determined by either the leakage inductance or the shunt capacitance acting alone; e.g. in a low-impedance circuit the series leakage reactance may be appreciable while the shunting effect of the capacitance could be negligible, while for

Wide-band transformers

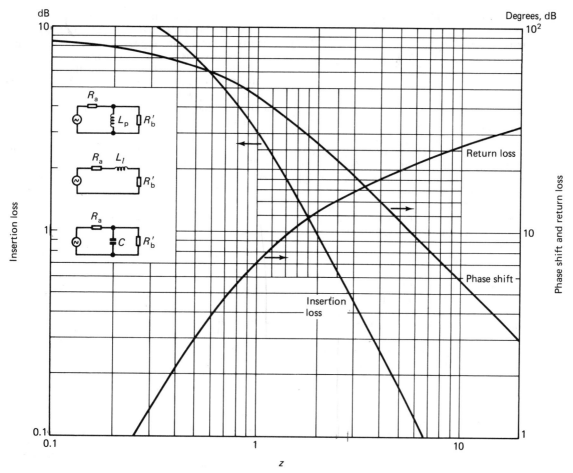

Figure 7.6 Insertion loss, return loss and phase shift due to shunt inductance, leakage inductance or shunt capacitance taken separately

Expressions for the independent variable, z

Element:	L_p	L_l	C
Insertion loss	$\omega L_p/R$	$(R_a+R'_b)/\omega L_l$	$1/\omega CR$
Return loss	$\omega L_p/R'_b$	$R'_b/\omega L_l$	$1/\omega CR'_b$
Phase shift	$\omega L_p/R$	$(R_a+R'_b)/\omega L_l$	$1/\omega CR$

The phase shift due to the shunt inductance is positive (leading) and for the other two elements it is negative

a high-impedance circuit the reverse could apply. When it is required to consider the the effect of either of these two reactances in isolation the following equations apply. The higher frequency droop due to leakage inductance is given by:

$$A_i = 10 \log_{10} \left\{ 1 + \left(\frac{\omega L_l}{R_a + R'_b} \right)^2 \right\} \quad \text{dB} \quad (7.17)$$

and due to shunt capacitance it is given by

$$A_i = 10 \log_{10} \ (1 + (\omega CR)^2) \quad \text{dB} \quad (7.18)$$

The corresponding return loss of the input impedance relative to R'_b is given by

$$A_r = 10 \log_{10} \left\{ 1 + \left(\frac{2R'_b}{\omega L_l} \right)^2 \right\} \quad \text{dB} \quad (7.19)$$

and

$$A_r = 10 \log_{10} \left\{ 1 + \left(\frac{2}{\omega CR'_b} \right)^2 \right\} \quad \text{dB} \quad (7.20)$$

and the corresponding phase shifts are given by

$$\tan \varphi = -\frac{\omega L_l}{R_a + R'_b} \qquad (7.21)$$

and

$$\tan \varphi = -\omega C R \qquad (7.22)$$

Figure 7.6 may again be used to express these functions, the abscissa being a scale of $(R_a + R'_b)/\omega L_l$ or $1/\omega CR$ for insertion loss and phase shift, or $R'_b/\omega L_l$ or $1/\omega CR'_b$ for return loss. C is the sum of C_1 and C'_2 in Figure 7.3; if the transformer has a turns ratio greater than about 3 only the shunt capacitance of the high-impedance side is significant and $C \simeq C_1$ or C'_2 depending on whether it is a step-down or a step-up transformer. This graph may be used in the early stages of design to determine maximum values of L_l and C acting individually.

At this stage experience may suggest that it will be either the leakage inductance or the stray capacitance that will limit the bandwidth, in which case the subsequent transformer design merely consists of meeting the leakage inductance or stray capacitance limit while at the same time providing adequate shunt inductance.

If, however, the performance requirements are stringent, e.g. a flat response over a wide bandwidth within a restricted volume, then it is probable that the combined effect of L_l and C will have to be considered. This is best done by synthesizing the high-frequency equivalent circuit from the performance requirements using wave filter techniques.* In this way the combined effect of L_l and C may often be turned to good account. The procedure may be used in reverse to analyse the response of a given equivalent circuit.

The step-up transformer will first be considered. In this case only C'_2 need be taken into account so, if the winding resistance can be neglected, the equivalent circuit is simplified to that shown in Figure 7.7. This is a simple half-section low pass filter having a nominal impedance:

$$\left. \begin{array}{l} R_o = \sqrt{\left(\dfrac{L_l}{C}\right)} \ \Omega \\[2mm] \text{and a cut-off frequency given by} \\[2mm] 2\pi f_c = \omega_c = \dfrac{1}{\sqrt{(L_l C)}} \quad \text{rad s}^{-1} \end{array} \right\} \qquad (7.23)$$

The mid-series† impedance is equal to

---

*The approach is based on an unpublished lecture given by J. H. Mole.
†Half-sections may be cascaded, series element to series element and shunt to shunt. Thus the series terminals of a half-section are referred to as mid-series and the shunt terminals are mid-shunt.

$R_o\sqrt{[1-(\omega/\omega_c)^2]}$, i.e. it is asymptotic to R_o at low frequencies but falls to zero at cut-off. Thus if it is desired to maintain a given degree of impedance matching over the widest possible frequency band it is better to design the filter such that the terminating resistance (R_a in this case) equals not R_o but a lower value, e.g. ρR_o where ρ is a nominal mismatch factor and is less than unity. Similarly the mid-shunt impedance is given by $R_o/\sqrt{[1-(\omega/\omega_c)^2]}$ i.e. it is asymptotic to R_o at low frequencies but rises to infinity at cut-off. In this case it is better to arrange that the terminating resistance (R'_b in practice) equals not R_o but a higher value, e.g. R_o/ρ. If the value of ρ is the same for both sides, the nominal primary and secondary mismatches will be equal and $R_a = \rho^2 R'_b$.

The effective loss of the circuit is given by

$$A_e = 10 \log_{10} \left\{ 1 + \left(\frac{\omega^2/\omega_c^2 - (1-\rho^2)}{2\rho} \right)^2 \right\} \ \text{dB} \qquad (7.24)$$

This function is presented graphically in Figure 7.7, with the mismatch factor ρ as a parameter. It will be noted that values of ρ greater than unity have been included. In practice only those equal to or less than unity will be considered in transformer design. The higher values may be useful in predicting the characteristic of a given transformer having arbitrary values of L_l and C. A scale of return loss has been added; this has been calculated from Eqn (7.10). It should also be noted that the graph is given in terms of effective loss, thus when the frequency approaches zero the effective loss is not zero unless $\rho = 1$, in which case $R_a = R'_b$ and the effective loss becomes equal to the insertion loss. However, the graph may easily be used for insertion loss. It may be shown that by taking the intercept at $\omega/\omega_c = 0$ as the zero insertion loss datum, the relative effective loss, positive or negative, read from the graph at other frequencies equals the insertion loss.

To obtain the shape of the response characteristic in the high frequency region, the intercept of the appropriate curve at $\omega/\omega_c = 0$ may be taken as equal to the mid-band loss. This is because the graph ignores factors affecting the response in other frequency regions, and both zero frequency and mid-band frequency may be regarded as far removed from the higher frequency region. Thus if it is required, for example, that the variation of attenuation between mid-band and f_2 should not exceed about 0.2 dB then a mismatch of $\rho = 0.8$ would be suitable and f_2 would correspond to $\omega/\omega_c = 0.84 = f_2/f_c$.

From R_a and ρ the value of R_o may be calculated, and from ω/ω_c and f_2 the value of ω_c may be obtained. Eqn (7.23) may then be used to calculate the required leakage inductance and shunt capacitance. When realizing the design it is necessary to adjust the actual values of L_l and C by adding inductance in series with

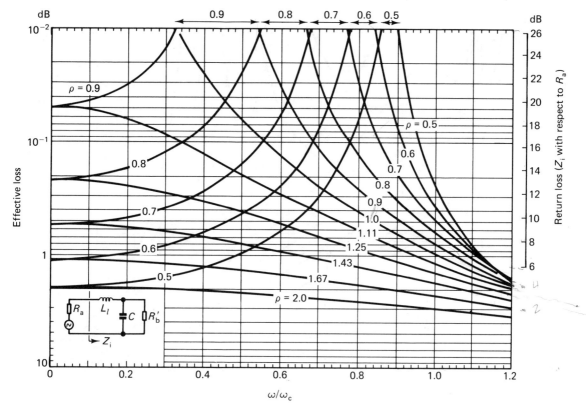

Figure 7.7. Transformer characteristics using the half-section low pass filter technique; effective loss and return loss due to a combination of L_l and C regarded as a half-section low pass filter with a mismatch ratio ρ. In the form shown, the graph applies to a step-up transformer in which C_1 (see Figure 7.3) may be neglected. Then:

$$R_a = \rho R_o \qquad R'_b = R_o/\rho \qquad R_a = \rho^2 R'_b$$
$$R_o = (L_l/C)^{1/2} \qquad \omega_c = (1/L_l C)^{1/2}$$

For the step-down transformer C'_2 (see Figure 7.3) may be neglected and C moves to the l.h.s. of L_l. The same graph may be used, but now:

$$R_a = R_o/\rho \qquad R'_b = \rho R_o \qquad R_a = R'_b/\rho^2$$

The expressions for R_o and ω_c remain as above
This graph may also be used for the corresponding half-section high pass filters, in which case it must be used in conjunction with Figure 7.9.

the primary and shunt capacitance to the secondary winding. The practical design must allow a margin for this adjustment.

Again, R_a and ρ may be used to calculate R'_b and from this the actual turns ratio is obtained i.e. $r = \sqrt{(R_b/R'_b)}$.

If the transformer has a large step-down ratio then the primary shunt capacitance will predominate and C must be moved to the other side of L_l. This does not affect the response, but now $R_a = R_o/\rho$ and $R'_b = \rho R_o$.

This mismatched filter technique is useful if the transformer must meet a severe attenuation or return loss requirement over the widest possible frequency band. If the requirement is not severe, e.g. if the permitted variation in attenuation up to f_2 is greater than 1 or 2 dB then it is usually better to use a simple design in which either L_l or C is, if possible, negligible and R_a equals R'_b.

If the transformer has a ratio approaching unity, the primary and secondary shunt capacitances will become comparable in magnitude. If there is a severe attenuation or return loss requirement then the full-section low-pass filter technique may be used. The circuit is shown in Figure 7.8; it consists of two half-section low-pass filters connected back to back. In the present context C is simply one of a pair of mid-shunt capacitances; in the realization, C_1 and C'_2 of Figure 7.3 must each be made equal to C. The filter section is mismatched at each end by a factor $\rho < 1$ but since both ends are mid-shunt there is no mismatch required between R_a and R'_b. This means that the attenuation may be expressed in terms of insertion

210 Wide-band transformers

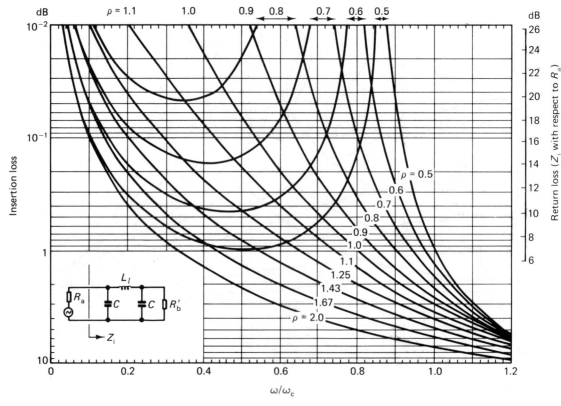

Figure 7.8 Transformer characteristics using the full-section low pass filter technique; insertion loss and return loss due to a combination of L_l and C regarded as a full-section low pass filter with a mismatch ratio ρ:

$R_a = R'_b = R_o/\rho$ $\qquad R_o = (L_l/2C)^{1/2}$ $\qquad \omega_c = (2/L_lC)^{1/2}$

This graph may also be used for the corresponding full-section high pass filters, in which case it must be used in conjunction with Figure 7.9

loss of the equivalent circuit and this insertion loss will approach zero as $\omega \to 0$.

$$R_a = R'_b = R_o/\rho$$

$$R_o = \sqrt{\left(\frac{L_l}{2C}\right)}, \quad \omega_c = \sqrt{\left(\frac{2}{L_lC}\right)} \qquad (7.25)$$

The insertion loss is given by

$$A_i = 10 \log_{10} \left\{ 1 + \left(\frac{\omega}{\rho \omega_c}\right)^2 \left(\frac{\omega^2}{\omega_c^2} - (1 - \rho^2)\right)^2 \right\} \text{dB} \qquad (7.26)$$

This function is presented graphically in Figure 7.8, with ρ as a parameter. Again only values of ρ less than unity are relevant in the present context but larger values have been included to enable transmission characteristics to be predicted. As before, a scale of return loss has been added.

The curves may be used in a similar manner to that described above for the single section filter. If the permissible insertion loss variation is for example less than about 0.2 dB then the curve of $\rho = 0.7$ would be satisfactory. The insertion loss would increase from the mid-band value by about 0.2 dB at $\omega/\omega_c = 0.4$, then return to the mid-band value at $\omega/\omega_c = 0.715$ before increasing again by about 0.2 dB at $\omega/\omega_c = 0.83$, this frequency corresponding to f_2.

It will be noted that if L_l and C have the same values respectively in the half-section and full-section equivalent circuits then the value of ω_c is larger by a factor of $\sqrt{2}$ in the latter case. Thus it is often advantageous to use a full-section even if the ratio of the transformer would suggest a half-section. In the practical realization the required values of L_l and C are obtained by the addition of external elements as previously mentioned.

The phase shift at cut-off, ω_c, is 90° (lagging) for the half-section and 180° for the full-section. It increases monotonically from zero to the cut-off value as the frequency rises through the pass band. The rate at which it approaches the cut-off value depends on the mismatch factor; if this is low then the phase shift rises relatively sharply.

Finally a word of caution must be added. It has been assumed that the resistance of the windings may be neglected. This is not always true. Particularly when the full-section low pass filter technique is used to extend the higher frequency region, the winding resistance, probably increased by eddy current effects, may cause the Q-factor of the circuit to be quite low. The curves given in Figure 7.8 are calculated on the assumption that the π-network consists of pure reactances; in practice the winding resistances are included in this network. This may have the effect of suppressing the zero attenuation peak in the characteristic and causing an additional droop at the higher frequency end of the band. Thus the expected performance may not be fully realized. This effect is less pronounced in the T-network (see Section 7.2.6) and half-section versions. Here the winding resistances may be considered as being absorbed in the terminating resistances leaving the inductive and capacitive elements as pure reactances.

7.2.6 Filter techniques in the lower frequency region

The filter techniques described above apply, with a few modifications, to the lower frequency region. The corresponding simple high pass filter circuits are shown in Figure 7.9 together with the essential equations. When considering the lower frequency region these equivalent circuits may be substituted for those in Figure 7.7 and 7.8 respectively. The graphs will give the lower frequency response in the same way as has been described for the higher frequency response except that the parameter of the frequency scale should now be ω_c/ω instead of ω/ω_c.

The required capacitance must be introduced in series with one or both windings (obviously if there is d.c. in one of the windings, only the half-section may be used. If the permitted variation of pass-band attenuation is small then for a given primary inductance the filter design extends the bandwidth. Alternatively, with the same qualification, it permits a given bandwidth to be obtained with the lowest value of primary inductance L_p. However if the permissible droop is greater than 1 or 2 dB the required characteristic is better obtained by using the simple design as this will result in a lower value of primary inductance.

The phase shift at cut-off, ω_c, is 90° (leading) for the half-section and 180° for the full-section. The remarks made in the previous section concerning rate of change of phase apply here also.

The high pass version of the filter technique does not appear to suffer the degradation due to winding resistance discussed at the end of the previous section. This is presumably because the circuit is mid-series connected to the source and load, and the winding resistances may be regarded as being absorbed by these terminations.

7.2.7 Example of the filter design technique

The following example will illustrate the filter technique outlined in the previous sections. A transformer is required to meet the following specification.

Source impedance, R_s	$= 600\,\Omega$
Load impedance, R_b	$= 1200\,\Omega$
Frequency, f_1	$= 300$ Hz
Effective loss at f_1	<0.1 dB w.r.t the mid-band loss.
Effective loss at mid-band	<0.3 dB
Frequency, f_2	$= 300$ kHz
Return loss relative to source impedance	>16 dB from f_1 to f_2

The transformation ratio is not large and the pass band requirements are quite severe. It might therefore be appropriate to use the full-section filter technique at both ends of the pass band. However as there is no margin available for any degradation of the characteristic in the region of f_2 due to the effects of winding resistance, the higher frequency characteristic will be based on a half-section filter, in which such degradation is less marked. The lower frequency region will use a full-section filter, thus the example will illustrate both techniques.

7.2.7.1 General

Turns ratio, $r = \sqrt{2}$ and $R_a = R'_b = 600\,\Omega$
$R = R_a R'_b/(R_a + R'_b) = 300\,\Omega$
$\omega_1 = 2\pi f_1 = 1885$ rad s^{-1}
$\omega_2 = 2\pi f_2 = 1.885 \times 10^6$ rad s^{-1}

7.2.7.2 Mid-band

It is assumed that the shunt resistance, R_p, due to the core loss may be neglected. From Figure 7.4 an effective mid-band loss of less than 0.3 dB will be obtained if

$$\frac{R_s}{R_a + R'_b} < 0.035, \text{ i.e. } R_s < 42\,\Omega$$

The corresponding return loss obtained from Figure 7.5 is approximately 30 dB.

7.2.7.3 Higher frequency region

When a half-section is used, the turns ratio does not in general equal $(R_b/R_a)^{1/2}$, i.e. there is a net mismatch.

212   Wide-band transformers

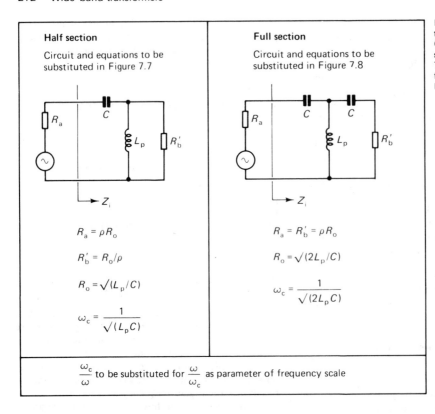

Figure 7.9 High pass filter technique in transformer design. Circuit and equations to be substituted in Figures 7.7 and 7.8 respectively to convert these figures to the corresponding high pass filter cases

However, since the lower frequency region will use a full-section filter it follows that the actual turns ratio will in fact equal $(R_b/R_a)^{1/2}$. Therefore, referring to Figure 7.7, ρ must equal unity for the higher frequency region. The curve for $\rho = 1$ crosses the 16 dB return loss line at $\omega/\omega_c \approx 0.55$ and falls monotonically with increasing frequency.

$$R_o = R_a/\rho = 600 = \sqrt{(L_l/C)}$$
$$\omega_c = 1.885 \times 10^6/0.55 = 3.43 \times 10^6 = \sqrt{(1/L_l C)}$$
$$\therefore \quad C = 486 \text{ pF} \quad L_l = 175 \text{ μH}$$

[Note that the return loss due to $L_l = 175$ μH in the absence of C may be obtained from Figure 7.6: $R_b'/\omega L_l = 600/1.885 \times 175 = 1.82$.

$\therefore$ Return loss would be 11.5 dB.

7.2.7.4 Lower frequency region

This time Figure 7.8 is used but it is modified in accordance with the r.h.s. of Figure 7.9. It must be remembered that the frequency scale is inverted, i.e. it must be read in terms of ω_c/ω. A curve is required such that its trough touches the 0.1 dB line (this neglects any margin for tolerances). As the frequency decreases from the value at the trough the characteristic passes through a point of zero attenuation and thereafter the attenuation increases monotonically, crossing the 0.1 dB line at a frequency which is put equal to f_1. Between f_1 and mid-band the variation of attenuation will not exceed 0.1 dB. Interpolating, a curve to meet this requirement has $\rho = 0.75$ and $\omega_c/\omega = 0.77$.

$$\therefore \quad R_o = 600/0.75 = 800 = \sqrt{(2L_p/C)}$$
$$\omega_c = 1885 \times 0.77 = 1450 = \sqrt{(1/2L_p C)}$$
$$\therefore \quad C = 0.862 \text{ μF} \quad L_p = 0.275 \text{ H}$$

Note that the effective loss due to $L_p = 0.275$ H if the series capacitances are not introduced is given by Figure 7.6:

$$\omega L_p/R = 1885 \times 0.275/300 = 1.73$$
$$\therefore \qquad \text{Effective loss} = 1.25 \text{ dB}$$

Alternatively the value of L_p required to obtain the specified effective loss at 300 Hz, in the absence of the series capacitances is obtained from

$$\omega L_p/R = 6.7, \therefore L_p \simeq 1 \text{ H}$$

However, in this particular example the provision

of two large-value capacitors may outweigh the advantage of the smaller inductance. A larger core may give a more satisfactory overall design.

7.2.7.5 Overall results

Figure 7.10 shows the complete equivalent circuit with the elements referred to the appropriate sides of the transformer. The calculated attenuation characteristic is also shown. This may be compared with the chararacteristic (shown by the broken line) which would be obtained if none of the capacitances had been introduced. It is seen that the effect of the filter design technique is to extend the relatively flat part of the pass band by sharpening the cut-off regions.

7.3 Low and medium frequency wide-band transformers

The transformers to be considered under this heading are those having the lower end of the pass band, i.e. f_1, below a few hundred kHz. Up to these frequencies the initial permeability of suitable manganese zinc ferrites is substantially constant so the design of such transformers may proceed as for an audio-frequency transformer. In general it will be necessary to use more than a few turns to obtain the required shunt inductance and the winding resistance will usually be a significant factor in the design. For these reasons it is usual to use a two-piece core embracing conventional windings on a coil former. The toroidal core is an alternative that is usually better suited to high-frequency transformers where relatively few turns are required.

The foregoing sections have considered the translation of the performance specification into element values in the equivalent circuit. The next stage of the design is to realize these values as parameters of a physical transformer. The relations between these elements and the design of the core and windings will now be discussed with particular reference to ferrite cores. An example of transformer design will then be given to illustrate the procedure.

7.3.1 The practical design in terms of the element values

7.3.1.1 Contribution of the core

The transformer core, by providing a low reluctance path for the magnetic flux ensures a high degree of coupling between primary and secondary windings. It is shown in text books on transformers that the coupling factor, k, defined as the proportion of flux due to one winding which actually links the other winding, is related to the shunt inductance and leakage inductance by the equation:

$$\frac{L_p}{L_l} = \frac{1}{2(1-k)} \quad (7.27)$$

The better the coupling, the nearer k approaches unity and the greater is the ratio L_p/L_l. Ignoring the limitations imposed by self capacitance, this ratio is proportional to the bandwidth of the transformer, a fact that follows from a consideration of Figure 7.6. The L_p/L_l ratio and other similar parameters will be discussed in more detail in the next section. From that discussion, other things being equal, the best transformer core is the one that gives the highest value of L_p for a given number of turns.

The relation between the inductance and the parameters of the core and winding is considered in detail in Chapter 4. The essential result is in Eqn (4.24) which for convenience may be expressed here as:

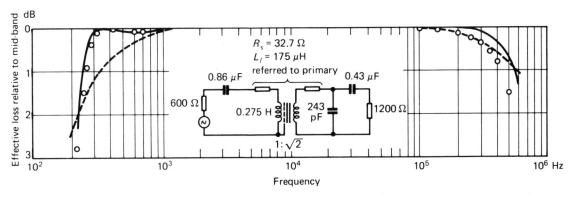

Figure 7.10 Example of wide-band transformer design using filter techniques
— calculated transmission characteristic
∘ ∘ ∘ values measured on a particular transformer to this design (see Section 7.3.2)
– – – calculated characteristic for circuit without capacitances

$$\frac{L_p}{N^2} = \mu_o \mu_e A_e / l_e = \frac{\mu_o \mu_e}{C_1} \quad \text{H} \tag{7.28}$$

where C_1 is the core factor expressing the core reluctance normalized with respect to permeability (see Eqn 4.9) and μ_e is the effective permeability of the core.

For a given magnetic material, a core having a low value of C_1 will give a high inductance for a given number of turns. However C_1, being a geometric factor, depends on the shape and size of the core and is thus related to other important design parameters such as winding resistance, leakage inductance and core volume. Thus the best value of C_1 is not necessarily the lowest.

The effective permeability, on the other hand, affects only the inductance for a given number of turns and the higher its value the better. At low audio frequencies cores of high permeability nickel iron alloy give the best performance. Strip wound toroids having permeabilities > 100 000 are obtainable, while conventional laminations stacks can provide effective permeabilities > 10 000. As the frequency increases, eddy currents within the alloy reduce the permeability and even when this reduction is combated by the use of very thin strip or laminations the permeability soon falls below the value obtainable with ferrites. The frequency of change-over depends on the initial permeability of the alloy and its thickness, and, of course, on the effective permeability of the ferrite core. Cost and convenience are also considerations. As the effective permeability of ferrite cores is progressively increased their use extends further into the audio frequency range. Although much of this chapter applies to wide band transformers in general, the core considerations will be confined to ferrite cores.

The effective permeability has been used in this discussion because ferrite transformer cores are often assembled from two or more pieces, and the residual gaps at the joints, however well the joints are made, reduce the permeability from the value intrinsic in the core material to an effective value. If the core has no gaps, e.g. a continuous toroid, then μ_e may be replaced by the material permeability, μ.

The relation between μ_e and μ is given in Eqn (4.26). In the present context, the gap length, l_g, is small so that the gap area may be taken as equal to the pole face area. Assuming a core of uniform cross-section, this expression is presented graphically in Figure 7.11 for various gap ratios, l_g/l_e. These ratios span the values that might be obtained in a two-part transformer core having mating surfaces made flat and smooth by grinding (the larger values) or by stress-free lapping (the smaller values), see also Sec-

(7.28) $\quad \dfrac{L_p}{N^2} = \dfrac{4\pi \mu_e A_e 10^{-9}}{l_e} = \dfrac{4\pi \mu_e 10^{-9}}{C_1} \quad \text{H}$

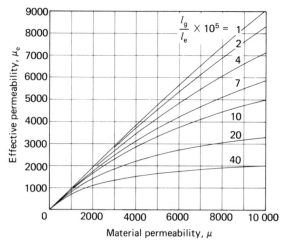

Figure 7.11 The effective permeability as a function of material permeability for several ratios of residual gap length to magnetic path length

tion 1.3.8 and Reference 100 of Chapter 1. It is clear that as the material permeability is made larger the greater is the degradation due to the residual gap.

While the concept of effective permeability is useful in assessing the merit of a transformer core, a more practical parameter is the inductance factor, A_L. This factor has already been introduced in connection with inductors (see Section 5.3). It relates the number of turns to the inductance thus:

$$L = A_L N^2 \quad \text{nH} \tag{7.29}$$

where A_L is the inductance in nH corresponding to one turn.

The specification for a ferrite transformer core invariably quotes a limit value of A_L. This value corresponds to the *initial* effective permeability, i.e. it relates to low flux density, low frequency, no magnetic polarization and usually a butt joint. It provides a ready means of calculating the number of turns required to give an inductance equal to or greater than a required minimum. In most simple transformer designs it does not matter by how much the shunt inductance exceeds the minimum value. However in a design involving the high pass filter technique an upper and lower limit must be placed on L_p and it is better to use a mean value of A_L.

If the amplitude is not small, so that there is an appreciable flux density in the core, then the permeability will be greater than the initial value (see Figure 3.4) and a higher inductance will be obtained. However, it is rarely possible to take advantage of this because wide-band transformers, even those operating at relatively high power levels, must retain their performance characteristics down to vanishingly small amplitudes. In fact high amplitudes might result in a greater number of turns being required in order to prevent excessive core loss, an aspect considered later in this section.

For most manganese zinc ferrites the parallel initial permeability, and therefore A_L, remain constant up to a few hundred kHz. Figure 3.11 shows typical permeability spectra; the higher the initial permeability the lower the frequency at which the fall of permeability occurs. As indicated at the beginning of Section 7.3 the low and medium frequency transformers at present being considered are those having f_1 within the substantially flat region of the permeability spectrum. The low frequency value of A_L will give a satisfactory design even if f_1 is situated at the upper extremity of the flat region and, in consequence, the pass band lies wholly in the region of permeability dispersion. This is because the shunt reactance (or impedance) is proportional to $\omega\mu_p$ and this continues to rise with frequency even when μ_p is falling due to dispersion phenomena (see the r.h. graphs of Figure 3.11 where the diagonal scale shows $\omega\mu'_p$ as a function of frequency).

Another possible dispersion in shunt inductance may arise from dimensional resonance. This phenomenon, which has been discussed in detail in Sections 2.2.4 and 4.2.3, arises from the establishment of standing electromagnetic waves across the core section. It may occur, in practice, when a manganese zinc ferrite core of relatively large cross-section is used at medium or high frequencies, e.g. > 20 kHz. Figure 4.7 relates the frequency of dimensional resonance to the least cross-sectional dimension of cores of typical manganese zinc ferrites. The resonant frequency is proportional to $\mu_e^{-1/2}$ so an air gap will move the resonance to higher frequencies. In the discussion of this phenomenon it was observed that the main effect of dimensional resonance on the impedance of the core is that the phase angle changes rapidly as a function of frequency while the modulus shows a relatively small variation. So unless it happens that the core is large enough for dimensional resonance to occur in the region of f_1 it will not normally cause any distortion of the transmission characteristic. Since the frequency of dimensional resonance increases as the core cross-section is reduced, and it is usually desirable for other reasons to use smaller cores at higher frequencies, there is a natural tendency to avoid dimensional resonance. However, if a design ever necessitates a manganese zinc core of abnormally large cross section then the possibility of dimensional resonance should be checked by reference to Figure 4.7.

If there is to be direct current flowing in one or more of the transformer windings so that there is a net polarization, this will reduce the material permeability to its incremental value and the simple procedure of using A_L may no longer be used. The most convenient procedure is that due to Hanna, in which an air gap is introduced into the magnetic circuit to minimize the effect of the direct current. This method has been discussed in the introduction to Figure 3.7 which gives Hanna curves for a variety of ferrites. The procedure is summarized here for convenience. A trial core of volume V and magnetic length l is selected. If L_p is the required inductance and I_o is the net polarizing current referred to the winding under consideration, then the value of $L_p I_o^2/V$ is calculated. Using the Hanna curves for a suitable transformer ferrite (see Table 3.1) the value of NI_o/l is obtained. Since I_o and l are known, the required number of turns, N, may be calculated. The curve is marked along its length by a scale of l_g/l and the value of this ratio at the intercept of the curve with the relevant co-ordinates enables the optimum air gap length l_g to be calculated. If the core is of non-uniform cross-section there is some uncertainty about the values to be assigned to l and V. In most cases the effective parameters l_e and V_e give reasonably accurate results. Even for pot cores these parameters are satisfactory. However when the core shape is complicated it is preferable to use a form of Hanna curve prepared specifically for the core type, if this is available. In this case the dimensional parameters of the core need not enter into the calculation. The curve gives $L_p I_o^2$ as a function of NI_o with gap or spacer thickness as a parameter. Figure 7.12 shows a family of these curves for typical wide band RM transformer cores as listed in Table 7.1 (see also Figure 9.12 which shows Hanna curves for a number of power transformer cores).

The air gap may, of course, be introduced into most cores by grinding back one of the pole faces. Some manufacturers supply transformer cores with a selection of standard gaps but since an optimum design usually requires a specific gap which might have any value within a rather large range it is more usual to obtain the required gap by inserting a paper or mica spacer. It must be remembered that, in the general Hanna curves, l_g is the total gap length, i.e. approximately twice the thickness of the spacer inserted between a pair of half cores, see also the caption of Figure 9.12.

In the foregoing, the core loss has been disregarded and in the majority of low and medium frequency transformers using ferrite cores the core loss is indeed negligible. At low amplitudes the residual loss and the eddy current loss in the core are the only possible contributions to the shunt loss resistance R_p. Considering the contributions separately, and assuming that the effect of the residual air gap is negligible, the residual loss component of R_p is given by Eqn (2.18)

$$R_p = \omega L_o \mu''_p \quad \Omega \quad (7.30)$$

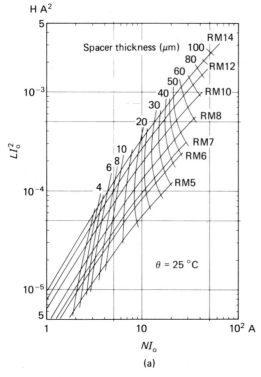
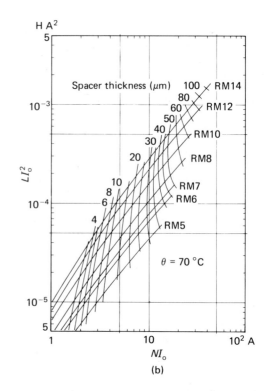

Figure 7.12 A family of Hanna curves for typical wide-band RM transformer cores as listed in Table 7.1. The core material is MnZn ferrite having $\mu \simeq 4500$; the gap data is in terms of the thickness of a spacer separating the core halves

where
$$L_o = \frac{\mu_o N^2}{C_1} \quad \text{H} \quad (7.31)$$

and μ_p'' is the loss component of the parallel complex initial permeability; it is given as a function of frequency for a number of ferrites in the r.h. graphs of Figure 3.11.

Multiplying and dividing the above equation by μ_p'

$$R_p = \omega L_o \left(\frac{\tan \delta_r}{\mu_p'} \right)^{-1} = \frac{\omega L_p}{\tan \delta_r} \quad (7.32)$$

where $(\tan \delta_r)/\mu_p'$ is for most practical purposes equal to the residual loss factor. This factor $(\tan \delta_r)/\mu$ is given as a function of frequency for a number of ferrites in Figure 3.12.

In the frequency range under consideration, $\tan \delta_r \ll 0.1$ for most manganese zinc ferrites. R_p will therefore be much greater than ωL and the attenuation that it causes will be small compared with the droop due to ωL_p. If the core has a residual air gap such that the effective permeability is μ_e it follows from Eqns (4.50) and (7.32) that the value of R_p is unchanged.

The component of the shunt resistance due to eddy current loss in the core may be obtained from Eqn (4.55)

$$R_p = \frac{\omega L_p}{\tan \delta_F} = \frac{\omega L_p \rho \beta}{\pi \mu_o \mu_e d^2 f} \quad \Omega \quad (7.33)$$

where the symbols are defined after Eqn (2.46).

This component is again negligible in most practical cases. Indeed if the frequency and core cross-sectional dimension, d, are such that the eddy current loss is significant then a check should be made (using Figure 4.7) to see whether dimensional resonance may occur. Eddy current core loss and dimensional resonance are described in greater detail in Sections 2.2.4 and 4.2.3.

Finally, if the amplitude is not negligibly small there may be a hysteresis loss contribution. For a given voltage amplitude, the flux density will increase

(7.31) $\quad L_o = \dfrac{4\pi N^2 10^{-9}}{C_1} \quad$ H

(7.33) $\quad R_p = \dfrac{\omega L_p}{\tan \delta_F} = \dfrac{\omega L_p \rho \beta 10^9}{4\pi^2 \mu_e d^2 f} \quad \Omega$

as the frequency decreases (see Eqn (4.45)) and so the hysteresis loss will become greater; if it becomes significant it will contribute to the lower frequency droop. Hysteresis loss is discussed in Sections 2.2.3 and 4.2.3.

If hysteresis data are available in the form of the loss tangent, $\tan \delta_h$, or the loss factor $(\tan \delta_h)/\mu$, then the form of Eqn (7.32) applies and the contribution to the shunt resistance is given by:

$$R_p = \omega L_o \left(\frac{\tan \delta_h}{\mu}\right)^{-1} = \frac{\omega L_p}{\tan \delta_h} \quad \Omega \quad (7.34)$$

assuming that $\tan \delta_h \ll 1$.

The hysteresis loss factor is given as a function of flux density for a number of ferrites in Figure 3.15. It is only necessary to calculate the effective flux density, $\hat{B}_e$, from Eqn (4.45) for a number of frequencies in the vicinity of f_1, obtain the corresponding values of $(\tan \delta_h)/\mu$ and use these to calculate R_p. Within the Rayleigh region, $\tan \delta_h$ may also be calculated from the hysteresis material constant η_B, see Eqn (2.63) or the hysteresis core constant η_i, see Eqn (4.44).

For large amplitudes, core loss density data such as given in Figure 3.19 may be used. If at a given frequency and flux density the power loss density is P_M and the core has a volume V_e

$$U^2/R_p = P_M V_e$$

$$\therefore \quad R_p = \frac{U^2}{P_M V_e} \quad (7.35)$$

If this core loss causes a significant increase in attenuation at f_1 allowance for this must be made by increasing the inductance. In higher-power transformers the hysteresis loss dominates the design. A given limit on R_p or total core loss determines P_M and hence places a limit on the flux density. For a given core operating at a given frequency and voltage, the e.m.f. Eqn (4.45) determines the minimum number of turns.

It is also possible for hysteresis to dictate the number of turns at low amplitudes. This might occur if there were a limit imposed on the permissible waveform distortion. This would limit $\tan \delta_h$ (see Section 4.2.4) and in consequence would restrict the flux density.

A final remark concerning the contribution of the core in transformer design. The transformer must usually have a satisfactory performance over a given temperature range. The permeability and losses may show appreciable variations within this range (e.g. see Figure 3.8 and 3.20) and, since the core normally has only a residual air gap, the inductance and shunt loss resistance vary by almost the same proportion. It is thus advisable to ensure that the design is based on the minimum A_L, or maximum loss density, occurring within the operating temperature range.

7.3.1.2 Relations arising from the geometry of core and winding

In the design of a transformer any core may, in principle, be used to obtain a sufficient shunt inductance; it is only necessary to provide enough turns. However, it is shown in Section 11.3 that the resistance of a winding on a given core is proportional to N^2, see Eqn (11.11). Therefore a core requiring a larger number of turns for a given inductance will give a relatively high winding resistance and this may cause excessive mid-band attenuation.

Since both the inductance, L_p, and the winding resistance, R_{dc}, are proportional to N^2, the ratio R_{dc}/L_p depends only on the effective permeability and on the geometry of the core and winding, see Eqn (11.12). This ratio may be taken as a figure of merit of a transformer core. In practice one transformer winding occupies only a fraction of the total winding space. It is shown in Section 11.3 that for a simple transformer the least total winding loss is obtained when the primary and secondary each occupy substantially half the total winding space. Therefore the resistance of the primary or secondary will be approximately twice the value that would be obtained for a full winding of the same number of turns, see Eqn (11.11). The resistance/inductance ratio for a winding occupying half the coil former, designated $(R_{dc}/L)_{1/2}$ will be twice the value given by Eqn (11.12). In terms of the inductance factor, A_L (in nH/turn²);

$$(R_{dc}/L)_{1/2} = \frac{2\rho_c l_w 10^9}{F_w A_w A_L} \quad \Omega H^{-1} \quad (7.36)$$

The mid-band and lower frequency specifications will determine limit values for $R_s \ (= R_1 + R_2')$ and L_p respectively. A core having

$$(R_{dc}/L)_{1/2} \leq R_1/L_p \quad (7.37)$$

will enable both specification requirements to be met. This assumes that eddy current losses in the winding are unimportant, which is usually true. However, if the transformer uses solid conductors at high frequencies and the d.c. winding resistance is near the limit, then the a.c. resistance should be checked by reference to Section 11.4.4.

After considering the mid-band performance relative to that at the lower frequency region, the next most important performance parameter is the bandwidth. It is often the case that the higher frequency cut-off is determined by the leakage inductance alone. This is particularly so when the circuit impedance is low or the transformation ratio is near unity. Under these circumstances the simple equivalent circuit will show a low maximum limit for the leakage inductance while the corresponding self capacitance limit may be so high that it clearly will not be a limiting factor in the design. When this is so the bandwidth is proportional to L_p/L_l. Indeed, it is easily shown from Figure 7.6 that if the source is matched to the load then for a 1 dB droop at both f_1 and f_2:

$$\frac{f_2}{f_1} = \frac{L_p}{L_l} \qquad (7.38)$$

Leakage inductance in transformers is considered in Section 11.7. Expressions are given which enable the leakage inductance of various winding arrangements to be calculated. A simplified expression for L_l in a transformer having two windings wound one above the other may be derived (see Eqns (11.58) and (11.60a)):

$$L_l = \mu_o N_1^2 \frac{l_w h_w}{3 b_w} \quad \text{H} \qquad (7.39)$$

where l_w is the mean turn length,
h_w is the total winding height of both windings,
b_w is the winding breadth,
and it is assumed that the inter-winding space is negligible
Therefore, combining this with Eqn (7.28)

$$\frac{L_p}{L_l} = \frac{3\mu_e b_w}{l_w h_w C_1} \qquad (7.40a)$$

The corresponding expression for the side-by-side winding arrangement is

$$\frac{L_p}{L_l} = \frac{3\mu_e h_w}{l_w b_w C_1} \qquad (7.40b)$$

The former arrangement, in which the windings are placed one over another, is usually preferred. In this case the lowest leakage inductance is obtained by using a broad shallow winding area.

The side-by-side arrangement favours a tall narrow winding space but this leads to a larger mean turn length, l_w, and increases R_{dc}/L_p.

Given an optimum core shape, the ratio $L_p L_l$ and hence the fractional bandwidth may be increased by increasing the effective permeability of the core. As observed in the previous section, a high effective permeability is the main attribute of a good transformer core. If the highest practical value of effective permeability has been achieved then L_p/L_l can only be further increased by reducing L_l. For a given core shape this is only possible by subdividing the windings into alternate sections of primary and secondary (see Section 11.7).

If the circuit impedance is high or the transformation ratio differs appreciably from unity, or if the pass band is towards the top of the frequency range at present under consideration, then self capacitance may limit the bandwidth at the higher-frequency end. Under these circumstances the simple equivalent circuit will have an easily achievable value of L_p/L_l, while the maximum limit of shunt capacitance will be low, e.g. < 50 pF. In the present context only the self capacitance of the windings is relevant; if there is shunt capacitance associated with the source or load this must be deducted from the overall limit to obtain the limit of winding self capacitance.

The calculation of winding self capacitance is considered in Section 11.6; self capacitance expressions are tabulated for most common winding configurations. If the transformation ratio differs appreciably from unity, only the self capacitance of the higher impedance winding is significant. Direct capacitance between windings is usually eliminated by means of an electrostatic screen. When windings consist of only a few layers, the capacitance of winding to screen must be taken into account.

From Section 11.6 the winding self capacitance is largely independent of the number of turns. For a simple multilayer winding it is shown in Eqn (11.56) that

$$C_s \propto \frac{b_w l_w}{h_w} \qquad (7.41)$$

where, in this case, h_w is the height of the winding under consideration (see Figure 11.1).

Thus the best winding geometry for minimum self capacitance is a tall narrow cross-section. This is the reverse of the shallow broad winding shape required for low leakage inductance and a low value of R_{dc}/L_p. However a tall narrow winding may be simulated in a shallow broad winding space by constructing the winding in narrow sections wound side-by-side, see expression 6 of Table 11.3 and Section 5.6.

For a simple two-winding unsectionalized transformer, Eqns (7.41) and (7.39) may be combined to give

$$L_l C_s \propto N_1^2 l_w^2 \qquad (7.42)$$

since h_w in the two equations, although different, are proportional. In Section 7.2.5 it was seen that L_l and C combine to form a low pass filter having a cut-off given by $\omega_c = (L_l C)^{-1/2}$. Although the upper limit of the transmission is usually less than the theoretical cut-off frequency, the value of ω_c may be taken as an indication of the limit of high frequency performance. From the above equation:

$$\omega_c \propto \frac{1}{N_1 l_w} \qquad (7.43)$$

i.e. for a given shunt inductance, ω_c is proportional to $\mu_e^{1/2}$ or $A_L^{1/2}$ and inversely proportional to the mean diameter of the winding. Therefore for the best high-

---

(7.39) $L_l = 4\pi 10^{-9} N_1^2 \dfrac{l_w h_w}{3 b_w}$ H

frequency performance the selected core should be the smallest that has a value of $(R_{dc}/L_p)_{1/2}$ adequate to ensure satisfactory mid-band and lower frequency transmission, and for that size it should have the highest possible A_L value. The winding cross-section affects ω_c only inasmuch as it affects the mean turn length. It follows that a wide shallow winding cross-section is preferable. Sectionalizing the windings either to reduce leakage inductance or self capacitance will raise the limit of ω_c.

7.3.1.3 Form of the core

For most wide-band transformer applications, the best core material is the one that provides the highest initial permeability at the lower end of the pass band, i.e. at f_1. Of the various types of ferrite available manganese zinc ferrite is invariably used for low and medium frequency transformer cores. As shown in Table 3.1, special high-permeability grades have been developed for this application.

Most ferrite transformer cores consist of two pieces assembled together with a butt joint. The residual air gap is made as small as possible by grinding (and often lapping) the mating surfaces so that they are flat and smooth. The magnetic properties of the transformer core and the effect of the residual gap are described earlier in this section.

Ferrite transformer cores are made in a variety of basically different shapes. For each shape it is possible to optimize the proportions to obtain the lowest value of $(R_{dc}/L)_{1/2}$, e.g. see Section 5.2. This is often done but in many designs practical considerations take precedence. Most important among these are the termination requirements. Telecommunication transformers are often wound with very fine wire and may have many terminations. The geometric arrangement of the terminals must usually be compatible with the standard printed circuit grid. These considerations lead to the use of a coil former having a relatively large number of fixed terminal pins arranged in a particular pattern. The core must often be designed round the coil former. In addition, the equipment practice may set a height limitation.

The first ferrite transformer cores were made in the form of double E cores to match the standard shapes that had long been in use for metal laminations. By matching these shapes, the ferrite cores carried the attendant restriction, i.e. the core cross-section was necessarily rectangular. It was soon realized that a pressed ferrite core need not be restricted in this way and E cores with round centre legs appeared. These had the advantage that it is easier to wind a cylindrical coil, and when this is done in production a much better control may be exercised over the leakage inductance and self-capacitance. This is particularly important if these elements are being used in a filter design technique.

With the greater versatility of core design offered by ferrites, shapes having other special advantages were introduced. The pot core shape with proportions optimized for minimum R_{dc}/L had already been developed for inductors. The minimum R_{dc}/L and the magnetic screening provided by these cores made them attractive as transformer cores and special ungapped, high permeability ranges were made available. The main disadvantage of the pot core is that the lead-out slots are not wide enough to accommodate a coil former having a sufficient number of fixed terminal pins.

The RM core, already described as an inductor core in Section 5.2, provides a more functional solution and is currently probably the most widely used core for wide band transformers. It combines compact shape, a coil former with fixed pins located on the printed circuit grid and good magnetic screening. Transformer versions using high permeability ferrite and having no central hole are readily available.

Another popular shape is the EP core, which is cubic with an offset winding annulus that breaks through the lower face of the core to allow access for the pinned flanges of the coil former. When mounted on a circuit board the winding axis is horizontal. Since the terminal pins are located on both flanges, single layer windings are easily terminated.

The simple toroidal core can provide potentially the greatest band width since it has no residual gap and, properly wound, will give minimal leakage inductance. Some aspects of toroidal cores for high-frequency wide-band transformers are considered in Section 7.4.3. In spite of its good performance, the application of the toroidal core is somewhat limited by the need for specialized winding machines.

Table 7.1 lists four current ranges of ferrite cores suitable for wide-band transformers and gives the physical dimensions of the cores and winding spaces together with typical A_L and $(R_{dc}/L)_{1/2}$ values. It also includes a parameter k_r which enables $(R_{dc}/L)_{1/2}$ to be quickly calculated where the given values of A_L or F_w differ from those assumed for the listed $(R_{dc}/L)_{1/2}$ values.

From the limit values of R_1 and L_p in the equivalent circuit, the maximum permissible value of $(R_{dc}/L)_{1/2}$ may be calculated. The core selected for a trial design should have an $(R_{dc}/L)_{1/2}$ value that does not exceed the permissible value of R_1/L_p. If the transformer must have extra windings or an unusually large allowance for insulation, or if a higher permeability core is justified, then the value of $(R_{dc}/L)_{1/2}$ must be obtained from k_r. If the transformer has d.c. magnetization so that the design involves the Hanna procedure then clearly the $(R_{dc}/L)_{1/2}$ criterion cannot be used. In such cases the trial design must be based on

Table 7.1 Properties of some transformer cores

Core type	Assy. vol. (mm³ × 10³)	Nominal core dimensions			Dimensional parameters of the core			
		A (mm)	B (mm)	C (mm)	l_c (mm)	A_c (mm²)	V_c (mm³ × 10³)	C_1 (mm⁻¹)
EE cores								
EF12.6	2.1	12.6	12.8	3.6	29.6	13.0	0.384	2.28
EF16	4.3	16.1	16.1	4.5	37.6	20.1	0.754	1.87
20/20/5	5.3	20.2	20.0	5.1	42.8	31.2	1.34	1.37
30/30/7	17.7	30.1	30.0	7.1	66.9	59.7	4.00	1.12
Pot cores								
7/4	0.22	7.25	4.15	–	10.0	7.0	0.070	1.43
9/5	0.44	9.15	5.3	–	12.5	10.0	0.125	1.25
11/7	0.80	11.1	6.5	–	15.9	15.9	0.252	1.00
14/8	1.66	14.1	8.35	–	20.0	25.0	0.50	0.80
18/11	3.42	17.9	10.55	–	25.9	43	1.12	0.60
22/13	6.25	21.6	13.4	–	31.6	63	2.00	0.50
26/16	10.5	25.5	16.1	–	37.2	93	3.46	0.40
30/19	16.9	30.0	18.8	–	45	136	6.1	0.33
36/22	27.4	35.5	21.7	–	52	202	10.6	0.264
42/29	52.9	42.4	29.4	–	69	265	18.2	0.259
RM cores*								
RM4	0.97	9.6	10.4	–	21.0	11.0	0.232	1.9
RM5	1.52	12.1	10.4	–	22.1	23.8	0.526	0.93
RM6R	2.61	14.5	12.4	–	27.5	38.0	1.04	0.73
RM6S	2.57	14.4	12.4	–	28.6	36.6	1.05	0.78
RM7	3.83	16.9	13.4	–	30.4	43.0	1.34	0.70
RM8	6.11	19.3	16.4	–	38.0	64.0	2.43	0.59
RM10	10.9	24.2	18.6	–	44.0	98.0	4.31	0.45
RM12	21.0	29.3	24.5	–	57.0	146	8.34	0.390
RM14	35.4	34.3	30.1	–	70.0	200	14.0	0.350
EP cores								
EP7	0.75	9.2	7.4	6.4	15.7	10.3	0.162	1.52
EP10	2.13	11.5	10.3	7.7	19.2	11.3	0.217	1.70
EP13	3.29	12.5	12.9	8.8	24.2	19.5	0.472	1.24
EP17	7.00	18.0	16.8	11.0	28.5	33.9	0.966	0.84
EP20	13.0	24.0	21.4	15.0	40.0	78	3.12	0.51
EP30	31.8	31.0	29.9	23.1	62.6	179	11.2	0.351

* with no central hole (except for RM4).

Notes (1) The data used in this table have been compiled with reference to the data sheets of a number of leading manufacturers. There are usually small variations between the different sources so the figures presented here will not agree exactly with any particular set of data sheets. The accuracy should be adequate for design purposes but where manufacturing specifications are involved reference should be made to manufacturer's data.

(2) The assembly volume is the volume of the least rectangular box that will just enclose the core plus coil former (excluding pins or tags).

(3) The inductance factors are typical nominal values based on a ferrite having a nominal initial permeability of 4300, e.g. *N30*, (courtesy of Siemens). Some of the listed cores are also available made of ferrites with much higher initial permeabilities but for the purpose of comparing core performance it is preferable to relate the data to a medium-permeability ferrite that is typical of those commonly used for wide-band transformers.

(4) The parameter k_r is the value of $(R_{dc}/L)_{\frac{1}{2}}$ normalized with respect to the winding copper factor, F_w, and the inductance factor, A_L. Therefore, for any given values of these parameters, $(R_{dc}/L)_{1/2} = k_r / A_L F_w$.

(5) The figures in the column headed $(R_{dc}/L)_{\frac{1}{2}}$ have been calculated from Eqn (7.36) in which F_w is put equal to 0.5 and A_L is the listed value. For the smaller cores, particularly if fine wire is to be used, an F_w value of 0.5 is somewhat optimistic and allowance should be made for this by choosing a core with a lower value of $(R_{dc}/L)_{\frac{1}{2}}$ or by using the parameter k_r.

Winding dimensions of coil formers				Inductance factor (typical) A_L (nH/turn²)	k_r $= \dfrac{2\rho_c l_w \times 10^9}{A_w}$ (Ω)	$(R_{dc}/L)_{1\,2}$ $= \dfrac{k_r}{A_l F_w}$ (Ω H⁻¹)	Standards
l_w (mm)	h_w min. (mm)	b_w min. (mm)	A_w min. (mm²)				
27.2	1.6	7.3	11.7	1000	78800	158.	DIN 41 985
34	2.3	9.9	22.8	1400	50500	72.2	
36	2.6	10.6	27	2500	45200	36.1	DIN 41 295
52	4.7	17.1	80	3300	22000	13.3	
14.6	1.0	2.05	2.05	2000	241000	241	
18.9	1.24	2.7	3.35	2500	191000	153	
22.7	1.50	3.4	5.10	3200	151000	94.2	
29.0	2.1	4.4	9.25	4200	106000	50.6	IEC 133
36.7	3.0	6.0	18.0	5600	69100	24.7	BSI 4061
44.5	3.5	7.75	27.1	7000	55600	15.9	(range 2)
52.6	4.0	9.55	38.2	9000	46600	10.4	DIN 41 293
62.0	4.8	11.3	54.3	10500	38700	7.37	
74.3	5.8	12.7	73.7	13500	34200	5.06	
86.2	7.8	17.7	138	13600	21200	3.11	
19.9	1.35	5.8	7.83	1700	86100	101	
25.0	1.98	4.95	9.8	3500	86600	49.5	
30.8	2.33	6.65	15.5	4300	67300	31.3	
30.7	2.33	6.45	15.0	4300	69300	32.3	IEC 431
35.8	3.05	6.95	21.2	5000	57200	22.9	DIN 41 980
42	3.4	8.95	30.4	5700	46800	16.4	
52	4.15	10.5	44	7600	40000	10.5	
61	4.95	14.6	72	8400	28700	6.83	
71	5.85	18.5	108	9500	22300	4.69	
17.9	1.20	3.4	4.1	2000	148000	148	
21.5	2.05	5.6	11.5	2000	63300	63.3	
23.8	1.83	7.6	13.9	2800	58000	41.4	
28.9	2.03	9.3	18.9	4300	51800	24.1	
40.8	2.75	12.3	33.8	6700	40900	12.2	
63.8	3.2	20.9	66.9	9600	32300	6.73	

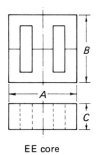

EE core

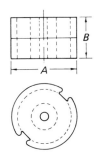

Pot core

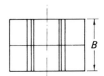

RM core

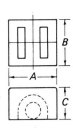

EP core

7.3.2 An example of transformer design procedure

The design procedure depends to some extent on the requirements in the specification. The procedure outlined below includes most of the steps in the design of a normal two-winding wide-band communication transformer. If there are few requirements and they are easily satisfied the more detailed steps may be omitted.

(1) Translate the specification into equivalent circuit element values as described in Section 7.2.
(2) If there is to be superimposed d.c. or if there are more than two windings the best core must be selected by experience or by trial designs. If these complications are absent, calculate R_1/L_p (reducing R_1 by perhaps 10% below the specification value to allow for manufacturing tolerances) and choose a core having $(R_{dc}/L)_{1/2}$ a little less than this calculated value.
(3) If the windings have to carry d.c., use the Hanna curves as described in Section 7.3.1. If not, use A_L for the core and calculate the number of turns required to obtain the shunt inductance, L_p (see Eqn (7.29)).
(4) From the turns ratio calculate the number of turns in the other winding(s).
(5) Calculate the winding area available for one of the windings; this will normally be half the total net winding area, A_w, after a deduction has been made for insulation. The conductor diameter may then be obtained as described in Section 11.2. In a similar way obtain the conductor diameter for the other winding(s).
(6) Calculate the winding resistance (see Section 11.3). Check that $R_1 \simeq R_2/r^2$ and that $R_1 + R_2/r^2$ is less than the permitted value by a suitable margin for tolerances. If not, either a better copper factor must be obtained or a core having a lower $(R_{dc}/L)_{1/2}$ ratio must be used.
(7) Using Eqns (7.30)–(7.35), the core loss may be checked to ensure that R_p is negligible over the bandwidth. If it is excessive, more turns or a larger core or one having less loss must be used.
(8) Assuming the simplest winding arrangement, calculate the leakage inductance from the expression in Figure 11.21. If the value is excessive try the effect of dividing the windings into alternately wound sections. If a sufficiently low value of L_l cannot be achieved by sectionalizing then a better core must be used, i.e. one having a higher L_p/L_l ratio; see Eqn (7.40a).
(9) Calculate the winding self-capacitance by the methods described in Section 11.6. If the total shunt capacitance due to source, load and windings is excessive, side-by-side sectionalizing must be tried.
(10) If a filter design technique is used, calculate the additional (external) inductance and capacitance that will be required.

These steps will now be exemplified by taking the transformer design already considered in Section 7.2.7 and continuing it to the practical realization.

The transformer has a ratio, secondary turns/primary turns, equal to $\sqrt{2}$. The main parameters referred to the primary (600 Ω) side are:

Open circuit inductance, $L_p = 0.275$ H
Total winding resistance, $R_1 + R_2/r^2 < 42$ Ω
Leakage inductance = 175 μH
Secondary shunt capacitance (referred to primary) = 486 pF

Voltage amplitudes are assumed to be small so that hysteresis may be neglected. It is also assumed that there is no direct current.

The primary and secondary windings will occupy approximately equal winding areas, so $R_1 \simeq R_2'$ and therefore $R_1 < 21$ Ω. Deducting 10% for tolerances, $R_1/L_p = 18.9/0.275 \simeq 69$ Ω/H. Referring to Table 7.1, suitable cores are the 14/8 pot core, the RM5 core and possibly the EP10 core. As the core is towards the smaller end of the size range and therefore the assumed value of F_w is probably optimistic, one of these cores with the lower value of $(R_{dc}/L)_{1/2}$ should be chosen. For the purpose of this example the RM5 core, $(R_{dc}/L)_{1/2} = 49.5$ will be used.

The number of turns required to ensure that the inductance will be greater than the specified value may be obtained from the limit A_L using Eqn (7.29). However, in the present example L_p is a filter element so it is better to use an average or nominal value of A_L. From Table 7.1 a nominal value for the RM5 core is 3500 nH/turn². Then the number of primary turns is given by

$$N_1 = (L_p/A_L)^{1/2} = (0.275 \times 10^9/3500)^{1/2} = 280$$

The available winding cross-sectional area for this winding may be taken as approximately half of the value of A_w given in the table, but as the area is rather small it is better to work in terms of the available winding height. For the RM5 core the minimum h_w is 1.98 mm; allowing 0.3 mm for insulation the net height is 1.68 mm of which half may be provisionally assigned to the primary winding, i.e. 0.84 mm. The minimum winding breadth is 4.95 mm so the provisional winding area is $0.84 \times 4.95 = 4.16$ mm². Referring to Section 11.2 and using a packing factor, F_p, of 0.87 in Eqn. (11.6), the maximum overall conductor diameter for the primary winding, if random winding is to be used, is theoretically,

$$d_o = (4.16 \times 0.87/280)^{1/2} = 0.114 \text{ mm}$$

If Grade 1 metric enamelled wire is to be used, Table B.1.2 of Appendix B shows that the next smaller wire gauge has $d_o = 0.110$ mm and a nominal copper diameter $d = 0.090$ mm. The preceding table gives the resistance per unit length, R_c, to be 2.687 Ω/m. The mean turn length from Table 7.1 is 25 mm. Therefore the resistance of the primary winding is, from Eqn (11.9)

$$R_1 = 280 \times 25 \times 10^{-3} \times 2.687 = 18.8 \text{ Ω}$$

Since in such a small winding the available area must be conserved it is advisable to calculate the actual height occupied by the chosen gauge of wire (which will be less than the height provisionally assumed) so that the surplus may be allocated to the secondary. The actual winding area is given by

$$A_w = (d_o^2 \times N_1)/F_p = 0.11^2 \times 280/0.87 = 3.89 \text{ mm}^2$$

So the actual winding height is $3.89/4.95 = 0.786$ mm. Therefore the height available for the secondary winding is $1.68 - 0.786 = 0.894$ mm and the area is $4.95 \times 0.894 = 4.43 \text{ mm}^2$.

The number of secondary turns is $N_2 = 280 \times \sqrt{2} = 396$ and the corresponding value of d_o is $(4.43 \times 0.87/396)^{1/2} = 0.987$ mm. From the same wire tables the next smaller wire has $d_o = 0.098$ mm, $d = 0.080$ mm and a resistance of 3.401 Ω/m. So the secondary resistance is

$$R_2 = 396 \times 25 \times 10^{-3} \times 3.401 = 33.7 \text{ Ω}$$

$$\therefore R'_2 = 33.7/2 = 16.9 \text{ Ω and}$$

$$R_s = 18.8 + 16.9 = 35.7 \text{ Ω}$$

This is just within the imposed limit of $0.9 \times 42 = 37.8$ Ω. It should be appreciated that in the design all dimensions have been put equal to the worst-case limits.

The mean value of R/L_p for the two windings is 65 and the actual winding copper factor is, from Eqn (11.4),

$$(280 \times \pi \times 0.09^2/4 + 396 \times \pi \times 0.08^2/4)/9.8 = 0.385$$

where 9.8 is A_w for the RM5 core

As predicted, the actual copper factor is significantly lower than that assumed for $(R_{dc}/L)_{1/2}$ in Table 7.1.

The requirements for the shunt inductance and series resistance having been satisfied, the effect of core loss may be checked. Hysteresis loss has been assumed negligible and the frequency is too low for significant eddy current loss in the core. The value of R_p may be estimated either from the residual loss factor or from μ_p'' for the appropriate ferrite. For instance, Figure 3.12.2–5 gives loss factors for a number of high permeability ferrites and Eqn (7.32) can be used to obtain R_p. For the purpose of this example the μ_p'' parameter will be used. Referring to Figure 3.11.3, the curve for ferrite T3 is suitable and yields the following figures:

at $f = 10$ kHz, $\mu_p'' = 6 \times 10^5$
at $f = 300$ kHz, $\mu_p'' = 1.5 \times 10^4$

From Eqn (7.30), $R_p = \omega\mu_o N^2 \mu_p''/C_1$, so inserting the above data and putting $N = 280$ and C_1 for the RM5 core $= 0.93 \text{ mm}^{-1} = 930 \text{ m}^{-1}$;

at $f = 10$ kHz, $R_p = 4$ MΩ
at $f = 300$ kHz, $R_p = 3$ MΩ

The lower of these values may be used to check the return loss by reference to Figure 7.5. $R_s/R'_b = 35.7/600 = 0.06$ and $R'_b/R_p = 600/(3 \times 10^6) = 2 \times 10^{-4}$. Clearly the core loss is so low that it may be neglected and the $R'_b/R_p = 0$ curve may be used. This yields a return loss of 31 dB when $R_s/R'_b = 0.06$. Thus the midband requirements are adequately met.

The leakage inductance must now be calculated; at first the simple winding arrangement shown in the l.h. diagram of Figure 11.21 will be assumed. If the windings occupy all the available winding space the parameters in the leakage inductance expression given in the caption to Figure 11.21 are as follows

$N = 280$ $\qquad \Sigma x \simeq 1.68$
$l_w = 25$ $\qquad \Sigma x_\Delta = 0.2$ (assumed)
$M = 1$
$Y = 4.95$

The expression then yields $L_l = 380$ μH; this value is too high. However, by using the second winding arrangement shown in Figure 11.21, i.e. a simple sandwich, the above figure may be divided by $M^2 = 4$, i.e. $L_l = 95$ μH. This is satisfactory; for the actual transformer it will be necessary to adjust the value to 175 μH by the addition of auxiliary (series) inductance.

It finally remains to check that the self capacitance of the windings is not excessive. For brevity only the shunt component of the self capacitance will be considered; the method used will be that outlined towards the end of Section 11.6.2.

The mean area of the layers is $l_w \times b_w = 25 \times 4.95 = 124 \text{ mm}^2$. From Figure 11.18 the effective thickness of the dielectric is

$$S_e = 1.26 d_o - 1.15 d$$

(Since there is no interleaving of insulation between layers, $\Delta_i = 0$.) Substituting the wire diameters for the primary, $S_e = 0.0351$ mm.

224 Wide-band transformers

From Eqn (11.55(a)), assuming that the effective permittivity of the dielectric is 4 and that the mean layer may be used, the capacitance between layers is

$$C_l = 0.00885 \times 4 \times 124/0.0351 = 125 \text{ pF}$$

The effective number of layers, $p = h_w/d_o$; for the primary winding, $p = 0.786/0.11 \simeq 7$. Therefore the shunt element of the self capacitance, obtained from expression 5 of Table 11.3 is:

$$C_a = \frac{4 \times C_l(p-1)}{3p^2} = \frac{4 \times 125 \times 6}{3 \times 49} = 20 \text{ pF}$$

The corresponding values for the secondary winding are:

$$C_l = 139 \text{ pF} \quad p = 9$$
$$C_a = 18 \text{ pF}$$

These capacitances are very small compared with the filter capacitances that are required. In practice they could be ignored and external capacitors of the specified values used to realize the circuit.

In the first edition, the transformer design was based on the 18 mm pot core because the A_L values then available were about half the present-day figures. At that time a model was made and its performance is relevant to the present example. It had the following properties referred to the 600 Ω side (the calculated values for that design are given in brackets to indicate the degree of agreement).

Open circuit inductance, L_p	= 299 mH	(275)
Primary winding resistance, R_1	= 15.5 Ω	(16.5)
Secondary winding resistance, R'_2	= 17.2 Ω	(16.9)
Leakage inductance, L_l	= 160 µH	(166)
Primary plus secondary self capacitance, $C_1 + C'_2$	= 106 pF	(109)

The transmission characteristic of this transformer was measured in the circuit shown in Figure 7.10; the plotted results are shown together with the theoretical characteristic. For comparison, the theoretical characteristic is given also for the circuit without capacitances, i.e. the characteristic that would have been obtained without recourse to filter technique (assuming L_l remains at 175 µH). It is seen that the measured high frequency droop is greater than that theoretically predicted. This is due to eddy current effects increasing the winding resistance at the upper end of the band. However, the return loss at 300 kHz was found to be 16.4 dB which is in accordance with the design value. The lower frequency region is in close agreement with the theoretical curve.

7.4 High-frequency wide-band transformers

Although there is no clear division of the frequency bands, high-frequency transformers may be considered as those having pass bands lying wholly above about 0.1 MHz. Above this frequency the permeability of suitable transformer ferrites begins to vary significantly with frequency and the loss angle begins to rise. The higher the operating frequency the more important it becomes to regard the core impedance as a complex function of frequency; simple constants, such as A_L used at low frequencies, become inappropriate.

Another important distinction is that, at high frequencies, the circuit impedances are usually lower so that generally lower shunt impedances are necessary. The required impedance may usually be obtained with few turns so, except at high powers, the winding resistance becomes a negligible factor in the design. It is no longer necessary to consider the $(R_{dc}/L)_{1/2}$ value of the core. Instead the main concern is that for a given shunt impedance at f_1 the leakage inductance and self capacitance shall be a minimum.

Much of the discussion concerning low and medium frequency transformers is relevant in the present context and indeed the design procedure set out in Section 7.3.2, adapted to take account of the complex permeability, could in principle be used for high-frequency transformers. The object of the present section is to consider alternative approaches and techniques that are specific to the design of ferrite-cored transformers intended to operate at these frequencies.

7.4.1. Contribution of the core

The shunt inductance, L_p, and the shunt resistance, R_p, see Figure 7.3, depend directly on the core permeability and losses respectively. At low and medium frequencies the permeability may be regarded as constant and the losses are usually negligible. At high frequencies the onset of dispersion phenomena such as ferromagnetic resonance invalidates these conditions, see Figure 3.11, and it is better to work in terms of the parallel complex permeability. From Eqn (2.18)

$$\left.\begin{aligned}\frac{L_p}{N^2} &= \frac{\mu_o \mu'_p}{C_1} \quad \text{H} \\ \frac{R_p}{N^2} &= \frac{\omega \mu_o \mu''_p}{C_1} \quad \Omega \\ \tan \delta_m &= \frac{\mu'_p}{\mu''_p}\end{aligned}\right\} \quad (7.44)$$

If the core has an air gap, the inductance is reduced in the ratio μ_e/μ'_p (see Section 4.2.2) and the shunt resistance is virtually unchanged (see Eqn (4.48)). In the present context the air gap is usually negligible or entirely absent.

For a transformer to have a specified limit of insertion loss or return loss, the modulus of the shunt impedance presented by the core must not be less than a certain value, this value depending on the loss tangent. If for a given ferrite core the value of $\tan \delta_m$ at f_1 is known then a minimum permissible value of L_p or R_p may be calculated and the required number of turns may be obtained from Eqn (7.44). This procedure may be simplified by normalizing some of the variables and preparing a graph giving the required number of turns for a given insertion loss or return loss as a function of frequency. Such curves apply only to the ferrite material from which the complex permeability data were obtained but they may be used for any core shape. The technique is an extension of that proposed by Maurice and Minns.[1] The main steps in the preparation of the graphs will now be indicated.

The shunt impedance Z of a transformer may be expressed in terms of R_p and Q-factor thus:

$$Z = \frac{j\omega L_p R_p}{R_p + j\omega L_p} = \frac{R_p}{1 - jQ} = \frac{R_p(1 + jQ)}{1 + Q^2} \quad \Omega \quad (7.45)$$

where $Q = R_p/\omega L_p$.

Applying this equation and that of Figure 7.2(c), the insertion loss introduced when Z is connected across a 1 Ω source and a 1 Ω load is given by

$$\begin{aligned}A_i &= 10 \log_{10}\left(1 + \frac{1}{R_p} + \frac{1 + Q^2}{4R_p^2}\right) \\ &= 10 \log_{10} F \quad \text{dB}\end{aligned} \quad (7.46)$$

where F represents the function in the brackets.

For a given insertion loss, F will have a particular value, e.g. F_1. Then

$$F_1 = 1 + \frac{1}{R_p} + \frac{1 + Q^2}{4R_p^2}$$

The solution of this quadratic equation is

$$R_p = \frac{1 + Q^2}{2\sqrt{\{1 - (1 - F_1)(1 + Q^2)\}} - 2} \quad \Omega \quad (7.47)$$

This gives the value of the resistive component, R_p, of impedance Z that will result in a given insertion loss, provided the value of the Q-factor is known. If the parallel components of the complex permeability have been determined for the ferrite under consideration then, at a given frequency, the Q-factor is given by

$$Q = \frac{\mu''_p}{\mu'_p}$$

and the value of R_p corresponding to N turns wound on a given core shape is

$$R_p = \frac{\omega \mu_o \mu''_p N^2}{C_1} \quad \Omega$$

Putting this equal to the value of R_p required to obtain the given insertion loss (Eqn (7.47)), the required number of turns may be calculated.

If the expression for R_p is normalized with respect to the core factor, C_1, by putting $C_1 = 1$ mm^{-1}, a normalized number of turns, N_a, may be calculated such that when wound on a core having $C_1 = 1$ mm^{-1} and connected across 1 Ω source and load resistances, the given insertion loss will be obtained. Thus

$$N_a^2 = \frac{R_p 10^3}{\omega \mu_o \mu''_p} \quad (7.48)$$

Figure 7.13 shows N_a as a function of frequency for four values of insertion loss and for three ferrites suitable for wide-band transformers. In applying these graphs to a practical situation where the actual core factor is C_1 and the parallel combination of source and load resistances is denoted by R, the actual number of turns required is given by

$$N = N_a\sqrt{(2RC_1)} \quad (7.49)$$

where C_1 is in mm^{-1}.

The corresponding graphs having return loss as parameter may be similarly prepared. The input impedance of the transformer, Z_i, equals the shunt impedance due to the core, Z, in parallel with the

---

$$(7.44) \begin{cases} \dfrac{L_p}{N^2} = 4\pi\mu'_p 10^{-9}/C_1 & \text{H} \\ \\ \dfrac{R_p}{N^2} = 4\pi\omega\mu''_p 10^{-9}/C_1 & \Omega \end{cases}$$

$$(7.48) \quad N_a^2 = \frac{R_p}{4\pi\omega\mu''_p 10^{-9}}$$

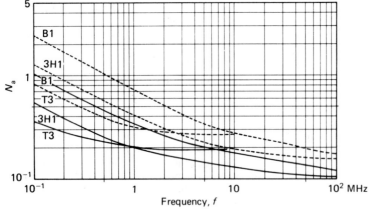

(a) —— $A_i = 3$ dB   --- $A_r = 15$ dB

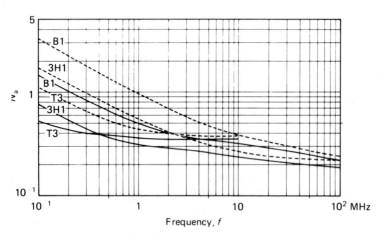

(b) —— $A_i = 1$ dB   --- $A_r = 20$ dB

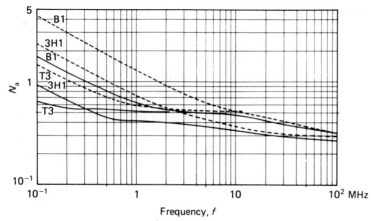

(c) —— $A_i = 0.5$ dB   --- $A_r = 25$ dB

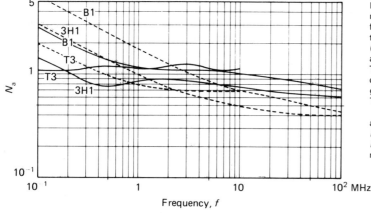

Figure 7.13 The normalized number of turns, N_a, as a function of frequency for three typical ferrites: MnZn ferrites T3 ($\mu_i=5000$) and 3H1 ($\mu_i=2100$) and NiZn ferrite B1 ($\mu_i=600$). The graphs are based on the complex permeability data given in Figure 3.11. See also Table 3.1.

The number of turns in an actual design is given by $N=N_a\sqrt{(2RC_1)}$, where $R=R_aR_b'/(R_a+R_b')$ and C_1 is in mm^{-1}.

(d) ——— $A_i = 0.1$ dB   - - - $A_r = 30$ dB

normalized load of 1 Ω, i.e.

$$Z_i = \frac{Z}{1+Z}$$

Combining this with Eqn (7.45) and the equation for return loss in Figure 7.2(d), the return loss is given by

$$A_r = 20 \log_{10} \left| \frac{Z_i + 1}{Z_i - 1} \right| = 10 \log_{10} \left(1 + \frac{4R_p(R_p + 1)}{1 + Q^2}\right)$$
$$= 10 \log_{10} F \quad \text{dB} \qquad (7.50)$$

where F is now the function defined by this equation.

As before, the particular value of F corresponding to a given return loss may be denoted by F_1 and the quadratic equation involving F_1, R_p and Q may be solved for R_p. This time

$$R_p = \frac{\sqrt{\{1 - (1 - F_1)(1 + Q^2)\}} - 1}{2} \quad \Omega \qquad (7.51)$$

Again this may be equated to the value of R_p obtained with N_a turns on a core of the given ferrite, the core factor being unity. The resulting graphs, relating to the same three grades of ferrite, are also shown in Figure 7.13. For the practical design, Eqn (7.49) again applies.

There are several things to notice about these graphs. Clearly the best material is that which provides the given performance at the required frequency with the least value of N_a. If the specification requires that the insertion loss should not exceed a given value, it is seen that the higher permeability MnZn ferrite is superior up to frequencies between 0.1 and 1 MHz depending on the insertion loss. The lower the specified insertion loss the lower is the cross-over frequency at which the MnZn inductor ferrite, 3H1, which has lower permeability and lower loss, becomes preferable. Above this cross-over frequency the inductor ferrite remains superior at frequencies up to at least 100 MHz since none of the 3H1 curves crosses the corresponding B1 curves. The NiZn ferrite B1 would therefore be preferable only when it was necessary to use a core with a very high resistivity. Curves for any lower permeability NiZn ferrites would lie above the B1 curves.

The frequency referred to is, of course, f_1, i.e. the frequency at the lower limit of the transmission band. Thus if a transformer were required, for example, to have less than 1 dB insertion loss between 5 and 200 MHz, or even 50 and 200 MHz, then the low loss inductor ferrite would be the best choice.

When the specification is in terms of return loss, the cross-over frequency between the higher permeability and the inductor MnZn ferrites is about 2 MHz and is insensitive to the return loss specification. Again the NiZn ferrite does not show superior performance in the frequency range considered.

It will be noticed that some of the curves relating to insertion loss show minima. This undulation occurs not because Z ceases to rise with frequency but rather because it changes from a mainly inductive impedance at the lower frequencies to a mainly resistive impedance at the upper end of the frequency range. If the design leads to a value of N_a in one of these troughs and this value of N_a is used to obtain the actual number of turns, the correct insertion loss will be obtained at f_1 but at higher frequencies, i.e. further into the pass band, the loss will increase to a shallow maximum. In fact, the idea of the lower frequency droop in the characteristic may have to be abandoned. A horizontal line drawn at the proposed value of N_a will indicate the variation of attenuation that will be obtained, the attenuation at any frequency being interpolated between the curves of constant insertion loss. The variation is usually not very great but in some cases it may be preferable to choose the highest value of N_a within the pass band of the transformer rather than the value at f_1.

The complex permeability data on which these graphs are based relate to the residual loss, i.e. eddy current and hysteresis loss are not included. The remarks in Section 7.3.1 regarding eddy current loss apply equally at high frequency. If a manganese zinc ferrite core is used, dimensional resonance is a possibility. The dimensional resonant frequency is related to the cross-sectional dimension of the core in Figure 4.7. Provided it occurs at a frequency well above f_1 it is unlikely to affect the pass band. Transformers having f_1 above 1 MHz will probably use very small cores and so avoid the effect of this resonance.

In most wide band high frequency transformers the signal levels are so small that hysteresis effects are negligible. The exceptions are the high power transformers considered in Chapter 9; see also Reference 2.

7.4.2 Relations arising from the geometry of core and winding

Before considering the particular core shapes that are suitable for high frequency transformers it will be useful to extend the discussion begun in Section 7.3.1. In that section general relations between core performance and geometry were derived for low and medium frequency transformer cores. At high frequencies, winding resistance ceases to be a significant factor and the winding may consist of only a few turns. In this case the winding breadth and height cease to be definite. Considering only the reactive component of the shunt impedance, Eqn (7.40(a)) yields, for an ungapped core, the proportionality

$$\frac{L_p}{L_l} \propto \frac{\mu_i}{l_w C_1} \propto \frac{A_L}{l_w} \qquad (7.52)$$

It is not valid similarly to invoke Eqn (7.42) to study the factors affecting $L_l C_s$ because that equation is based on the self capacitance of a multilayer winding; in the present context the winding consists of a few turns, probably wound in a single layer against a screen. The self capacitance may be estimated from expression 1 of Table 11.3 or it may be better to use Medhurst's data (see Figure 11.19). If the former expression is used:

$$C_s = \frac{C_m}{3} \propto b_w l_w$$

According to Medhurst, C_s/l_w is a function of $b_w/$(winding diameter) and is approximately constant in the region where $b_w/$(winding diameter) = 1. Thus from either consideration it is sufficient for the present purpose to conclude that the self capacitance is proportional to l_w.

The leakage inductance is proportional to $N^2 l_w / b_w$ so the factor determining the upper frequency limit of a core, i.e. $L_l C_s$, is related to the geometry by the following proportionality

$$L_l C_s \propto N^2 l_w^2 \qquad (7.53)$$

These proportionalities provide a useful guide to core design or selection. From the proportionality of 7.52 it is clear that for greatest bandwidth it is desirable to use the highest available permeability and to proportion the core to minimize $l_w C_1$. From 7.53, the highest cut-off frequency will be obtained by using the highest permeability as this will result in the smallest number of turns. Given a core shape and a permeability, it is easily shown that for a given inductance $N^2 l_w^2$ is proportional to the linear dimensions of the core, so it is best to use the smallest practical core size.

In this discussion the permeability of the material has been regarded as a measure of merit and the core loss has been ignored. However, in general the criteria of the previous section do, of course, still apply, i.e. the better core material is that which yields the lower value of N_a in Figure 7.13.

7.4.3 The form of the core

High-frequency transformers may be designed using any of the conventional core shapes, e.g. small pot cores or RM cores are often used. However, because the required number of turns tends to be small at the higher frequencies, the toroidal core shape becomes attractive. It is useful to consider the optimum shape of the toroidal core. In the previous section it was found that for the widest bandwidth $l_w C_1$ should be a minimum. If the inner and outer diameters of the toroid are d_1 and d_2 respectively and the axial thickness is h, then assuming the turns lay flat against the core the turn length, l_w, is given by

$$l_w = d_2 - d_1 + 2h$$

The core factor, C_1, for such a toroid is given by Eqn (4.6):

$$C_1 = \frac{2\pi}{h \log_e (d_2/d_1)} \qquad (7.54)$$

$$\therefore \quad l_w = d_2 - d_1 + \frac{4\pi}{C_1 \log_e (d_2/d_1)} \qquad (7.55)$$

If d_1 and C_1 are constant, l_w is a minimum when

$$C_1 d_1 = \frac{4\pi}{(d_2/d_1)[\log_e (d_2/d_1)]^2} \qquad (7.56)$$

and from Eqn (7.54)

$$C_1 h = \frac{2\pi}{\log_e (d_2/d_1)} \quad (7.57)$$

Figure 7.14 shows d_2/d_1 as a function of $C_1 d_1$ obtained from Eqn (7.56) and the corresponding value of $C_1 h$ obtained from Eqn (7.57). Thus from the required value of C_1, assuming a value of d_1, the value of d_2 and h may be found. The value of d_1 must be as small as possible. This follows from Eqn (7.54); C_1 is equally reduced by increasing d_2 or reducing d_1 in the same proportion, but if $d_2/d_1 \gg 1$, an increase in d_2 increases l_w far more than a proportionate decrease of d_1. The only limit to the smallness of d_1 is in the necessity to thread the windings through the hole. Since there are to be few turns and the wire diameter may be small, d_1 may in practice be in the region of 2 mm. This results in a toroid that is quite different in shape from the usual toroidal core; it is a short cylindrical core with a small axial hole, i.e. a bead.

The upper part of Table 7.2 lists some examples obtained from the graphs. It is seen that as C_1 is reduced, d_2 and the overall volume increase rapidly. The bandwidth, which is inversely proportional to $l_w C_1$, increases with volume but the rate of improvement rapidly diminishes as the volume becomes larger. A given transmission requirement at f_1 determines L_p

$$\therefore\ L_p \propto \frac{N^2}{C_1} = \text{const.}$$

Substituting this in Eqn (7.53)

$$L_l C_s \propto l_w^2 C_1 \quad (7.58)$$

From this proportionality and the figures in Table 7.2 it is seen that as d_2 increases $L_l C_s$ increases and the upper cut-off frequency is lowered. Thus although an increase in volume will somewhat improve the bandwidth it will move the pass band to a lower frequency. If transmission at the highest possible frequency is required then for a given value of d_1 it is better to use the smallest practicable core; the limitation being the number of turns that it is feasible to apply. The improvement resulting from reducing d_1 is demonstrated in the table. Considerable licence may be used in fixing the outside dimensions of the core since the proportions of these yield a very flat optimum.

There is a practical limit to how far C_1 and d_1 may be reduced. However, a further improvement may be obtained by using two toroids side by side as shown in the right hand diagram of Figure 7.14; a single turn would thread both holes. In practice a single core having two holes would be used; the full line shows a possible profile. Optimum proportions may again be found. The turn length is now given by

$$l_w = 2(d_2 - d_1 + h)$$

and the core factor assuming two toroids is

$$C_1 = \frac{\pi}{h \log_e (d_2/d_1)}$$

The result is that l_w is a minimum when

$$C_1 d_1 = \frac{\pi}{(d_2/d_1)[\log_e (d_2/d_1)]^2} \quad (7.59)$$

Figure 7.14 also gives d_2/d_1 and $C_1 h$ as functions of $C_1 d_1$ for this two-hole core and the lower part of Table 7.2 lists several examples. The advantages of the two-hole core are apparent. The above conclusions concerning core size apply also to this version. A possible disadvantage of this core is that each turn has to be threaded twice through a hole but since the number of turns is not large this is not a serious limitation.

7.4.4 The windings

In the foregoing discussion of core shape it has been

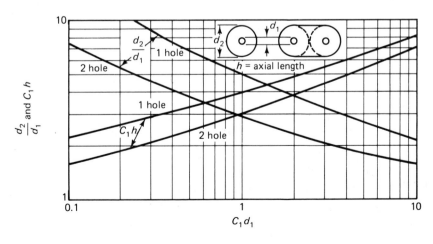

Figure 7.14 Optimum toroidal cores for high-frequency transformers

Table 7.2 Examples of 1 hole and 2 hole high-frequency wide-band transformer cores having optimum proportions (see Figure 7.14). Dimensions in millimetre units.

1 hole core

d_1	C_1	$C_1 d_1$	$\dfrac{d_2}{d_1}$	$C_1 h$	d_2	h	l_w	$l_w C_1$	Overall volume
2	0.8	1.6	4.02	4.5	8.0	5.6	17.3	13.8	285
2	0.4	0.8	5.45	3.7	10.9	9.3	27.4	10.95	860
2	0.2	0.4	7.6	3.1	15.2	15.5	44.2	8.85	2800
1	0.8	0.8	5.45	3.7	5.5	4.7	13.9	11.1	111

2 hole core

d_1	C_1	$C_1 d_1$	$\dfrac{d_2}{d_1}$	$C_1 h$	d_2	h	l_w	$l_w C_1$	Overall volume
2	0.8	1.6	2.45	3.5	4.9	4.37	14.5	11.6	188
2	0.4	0.8	3.1	2.8	6.2	7.0	22.4	9.0	480
2	0.2	0.4	4.02	2.25	8.04	11.3	37.7	7.55	1300
2	0.1	0.2	5.45	1.85	10.9	18.5	54.8	5.48	3900
1	0.1	0.1	7.6	1.55	7.6	15.5	44.2	4.42	1600

assumed that factors such as the turns ratio and the winding arrangement have been constant. With conventional core shapes it is often necessary to use rather elaborate windings consisting, perhaps, of copper foil. However, with the core shapes described in the previous section the windings may be very simple. The requisite number of turns of a convenient size of enamelled copper wire may be used and no special skill is required in the winding (or rather threading) operation.

Sometimes toroidal transformers are designed using a transmission line technique,[3-6] the primary and secondary windings being twisted together to form a transmission line. Such techniques may extend the upper limit of the frequency range. However, using the principles described in the foregoing sections to maximize the bandwidth, by optimizing the design at the lower transmission frequency and so minimizing the number of turns and the total length of conductor, can be equally or even more effective.

Windings requiring few turns may give rise to problems in obtaining the correct turns ratio. Indeed the minimum number of turns may in practice be determined by the required ratio, e.g. an impedance ratio of 4 may be obtained by windings having 1 and 2 turns respectively whereas for an impedance ratio of approximately 3 the least numbers of turns are 4 and 7 respectively.

To minimize leakage inductance, the primary and secondary should be wound in close proximity. If auto transformer winding (see Section 11.7.2) is possible it will give the lowest possible leakage inductance. If not, bifilar winding using a twisted pair may be used.

If the transformer must provide a large impedance transformation the higher impedance may result in a self capacitance limitation. The situation is made more difficult if an electrostatic screen between windings is required. These factors may prevent the use of twisted pair windings. A low value of leakage inductance may be obtained without close proximity of the windings if a closely fitting screen is used. This envelopes the core except for an annular gap to prevent a short circuit. It provides, in effect, a tight single turn coupling between primary and secondary. The windings may be placed over the screen and in the case of a single hole core may be on diametrically opposite sides of the toroid. In practice such a screen could be provided by silvering the surface of the ferrite, the annular gap being provided, for example, by subsequently grinding a chamfer at one end of the hole. The direct silvering technique can only be successfully applied to high resistivity nickel zinc ferrite; for manganese zinc ferrite an insulating barrier must first be applied so that the lower resistivity ferrite does not short circuit the screen. If a high degree of screening is required, the winding with the fewest turns could be wound with small diameter screened or coaxial wire.

In the majority of designs, however, adequate performance may be obtained by using the simplest winding arrangement. If a nickel zinc ferrite core is used, the core resistivity is high enough for chance electrical contact between the windings and core to be ignored. However, if the lower resistivity manganese zinc ferrite is used, insulation, perhaps in the form of lacquer, must be provided.

7.4.5 An example of high-frequency transformer design

Suppose a transformer is required to provide isolation

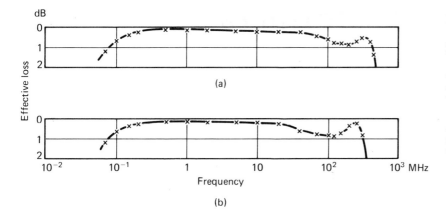

Figure 7.15 Measured transmission characteristics of high-frequency transformer examples using the first core in Table 7.2.
(a) MnZn core (μ_i=2100), 6:6 turns, twisted enamelled copper wire, matching 50 Ω with phase inversion;
(b) NiZn core (μ_i=600), 11:11 turns of twisted enamelled copper wire, matching 50 Ω with phase inversion

and phase inversion between two 50 Ω circuits, with less than 1 dB insertion loss between 100 kHz and 100 MHz.

From the list of optimum one hole cores in Table 7.2 the first will be used in this example. For this core $d_1 = 2$ mm, $d_2 = 8$ mm, $h = 5.6$ mm and $C_1 = 0.8$ mm^{-1}.

Referring to Figure 7.13(b) the higher permeability MnZn ferrite would give the best performance. However, at the time of writing, the above core was available only in the inductor grade ferrite (3H1) so this ferrite is used as the basis of the design. The normalized number of turns required to meet the insertion loss specification is approximately 0.8. Therefore, the actual number of turns is given by Eqn (7.49) in which $R = 25$ Ω. Thus

$$N = 0.8\sqrt{(2 \times 25 \times 0.8)} \simeq 5$$

therefore, allowing a margin for tolerances, each winding may consist of 6 turns.

There is no simple analytical means of estimating the performance at the upper end of the frequency band; the use of an optimum core will ensure that, within the limits imposed by the core size and material, the best performance will be obtained. A transformer was made according to this design and the measured transmission characteristic is shown in Figure 7.15(a).

Had the higher permeability MnZn ferrite been used, then $N_a \simeq 0.5$ and the number of turns would have been reduced to 4. This would have resulted in the same lower frequency characteristic but the bandwidth would have been extended at the higher frequency end.

If, to achieve adequate insulation, it were necessary to specify a nickel zinc ferrite core then from Figure 7.13, N_a would be raised to 1.4 and the actual number

of turns would have to be raised from 6 to 10 or 11 to make the design equivalent to the previous one. A transformer was made according to this design using a nickel zinc ferrite corresponding to that quoted in Figure 7.13.

The measured transmission characteristic is shown in Figure 7.15(b). It is seen that the lower permeability has had the effect of lowering the upper cut-off frequency because more turns were necessary.

7.5 References and bibliography

Section 7.4

1. MAURICE, D. and MINNS, R. H. Very-wide band radio frequency transformers, *Wireless Engr.*, **24**, 168 (Part 1) and 209 (Part 2) (1947)
2. HILBERS, A. H. Design of high-frequency wideband power transformers, *Electron. Appl. Bul.*, **32**, 44 (1973–74)
   BERG, R. S. and HOWLAND, B. Design of wide-band shielded toroidal r.f. transformers, *Rev. scient. Instrum.*, **32**, 864 (1961)
   O'MEARA, T. R. Wide-band transformers and associated coupling networks, *Electro-Technology*, **70**, no. 3, 80 (1962)
3. TALKIN, A. I. and CUNEO, J. V. Wide-band balun transformer, *Rev. scient. Instrum.*, **28**, 808 (1957)
4. RUTHROFF, C. L. Some broad-band transformers, *Proc. Inst. Radio Engrs*, **47**, 1337 (1959)
5. MATICK, R. E. Transmission line pulse transformers—theory and applications, *Proc. IEEE*, **56**, 47 (1968)
6. TITTERINGTON, R. G. The ferrite-cored balun transformer, *Radio Commun.*, 216, March (1982)

General

GROSSNER, N. R. *Transformers for electronic circuits*, McGraw-Hill Book Co. (1983)

8 Pulse transformers

8.1 Introduction

Pulse transformers, like wide-band transformers, are required to transmit energy that is distributed over a wide frequency spectrum. Whereas the performance of the wide-band transformer is specified in terms of this spectrum, the performance of the pulse transformer is specified in terms of its effect on the shape of a pulse of current or voltage, this shape being expressed as a function of time. The shape of the voltage pulse is usually assumed to approximate to a rectangle. The transformer must transfer the pulse of energy from the primary circuit, which is often the pulse generating circuit, to the secondary, or load circuit without excessive shape distortion. In common with other types of transformer it must of course perform one or more of the following functions:

- accurate voltage or current transformation,
- isolation,
- reversal of polarity,
- balance or unbalance to earth,
- impedance matching.

It is clear that the function of a pulse transformer is similar to that of a wide-band transformer. Indeed it is possible to relate pulse distortion to bandwidth; this aspect is discussed in Section 8.4. However, the most satisfactory procedure in designing a pulse transformer is to determine the elements of the equivalent circuit from the permissible pulse distortion. It is mainly in this respect that the two transformers types differ. This chapter is therefore principally concerned with the method of determination of the equivalent circuit elements. The basic transformer concepts and the practical design to realize a given equivalent circuit are common to both types of transformer; they have been adequately covered in the previous chapter and need not be repeated.

Pulse transformers range in application from low power, e.g. use in data handling circuits, to high power, e.g. the pulsing of radar transmitters. The switched mode power supply transformer, considered in the next chapter, is a type of pulse transformer in which the duty cycle approaches unity. However, its design is determined primarily by power transfer not pulse distortion so it only marginally enters the scope of this chapter.

In the design of high-power pulse transformers, heat transfer is a major consideration and the best core material is that which provides the widest excursion of flux density. In practice this implies a high saturation flux density. For this reason the iron alloys such as grain oriented silicon steel or the nickel iron alloys are usually chosen as cores for very-high-power pulse transformers, e.g. for high power radar transmitters. The saturation flux density of ferrites is typically only 20% of the figure available for the best alloys. Moreover, because the Curie point of ferrites is much lower than for alloys, the saturation flux density of ferrites is appreciably reduced at the temperatures at which the higher power transformers normally operate. For these reasons ferrite cores are used only in pulse transformers intended for low or medium powers, and the scope of this chapter is restricted accordingly. Much of the treatment is, however, quite general.

Some useful terms and their definitions are illustrated in Figure 8.1. In the upper part of this figure the ideal form of a rectangular or trapezoidal pulse is shown.

Such pulses are usually identically repeated at regular time intervals (t_P) and the number of pulses per second (pps) is termed the Pulse Repetition Frequency (P.R.F.). If P_P is the power during the pulse then the average power is

$$P_{av} = P_P t_d / t_P \qquad (8.1)$$

where t_d is the pulse duration. In high power applications t_d/t_P is very small so it is possible to achieve very high peak power without the necessity of handling high average powers. In low power applications also the ratio t_d/t_P is usually small but there are cases where it approaches 0.5 and the pulse train becomes a square wave.

When an ideal voltage pulse is applied to the primary of a transformer, the finite reactances associated with the transformer cause the pulse appearing across the loaded secondary to be distorted. A typical secondary waveform is shown in the lower half of

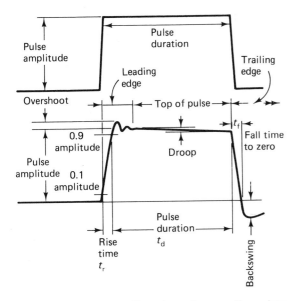

Figure 8.1 Rectangular pulses, ideal and distorted, illustrating some of the terms used

t_p = time between leading edges of consecutive pulses

Figure 8.1. A pulse transformer specification, in addition to specifying source and load impedances, turns ratios, voltages, pulse durations, etc., will define the limits of permissible pulse distortion in terms of rise time, overshoot, droop, fall time, back-swing, etc. The object of the transformer design is to achieve the required energy transformation while keeping all the elements of the transformer equivalent circuit within the limits imposed by the pulse distortion requirements.

In practice, the leading and trailing edges of the pulse usually have short durations relative to the top of the pulse. It is clear that under such circumstances these three parts of the pulse have little influence on one another and may therefore be considered separately. This principle enables the general transformer equivalent circuit, e.g. Figure 7.3, to be simplified so that it includes only those elements which influence the part of the pulse being considered. There are two main cases, the step-up transformer in which the primary shunt capacitance may be neglected and the step-down transformer in which the secondary shunt capacitance may be neglected.

The following sections study the influence of the main elements of the equivalent circuit on the leading edge, the top of the pulse, and the trailing edge respectively. The approach and presentation are an extension of the conventional treatment to be found in the literature, e.g. Wilds[1]. It is shown how the relations so derived may be used to calculate the permissible values of these elements. After noting some relation between pulse transformer and wide-band transformer design, the properties of the core material are considered with particular reference to pulse permeability. Finally some practical aspects of pulse transformer design are considered.

8.2 Pulse distortion in terms of the equivalent circuit elements

8.2.1 The leading edge of the pulse

It is usual to suppose that the pulse applied to the primary of the transformer is a simple step function, i.e. it may be regarded as switching a source of d.c. directly across the primary terminals. In practice, of course, the pulse source will give a finite rise time, t_1, and it may be shown that this will combine with the actual rise time, t_r, contributed by the transformer, to give an overall rise time

$$t = (t_1^2 + t_r^2)^{1/2}$$

If the resulting pulse is observed with an oscilloscope having a rise time, t_2, the observed pulse will have a rise time

$$t = (t_1^2 + t_2^2 + t_r^2)^{1/2} \qquad (8.2)$$

Knowing the value of t_1 and t_2 the necessary correction may be made.

It may be assumed that during the rise time there is negligible build-up of magnetizing current in the transformer shunt impedance, see Figure 7.3, so this arm of the equivalent circuit may be ignored. Further, if the transformer has a reasonably large turns ratio the shunt capacitance on the low impedance side may be ignored. This means that one of two simplified equivalent circuits, Figure 8.2, may be used depending on whether the transformer steps up or down. In this and the following equivalent circuits the ideal transformer is omitted and the analysis is referred to the primary circuit.

Clearly the solution of the transient problem will yield an aperiodic result or an oscillatory result depending on the relative values of the resistive and reactive elements. The aperiodic solution for the step-up case is as follows:

$$\frac{u'_b}{E_a} = \frac{\eta}{1+\eta}[1 - \exp(-2\pi S_r \sigma_r)\{\cosh(2\pi S_r(\sigma_r^2 - 1)^{1/2})$$
$$+ (1 - 1/\sigma_r^2)^{-1/2}\sinh(2\pi S_r(\sigma_r^2 - 1)^{1/2})\}] \qquad (8.3)$$

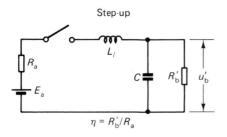

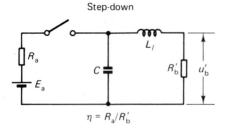

Figure 8.2 Simplified equivalent circuit used in leading edge analysis of a step-up and step-down pulse transformer. All the elements shown are referred to the primary side; the ideal transformers have been omitted.

E_a is the generator e.m.f.   V
R_a is the source resistance   Ω
R'_b is the load resistance   Ω
L_l is the leakage inductance   H
C is the total shunt capacitance (transformer plus external circuit)   F
u'_b is the instantaneous voltage across load   V

where u'_b is the instantaneous load voltage referred to the primary

$$\sigma_r = a/\sqrt{b}$$

$$S_r = \frac{t\sqrt{b}}{2\pi} \quad (t = \text{time in seconds after switch has closed})$$

and $\quad a = \frac{1}{2}\left(\frac{R_a}{L_l} + \frac{1}{CR'_b}\right)$

$b = \frac{1}{L_l C}(1 + 1/\eta)$

where $\eta = R'_b/R_a$

When $\sigma_r < 1$ the circuit is oscillatory and the solution becomes

$$\frac{u'_b}{E_a} = \frac{\eta}{1+\eta}[1 - \exp(-2\pi S_r \sigma_r)\{\cos(2\pi S_r(1-\sigma_r^2)^{1/2}) + (1/\sigma_r^2 - 1)^{-1/2}\sin(2\pi S_r(1-\sigma_r^2)^{1/2})\}] \quad (8.4)$$

where the parameters have the meanings given above.

For the step-down case the same result applies except that R_a and R'_b must be interchanged, i.e. for step-down

$$\left.\begin{array}{l} a = \frac{1}{2}\left(\dfrac{R'_b}{L_l} + \dfrac{1}{CR_a}\right) \\[2mm] b = \dfrac{1}{L_l C}(1 + 1/\eta) \end{array}\right\} \quad (8.5)$$

where now $\eta = R_a/R'_b$

Figure 8.3 shows the response relative to the steady state amplitude plotted against S_r with σ_r as a parameter.

The factor S_r may be regarded as a time factor and σ_r is in effect a damping factor. Looking at these factors in more detail

$$S_r = \frac{t\sqrt{b}}{2\pi} = \frac{t}{2\pi}\left(\frac{1+1/\eta}{L_l C}\right)^{1/2} \quad (8.6)$$

$$\therefore \quad (L_l C)^{1/2} = \frac{t}{2\pi S_r}(1 + 1/\eta)^{1/2} \quad (8.7)$$

In Figure 8.1 the rise time is shown as the time required for the output voltage to rise from 0.1 to 0.9 of its full amplitude. Sometimes other fractions are used. Whatever amplitude fractions are used, Figure 8.3 enables the rise time to be obtained in terms of S_r for any given value of σ_r.

Figure 8.4 gives S_r as a function of σ_r for two specific amplitude fractions often quoted in practice, namely, 0 to 0.95 and 0.1 and 0.9. The transformer specification will require that the rise time, e.g. from 0.1 to 0.9 of full amplitude, shall be less than a certain value. This value may be substituted for t in Eqn (8.7). From Figure 8.3 the rise time characteristic appropriate to the design requirements, e.g. slow aperiodic or fast oscillatory, may be chosen and so the design value of σ_r may be determined. The corresponding value of S_r obtained from Figure 8.4 may then be substituted in Eqn (8.7) which will give the maximum value of $L_l C$.

The damping factor expression may also be manipulated[2] to yield information about L_l and C. For the step-up case:

$$\sigma_r = \frac{a}{\sqrt{b}} = \frac{1}{2}\left(\frac{R_a}{L_l} + \frac{1}{CR'_b}\right)\left(\frac{L_l C}{1+1/\eta}\right)^{1/2}$$

$$= \frac{1}{2}(1+\eta)^{-1/2}\left[\left(\frac{\eta C R_a^2}{L_l}\right)^{1/2} + \left(\frac{L_l}{\eta C R_a^2}\right)^{1/2}\right] \quad (8.8)$$

$$\left.\therefore \quad \sigma_r = \frac{1}{2}(1+\eta)^{-1/2}\left(\frac{1}{\sqrt{x}} + \sqrt{x}\right) \\[2mm] \text{where} \quad x = \dfrac{L_l}{R_a R'_b C} \right\} \quad (8.9)$$

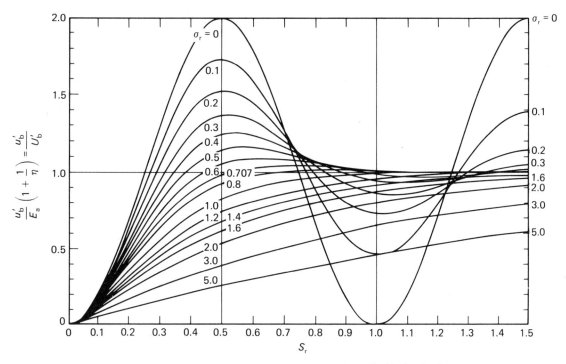

Figure 8.3 Leading edge response; relative amplitude as a function of the time factor S_r with damping factor σ_r as a parameter

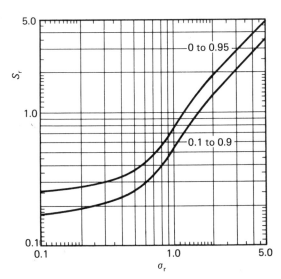

Figure 8.4 Time factor S_r for the load voltage to rise between given fractions of the final amplitude, plotted as a function of σ_r (after Wilds[1])

This quadratic equation applies also to the step-down case when the appropriate value of η is used. If the load-source mismatch factor, η, and the terminating impedances are known then the limit value of σ_r will restrict the choice of L_l/C according to the above equation. The value of σ_r is usually limited by the overshoot requirements of the specification, e.g. if there must be no overshoot then σ_r must not be less than unity. If a certain amount of overshoot is tolerable, Figure 8.3 will indicate the minimum value of σ_r. More conveniently Figure 8.5 shows σ_r as a function of overshoot.

In practice the limit value of σ_r may not yield real roots to Eqn (8.9); for this reason and others to be discussed later the overshoot limit forms only one of the boundary conditions for σ_r.

Eqn (8.9) may be solved for $\sqrt{x}$:

$$\sqrt{x} = \sigma_r(1+\eta)^{1/2} \pm (\sigma_r^2(1+\eta) - 1)^{1/2} \qquad (8.10)$$

For x to be real, $\sigma_r^2(1+\eta)$ must be greater than unity, therefore when $\eta = 1$ (i.e. $R_a = R'_b$), $(\sigma_r)_{\min} = 0.707$. Referring to Figure 8.3 it will be seen that a curve has been drawn for $\sigma_r = 0.707$; the corresponding overshoot is 4.5%. Therefore a transformer that perfectly matches the load to the source cannot have more than 4.5% overshoot whatever values the elements of the equivalent circuit are given. On the other hand, a designer wishing to increase the rise time at the expense of overshoot cannot go far in this direction if the transformer provides impedance matching. However, pulse transformer ratios are

236 Pulse transformers

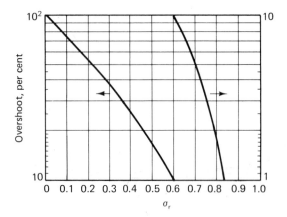

Figure 8.5 Overshoot as a function of the damping factor σ_r

often fixed by the required voltage or current ratios without reference to the impedances, so the mismatched condition is more common than with wide band transformers.

Eqn (8.9) is expressed graphically in Figure 8.6. To provide real roots, and therefore real values of L_l/C, a horizontal line drawn through the actual value of σ_r must touch or cut the curve for the appropriate value of η, i.e. the actual value of σ_r must be equal to or greater than the value on the η curve at $x=1$. If the minimum value of σ_r permitted by the overshoot specification equals the value at $x=1$ on the curve for the appropriate value of η then, since σ_r may have any larger value, the roots may have any values, i.e. there is no restriction on L_l/C.

If the overshoot specification allows σ_r to be less than the value at $x=1$ on the curve for the appropriate value of η, this must be disregarded; the minimum value of σ_r is then set by the need for real roots. The situation is then the same as the previous paragraph, $(\sigma_r)_{min}$ is the value at $x=1$ and there is no restriction on L_l/C.

The other possibility is that the overshoot specification may require $(\sigma_r)_{min}$ to exceed the value at $x=1$ (on the curve for the appropriate value of η) by a certain amount. Since σ_r may have any value larger than $(\sigma_r)_{min}$ it follows that in this case there are two possible zones for x (and L_l/C). This may be illustrated on Figure 8.6 by the example $(\sigma_r)_{min} = 0.7$, $\eta = 3$; x may be greater than 5.7 or less than 0.177.

Thus the rise time specification places a maximum limit on $L_l C$ while the overshoot specification and/or the solution of Eqn (8.9) may place some restrictions on the value of L_l/C. The actual leakage inductance and shunt capacitance cannot be evaluated until other parts of the pulse have been considered and any further restrictions have been taken into account.

Before leaving the leading edge analysis some mention should be made of the unity-ratio transformer. This is used occasionally as an isolation or a polarity reversing transformer and a solution to the transient equations is considered to be too lengthy to be justified. The difficulty arises because primary and secondary shunt capacitances are commensurate and thus an additional element must be included in the circuit for analysis. The following procedure is similar to that proposed by Fenoglio et al.[3]

(1) As a first approximation the primary shunt capacitance is ignored and the rise-time and overshoot may be obtained from the solution to the step-up transformer case.
(2) The rise time of the voltage across the primary capacitance may be simply calculated on the assumption that it alone constitutes the load on the source.

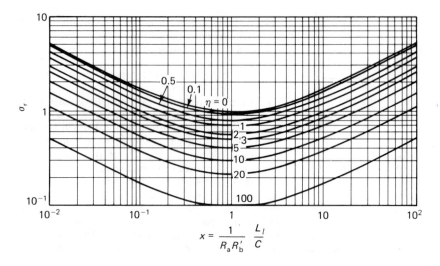

Figure 8.6 σ_r as a function of x $(=[1/R_a R_b'][L_l/C])$ with the mismatch factor η as a parameter. (Courtesy Postma[2])

(3) The total rise time may be obtained by the root of the sum of the squares relationship (see Eqn (8.2)). The overshoot may be taken as approximately that obtained in (1).

8.2.2 The top of the pulse

When the pulse has attained its full amplitude and the leading edge transient has died away, the leakage inductance and shunt capacitances have little further effect within the pulse duration; ideally there is a constant current through the leakage inductance and a constant voltage across the capacitance. The equivalent circuit then reduces to the core impedance shunting the load impedance. Usually the core loss may be neglected so the core impedance may be regarded as due only to the open circuit inductance, L_p.

If during the pulse the inductance and the load voltage both were constant, the magnetizing current, i.e. the current in L_p, would rise linearly with time. Assuming a finite source impedance, the magnetizing current would cause a progressive fall of load voltage. This is referred to as droop, see Figure 8.1. For a practical core, the inductance depends on the magnetizing current so strictly a transient analysis based on a constant inductance is not entirely valid. However, to avoid complication, it is usual to base the analysis on a constant inductance and when it is necessary to realize the inductance an appropriate value of permeability is taken. This will be discussed at greater length in Section 8.5.

Figure 8.7 shows the simplified equivalent circuit, the analysis of which will enable the droop to be expressed in terms of the element values. This circuit applies to all transformation ratios. The solution is

$$\frac{u'_b}{E_a} = \frac{\eta}{1+\eta} \exp\left(-\frac{tR}{L_p}\right) \quad (8.11)$$

where $R = \dfrac{R_a R'_b}{R_a + R'_b}$

This function is plotted in Figure 8.8. A derived function relating percentage droop to the element values is also useful. If D denotes the percentage droop at the end of a time substantially equal to t_d then

$$\frac{u'_b(1+1/\eta)}{E_a} = 1 - D/100 = \exp\left(-\frac{t_d R}{L_p}\right)$$

$$\therefore \quad \frac{L_p}{t_d R} = -\frac{1}{\log_e(1-D/100)} \quad (8.12)$$

Figure 8.9 shows $L_p/t_d R$ as a function of D. From this graph the minimum value of L_p may be obtained for any specified droop and pulse duration.

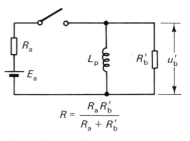

$$R = \frac{R_a R'_b}{R_a + R'_b}$$

Figure 8.7 Simplified equivalent circuit used in top of the pulse analysis of a pulse transformer. All the elements shown are referred to the primary side; the ideal transformer has been omitted. L_p=inductance of primary winding, in henries

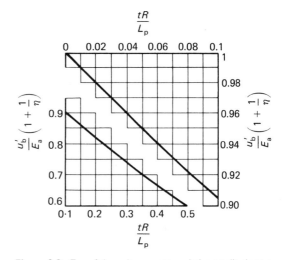

Figure 8.8 Top of the pulse response; relative amplitude as a function of tR/L_p

8.2.3 Trailing edge of the pulse

At the end of the pulse the source voltage is removed and the energy stored in the magnetic core and the capacitances is returned to the circuit to give a trailing edge transient. If there are limitations on the shape of the trailing edge transient they may over-ride the leading edge and pulse top requirements in determining the values of the circuit elements. On the other hand it is often found that the shape of the trailing edge is of no consequence and this greatly simplifies the design.

Figure 8.10 shows the equivalent circuit for analysis. Most references show the source as being open-circuited at the end of the pulse. While this may often be true of high power transformers there are many cases where low power transformers remain connected to the source impedance after the end of the pulse. Therefore to make the results general the

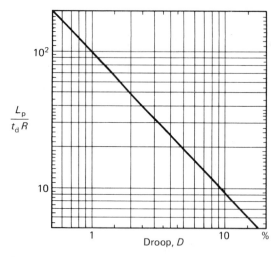

Figure 8.9 L_p/t_dR as a function of pulse droop

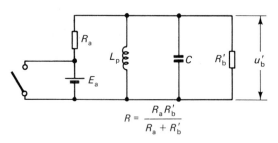

$$R = \frac{R_a R'_b}{R_a + R'_b}$$

Figure 8.10 Simplified equivalent circuit used in trailing edge analysis of a pulse transformer. All the elements shown are referred to the primary side; the ideal transformer has been omitted. C=total shunt capacitance, in farads

source e.m.f. is shown as short-circuited and R_a may be given the value actually remaining after the end of the pulse. So that the results have the same form as those quoted in the literature let

$$R = \frac{R_a R'_b}{R_a + R'_b} \quad \text{as before.}$$

The solution again assumes that the element values remain constant with time and the core loss is again neglected (the latter assumption is reasonable provided the external resistances are not disconnected; if they are, the core loss becomes the principal means of dissipating the energy). The transformation ratio has no effect on the solution. The aperiodic trailing edge transient is given by:

$$\frac{u'_b}{U'_b} = \exp(-2\pi S_f \sigma_f)\{\cosh[2\pi S_f(\sigma_f^2 - 1)^{1/2}] \\ - (1 + 2\chi)(1 - 1/\sigma_f^2)^{-1/2}\sinh[2\pi S_f(\sigma_f^2 - 1)^{1/2}]\} \quad (8.13)$$

where U'_b = voltage across load at end of pulse (referred to primary)

$$\sigma_f = a/\sqrt{b}$$

$$S_f = \frac{t\sqrt{b}}{2\pi}$$

$$\chi = \frac{t_d R}{L_p}$$

and $\quad a = \dfrac{1}{2CR}$

$$b = \frac{1}{L_p C}$$

When $\sigma_f < 1$ the circuit is oscillatory and the trailing edge transient is given by

$$\frac{u'_b}{U'_b} = \exp(-2\pi S_f \sigma_f)\{\cos[2\pi S_f(1 - \sigma_f^2)^{1/2}] \\ - (1 + 2\chi)(1/\sigma_f^2 - 1)^{-1/2}\sin[2\pi S_f(1 - \sigma_f^2)^{1/2}]\} \quad (8.14)$$

Note that the response is relative to the load voltage at the end of the pulse. If it is required in terms of the pulse amplitude a correction must be made to allow for the droop. Normally the droop is small and this correction may be neglected.

Again S_f is a time factor

$$S_f = \frac{t\sqrt{b}}{2\pi} = \frac{t}{2\pi}(L_p C)^{-1/2} \quad (8.15)$$

thus $L_p C = \left(\dfrac{t}{2\pi S_f}\right)^2 \quad (8.16)$

As before σ_f may be regarded as a damping factor:

$$\sigma_f = a/\sqrt{b} = \frac{1}{2R}\left(\frac{L_p}{C}\right)^{1/2} \quad (8.17)$$

$$\therefore \frac{L_p}{C} = (2\sigma_f R)^2 \quad (8.18)$$

Just as, for the leading edge, S_r and σ_r yielded values for the product and quotient of L_l and C, so for the trailing edge S_f and σ_f determine the values of the product and quotient of L_p and C.

The factor χ is the ratio of the magnetizing current in the shunt inductance at the end of the pulse, to the load current*. Thus it may be regarded as an indication of the stored magnetic energy.

*This is true only if $R_a = \infty$, otherwise it is the ratio of the final magnetizing current to E/R.

Figures 8.11 to 8.15 show the trailing edge responses calculated from Eqns (8.13) and (8.14) with σ_f and χ as parameters. From these results two additional graphs may be derived. Figure 8.16 shows the particular value of S_f for the response to fall to zero, denoted $(S_f)_o$, plotted as a function of χ with σ_f as a parameter. Figure 8.17 shows χ plotted against σ_f with the backswing (percentage of U'_b) as a parameter. These last two graphs contain all the important information required from the actual trailing edge response curves.

Referring to the various responses shown in Figures 8.11 to 8.15 it is interesting to consider the net area under the trailing edge and backswing curves. If either of Eqns (8.13) or (8.14) is integrated over the time period from the closure of the switch (Figure 8.10) to infinity the result is

$$\int_0^\infty \frac{u'_b}{U'_b} dt = -t_d$$

thus $\int_0^\infty \frac{u'_b}{R'_b} dt = -\frac{U'_b t_d}{R'_b}$

Or $\int_0^\infty i \, dt = -It_d$

where I is the current in the load resistor at the end of the pulse, i.e. immediately before the trailing edge starts, and i is the instantaneous load current thereafter. In words; the net ampere seconds in the load during the backswing exactly balances the ampere seconds in the load during the pulse. The analysis assumes zero droop but the result is generally true; the positive and negative areas of the pulse balance, i.e. the transformer does not transmit a d.c. level. If the shunt inductance is made very large so that the amplitude of the backswing is very small, the time constant becomes very large and the balance is maintained.

If a subsequent pulse follows the first before its trailing edge transient has completely decayed, the second pulse will cut off a part of the backswing area and disturb the balance. But the second pulse will be displaced (e.g. downwards) by the residual current of the first pulse. Succeeding pulses will be similarly displaced until the area balance is restored. Thus as the mark/space ratio is increased towards unity the zero-current line moves upwards relative to the pulse wave-form and approaches a level approximately midway between the maximum positive and negative excursions, always maintaining equal areas.

8.3 Determination of the element values

In the preceding sections the relations between the element values and the pulse distortion have been expressed in graphical form. However, since these relations usually involve more than one circuit element and as each circuit element is the subject of several, often conflicting, requirements, the treatment so far does not lend itself to transformer design other than by the trial and error method of calculation.

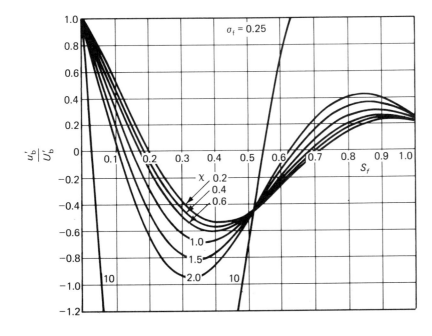

Figure 8.11 Trailing edge response; relative amplitude as a function of the time factor S_f with $\sigma_f=0.25$ and χ as a parameter

240 Pulse transformers

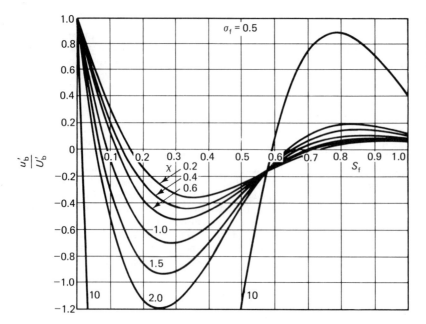

Figure 8.12 Trailing edge response; relative amplitude as a function of the time factor S_f with $\sigma_f=0.5$ and χ as a parameter

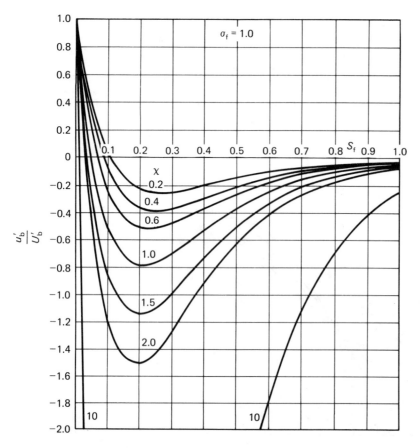

Figure 8.13 Trailing edge response; relative amplitude as a function of the time factor S_f with $\sigma_f=1.0$ and χ as a parameter

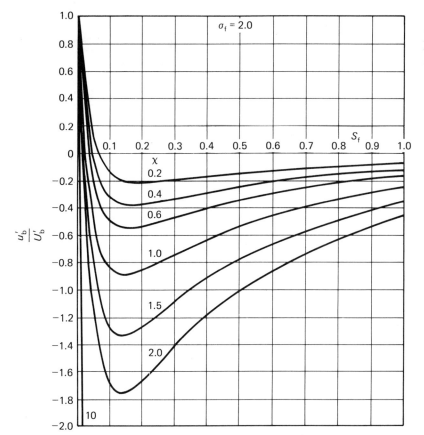

Figure 8.14 Trailing edge response; relative amplitude as a function of the time factor S_f with $\sigma_f = 2.0$ and χ as a parameter

If the pulse transformer is fully specified, complete design can be a long process. Modern transient analysis CAD can reduce much of the labour involved in the design and can be of great value to the designer who is regularly concerned with pulse transformers having stringent requirements. However, powerful CAD is not always available, and even when it is, it is useful to have access to a descriptive manual procedure that may be readily adapted to varying requirements. Such a procedure is set out in the following section. It is based on that described by Wilds[1] but is somewhat less comprehensive and more direct. If, as is often the case, the trailing edge is not specified in detail, some or all of the latter part of the procedure may be omitted. After the steps in the procedure have been given in general terms they are illustrated by means of a worked example in Section 8.3.2.

It often happens that a simple pulse transformer is required quickly for a low-power circuit. If the specification is confined to the bare necessities, a formal design procedure is not justified. To meet this situation a rapid, rather empirical, method may be adopted. Such a method is described in Section 8.3.3.

8.3.1 A complete design procedure

A pulse transformer specification will list some or all of the following requirements: (the figures in brackets refer to the example to be worked in the next section)

Primary pulse amplitude	(6V)
Pulse duration, t_d	(4 μs)
Source resistance, R_a	(50 Ω)
Load resistance, R_b	(600 Ω)
Shunt capacitance of the load	(150 pF)
Turns ratio, r	(2)
Maximum rise time, t_r from 10 to 90% amplitude	(0.1 μs)
Maximum overshoot	(5%)
Maximum droop, D	(5%)
Maximum fall time to zero amplitude, t_f	(1 μs)
Maximum backswing	(10%)

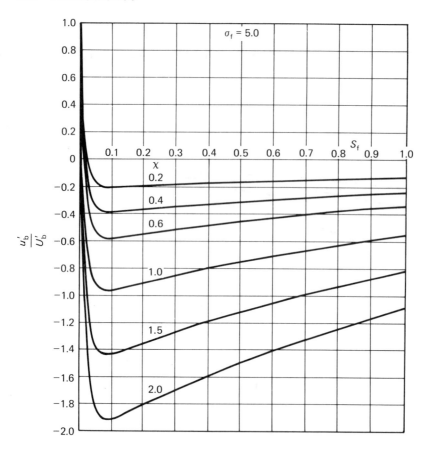

Figure 8.15 Trailing edge response; relative amplitude as a function of the time factor S_t with $\sigma_f = 5.0$ and χ as a parameter

Whether trailing edge oscillations are permissible (No)
Type of winding, e.g. isolation, phase reversing, autotransformer, etc.

First the load-source mismatch factor η must be calculated. For a step-up transformer

$$\eta = \frac{R'_b}{R_a} = \frac{R_b}{R_a r^2}$$

From the overshoot limitation and Figure 8.5, $(\sigma_r)_{min}$ is obtained. This is the minimum value of σ_r assuming overshoot is the only restriction (see Section 8.2.1). A horizontal line may now be drawn on Figure 8.6 at $\sigma_r = (\sigma_r)_{min}$. If it intersects the curve for the given value of η in two places then x must lie outside the range enclosed by these two intersections. It is usually preferable to choose the lower region because, if the load capacitance is appreciable, it is easier to make L_l/C small than large. This procedure will be assumed. Therefore $L_l/CR_a R'_b$ may, so far, have any value below that at the lower intersection of $(\sigma_r)_{min}$ and the appropriate η curve. As previously explained, if the $(\sigma_r)_{min}$ line does not intersect the appropriate η curve then the value of σ_r at the minimum of that curve must be taken as $(\sigma_r)_{min}$ and there is no restriction on $L_l/CR_a R'_b$.

The rise time must now be considered. Figure 8.4 will show, for any value of σ_r the corresponding value of S_r for typical fractions of the leading edge used to specify the rise time. This value may be designated $(S_r)_{min}$. If for any value of σ_r the factor S_r is greater than $(S_r)_{min}$, it merely means that more of the leading edge will be traversed in a given time so that the rise time will be shorter than specified. For any value of $(S_r)_{min}$, Eqn 8.7 will give a maximum value of $L_l C$.

A simple table now now be constructed, e.g. Table 8.1.

Col. 1 σ_r in progressive steps starting at $(\sigma_r)_{min}$. Since in the final design σ_r will not differ greatly from $(\sigma_r)_{min}$, three or four values will usually be sufficient.

Col. 2   Corresponding to the value of σ_r in col. 1, values of $L_l/CR_a R'_b$ from Figure 8.6.

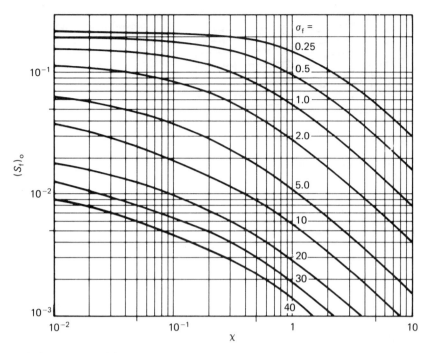

Figure 8.16 Time factor for amplitude to fall to zero, $(S_t)_0$, as a function of χ with σ_f as a parameter

Col. 3 $(L_l/C)^{1/2}$ from col. 2.
Col. 4 $(S_r)_{min}$ from σ_r and Figure 8.4.
Col. 5 $(L_lC)^{1/2}_{max}$ from $(S_r)_{min}$ in col. 4 and Eqn (8.7), i.e. $(L_lC)^{1/2}_{max} = (t_r/2\pi(S_r)_{min})(1 + 1/\eta)^{1/2}$
Col. 6 $(C)_{max} = (L_lC)^{1/2}_{max} \times (L_l/C)^{-1/2}$
Col. 7 $(L_l)_{max} = (L_lC)^{1/2}_{max} \times (L_l/C)^{1/2}$

If there is no trailing edge specification, cols. 6 and 7 give the range of values within which C and L_l may be chosen. For each value of σ_r, C and L_l may not be larger than the corresponding maximum values because this would violate $(L_lC)_{max}$ and the permitted rise time would be exceeded. They may have smaller values provided L_l/C does not pass outside the range permitted by the $(\sigma_r)_{min}$ line in Figure 8.6. If $(\sigma_r)_{min}$ is greater than the value at $x = 1$ then this means that $(L_l/C)^{1/2}$ must not be greater than the first value in col. 3; if $(\sigma_r)_{min}$ equals the value at $x = 1$ then there is no restriction on L_l/C provided $(C)_{max}$ and $(L_l)_{max}$ are not exceeded.

Again in the absence of a trailing edge specification, the minimum value of the shunt inductance $(L_p)_{min}$ may easily be obtained from the permitted droop, D, and Figure 8.9. The transformer design then involves choosing a core and winding configuration to satisfy the limits of shunt inductance, leakage inductance and shunt capacitance (part of which may be external to the transformer). In practice the design should allow margins for manufacturing tolerances.

The additional restrictions on L_l, C and L_p which arise from the specification of the trailing edge response will now be considered. First the least value of the damping factor, i.e. $(\sigma_f)_{min}$ must be decided. If oscillations are not permitted then $(\sigma_f)_{min} = 1$; if a certain amplitude of oscillation is acceptable $(\sigma_f)_{min}$ may be less than unity and its value may be estimated from Figures 8.11 and 8.12 by assuming a provisional value of χ. Following a procedure similar to that for the leading edge, a table may now be constructed for a range of values of σ_f, e.g. Table 8.2.

Col. 1 σ_f in progressive steps starting at $(\sigma_f)_{min}$.
Col. 2   Corresponding to the values of σ_f in col. 1, values of $(\chi)_{max}$ from Figure 8.17 and the specified backswing. It may happen that for the lower values of σ_f there are no corresponding values of $(\chi)_{max}$; it follows that these values of σ_f are inadmissible.
Col. 3 $(L_p)_{min}$ from $(\chi)_{max} = t_d R/(L_p)_{min}$. It will be found that $(L_p)_{min}$ becomes substantially constant as σ_f increases, and a value of $(L_p)_{min}$ may be taken that is valid over a reasonable range of σ_f values. This value may be compared with the $(L_p)_{min}$ calculated earlier from the droop specification. A value of L_p may be provisionally chosen such that it is larger, by a reasonable margin, than the value of $(L_p)_{min}$ obtained from either criterion. The corresponding value of χ may be calculated from $\chi = t_d R/L_p$. This value of χ is provisionally fixed, i.e. it is independent of

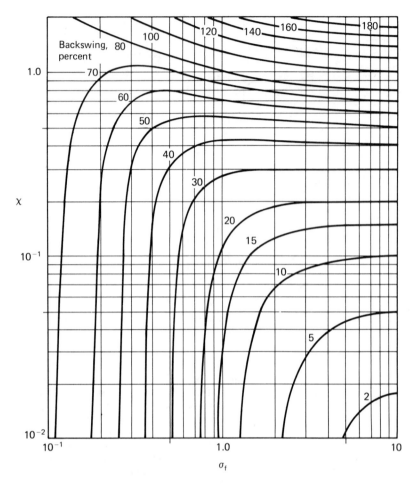

Figure 8.17 χ as a function of σ_f with backswing (percentage of U_b') as a parameter

Col. 4 $(S_f)_o$, i.e. the value of S_f corresponding to the fall time, from Figure 8.16, σ_f and χ,
Col. 5 $(L_pC)_{max} = \{t_f/2\pi(S_f)_o\}^2$; see Eqn (8.16),
Col. 6 $(C)_{max} = (L_pC)_{max}/L_p$,
Col. 7 $C = L_p/(2\sigma_f R)^2$; see Eqn (8.18).

The values of $(C)_{max}$ in col. 6 are the maximum values of C for the specified maximum fall time. They increase in value from the top to the bottom of the column. But because L_p has been fixed there is a value of C corresponding to each value of σ_f; see Eqn (8.18). These values of C are calculated in col. 7; they decrease in value as σ_f increases. Figure 8.18 indicates the variation of $(C)_{max}$ and C with σ_f. If any value of C is chosen from col. 7 then σ_f is fixed and col. 6 gives the maximum permissible value of C to meet the fall time requirement; if both criteria are to be observed then clearly the permissible values of σ_f are those for which $C \leq (C)_{max}$ and the maximum value of C equals $(C)_{max}$ at the cross-over point.

The value of $(C)_{max}$ obtained from the trailing edge requirements may now be compared with the range of values obtained in the table of leading edge parameters and a suitable limit value may be chosen, e.g. if the largest value of $(C)_{max}$ in the leading edge table is less than $(C)_{max}$ obtained from the trailing edge table then the former value would be selected as the limit. In practice the actual value chosen would be lower than this by a suitable margin to allow for manufacturing tolerances. A check should be made that it is a reasonable value when the shunt capacitance associated with the load is taken into account. The corresponding maximum value of leakage inductance may now be calculated; in most cases $(L_l/C)^{1/2}$ must not exceed the value corresponding to $(\sigma_r)_{min}$ at the top of col. 3 of the first table (see the discussion following the description of this table).

The element values having been determined, all the response parameters should be checked to ensure that

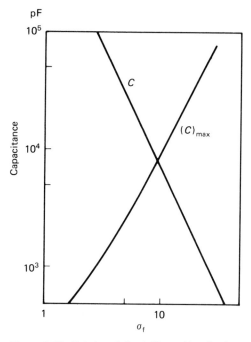

Figure 8.18 Variation of C and $(C)_{max}$ with σ_r for the worked example

they do not lie outside the limits given in the specification.

8.3.2 An example of design

The foregoing procedure will now be illustrated by a worked example of a step-up pulse transformer, the specification values being the figures in brackets in the list at the beginning of the previous section.

The load-source mismatch factor is

$$\eta = R_b/R_a r^2 = 600/50 \times 4 = 3$$

Other useful values are

$$R'_b = 600/4 = 150\,\Omega$$

$$R = R_a R'_b/(R_a + R'_b) = 37.5\,\Omega$$

$$R_a R'_b = 7500$$

Referring to Figure 8.5, a 5% overshoot will not be exceeded if $(\sigma_r)_{min} = 0.7$. A line drawn at this value on Figure 8.6 intersects the $\eta = 3$ curve at $x = 0.177$ and 5.7. Assuming that it is preferable to use the lower region then x must be less than 0.177.

The first table may now be constructed as described in the previous section, see Table 8.1.

For each value of σ_r, the values of C and L_l may not be larger than those given in columns 6 and 7 respectively because this would violate $(L_l C)^{1/2}_{max}$ and lead to an excessive rise time. They may have smaller values provided $(L_l/C)^{1/2}$ is not greater than 36.4; if this value is exceeded then σ_r will be less than 0.7 and the overshoot limitation will be exceeded. Thus 740 pF and 1.97 µH are not permissible $((L_l/C)^{1/2} = 51.6)$ but 740 pF and 0.98 µH would be acceptable $((L_l/C)^{1/2} = 36.4)$.

If, on the other hand, L_l is made disproportionately small so that $(L_l/C)^{1/2}$ is much smaller than the value corresponding to the minimum value of σ_r the damping factor will be larger than necessary. Provided the value of C does not exceed the limit, the resulting rise time will not be greater than the specified value. However, because the damping factor σ_r is greater than the minimum value, the rise time will be greater than it need be (see the effect of σ_r in Figure 8.3). If C can be reduced so that $(L_l/C)^{1/2}$ approaches the value corresponding to $(\sigma_r)_{min}$ the rise time will be decreased, and the leading edge response will be optimized well within the specification. However, in practice it is often quite difficult to meet the $(L_l)_{min}$ limit and the question of the disproportionately low value of L_l does not arise.

Returning to the example, the top of the pulse must now be considered. The permitted droop is 5%. From Figure 8.9, $L_p/t_d R = 19.5$.

$$(L_p)_{min} = 19.5 \times 4 \times 10^{-6} \times 37.5 = 2.93\,\text{mH}$$

The backswing limitation may modify this value.
The second table may now be constructed, see Table 8.2.

Table 8.1 Calculation of leading edge parameters

1	2	3	4	5	6	7
σ_r	$x = L_l/CR_a R'_b$ (from Figure 8.6)	$(L_l/C)^{1/2}$ (from col. 2)	$(S_r)_{min}$ (from Figure 8.4)	$(L_l C)^{1/2}_{max} = \left(\dfrac{t_r}{2\pi(S_r)_{min}}\right)(1+1/\eta)^{1/2}$	$(C)_{max} = (L_l C)^{1/2}_{max} \times (L_l/C)^{-1/2}$	$(L_l)_{max} = (L_l C)^{1/2}_{max} \times (L_l/C)^{1/2}$
0.7	0.177	36.4	0.34	5.4×10^{-8}	1480 pF	1.97 µH
1	0.072	23.2	0.54	3.4×10^{-8}	1470	0.79
1.5	0.0296	14.9	0.94	1.96×10^{-8}	1320	0.29
2.0	0.0161	11.0	1.32	1.39×10^{-8}	1260	0.15

Table 8.2 Calculation of the trailing edge parameters

1	2	3	4	5	6	7
σ_f	$(\chi)_{max}$ (from Figure 8.17 and backswing)	$(L_p)_{min} = t_d R/(\chi)_{max}$ (mH)	$(S_f)_o$ (from σ_f and χ in Figure 8.16)	$(L_pC)_{max} = \left(\dfrac{t_f}{2\pi(S_f)_o}\right)^2$	$C_{max} = \dfrac{(L_pC)_{max}}{L_p}$	$C = \dfrac{L_p}{(2\sigma_f R)^2}$
1	—	—	Provisionally choose $L_p = 4$ mH	0.15	1.13×10^{-12}	280 pF	710 000 pF
2	0.067	2.14	$\therefore \chi = \dfrac{4 \times 10^{-6} \times 37.5}{4 \times 10^{-3}}$	0.1	2.53×10^{-12}	630	180 000
5	0.095	1.5	$= 0.0375$	0.05	10×10^{-12}	2 500	29 000
10	0.1	1.5		0.027	35×10^{-12}	8 800	7 100
20	0.1	1.5		0.013	150×10^{-12}	38 000	1 800
30	0.1	1.5		0.009	310×10^{-12}	78 000	790

The requirement that the trailing edge must not be oscillatory limits σ_f to values equal to or greater than one. The maximum fall time is 1 µs and the maximum backswing is 10%.

An inspection of the tables shows that the maximum value of the shunt capacitance is determined by leading edge requirements; the crossover capacitance in the second table is outside the range permitted by the first table. A value of 1480 pF would be possible but it would be better to choose a lower value, e.g. 1000 pF to allow some margin for manufacturing tolerances. It must not be made too low because (1) when referred to the secondary it must not approach too closely the load capacitance and (2) the maximum leakage inductance must be reduced in proportion to satisfy the $(\sigma_r)_{min}$ requirements (i.e. the overshoot) and even with $L_l = 1.97$ µH the ratio L_p/L_l is rather large. In practice it will normally be necessary to add external capacitance to achieve the design value and so to avoid excessive overshoot. In the present example the total secondary shunt capacitance must be $1000/4 = 250$ pF of which 150 pF exists across the specified load. Taking $C = 1000$ pF, the maximum value of the leakage inductance is $L_l = 1.97 \times 1000/1480 = 1.33$ µH. It may be less than this without invalidating the design.

Figure 8.19 shows the element values in the equivalent circuit and Table 8.3 shows the calculated values of all the main parameters based on these element values.

The practical realization of a low-power pulse transformer having the required transformation ratio and the calculated element values is almost identical to the realization of a wide-band transformer. This has been discussed in detail in Section 7.3. Basically it consists of choosing a core and a winding configuration that will give the required shunt and leakage inductance with an acceptable value of self capacitance. For most purposes the number of turns may be calculated from the initial effective permeability, or from A_L. If the flux density in the core is not negligible

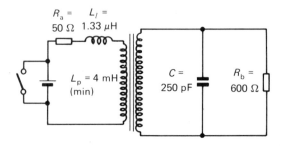

Figure 8.19 Element values for the worked example

the pulse magnetization may lower the permeability so that the initial permeability is no longer the appropriate parameter. This aspect is considered in Section 8.5.

8.3.3 A rapid design procedure

It often happens that a small pulse transformer having a simple specification is required quickly for experimental purposes. The following method, which is partly experimental, will often provide a satisfactory transformer.

1. From the required pulse duration, permissible droop and circuit impedance, R, obtain $(L_p)_{min}$ from Figure 8.9.
2. Select a suitable core, and using the published value of A_L (see for example Table 7.1) determine the number of primary turns.
3. Calculate the number of turns on the other winding(s) from the required turns ratio(s) and wind the transformer using a reasonably low leakage inductance configuration, e.g. single layer windings (see Figure 11.21).
4. Measure the pulse distortion of the transformer in the circuit. If the leading edge transient is oscillatory this is due to the load not being matched to the

Table 8.3

Parameter	Remarks	Calculated value	Spec. value
x	$= L_l/CR_aR_b'$	0.177	
σ_r	from Figure 8.6	0.7	
Overshoot	corresponding to σ_r	5%	5% max
S_r (10-90%)	from Figure 8.4	0.34	
Rise time (10-90%)	from Eqn (8.6)	0.067 µs	<0.1 µs
L_p/t_dR		26.7	
Droop	from Figure 8.9	3.7%	<5%
$\bar{\sigma}_f$	from Eqn (8.17)	26.7	>1 for no oscillation
χ	$= t_dR/L_p$	0.0375	
$(S_f)_o$	from Figure 8.16	0.01	
Fall time to zero	from Eqn (8.15)	0.13 µs	<1 µs
Backswing	from Figure 8.17	3.7%	<10%

source. It may or may not be possible to improve the matching. If the overshoot is excessive, shunt capacitance may be added; this will reduce the overshoot but increase the rise time. If it is not possible to meet the leading edge and pulse top requirements by this simple approach then some or all of the full design procedure of Section 8.3.1 must be followed.

8.4 Pulse distortion in relation to bandwidth

Any transformer connected between a source and load may be represented by the elements of the equivalent circuit. From these elements it is possible to calculate the bandwidth for the transmission of sinusoidal currents and, separately, the distortion that occurs when a given pulse is transmitted. Clearly these two aspects of the performance of the transformer are related, indeed some designers attempt to translate a pulse distortion specification into an approximately equivalent bandwidth in order to visualize the requirement more easily.

Wilds[4] has analysed the relation between the two aspects of performance to discover how feasible it is to predict the pulse distortion from the transmission bandwidth. He shows that it is possible if only the leading edge and the pulse top are of interest. To predict the leading edge response it is necessary to know not only the attenuation/frequency relation at the upper end of the transmission band but also the corresponding phase-shift/frequency response. Since for a given transformer the latter is usually not known, he points out that there is little to be gained from the analogy; it would be better to measure the principal elements of the equivalent circuit of a given transformer and then use these values directly to predict the pulse distortion using functions such as those given earlier in this chapter.

However, the droop of the pulse top is very simply related to the attenuation at the lower end of the transmission band and, as it may be useful in certain circumstances, this relation is now derived.

From Eqn (7.14) or Figure 7.6, the attenuation due to the shunt inductance, L_p, is 3 dB at $\omega = \omega_1$ when

$$\frac{\omega_1 L_p}{R} = 1$$

Let $L_p = \mu_i L_o$ where L_o is defined in Eqn (2.14) and μ_i is the initial permeability, assuming low amplitude signals.

$$\therefore \mu_i L_o = \frac{R}{\omega_1} \quad \text{H} \tag{8.19}$$

From Eqn (8.12) the pulse top droop is related to L_p by the following expression

$$\left. \begin{array}{l} L_p = -\dfrac{t_dR}{\log_e(1-D/100)} \quad \text{H} \\ = \mu_p L_o \end{array} \right\} \tag{8.20}$$

where μ_p is the pulse permeability, i.e. that value of permeability which characterizes the behaviour of the material under pulse conditions, (see Section 8.5). Combining these equations, the pulse droop of a

248 Pulse transformers

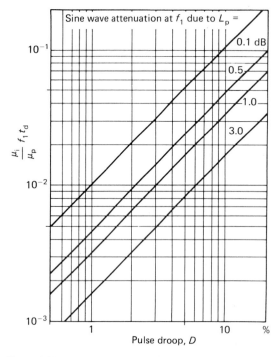

Figure 8.20 Pulse droop in time t_d related to the frequency f_1 at which the low frequency sine wave attenuation of the transformer, due to L_p, has a given value

given transformer may be expressed in terms of the frequency at which the insertion loss due to the shunt inductance reaches 3 dB:

$$\frac{\mu_p}{\mu_i} = -\frac{\omega_1 t_d}{\log_e(1 - D/100)}$$

or $f_1 t_d = -\dfrac{\mu_p}{\mu_i} \cdot \dfrac{\log_e(1-D/100)}{2\pi}$ (8.21)

Similar expressions may be written for other values of insertion loss due to L_p, e.g. 0.1, 0.5 and 1dB at f_1. The results may be shown graphically as in Figure 8.20. It is interesting to note that a 5% droop on a 4 μs pulse corresponds to a 3 dB insertion loss at 2 kHz, assuming that $\mu_p/\mu_i = 1$. This example shows that the energy spectrum of a typical pulse extends down to a surprisingly low frequency. It also shows that even this limited correspondence between pulse and wide-band performance is only possible if the ratio μ_p/μ_i is known. The pulse permeability is considered in the next section.

8.5 Pulse permeability

As for wide-band transformers, the core of a pulse transformer must provide sufficient shunt inductance with a reasonable number of turns on the winding. Ignoring the effect of any residual air gap, the inductance at low amplitude alternating currents is given by

$$L_p = \frac{\mu_0 \mu_i N^2 A_e}{l_e} \quad \text{H} \tag{8.22}$$

The problem is to find a substitute for the initial permeability, μ_i, which will give the appropriate value of L_p under pulse conditions. If the flux density, B, were directly proportional to the field strength, H, over the excursions considered then the permeability and inductance would be constant. If a constant direct voltage U were applied to such an inductance then by definition

$$U = L_p \frac{di}{dt} \quad \text{V} \tag{8.23}$$

From Eqn (2.11),

$$U = -e = NA\frac{dB}{dt} \quad \text{V} \tag{8.24}$$

The current and flux density would rise linearly with time. When the current reaches a value I the energy stored in the core is

$$\tfrac{1}{2}L_p I^2 = V \int_0^{\Delta B} H \, dB \quad \text{J} \tag{8.25}$$

where ΔB is the change in flux density corresponding to I, and V is the core volume. This assumes a uniform distribution of flux density; if this is not so the r.h.s. must also be integrated with respect to volume.

In a practical transformer there is in general a non-linear relation between B and H, and the inductance L_p is a function of the instantaneous amplitude. From Eqns (8.23) and (8.24), by integration,

$$\int_0^{t_d} U\,dt = \int_0^{\Delta I} L_p \, di = NA \int_0^{\Delta B} dB \tag{8.26}$$

It may usually be assumed that during the interval t_d (the pulse duration) the variation of U is negligible (i.e. the droop is small) so this integration may be written

---

(8.22) $L_p = 4\pi\mu_i N^2 A_e 10^{-9}/l_e$ H

(8.24) $U = NA\dfrac{dB}{dt} \times 10^{-8}$ V

(8.25) $\tfrac{1}{2}L_p I^2 = \dfrac{V}{4\pi}\int_0^{\Delta B} H\,dB \times 10^{-7}$ J

(8.26) $\int_0^{t_d} U\,dt = \int_0^{\Delta I} L_p\,di = NA\int_0^{\Delta B} dB \times 10^{-8}$ V s

$$Ut_d = L_p \Delta I = NA\Delta B \quad \text{V s (Wb)} \qquad (8.27)$$

where L_p is now an effective value, defined by this equation.

Now
$$\Delta I = \frac{\Delta H l}{N} \quad \text{A} \qquad (8.28)$$

Therefore the effective value of L_p is given by

$$L_p = \frac{N^2 A}{l} \cdot \frac{\Delta B}{\Delta H} = \frac{N^2 A}{l} \cdot \mu_o \mu_p \quad \text{H} \qquad (8.29)$$

This value of inductance may be used in the equivalent circuit of the pulse transformer; it will give the same value of magnetizing current and magnetic energy at the end of the pulse as the actual, varying, inductance. In the above equation the pulse permeability is introduced and it is defined by

$$\mu_p = \frac{\Delta B}{\mu_o \Delta H} \qquad (8.30)$$

which is formally the same as the definition of amplitude permeability in the a.c. case. Therefore in calculations of inductance and number of turns for pulse transformers the normal expressions may be used, e.g. Eqn (8.22), but μ_p is substituted for μ_i.

Before discussing the nature of $\Delta B / \Delta H$ for pulsed magnetic cores it is necessary to be able to evaluate the magnitude of the change of flux density. From Eqn (8.27)

$$\Delta B = \frac{Ut_d}{NA} \quad \text{T} \qquad (8.31)$$

where, for small droops, U may be taken as the average voltage across the winding during the pulse. If the cross-sectional area A is not uniform then ΔB will depend on the cross-section taken. In practice, provided the non-uniformity is not large, a mean value of A may be used or, if known, the effective value A_e.

In the majority of small pulse transformers used in electronic equipment, e.g. data handling transformers, the value of ΔB is less than 10 mT (100 Gs). At these low flux densities the core material is operated substantially at the origin of the B–H loop and, for a ferrite, the pulse permability as defined by Eqn (8.30) may be taken as equal to the initial permeability without significant error. In illustration, the example transformer design considered earlier in this chapter may be used. The relevant parameters are: pulse duration, $t_d = 4$ μs, the primary inductance, $L_p = 4$ mH and the primary pulse amplitude = 6V. A suitable core for such a transformer might be the 14 × 8 pot core listed in Table 7.1. Assuming that the minimum inductance factor, A_L, is 3400 nH/turn², then

$$N = (L/A_L)^{\frac{1}{2}} \simeq 35 \text{ turns}$$

For this core, $A_e = 25.0$ mm². Therefore from Eqn (8.31)

$$\Delta B = \frac{6 \times 4 \times 10^{-6}}{35 \times 25.0 \times 10^{-6}} \simeq 30 \text{ mT (300 Gs)}$$

This flux density is attained after 4 μs, i.e. $dB/dt = 7.5$ mT/μs. This is equivalent to the maximum rate of change of flux density in a sine wave of frequency 200 kHz and $\hat{B} = 6.0$ mT (60 Gs). So it is seen that in such a transformer the pulse magnetization is not very great or rapid; eddy current and permeability dispersion effects may usually be ignored and initial permeability may be used in the design.

It is appropriate now to consider pulses of larger energies. If the flux density excursion becomes appreciable then the core material magnetization will no longer be confined to a region near the origin of the B–H loop. It becomes very difficult to predict, in general terms, over which minor B–H loop the core will operate. Until this loop has been established the pulse permeability ($\propto \Delta B/\Delta H$) cannot be determined. The pulse permeability may determine the number of turns on the primary winding and this in turn affects the flux density excursion (see Eqn (8.31)). Thus there is a closed loop of relations which can only be made to yield the best design conditions by a process of preparing a table of parameters such as $N_1, \Delta B, \mu_p$ and L_p and interpolating for the most suitable design.

Melville[5] has described in detail various modes of magnetization in ferromagnetic cores; only a brief and simplified account of his description will be given here. In the present context it is first assumed that effects depending on the rate of change of flux and on the velocity of propagation within the ferrite may be ignored, i.e. the magnetization is similar to that occurring during the determination of B–H loops with direct current. This assumption is probably valid for most ferrite pulse transformers; the limitations will be discussed later.

Figure 8.21 shows the B–H curve traced in a ferrite core, initially demagnetized, when a succession of unidirectional pulses are applied. The interval

---

(8.27) $Ut_d = L_p \Delta I = NA\Delta B \times 10^{-8}$ V s

(8.29) $L_p = \frac{4\pi N^2 A \Delta B}{l \Delta H} \times 10^{-9} = 4\pi N^2 \frac{A}{l} \times \mu_p \times 10^{-9}$ H

(8.30) $\mu_p = \frac{\Delta B}{\Delta H}$

(8.31) $\Delta = \frac{Ut_d}{NA} \times 10^8$ Gs

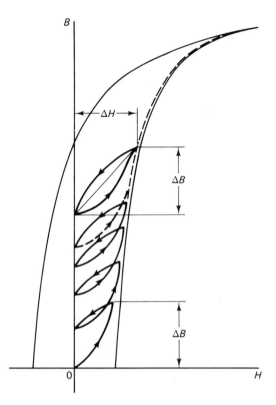

Figure 8.21 The B–H curves of an initially demagnetized ferrite core, when a succession of identical unidirectional voltage pulses are applied to a magnetizing winding. (Courtesy Melville[5])

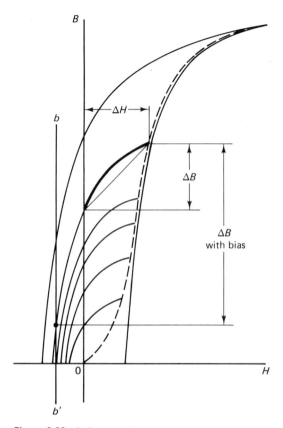

Figure 8.22 A diagrammatic representation of a series of B–H loops showing how the minor loop corresponding to a series of flux density excursions ΔB may be determined

between pulses is sufficient to ensure that the flux changes due to one pulse have finished before the next pulse arrives. Each pulse is of such an amplitude and duration that the change of flux is ΔB (see Eqn (8.31). During each pulse the flux density having increased by ΔB at the end of the pulse, returns to a residual value on the B-axis. A series of unclosed minor loops will be described until the curvature of the B–H loop permits a fall of flux density equal to ΔB and a closed minor loop is achieved.

If for a given ferrite there is available a series of B–H loops of successively increasing size such as shown diagrammatically in Figure 8.22, the minor loop corresponding to a given ΔB may readily be determined. A loop is found in which the flux density change from the tip to the residual value is ΔB; this section of the curve coincides with the upper arm of the minor loop, and the slope of the line joining the tip to the residual value gives the pulse permeability.

For a given material, the maximum available flux density excursion is limited by shape and size of the B–H loop and this limits the energy that may be transmitted by a given size of transformer core made from this material. However there are two ways in which the flux density excursion may be increased beyond the value obtainable in the simple case just considered. The first is to introduce an air gap so that the loop is sheared as described in Section 4.2.2. This reduces the residual flux density for a given peak flux density and so increases ΔB as shown in Figure 8.23. The pulse permeability will generally be reduced to the corresponding effective value, so this method is only applicable where a design is limited by ΔB rather than pulse permeability (or droop, or back-swing).

The other method is to introduce some reverse magnetic field so that between pulses the material is returned to a point on the curve to the left of the B-axis. This may be done by applying a negative reset pulse between the main pulses or by providing a d.c. or permanent magnet bias to the core[6]. A reset pulse will set the core to a point on the B-axis below the normal residual flux density, indeed a negative value is quite feasible. A steady field or current bias will

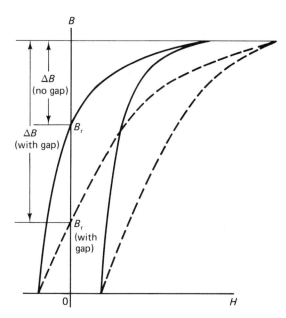

Figure 8.23 The increase of available flux density excursion due to shearing the B–H loop by the introduction of an air gap

result in the flux density returning to the value where the minor loop cuts the bias line (b–b' in Figure 8.22). A substantial increase in ΔB may be achieved in this way.

Sets of B–H loops for some ferrites suitable for pulse transformers are shown in Figure 3.2. Using these loops and the above method, the operating minor loops may be determined and the pulse permeability may be scaled off, see Figure 3.2.4. This procedure will give sufficiently accurate results for most ferrite pulse transformers.

Eddy currents, which have such a pronounced effect in laminated alloy cores, cause relatively little disturbance of the magnetization loop in ferrites. Instead the velocity of propagation within the ferrite must be considered. It was shown in Section 2.2.4 that the velocity of propagation of the magnetization in a manganese zinc ferrite is about 3×10^4 m/s. If it is assumed that the surface change of flux density must reach the centre of the core by the shortest route in a time that is less than a tenth of the pulse duration, t_d, then it may readily be shown that the least cross-sectional dimension of the core, b, must be less than about $6t_d$ where b is in mm and t_d in μs.

Before leaving this discussion of cores for higher power ferrite pulse transformers some mention must be made of core losses. During the pulse the flux density rises steadily from its residual or bias value to its maximum value, a total change of ΔB. At the end of the pulse the flux will decay back to the starting value at a rate which depends on χ and σ_f (see Figures 8.11 to 8.15); dB/dt is proportional to the voltage amplitude. Thus a minor B–H loop will be traversed and the rates of change of flux density will not normally be high even if the voltage pulse is very square. The magnetic core loss depends to some extent on dB/dt but a reasonable estimate may be made from the area of the minor loop. Since there is an infinite variety of possible minor loops it is not usually possible to give data on this subject; one approach would be to estimate the fraction of the full B–H loop that is enclosed and take this fraction of the energy loss per cycle of the full loop. However, in Section 9.3 and in the caption to Figure 3.19 experimental evidence is quoted which shows that for some typical power ferrites the magnetic loss at a given frequency depends on the total flux density excursion and is not substantially affected by the position of the minor loop. Where this is so, the core loss (per cycle) may be estimated from the loss curves relating to symmetrical sinusoidal excitation. The magnetic loss energy is dissipated at the pulse repetition rate and therefore the power loss may be calculated. At pulse repetition rates below 5 kHz this loss will not be appreciable in manganese zinc ferrite unless the magnetization approaches the full B–H loop.

Eddy current core loss could be significant at higher pulse energies. It is easily shown that for a core having a resistivity ρ and a circular cross-sectional area A the eddy current loss density when the flux density changes linearly by ΔB in time t is given by

$$P_F = \frac{(\Delta B)^2 A}{8\pi \rho t^2} \quad \text{W m}^{-3} \text{ during time } t \qquad (8.32)$$

Taking a rather extreme example, suppose $\Delta B = 0.1$ T (1000 Gs), $A = 10^{-4}$ m² (1 cm²), $\rho = 1$ Ωm (100 Ω cm) and $t = t_d = 1$ μs, then P_F is 4×10^4 Wm^{-3} or 40 mWcm^{-3}. The mean dissipation would be much less than this because normally the duty cycle (mark/space ratio) is small. Thus even taking into account both rising and decaying flux density, the eddy current core loss is not likely to be serious.

Winding losses under pulse conditions present, in a general case, a somewhat complex problem and the reader is referred to Glasoe[7] for a detailed treatment and also to Jongsma[8] who gives some relations between pulse and sine wave eddy current loss in windings. However a useful upper limit to the effective resistance may be obtained by using the results for the high frequency resistance of windings given in Section 11.4.4 and putting

$$f = \frac{1}{2t_d} \qquad (8.33)$$

(8.32) $P_F = \dfrac{(\Delta B)^2 A}{8\pi \rho t^2} \times 10^{-16}$ W cm^{-3}

252 Pulse transformers

The loss in the winding resistance occurs only during the pulse and therefore the winding loss so obtained must be reduced by the duty cycle ratio.

Summarizing the characteristics of ferrite cores at higher pulse powers, the achievement of a sufficient flux density excursion is the main limitation. Polarization or gapping can improve the performance in this respect but it cannot approach that of the best alloy or amorphous metal cores when the application involves very high power and the PRF is low. On the other hand, core losses are unlikely to be a limitation in the use of ferrite cores.

8.6 Practical considerations

8.6.1 General

As previously observed, the realization of a low power pulse transformer is similar to the realization of a wide-band transformer and much of Section 7.3 applies. In low power pulse transformer design there is less emphasis on winding resistance because this has little effect on the shape of the transmitted pulse. However, winding resistance will reduce the pulse amplitude and if this is an important aspect of performance it may easily be taken into account by using the following equation:

$$U'_b = \left(\frac{E_a R'_b}{R_a + R'_b + R_1 + R'_2} \right) \qquad (8.34)$$

where R_1 is the resistance of the primary winding and R'_2 is the resistance of the secondary winding referred to the primary.

If there is a pulse attenuation limit the corresponding limit of winding resistance may thus be readily found. This could lead to the evaluation of $(R_{dc}/L)_1$, i.e. the resistance/inductance ratio for one of the windings, e.g. the primary. Table 7.1 could then be used for the selection of a suitable ferrite core.

Another important design parameter is the ratio of shunt inductance to leakage inductance, L_p/L_l. It has been seen that the energy spectrum of a pulse transformer can be very wide and this tends to lead to large values for L_p/L_l. This ratio has been discussed in Section 7.3.1 It is often the case that a large value of L_p/L_l is more important than a low value of $(R_{dc}/L)_1$. This tends to lead to a winding that occupies only a small fraction of the available winding height, a shallow winding giving low leakage inductance at the expense of a higher resistance.

Whichever criterion is used in selecting a core, the effective permeability (or the inductance factor) should be as high as possible. The core shapes listed in Table 7.1 provide a suitable selection; where very high values of L_p/L_l are required, high permeability toroids may provide the best solution.

8.6.2 A practical example of a low-power pulse transformer

To illustrate the realization of a low power pulse transformer, the theoretical design considered in Section 8.3.2 will be used. Figure 8.19 shows the equivalent circuit and Table 8.3 compares the predicted performance with the specification.

Due to the severe specification, the ratio L_p/L_l is required to be rather high (>3000) so the transformer is not easy to realize. It was proposed to use a ferrite pot core and to select a size of core that would allow the low impedance winding to be accommodated in one layer using a reasonable conductor diameter, thus minimizing the leakage inductance.

The core chosen is the 18/11 pot core from Table 7.1; the winding breadth, b_w, is 6.0 mm and the mean turn length, l_w is 36.7 mm. An inductance factor $A_L = 2500$ nH/turn^2 is assumed, (in practice much higher values are now available).

Required primary inductance > 4 mH
∴ Number of turns on the primary winding = 40
∴ Maximum overall conductor diameter = 6.0/40
 = 0.15 mm

The conductor selected is 0.112 mm Grade 1 enamelled wire, $d_o = 0.134$ (see Appendix B, Table B.1.2). The secondary consists of two layers, each identical to the primary, placed one above and one below the primary as shown in Figure 8.24

From Figure 11.21 the leakage inductance referred to the primary is given by

$$L_l = 4\pi 10^{-4} \frac{N_1^2 l_w}{4 b_w} \left(\frac{\Sigma x}{3} + \Sigma x_\Delta \right) \quad \mu H$$

$l_w = 36.7$ mm

$b_w = 6.0$ mm

$\Sigma x = 3 \times 0.112$ mm

$\Sigma x_\Delta = 2 \times 0.1$ mm + 0.02 due to the enamel covering

Figure 8.24 Winding arrangement of the example pulse transformer. Conductor: 0.112 mm diameter, Grade 1 enamelled copper wire. Screen: 0.05 mm copper foil between 0.025 mm paper

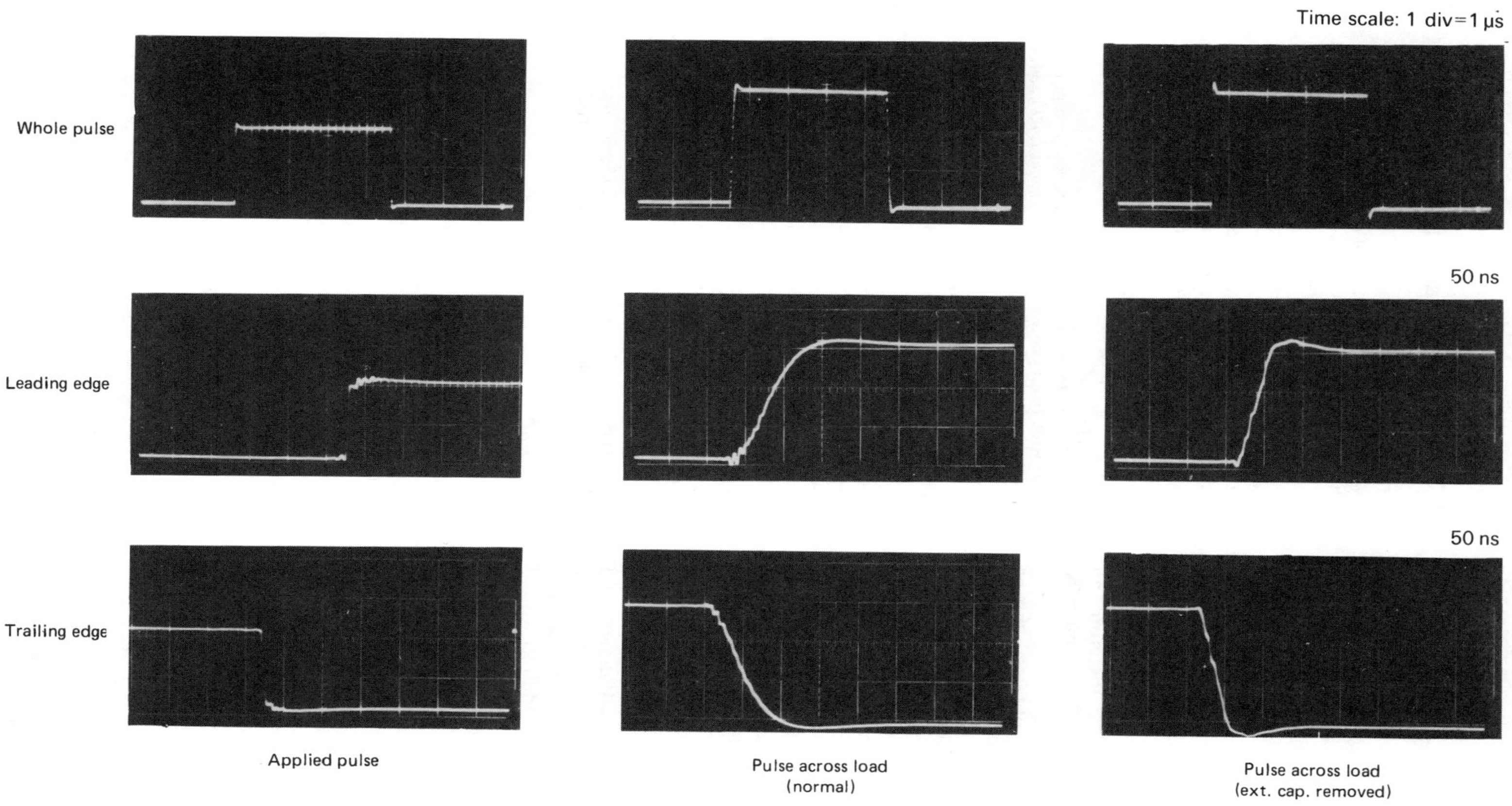

Time scale: 1 div = 1 µs

50 ns

50 ns

Pulse across load
(ext. cap. removed)

Pulse across load
(normal)

Applied pulse

Whole pulse

Leading edge

Trailing edge

Figure 8.25 Oscillograms showing the performance of the example pulse transformer (see Section 8.6.2). Net distortion:

	Normal	External capacitance removed
Rise time	62 ns	30 ns
Overshoot	4%	6%
Droop	1.5%	1.5%
Fall time	60 ns	30 ns
Backswing	4%	6%

$$\therefore L_l = 4\pi 10^{-4} \frac{40^2 \times 36.7}{4 \times 6} (0.112 + 0.22)$$

$$\simeq 1 \mu H$$

A model of this transformer was constructed and had the following measured properties.

Primary inductance = 4.3 mH

Leakage inductance = 1.32 μH

Self capacitance referred to the secondary side = 58 pF

The design shunt capacitance referred to the secondary side is $1000/4 = 250$ pF, therefore an eternal shunt capacitance of about 190 pF must be added.

Figure 8.25 shows oscillograms obtained during pulse tests with the transformer connected between a 50 Ω source and a 600 Ω load. It is seen that with the proper external capacitor the measured pulse distortions are close to the design values, allowing for the difficulty of measuring small deviations on oscillograms. In particular the use of the initial permeability in calculating the shunt inductance is shown to be justified.

The interwinding screen was included in case the interwinding capacitance (otherwise present) should give rise to a spike on the leading edge of the output pulse. The likelihood of such a spike in the absence of a screen depends on the time constants involved; when the screen was disconnected in the above example the leading edge waveform was unchanged. The presence of the screen makes the attainment of a low leakage inductance more difficult. An alternative design of transformer, without a screen, was also made. This consisted of three equal windings wound trifilar on a 14 mm pot core. There was no sign of a leading edge spike. The leakage inductance was much less (0.36 μH) but this reduction of leakage inductance by itself led to a somewhat longer rise time (0.075 μs) because it increased the damping factor, σ_r (see Table 8.1, Figure 8.6 and Figure 8.3) and this was only partly offset by a decrease in $(L_l C)^{1/2}$ (see Eqn 8.7). The new value of σ_r was 1.2 and the observed overshoot was zero as would be expected. A reduction of C to restore the ratio L_l/C would have decreased the rise time significantly below the value obtained with the first version.

8.7 References and bibliography

Section 8.1
1. WILDS, C. F. Determination of core size in pulse transformer design, *Electronic Engng,* **33,** 566 (1961)

Section 8.2.1
2. POSTMA, W., unpublished report
3. FENOGLIO, P., PECK, C. W., RICHARDSON, F. R., LORD, H. W. and BOYAJIAN, A. High power, high voltage pulse transformer design criteria and data, *General Electric Contract report, Req. No. NWK 60F200* (1953)

Section 8.3
1. WILDS, C. F., see Reference 1

Section 8.4
4. WILDS, C. F. Pulse transformers, frequency response, and wide band transformers, *Electronic Engng,* **34,** 608 (1962)

Section 8.5
5. MELVILLE, W. S. The measurement and calculation of pulse magnetization characteristics of nickel irons from 0·1 to 5 microseconds, *Proc. Instn elect. Engrs,* **97,** Part 2, 165 (1950)
6. RUSCHMEYER, K. Höher ausnutzbare Spulen durch permanentmagnetische Vormagnetisierung, *Valvo Berichte,* Hamburg, 1, Nov. (1986)
7. GLASOE, G. N., et al., see General
8. JONGSMA, J. High-frequency ferrite power transformer and choke design; Pt. 3, Transformer winding design, *Philips Elcoma Publication,* 40011, Sept. (1982)
   BROWN, D. R., BUCK, D. A. and MENYUK, N. A comparison of metals and ferrites for high speed pulse operation, *Trans. Am. Inst. elect. Engrs,* **73** Part 1, 631 (1955)
   SHARPLES, T. K. Saturation and gain characteristics of pulse transformers using various core materials, *Trans. I.E.E.E.,* **CP11,** 317 (1964)

Section 8.6
WILDS, C. F. Graded windings for pulse transformers. *Proc. Instn elect. Engrs,* **109** Part C, 589 (1962)

General
GLASOE, G. N. and LEBACQZ, J. V. (Eds) *Pulse generators,* MIT Radiation Lab. Series, McGraw-Hill Book Co, New York (1948)
BADY, I. Measurements of parameters that determine the front edge response of step-up transformers, *Conv. Rec. Inst. Radio Engrs,* **3,** Part 10, 26 (1955)
GROSSNER, N. R. Pulse transformer circuits and analysis, *Electro-Technology,* **69,** no. 2, 71 (Part 1), and no. 3, 80 (Part 2) (1962)
GROSSNER, N. R. *Transformers for electronic circuits,* 2nd Ed., McGraw-Hill Book Co. New York (1983)

9 Power transformers and inductors

9.1 Introduction

This chapter is concerned with transformers and inductors used, in the broadest sense, for power conversion. They may be characterized as having performance that is limited by temperature rise or magnetic saturation. Although the emphasis is mainly on the use of ferrite cores in this application, much of the treatment is not dependent on the type of core material and can be applied generally.

As indicated in Chapter 1, ferrites found their first power application as cores for the output transformers in the line time base and e.h.t. generation circuits of domestic television receivers. Initially, the operating frequency (in the UK) was about 10 kHz and this was subsequently increased to the present line frequency of approximately 16 kHz.[1] All the initial ferrite material and core developments for power use were designed to meet the requirements of this specific application. They were stimulated by the very large and sustained market for these cores. A wide range of other, smaller scale, power applications followed; examples are transformers supplying typically 2 kW at 30 kHz for ultrasonic power transducers, and matching transformers delivering 5 kW to transmitter antennas over a frequency band 2 to 30 MHz.

By the early 1970s the development of electronically switched power conversion circuits, generally referred to as Switched Mode Power Supplies (SMPS), was creating a demand for a new generation of power transformers, inductors and chokes. This in turn stimulated the development of ferrite materials and cores to meet the new requirements.[2] The switching frequency was, at first, generally chosen to be about 25 kHz. This was high enough to ensure that any acoustic emissions were above the audible range but was constrained by the limited frequency capability of the semiconductor devices and capacitors, and the availability of suitable components for the control circuits.

From that time until the present, the switching frequencies have steadily increased as the performance of the semiconductors and capacitors has been improved. The urge to increase switching frequencies arises from resulting substantial reductions in size and weight, not only of the wound components but also of the capacitors and other components.

Relative to the conventional mains-frequency power supplies, using transformers and chokes with heavy laminated silicon iron cores, the SMPS techniques increasingly offered economic and performance advantages. These were at first confined to the professional equipment sector but they are now being widely applied, e.g. in home computers and other consumer electronic equipment. The consequent demand for ranges of ferrite cores specifically designed for this application rapidly grew and has now exceeded that for all other industrial ferrite applications.[3] In view of the dominant position of ferrite cores for SMPS, this chapter will give precedence to this application. However, in the earlier sections the treatment will be quite general, recognizing that there still remains a wide diversity of power applications for ferrite cores.

In general, power transformers may be required to transmit a sinusoidal wave form at a single frequency, e.g. in a power distribution network, or over a narrow band of frequencies such as that required by an ultrasonic generator, or over a wide band as in a communication system. In the SMPS application the applied voltage wave form is, in general, rectangular, asymmetrical and of varying pulse width; the resulting flux wave forms are approximately triangular.

Single-frequency and narrow-band transformers will sometimes have a regulation specification that will limit the winding resistance and perhaps also the leakage inductance. Such special requirements will augment or may even take precedence over the temperature rise or saturation constraints. Wide-band power transformers will normally have to meet a return loss requirement over a given frequency band, so the design of such transformers is a combination of the principles discussed in Chapter 7 and the design procedure for single-frequency power transformers. Normally a wide-band transformer would be initially designed as a single-frequency transformer operating at the lower frequency of the transmission band. The design would then be checked, again as a single-frequency transformer, at the upper frequency of the band. Having achieved a satisfactory design as a power transformer, the wide-band transmission specification, e.g. return loss, would then be considered and the design (involving shunt and leakage inductance, etc.) would be developed accordingly.

At the higher frequencies, power transformer performance is limited mainly by the core loss. As the design frequency decreases then, for a given temperature rise, the flux density at which the core may be operated becomes higher until a frequency is reached below which the design is limited by saturation. For a silicon iron core, the maximum working flux density is typically $1.5\,T$ ($1.5 \times 10^4\,Gs$) and the saturation value may be $1.8\,T$; for 50/50 nickel iron the saturation is about $1.5\,T$ and for 80/20 (high permeability) nickel iron it is about $0.8\,T$. From Chapter 3 it is seen that for a typical power ferrite the maximum working flux density is about $0.32\,T$ ($3200\,Gs$) at $100°C$. Therefore at mains power frequencies, e.g. 50 and 400 Hz, the transformer steels have a clear advantage over ferrites. However, at higher frequencies where loss considerations dominate the design, the properties of ferrites, particularly the very low eddy current losses, give the ferrite core an undisputed superiority. Even the advent of amorphous metal magnetic materials poses no immediate threat.[4,5]

Heat transfer considerations are common to the design of most wound components used in power applications and the relevant theory and data are presented in the next section. This is followed by a general discussion of some basic aspects of the electrical and magnetic design of power transformers and inductors.

The next section extends the discussion in more detail to transformers and inductors having operating frequencies that are not so high that eddy current losses in the windings are significant. In particular this section deals with regulation, saturation and loss limited transformers and introduces parameters and design procedures to aid core selection. It also considers chokes and energy storage inductors. This section continues with a review of commercially-available core types together with their performance and design parameters. The section concludes with a worked example.

The final section considers higher frequency power transformers with particular attention to the design of windings to minimize their a.c. resistance.

9.2 Heat transfer and temperature rise

9.2.1 General

This subject is treated extensively in text books on heat and elsewhere and the reader is referred to the literature for a more detailed study than is possible here.[6-9] This section will be confined to interpreting some of the more straightforward results in terms that are applicable to transformer design.

A part of the power applied to the primary of a transformer is dissipated in the form of heat due to losses in the windings, in the core and in the dielectric materials that insulate the winding. The latter loss is usually negligible but may become significant in high voltage transformers. These losses cause a rise in temperature, the equilibrium temperature being reached when the heat lost balances the heat produced in the transformer. Since the production of heat per unit volume will in general vary from one part of the transformer to another and because the thermal conductivities involved are finite, the temperature, and therefore the rate of heat transfer, will in general vary from place to place on the surface of the transformer.

From the engineering point of view the temperature is important mainly because of its effects, usually deleterious, on the properties of the materials composing the transformer, e.g. increase of copper resistance, possible increase of core losses, reduction of the saturation flux density of the core material and degradation of the insulating materials. Since these effects concern the interior of the transformer it is the interior temperature that is important and this will be higher than the surface temperature by an amount depending on the thermal conductivities of the transformer materials and the distribution of the heat sources.

Sometimes the designer is more concerned with the heat dissipation than the temperature rise, e.g. the efficiency of the transformer may be important or its contribution to the ambient temperature inside an equipment may be a design criterion. At equilibrium, the calculation of the rate of heat dissipation usually presents no problems of heat transfer because it equals the total power loss. The present discussion is therefore concerned mainly with the temperatures attained by the transformer.

As indicated above, the surface temperature is a function of the rate of production of heat, i.e. power loss, and the heat dissipating properties of the transformer assembly. The heat will in general be transferred to the surroundings by means of conduction, convection and by radiation, and each of these processes may be natural or may be artificially aided. The way in which the total heat transfer is proportioned between these means of dissipation will vary widely from one design to another.

Enough has been said to illustrate the complexity presented by a proper analysis of the heat transfer problem. Computer programs[10,11] are available for the solution of two or three-dimensional heat transfer problems by numerical analysis and with good accuracy. However, in the majority of transformer designs great accuracy is not normally justified and in such cases satisfactory estimates may be obtained by making some simplifying assumptions and using empirical heat transfer formulae or data.

In the extreme case these lead to the rule of thumb that a heat loss of about 300 µW per square millimetre of surface (0.2 W in^{-2}) is a reasonable design figure. The design of a highly rated power transformer usually requires a more reliable guide. Accordingly data will now be given that will enable most transformer temperature problems to be solved with reasonable accuracy. Since the transformers under discussion are not large by power engineering standards, they attain equilibrium temperature in a relatively short time and it is this equilibrium temperature that is of main concern. The transient heating and cooling problem arises when the operation is intermittent or when the transformer is so large that the thermal time constant is too long for convenient measurement of the equilibrium temperature.[12] The transient problem is dealt with in most text books on power transformers.

A transformer is usually cooled mainly by convection (forced or natural) assisted by radiation. The amount of cooling by conduction is significant only if special means are adopted to make it so, e.g. some transformers are equipped with special heat conductors and heat sinks so that conduction is the principal means of cooling. However the majority of transformers are not designed in this way so attention will be devoted mainly to the other two types of cooling.

A transformer may be operated in contact with surrounding air or it may be contained in an oil filled vessel. In either design the transfer of heat depends ultimately on the radiation and convection from the surface in contact with the air (incidental conduction and the possibility of water cooling being ignored). The convection may be natural, or forced air cooling may be provided. The treatment below starts with the cooling of a body in air.

If the transformer is contained in an oil-filled vessel then the heat must be transferred from the transformer surface to the volume of oil before being dissipated to the surrounding air. This process is next considered. Finally attention is given to the passage of heat from the interior of the transformer to its surface. Applying these calculations in the above order enables the complete heat transfer problem to be solved in principle. In practice the procedure is reasonably reliable unless some factor is present which either invalidates the assumptions, e.g. significant conduction, or complicates the calculation, e.g. an unusual winding arrangement.

Table 9.1 The emissivities of various surfaces (after King[13])

Surface	Emissivity E
Polished aluminium and brass	0.05
Clean, smooth iron and steel	0.2
Polished silver	0.025
Oxidized aluminium	0.1–0.2
Oxidized brass	0.25–0.6
Oxidized iron and steel	0.6–0.9
Metallic paint	0.5
Non-metallic solids such as porcelain, glass, paper and non-metallic paints of any colour	0.9–0.95

9.2.2 Radiation

This is taken first not because it is necessarily the most important means of cooling but because it does not depend on the ambient air conditions.

The rate of dissipation of heat from a surface follows from the Stefan–Boltzmann law and is given by:

$$P_{rad} = 5.67 \, 10^{-8} \, E(T_s^4 - T_o^4) \quad \mu W \, mm^{-2} \quad (9.1)$$

where E is the emissivity of the surface

T_s is the surface temperature of the transformer in K

T_o is the temperature of the surrounding objects in K

The values of emissivity for a number of surfaces are given in Table 9.1.

In subsequent calculations involving radiation, T_o will be assumed to be 300K (27°C) and E will be taken as 0.95. These data have been used to calculate the broken curve in Figure 9.1; this expresses the relation given in Eqn (9.1) and may be used to find the heat lost by radiation in most practical situations. Radiation loss under other conditions may be readily obtained from the above equation.

9.2.3 Convection

Convection is the removal of heat by the passage of a fluid over a surface.

The fluid in close proximity to the surface may be regarded as a stagnant film transmitting heat by conduction to the main body of the fluid which due to its movement may be regarded as a constant temperature heat sink. There is, of course, a gradual transition between the stagnant film and the main body of the fluid. The rate of heat dissipation depends on the properties of the fluid and the difference in temperature between the surface and the fluid; it does not depend on the fine texture or the colour of the surface.

In free convection the movement of the fluid arises

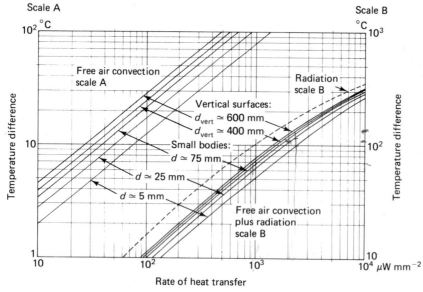

Figure 9.1 Heat transfer by free air convection and radiation. The temperature difference between the surface and the surrounding air as a function of the rate of heat transfer for large vertical surfaces (heights > 0.6 m or 24 in), and small bodies having various values of d in mm where d is the diameter for long horizontal cylinders or $d = d_{hor} d_{vert}/(d_{hor} + d_{vert})$ for more compact bodies in which the horizontal length d_{hor} becomes comparable with the vertical height d_{vert}. For small spheres d = radius. The l.h. group of curves is for free air convection, the broken curve is for radiation assuming an emissivity of 0.95 and an ambient temperature of 27°C, and the r.h. group is for the sum of the convection and radiation. Based on data given by Montsinger[14] and King[15]

from its reduced density when heated. It gathers heat as it rises over the heated surface and subsequently loses it to the surrounding fluid or containing vessel, becomes more dense as it cools and descends, perhaps to recycle past the heated surface indefinitely. Clearly the rate of heat transfer per unit area of surface will be influenced by the geometry of the surface, e.g. a vertical surface will lose heat by free convection at a rate different to that from a horizontal surface.

In forced convection the fluid is passed over the surface at a velocity determined only by the performance of the pump or blower. Thus the rate of cooling is less dependent on the geometry of the surface but obviously depends on the fluid velocity.

9.2.3.1 Free convection in air

Two formulae are quoted from the literature; they are based on laminar air flow which is the mode of convection normally applicable to heated objects of the size common in electronic equipment. For a long horizontal cylinder in air at atmospheric pressure the following formula is derived from one quoted by King[15]

$$P_{conv} = 6.3 \frac{\theta^{1.25}}{d^{0.25}} \quad \mu W\, mm^{-2} \tag{9.2}$$

where P_{conv} is the rate of heat transfer by convection,
θ is the difference in temperature in °C between the surface and the air,
d is the diameter of the cylinder in mm.

When the diameter exceeds 150 to 200 mm (6 to 8 in) the rate of heat transfer per unit area tends to become independent of the diameter, and the body may be treated as a number of large plane surfaces. A formula is given below for the heat transfer from a large vertical surface. Before considering this however it should be mentioned that Eqn (9.2) may be extended to apply to all bodies having dimensions that are small enough to affect the convection per unit area, i.e. linear dimensions of less than about 300 mm (12 in). This is done by giving d an effective value depending on the horizontal dimension and the vertical dimension of the given body, d_{hor} and d_{vert} respectively.[16] Then

$$\frac{1}{d} = \frac{1}{d_{hor}} + \frac{1}{d_{vert}} \tag{9.3}$$

Thus for a sphere $1/d = 2/\text{diameter}$, therefore d equals the radius of the sphere. This method is applicable to most power transformers used with electronic apparatus.

For vertical surfaces having a height d_{vert} (in mm) of less than about 1000 mm, King[17] gives a formula which may be stated as follows:

$$P_{conv} = 7.54 \frac{\theta^{1.25}}{d_{vert}^{0.25}} \quad \mu W\, mm^{-2} \tag{9.4}$$

θ is defined in Eqn (9.2).

From the upper side of a plane horizontal surface the rate of heat loss will be 15 to 25% more than for a vertical surface and for a similar downward facing surface the rate of heat loss will be less and will decrease as the surface becomes larger; a reduction of 33% relative to the vertical surface is quoted by King.

If the air pressure is n atmospheres then both of the heat transfer expressions, i.e. the r.h.s. of Eqn (9.2) and (9.4), must be multiplied by $\sqrt{n}$, thus natural convection at high altitudes is much less effective.

Figure 9.1 shows the surface temperature rise above ambient as a function of the rate of heat transfer at normal atmospheric pressure. The left hand group of curves are for free convection cooling of horizontal cylinders or small bodies in accordance with Eqn (9.2) and for large vertical surfaces in accordance with Eqn (9.4). The broken curve is for radiation cooling under the conditions mentioned in connection with Eqn (9.1). The right hand group of curves are for free convection plus radiation and, neglecting conduction losses, represent the equilibrium temperatures of typical bodies cooling under normal atmospheric conditions.

For free convection at normal atmospheric pressures the thickness of the stagnant film referred to above is about 2–3 mm. This must be remembered when considering clearances. If the space between the surface of the winding and the outside leg of a shell type core is less than 10 mm then the convection cooling from these surfaces will be less than the calculated value. If the spacing approaches 3 mm it may be assumed that no cooling by free convection will occur in that space and the areas concerned must be disregarded in calculating the total cooling surface of the transformer. Similarly any surface which clears an adjacent obstruction by less than about 5 mm will lose relatively little heat by free convection. If the transformer or its case is fitted with cooling fins these should be at least 12 mm (1/2 in) apart if they are to contribute to cooling by free convection. In fact the effective area of such fins may be estimated by assuming that their surfaces (but not their exposed edges) and the surfaces from which they project are covered by a 6 mm thickness of conductive material. The derived surface including the (now thickened) exposed edges of the fins may be used to calculate the effective area.

9.2.3.2 Forced convection in air

For air flowing over a plane surface King[18] gives a formula from which the following expressions may be obtained

$$P_{conv} = (4.54 + 4.1v)\theta \quad \mu W\,mm^{-2} \quad (9.5)$$

where v is the velocity of the air stream at a distance of about 25 mm from the surface, in m s^{-1}, and

         θ is the temperature difference between the surface and the air in °C

or $P_{conv} = (4.54 + 1.25v)\theta \quad \mu W\,mm^{-2}$

where the symbols are as before except that v is in ft s^{-1}

Figure 9.2 shows the relative surface temperature as a function of the rate of heat transfer due to forced air cooling. The amount of radiation cooling is usually relatively small but if it is significant it may be added from the data given in Figure 9.1.

9.2.3.3 Free convection in oil

In an oil-immersed transformer the heat is normally

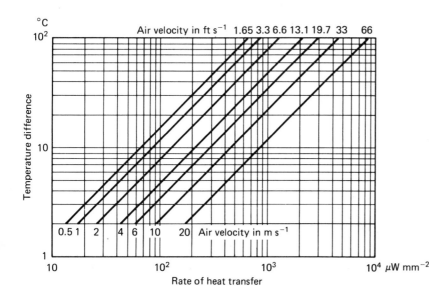

Figure 9.2 Heat transfer by forced air convection. The temperature difference between the surface and the surrounding air as a function of the rate of heat transfer for various air velocities. These data apply to air flow parallel to a smooth surface. If the surface is rough the transfer may be 10 to 40% greater; if the air meets the surface at an oblique angle the transfer is somewhat less. Based on data quoted by King[18]

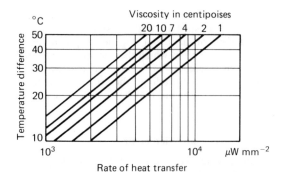

Figure 9.3 Heat transfer by free convection in oil. The approximate temperature difference between a large dissipating surface and the average temperature of the surrounding oil, shown as a function of the rate of heat transfer with the oil viscosity as a parameter. The viscosity is the value corresponding to a temperature that is the mean between that of the dissipating surface and the bulk of the oil. Based on data quoted by King[19]

transferred from the transformer surface to the walls of the containing vessel by free convection. The rate of heat transfer per unit area depends on the viscosity and other properties of the oil and also on the temperature difference, θ, between the transformer surface and the main body of the oil. In practice the main body of the oil may be assumed to be at a temperature that is not much higher than that of the walls of the containing vessel when these are cooled by air convection. King[19] quotes results obtained by Montsinger and Metrovick, and suggests that these results may be approximately represented by

$$P_{conv} = \frac{112\,\theta^{1.25}}{v^{0.4}} \quad \mu W\,mm^{-2} \quad (9.6)$$

where v is the oil viscosity in centipoises measured at a temperature that is a mean between that of the dissipating surface and the bulk of the oil.

The expression is shown graphically in Figure 9.3.

The remarks made concerning the effect of the stagnant layer between the surface and the convection currents apply here also except that the effective skin thickness is about 1 to 1.5 mm. Thus if convection cooling is to be assisted by providing vertical ducts in the core or winding these should have least dimensions of not less than 3 mm. If this limitation is observed then the surface area of the duct may be added to the total cooling area.

9.2.4 Conduction

The rate of heat flow perpendicular to a lamina of area A, thickness δx and thermal conductivity λ when the temperature difference is $\delta\theta$ is given by $\lambda A \delta\theta/\delta x$

or $\quad P_{cond} = \dfrac{\lambda \delta \theta}{\delta x} \quad \mu W\,mm^{-2} \quad (9.7)$

where P_{cond} is the rate of heat transfer by conduction per unit area
δx is in mm
$\delta\theta$ is in °C
λ is in $\mu W\,mm^{-1}\,°C^{-1}$.

This may conveniently be regarded as the 'Ohm's law' of heat flow by conduction:

$$P_{cond} = \frac{\delta\theta}{R_T} \quad \mu W\,mm^{-2} \quad (9.8)$$

where R_T is the thermal resistance $= \delta x/\lambda$ and has the units: $°C.mm^2\,\mu W^{-1}$.

Given the thermal resistance of a layer through which a uniform perpendicular heat transfer is occurring, the temperature difference across the layer may be readily calculated. Where the heat flow is divergent, P_{cond} is not constant across the thickness and the temperature drop must be obtained by integration.

The thermal conductivities of some commonly used transformer materials are given in Table 9.2.

Pursuing the electrical analogy, a complex arrangement of heat sources and heat transfer paths may be represented by an equivalent electrical circuit of current sources and resistances (e.g. see Figure 9.5). In the transient case capacitors would be included to represent thermal capacitances.

Before considering an example to illustrate the equivalent electrical circuit it will be convenient to consider the heat transfer in a homogeneous ferrite core. Figure 9.4 shows the cross-section of a cylindrical core of radius r and axial length l. The power converted to heat in unit volume of the ferrite, P_m, is assumed to be uniform and therefore at thermal equilibrium the heat passing through an elementary surface of radius x is

$$P_m \pi x^2 l \quad W$$

The heat flow per unit area is

$$\frac{P_m \pi x^2 l}{2\pi x l} = \frac{P_m x}{2} = P_{cond}$$

Therefore from Eqn (9.7)

$$d\theta = \frac{P_m x}{2}\,\frac{dx}{\lambda}$$

Integrating from $x = 0$ to $x = r$, the temperature difference between the centre and the surface is

$$\theta_{cs} = \frac{P_m r^2}{4\lambda} = \frac{P_m d^2}{16\lambda} \quad °C \quad (9.9)$$

If λ is in $\mu W\,mm^{-1}\,°C^{-1}$ and the diameter d is in mm then P_m is in $\mu W\,mm^{-3}$ or $mW\,cm^{-3}$.

For normal sizes of ferrite cores operating at typical frequencies and flux densities this conductive

Table 9.2 Thermal conductivities of some materials commonly used in power transformers

Material	Thermal conductivity λ (μW mm^{-1} °C^{-1} at 20°C)
Metals	
Aluminium, 90% pure	220×10^3
Brass	111×10^3
Copper	385×10^3
Lead	38×10^3
Silver	410×10^3
Steel	43×10^3
Insulating materials	
Air (still)	25.8
	(30 at 77°C)
Alumina ceramics	$(10-30) \times 10^3$
Cotton cloth, varnished	160–220
Bitumen	420–700
Epoxy resin, cast without filler	170–210
Epoxy resin, cast with silica filler	400–850
Ferrite	3500–4300
Glass	720–1200
Glass bonded mica	380–600
Mica	400–600
Paper, dry	50–140
Paper board, varnished	250
Phenol formaldehyde (Bakelite etc)	
moulded without filler	125–250
moulded with wood flour filler	170–350
paper laminate	240–260
fabric laminate	320
Polyamide resin (Nylon etc)	
moulded without filler	220–280
moulded with glass fibre filler	210–280
Polycarbonate, moulded without filler	190
Polyester resins, cast without filler	170–260
moulded with glass fibre filler	420–670
glass cloth laminate	210
Polyethylene, moulded without filler	330–520
Polystyrene, moulded without filler	40–140
Polytetrafluoroethylene (PTFE)	
moulded without filler	250
Polyvinychloride (PVC) moulded without filler	125–300
Rubber	130–280
Silicone rubber	210–400
Silk, varnished	160
Wax, paraffin	230
microcrystalline	200–250

Note In the literature thermal conductivities are often quoted in the units: Btu h^{-1} ft^{-2} °F^{-1} in. Multiply the figures so quoted by 144 to convert to μW mm^{-1} °C^{-1}.

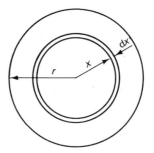

Figure 9.4 Cross-section of a homogeneous cylindrical ferrite core

temperature drop is not large. Where it is large then Eqn (9.9) becomes invalid since both P_m and λ are functions of temperature and this should be allowed for in the integration.

For cores with cross-sections approximating to a square, Eqn (9.9) may be used with reasonable accuracy if the effective diameter d is taken as the diameter of the circle having the same area, A, as the rectangular core, i.e. $d = (4A/\pi)^{1/2}$.

A similar treatment of ferrite cores having cross-sections with rectangular dimensions $b \times d$ where $b \gg d$, i.e. relatively thin slabs of ferrite, yields a temperature difference between the central plane and the parallel surfaces given by:

$$\theta_{cs} = \frac{P_m d^2}{8\lambda} \quad °C \qquad (9.10)$$

In the derivation of this expression it is assumed that the heat flow is normal to the plane of the slab; for a slab of finite thickness and limited superficial dimensions some heat will flow towards the edges and this will reduce the central temperature.

A comparison of the last two equations shows that for an infinite slab the temperature at the centre is twice that for a cylinder having a diameter equal to the slab thickness. In practice a general rectangular core will have an intermediate value.

9.2.5 An example of a heat transfer calculation

An example illustrating the calculation of the temperatures within a transformer will now be considered. Figure 9.5 shows a cross-section of a cylindrical ferrite core carrying two windings. Details of the dimensions, currents and core loss are also given in the figure. Alongside the cross-section of the transformer is the equivalent electrical circuit of the heat transfer path. The current sources are analogous to heat sources and the resistances are analogous to parts of the heat transfer path having relatively low conductivity. The core loss is represented by a distributed current source and the thermal resistance of the windings is neglected. It is assumed that the heat flow is radial and parallel to the plane of the diagram. This is a reasonable assumption in the case of these windings since they are wide and shallow so that the heat loss through the ends may be neglected. Considering the core, the part with which this example is concerned may be regarded as one limb of a closed core, the closing limbs carrying no windings. The temperature of these limbs may be separately calculated and will clearly be less than that of the limb carrying the windings. Thus there will in practice be some flow of heat axially along the core. Three-dimensional flow of heat presents a complex problem so the axial flow is neglected here; this will make the answer somewhat pessimistic.

Proceeding with the calculation in simple stages, the heat generated at each of the main sources is first deduced.

Heat generated in the core =
$150 \times 10^{-6} \times (\pi/4) \times 20^2 \times 50 = 2.36$ W

Heat generated in the primary winding:

Mean turn length $\simeq (20 + 1.8)\pi = 68.5$ mm
∴ Total length of the conductor =
$50 \times 68.5 = 3430$ mm
For 0.800 mm diameter copper wire the specific resistance,* $R_c = 0.034$ Ω m^{-1}
∴ Resistance $= 3.43 \times 0.034 = 0.117$ Ω
∴ Heat generated in the primary =
$I^2 R = 4 \times 0.117 = 0.47$ W

Heat generated in the secondary winding:

Mean turn length $\simeq 75.4$ mm
∴ Total length of the conductor $= 7540$ mm
$R_c = 0.121$ Ωm^{-1}
∴ Resistance $= 7.54 \times 0.121 = 0.912$ Ω
∴ Heat generated in the secondary =
$1 \times 0.912 \simeq 0.91$ W

The total rate of heat flow through R_3 is

$$\frac{2.36 + 0.47 + 0.91}{80.0 \times 50} = 0.00094 \text{ W mm}^{-2}$$

$$= 940 \text{ μW mm}^{-2}$$

where 80.0 = mean circumference of R_3 in mm.

The thermal resistance of $R_3 = \frac{1.0}{160}$

$$= 0.00625°\text{C mm}^2 \text{ μW}^{-1}$$

Therefore the temperature drop across R_3, given by Eqn (9.8), is

$\theta_3 = 0.00625 \times 940 = 5.9°$C

*See Appendix B, Table B.1.1.

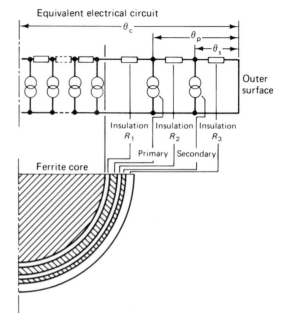

Figure 9.5 Diagram for the example calculation of temperature distribution in a ferrite cored transformer, showing the cross-section of the core and windings and the equivalent electrical circuit of the heat transfer path. The following data are used in the example:

Ferrite core	Axial length:	50 mm
	Diameter:	20 mm
	Power loss in core:	150 μW mm^{-3}
	Thermal conductivity:	4000 μW mm^{-1}°C^{-1}
Windings	Primary winding consists of 50 turns of 0.80 mm dia. enamelled copper wire and carries an r.m.s. current of 2 A. Secondary winding consists of 100 turns of 0.425 mm dia. enamelled copper wire and carries an r.m.s. current of 1 A. Axial length of windings=50 mm	
Insulation	Varnished cotton, λ=160 μW mm^{-1}°C^{-1} Insulation thicknesses: R_1, 0.5 mm R_2, 0.5 mm R_3, 1.0 mm	

The rate of heat flow through R_2 is less than that through R_3 by the heat generated in the secondary winding. Its value is

$$\frac{2.36 + 0.47}{72.6 \times 50} = 0.00078 \text{ W mm}^{-2}$$

$$= 780 \text{ μW mm}^{-2}$$

where 72.6 = mean circumference of R_2 in mm.

The thermal resistance of $R_2 = \frac{0.5}{160}$

$$= 0.00312°\text{C mm}^2 \text{ μW}^{-1}$$

Therefore the temperature drop across R_2 is

$$\theta_2 = 0.00312 \times 780 = 2.4°C$$

Similarly the temperature drop across R_1

$$\theta_1 = 0.00312 \times \frac{2.36 \times 10^6}{64.4 \times 50} = 2.3°C$$

The temperature drop between the centre and the surface of the core is given by Eqn (9.9)

$$\frac{P_m d^2}{16\lambda} = \frac{150 \times 400}{16 \times 4000} \simeq 0.9°C$$

Therefore the temperatures relative to the coil surface are

θ_s = relative temperature of secondary = + 5.9°C

θ_p = relative temperature of primary = + 8.3°C

θ_c = relative temperature of centre of core = + 11.5°C

Assuming the surface of the winding is being cooled by free air convection and radiation, reference to Figure 9.1 shows that the temperature of the surface when the dissipation is 940 µW mm^{-2} (= the rate of heat flow through R_3) is about 60°C above ambient. The curve for $d = 25$ mm has been used because the present case may be regarded as a long cylinder, it being assumed that there is no heat loss at the ends. The actual internal temperatures relative to ambient are:

Secondary	65.9°C
Primary	68.3°C
Centre of core	71.5°C

It is seen that, in this example, the heat transfer at the surface accounts for the major temperature rise. The temperature gradient within the ferrite core is negligible. Had the insulation been thicker or if the windings had consisted of many layers interleaved with paper the internal temperature drop would have been greater. However the example serves to show the temperature distribution in a typical ferrite power transformer.

9.2.6 Thermal resistance

The type of heat transfer calculation given in the previous section can be very useful in assessing the relative importance of the various elements of the radial heat transfer path in terms of temperature drop. However, as it is only a two-dimensional calculation, it neglects the heat flow in the axial direction and, for a typical power transformer, this can be a very significant proportion of the total. In practice, ferrite core manufacturers usually publish values of the thermal resistance for their ranges of power transformer cores. The thermal resistance of a transformer may be defined as:

$$R_{th} = \theta/P_{tot} \quad °C\,W^{-1} \qquad (9.11)$$

where P_{tot} is the total power dissipated in the core and the windings, and θ is the temperature rise above ambient measured at some defined internal point. This point should be at or near the point of maximum temperature, referred to as the 'hot-spot'. The actual hot-spot temperature is generally at the centre point of the axis of the limb carrying the windings. It can be seen by reference to the example in Section 9.2.5 that the temperature drop between the axis and the surface of a quite large cylindrical core is only of the order of 1°C, so it is usual to measure the temperature at the surface of the limb carrying the winding, at a point mid-way along the length of the coil former.

A number of factors influence the result. Atmospheric pressure affects the convection as indicated in Section 9.2.3; measurements are invariably performed at normal atmospheric pressure so correction by a factor equal to the square root of the pressure ratio must be applied if, for example, the transformer is to be operated unpressurized at high altitude.

Another factor is the distribution of the heat sources. In measurements on the EC range of cores undertaken by the author, half of the total power was supplied to the core at a suitable frequency via a virtually loss-free winding and the other half, representing the winding loss, was supplied by direct current to a non-inductive winding occupying most of the winding space, thus simulating the practical situation. Bracke[20] measured thermal resistance with various power distributions and concluded that the differences were within the experimental error. He therefore used the simpler method of supplying power only as d.c. to a suitable winding. He observed a dependence on the width of the winding, according to whether or not an allowance was made for creepage distance (see Section 9.4.4).

The position and method of mounting will also have some effect. It is usual to mount the transformer assembly on a printed circuit board in the position dictated by the terminating pins on the coil former. The difference between having the axis vertical or horizontal is probably small. Heat sinks are not normally used when measuring thermal resistance.

Finally, there are some less obvious factors. The example in Section 9.2.5 shows that the major temperature drop is at the outer surface/air interface. The temperature drop between the ferrite and this outer surface is typically about 5°C for an average SMPS transformer but it is strongly influenced by the presence of insulating layers. Of these, the annular air space between the ferrite core and the internal dimension of the coil former, which can be in the range 0.3 to 0.6 mm, can give a temperature drop that might be between 3 and 6°C. Clearly there could be some advantage in reducing the thermal resistance of this annular gap.

Hess and Zenger[21] show that the thermal resistance is significantly dependent on the thermal conductivity of the ferrite, even when the hot-spot temperature is measured on the surface of the ferrite as described above. They demonstrate that the thermal conductivity of ferrite depends on its (mass) density and the level of calcium at the grain boundaries, and underline the importance of these properties in power ferrites.

To put the estimation of thermal resistance into perspective, the following figures relate to the ETD 39 core (see Table 9.3). The cooling surface area of the exposed core and winding, assuming geometric surfaces, is estimated to be 5930 mm². From Figure 9.1, for a uniform surface temperature 35°C above ambient, the rate of cooling due to convection and radiation combined is 460 µW/mm². Therefore the power dissipation is $460 \times 10^{-6} \times 5930 = 2.73$ W and $R_{th} = 35/2.73 = 12.8$°C/W. Table 9.3 quotes $R_{th} = 15$°C/W for the ETD 39 core with a full-breadth winding, but this refers to the hot-spot temperature which, for a dissipation of 2.73 W, would be $15 \times 2.73 = 41$°C. The difference is the temperature drop between the ferrite core and the surface.

Clearly the calculation of the thermal resistance from the cooling surface area is a useful guide in the absence of data from direct measurement, but where such direct data is available it is to be preferred.[20, 22, 23] Table 9.3 lists manufacturers' published values of thermal resistance for the most widely used ranges of ferrite power cores.

9.2.7 Thermal shunts and thermal isolation

A thermal shunt is a means of making the cooling due to conduction a significant if not a major factor. One means of doing this is to use a coil former made of a metal such as aluminium, provided, of course, with a slit to prevent the occurrence of a shorted turn. The flange of the coil former might be fastened to a heat sink such as a chassis, a panel or a large cooling fin, again taking care to avoid an electrical short circuit. The temperature of the heat sink would depend on the heat flowing into it from the thermal shunt and on its rate of cooling due to convection, etc. This may be readily obtained from the heat transfer graphs shown earlier in this section. Usually the heat sink temperature will be near that of the equipment ambient so a very effective heat transfer path is provided.

The equivalent circuit of Figure 9.5 may be modified to illustrate the resulting analytical problem. The reference temperature in this case must be the ambient temperature and not that of the coil surface,

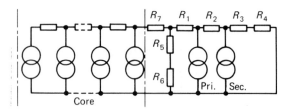

Figure 9.6 Modification of equivalent circuit shown in previous figure so that it includes a thermal shunt

so the closing short circuit must be replaced by a resistance, R_4, which represents the thermal resistance of the surface of the transformer, i.e. $R_4 = \theta/P_{tot}$ °C mm² µW^{-1} where θ is the temperature difference between the surface and the ambient, and P_{tot} is the total rate of heat transfer from the surface in µW mm^{-2}. This resistance may be estimated from the appropriate heat transfer graph. Then a shunt resistance, R_5, is connected from a point a little to the left of the primary winding. This resistance represents the thermal resistance of the shunt and would normally have quite a low value. In series with it would be connected the thermal resistance, R_6, representing the heat transfer of the heat sink. If this is being cooled by air having the same temperature as that cooling the transformer surface this resistance, R_6, would be connected to the common potential at the bottom of the circuit. Figure 9.6 shows the resulting circuit; R_1 now represents the thermal resistance between the primary winding and the coil former, and R_7 the thermal resistance between the coil former and the core. Ordinary network analysis will yield the temperatures of the various points.

An alternative or additional means of reducing winding temperature is to isolate the heat coming from the core and prevent it from entering the windings. This may be done, for example, by making a sufficiently large gap between core and coil former so that all the core heat is carried away by direct convection. This situation could be represented in Figure 9.6 by making R_7 open circuit and connecting a thermal shunt from the core surface. The numerical value of this shunt would depend on the rate of heat transfer from the core surface.

Further ways of dealing with the heat transfer problem are to be found in the literature. Sufficient has been included here to enable the designer to obtain a reasonably accurate estimate of the equilibrium temperatures within a proposed transformer.

9.3 General electrical design considerations

The basic relation is the induction equation. Various

forms of this relation have been introduced in earlier chapters but it is convenient to gather them together here.

From Eqn (2.12) the average e.m.f. during a change of flux density, ΔB, is given by

$$\bar{E}\Delta t = -NA\Delta B \quad \text{V s} \qquad (9.12)$$

For the purpose of transformer design the applied voltage, U, can usually be taken as equal and opposite to the e.m.f. If, in the case of square waves (rectangular pulses), U is constant for the pulse duration, t_d, the flux density excursion is given by

$$\Delta B = B_{p-p} = \frac{Ut_d}{NA} \quad \text{T} \qquad (9.13)$$

or the flux excursion is

$$\Phi_{p-p} = \frac{Ut_d}{N} \quad \text{Wb} \qquad (9.14)$$

If the square waves are symmetrical about zero, the peak values of the flux density and flux are given by

$$\hat{B} = \frac{\hat{\Phi}}{A} = \frac{Ut_d}{2NA} \quad \text{T} \qquad (9.15)$$

and t_d can be replaced by $1/2f$ (see also Eqns (4.58)–(4.60)).

For sinusoidal excitation at a frequency $f = \omega/2\pi$,

$$\hat{B}_e = \frac{\hat{\Phi}}{A_e} = \frac{\sqrt{2}U}{\omega NA_e} \quad \text{T} \qquad (9.16)$$

where $\hat{B}_e$ is the peak value of the effective flux density,
A_e is the effective core cross-sectional area,
U is the r.m.s. value of the voltage.

When the magnetic loss is being considered, the effective flux density and the effective core area, as defined in Section 4.2.1., will generally be used although, for high flux densities, the assumption on which their derivation depends, i.e. core loss $\propto \hat{B}^3$, is usually not very accurate. When any of the above induction equations is being used in connection with saturation or the degree of magnetization then the appropriate value for the area is the minimum cross-sectional area of the core.

At this point it is useful to discuss qualitatively some aspects of the induction equation in relation to a given source voltage, U, and a given core. The context of this discussion is the general case of a power transformer with sinusoidal excitation and perhaps a wide band specification. However, much of the discussion is relevant to more specialized applications such as switched mode power supplies.

Since the core losses rise with frequency and flux density, the flux density at high frequencies is usually limited to relatively low values by the permissible core loss. At such frequencies the necessary low flux density may be obtained with relatively few turns. If the number of turns required to obtain the flux density is too small to provide sufficient shunt impedance then N must be suitably increased; this will further reduce the flux density and both the reactive and core loss components of the magnetizing current will be reduced.

For a given voltage and core, $\hat{B}_e N \propto 1/f$. The core loss expressed in watts is approximately proportional to $f^m \hat{B}_e^n$ where m is typically 1.3 and n lies between about 1.9 and 3 (see Figure 3.19 and Eqn (9.17)). Thus for a given core loss, $\hat{B}_e \propto 1/f^{m/n}$, i.e. approximately $1/f^{1/2}$. It follows that as the frequency is decreased, both the flux density and the number of turns will increase if the core loss is maintained at a specified value. Under these conditions the design is core loss limited. In such a design it may be advantageous to obtain a certain ratio between core loss and copper loss in order to achieve optimum performance with respect to efficiency or volume.

At lower frequencies the flux density will approach a limit determined by the need to avoid magnetic saturation. Generally this limit will be appreciably below the maximum flux density for the given material advised by the manufacturer, the margin being necessary to allow for voltage tolerances and transients that might otherwise lead to the core becoming saturated. At such frequencies the design is saturation limited. At progressively lower frequencies the core loss decreases and the winding loss rises because N must now rise in inverse proportion to f in order to keep $\hat{B}_e$ constant. The total power loss, or temperature rise is still the factor limiting the throughput.

It is desirable, if possible, to choose a core size that results in a design in which the flux density determined by loss considerations coincides with the value arising from the need to avoid saturation; in such a design the core material is more fully utilized.

At even lower frequencies the winding resistance or

(9.12) $\quad \bar{E}\Delta t = -NA\Delta B \times 10^{-8} \quad \text{V s}$

(9.13) $\quad \Delta B = B_{p-p} = \dfrac{Ut_d}{NA} \times 10^8 \quad \text{Gs}$

(9.14) $\quad \Phi_{p-p} = \dfrac{Ut_d}{N} \times 10^8 \quad \text{Mx}$

(9.15) $\quad \hat{B} = \dfrac{\hat{\Phi}}{A} = \dfrac{Ut_d}{2NA} \times 10^8 \quad \text{Gs}$

(9.16) $\quad \hat{B}_e = \dfrac{\hat{\Phi}}{A} = \dfrac{\sqrt{2}U}{\omega NA_e} \times 10^8 \quad \text{Gs}$

leakage inductance will become large enough to cause unacceptable regulation. Below this frequency the design is regulation limited and temperature rise ceases to be a restriction. Regulation limitation is not often encountered in transformer designs using ferrite cores.

The shunt inductance has been mentioned above. This is often a major parameter in transformer design. If it is too low, the magnetizing current will be too great; in a wide-band transformer this will cause excessive attenuation or insufficient return loss at the lower end of the transmission band (see Section 7.2.4). Except for tuned transformers and certain types of SMPS transformers that operate in an energy-storage mode, the shunt inductance is usually required to be greater than a specified value. However, excessive inductance may lead to leakage inductance or winding resistance that is too high for other specification requirements to be met.

The induction equation is sometimes combined with the expression for inductance to give a relation that is independent of the number of turns. This is not a very satisfactory procedure because the result can be misleading. If the flux density and the inductance are the main considerations in a design then it is probably better to consider them separately. Where the flux density is the limiting parameter, due to core loss or saturation then, for a given core, the minimum number of turns is determined, for example, by Eqn (9.16). The shunt inductance may then be calculated, e.g. from Eqn (4.19). Since it was postulated that the design was flux density limited, the inductance will be greater than the minimum required to meet the magnetizing current or transmission band specification. The leakage inductance and resistance may then be estimated (see Chapter 11); if they are excessive it may be necessary to use a larger core. If, on the other hand, the design is inductance limited the required shunt inductance will determine the number of turns and the resultant flux density will be acceptable. At the start of a design it may not be known which is the limiting parameter so the number of turns is calculated by both criteria and the larger number is used.

In connection with these basic calculations it should be noted that the core permeability is a function of flux density and in power transformers the amplitude permeability at the maximum working flux density may be appreciably larger than the initial permeability (see Figure 3.4). Which value of permeability is used depends on whether the transformer is to be operated only at the rated voltage or whether it must give a satisfactory performance at voltages down to zero. If the latter requirement applies the initial permeability must be used.

While, for many years, ferrite cored power transformers have been used to handle a variety of waveforms, notably the line scan waveform in the domestic television receiver and the switching waveforms used in SMPS, it has until recently been the practice to express the magnetization and power loss performance in terms of a symmetrical sinusoidal excitation. Due to the non-linearity of the B–H loop, the flux density and field strength cannot both be sine waves. For the purpose of material characterization it is usual to specify that the flux density is sinusoidal. This corresponds to the applied voltage being sinusoidal, which is usually true in sine wave operation and, when core properties are being measured, it generally simplifies the measuring technique. Such data is convenient for the specification and comparison of materials and can, in fact, be used with reasonable accuracy in a wide range of applications of power ferrites. It is, of course, directly applicable to the design of transformers involving sinusoidal waveforms.

In the SMPS application the applied voltage is generally in the form of rectangular pulses. These may be symmetrical or asymmetrical, with arbitrary ratios of forward/reverse periods and often with periods of zero voltage between the pulses. Figure 9.8 illustrates some basic waveforms in idealized representation. The thicker lines indicate the voltage waveforms and the thinner lines represent the variation of magnetizing current and the flux density, the effects of hysteresis and saturation having been omitted for simplicity.

The simplest form is the symmetrical square wave as shown by the thicker line in Figure 9.7(a). As far as the core material is affected, the difference between the sinusoidal and the square wave excitation is that in the latter the magnetization follows a triangular wave instead of a sine wave. This makes rather little difference to the magnetization performance; the linear variation of flux density actually results in somewhat lower eddy current core loss, see Section 4.2.3.

Another common waveform is the unidirectional pulse excitation shown in Figure 9.7(b). If the circuit resistance were zero, the current would rise linearly during the initial pulse ($di/dt \propto U$) and, when the voltage falls to zero, the current attained at the end of the initial pulse would persist. During the next voltage pulse the current would rise again by the same amount (ignoring saturation).

In practice the circuit will have finite resistance and at the end of each voltage pulse the current will fall exponentially, generally becoming substantially zero before the ensuing pulse. This fall induces a reverse e.m.f. as shown by the broken curve of Figure 9.7(c). Although this mode of operation is quite common it presents two possible disadvantages; all the energy stored in the core at the end of the pulse is dissipated in the circuit resistance, and the reverse voltage spike could lead to difficulties.

When it is necessary to avoid these disadvantages, a

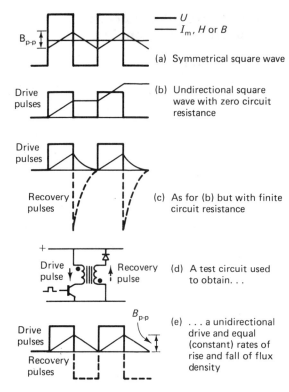

Figure 9.7 Voltage and flux density (or magnetizing current) relations for square wave excitation, assuming a linear B–H curve. (Courtesy Research Studies Press)

suitable energy-recovery circuit can be used. Such a circuit is illustrated in Figure 9.7(d). The core is provided with an additional winding which returns the magnetic energy to the supply via a diode. As soon as the drive pulse ends, the diode conducts and connects this winding to the supply rail which, being at a constant voltage, constrains the magnetization to decrease linearly. In the arrangement shown, the two windings have equal numbers of turns so the magnetization wave form is symmetrical as shown in Figure 9.7(e). Such an energy-recovery arrangement is particularly useful in the measurement of the pulse properties of cores. It reduces the power that has to be supplied by the pulse generator and, since the rate of change of magnetization has constant magnitude, the calculation of eddy current loss is much simpler.

When hysteresis is taken into account, the unidirectional voltage pulses initially result in asymmetrical current waveforms as the magnetization climbs from the origin of the B–H loop to settle into the minor loop that will support stable operation. This is illustrated in Figure 8.21. In practice the circuit configuration, or the presence of an air gap, will often result in the lower tip of the minor loop being very near the origin of the B–H axis so that the effective residual flux density is quite small.

Clearly, the unidirectional mode of operation results in a much smaller available flux density excursion and to this extent the core is under-utilized. However, at the higher frequencies being used for SMPS, the smaller flux density excursion is often acceptable or even necessary to limit the core loss, so this mode of operation is quite common.

In view of the importance of square wave operation, performance data relating to power ferrite materials and specific cores are increasingly being presented in forms that are directly applicable to such operation. In particular, as the flux excursion is no longer necessarily symmetrical, the expression of magnetization and magnetic loss in terms of peak flux density is not appropriate. Instead, the peak-to-peak excursion is becoming more usual as a parameter. Further, for the characterization of specific cores, where the cross-sectional area may be non-uniform, the use of flux (equals volt seconds per turn) is more directly applicable.

The magnetic loss has been defined in Section 2.3 as the material loss at a given frequency and flux density when the bulk eddy current loss has been eliminated. The magnetic loss (volume) density, P_M, is the material property as given in the graphs of Figure 3.19. For a typical power ferrite it can be represented by the formula:

$$P_M = k f^{1.3} \hat{B}_e^{2.5} \tag{9.17}$$

It is noted in the caption to Figure 3.19 that Annis,[24, 25] who made extensive measurements on power ferrites, found that for a given ferrite the values of k obtained from measurements with sine wave excitation also hold approximately for symmetrical and asymmetrical (including unidirectional) square waves provided the total flux density excursion is the same in each case. It follows that one set of loss curves will generally suffice provided $\dot{B}_{p-p}$ (or Φ_{p-p}) is used as a parameter.

It is of interest at this point to consider the relative magnitudes of the magnetic and eddy current core losses. For some particular grades of power ferrites, Figure 3.19 gives a specific version of Eqn (9.17):

$$P_M = 4.23 \times 10^{-6} f^{1.3} \hat{B}_e^{2.5} \quad \mu W\,mm^{-3} \tag{9.18}$$

where f is in kHz, B_e is in mT and the parameters relate to measurements at 100°C.

From Eqn (2.46), the eddy current loss (volume)

(9.18) $P_M = 1.34 \times 10^{-8} f^{1.3} \hat{B}_e^{2.5}$ mW cm^{-3}

where f is in kHz and B is in Gs.

density, for symmetrical sine wave excitation, is given by:

$$P_F = \frac{(\pi f \hat{B} d)^2}{\rho \beta} \quad \text{W m}^{-3}$$

Following the derivation of Eqn (4.61), if this expression is applied to a cylinder of diameter d, the cross-sectional area $A = \pi d^2/4$ and $\beta = 16$,

$$P_F = \frac{\pi f^2 \hat{B}^2 A}{4\rho} \quad \text{W m}^{-3}$$

This equation may be used to estimate the approximate eddy current core loss density in a non-uniform core if the effective values of core area, A_e and the flux density, $\hat{B}_e$ are used. Then:

$$P_F = \frac{\pi f^2 \hat{B}_e^2 A_e}{4\rho} \times 10^{-9} \quad \mu\text{W mm}^{-3} \tag{9.19}$$

where f is in kHz, B_e is in mT, A_e is in mm² and ρ is in Ω m.

The error in using A_e will generally be much smaller than the error resulting from the uncertainty in the value of the ferrite resistivity, an uncertainty that can sometimes be quite large. From Eqns (9.18) and (9.19)

$$\frac{P_F}{P_M} = 1.86 \times 10^{-4} \frac{A_e f^{0.7}}{\rho \sqrt{\hat{B}_e}} \tag{9.20}$$

the quantities having the same units as in Eqn (9.19).

This equation provides a convenient means of estimating the relative importance of the eddy current core loss in any projected design. As an example, if $A_e = 100 \text{ mm}^2$, $\rho = 1 \Omega\text{m}$, $f = 100 \text{ kHz}$ and $\hat{B}_e = 100 \text{ mT}$, then $P_F/P_M = 0.047$, i.e. about 5%.

For further discussion of eddy current core loss see Section 2.2.4 and 4.2.3.

The most reliable way of establishing the performance of a particular ferrite core is to use the manufacturer's data sheet for that core, where guaranteed limits apply and the total core loss data will include eddy current loss, see for example Reference 26.

Figure 9.8 shows typical manufacturer's data relating to a specific core; (a) gives the magnetization curve and (b) gives the loss curves both in terms of peak flux and peak-to-peak flux. The latter apply to all the forms of excitation mentioned above.

(9.19) $P_F = \dfrac{\pi f^2 \hat{B}_e^2 A_e}{4\rho} \times 10^{-7} \quad \text{mW cm}^{-3}$

where f is in kHz, B_e is in Gs, A_e is in cm² and ρ is in Ω cm.

(9.20) $\dfrac{P_F}{P_M} = 5.88 \dfrac{A_e f^{0.7}}{\rho \sqrt{\hat{B}_e}}$

the quantities having the same units as in Eqn (9.19).

9.4 Low-frequency power transformers and chokes

9.4.1 General

The transformers and chokes to be considered under this heading are those for which the winding resistance may be taken as substantially equal to the d.c. value. The frequency at which this ceases to be true depends on a number of factors, particularly the load currents and the way in which the windings are designed to reduce the effects of eddy currents. For a large (e.g. 1 kW) transformer with a low voltage output, eddy currents may affect the winding resistance at frequencies as low as 3 kHz. The special considerations that influence the design of higher frequency transformers (which include most SMPS transformers) are the subject of Section 9.5. However, a number of basic relations derived here in Section 9.4

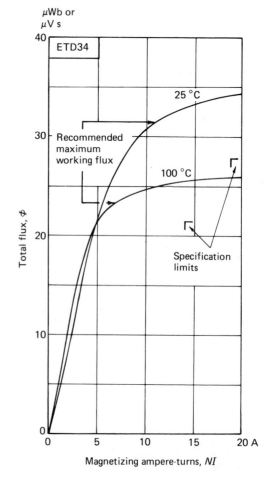

(a)

will, with suitable adaptation, be relevant to these higher frequency designs.

The maximum permissible operating temperature is the principal constraint and starting point in most power transformer design, particularly for low-frequency transformers. As indicated in Section 9.2.1, this temperature depends primarily on the materials used in the construction. In fact, the maximum temperature at which these materials can be operated has not changed much in recent years and 100°C has become the normal design figure. The next most important parameter is the ambient temperature; this is often specified as 60°C. The difference between these temperatures is the permissible temperature rise, which for a given core with a given thermal resistance, determines the maximum power dissipation, P_{tot}.

$$P_{tot} = P_w + P_c \quad \text{W} \tag{9.21}$$

where P_w is the winding loss and P_c is the core loss.

If either the core or winding loss is limited by some special constraint, then the other may assume the value obtained from this equation. When neither is individually limited, their ratio may be adjusted for maximum efficiency.

It is often stated that for maximum efficiency in a transformer the core loss should equal the winding losses. This arises from the following argument based on the equivalent circuit shown in Figure 9.9. Assuming that

$$R_1 + R_2' \ll R_b' \ll |L_p \text{ and } R_p \text{ in parallel}|,$$

the winding and core losses are given by:

$$\left. \begin{array}{l} \text{winding loss} = P_w = I_1^2 (R_1 + R_2') \quad \text{W} \\ \\ \text{core loss} = P_c = P_m V_e = U_1^2 / R_p \quad \text{W} \end{array} \right\} \tag{9.22}$$

where P_m is the total core loss density and V_e is the effective core volume (see Section 4.2.1). The power in the load is

$$P_o = I_1^2 R_b' = U_1^2 / R_b'$$

$$\therefore \frac{P_w + P_c}{P_o} = \frac{R_1 + R_2'}{R_b'} + \frac{R_b'}{R_p} \tag{9.23}$$

This expression for the ratio of the total power loss to the output power may be differentiated with respect to R_b' to obtain the condition for minimum power loss. This condition is $(R_1 + R_2')/R_b' = R_b'/R_p$, i.e. the value of load resistance is such that the core loss equals the winding loss. The result is not valid generally because if the output power is constant the applied voltage will depend on R_b' and so, consequently, will the flux density. Under these conditions R_p, the core loss resistance, will not, in general, be constant.

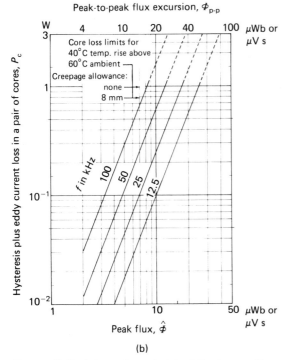

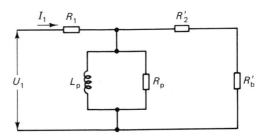

Figure 9.8 Typical manufacturer's data relating to a specific core (the ETD34 core): (a) magnetization curves in terms of total flux in the core; (b) total core loss as a function of peak-to-peak total flux excursion (upper scale), applicable to square or sine wave excitation, either symmetrical or asymmetrical, and as a function of the peak total flux (lower scale), applicable only to symmetrical square or sine wave excitation. The core loss limit lines are set at $P_{tot}/(1+n/2)$ where $n=2.5$ (see Eqn (9.37)), i.e. they assume optimum diversion of losses between core and windings

Figure 9.9 Simplified equivalent circuit of a power transformer, the elements referred to the primary side.

U_1 is the primary voltage
R_1 is the primary winding resistance
I_1 is the primary current
R_2' is the secondary winding resistance
R_b' is the load resistance
L_p is the shunt inductance of primary
R_p is the shunt core loss resistance

The more practical situation is the design of a transformer on a given core to give minimum total power loss when delivering a given power into a given load. The number of primary turns, N_1, is the variable. If N_1 is large the winding loss will be high and the flux density and core loss will be low; if N_1 is small the reverse applies. The problem is to find the correct ratio of core and winding loss for minimum total loss. It is assumed that the winding fills the winding space, the copper factor is constant and eddy current effects in the winding are negligible. If R_s represents the total winding resistance referred to the primary winding then, making the same assumptions as in the previous analysis:

$$P_w = I_1^2 R_s = \frac{U_1^2 R_s}{(R_b')^2} \quad \text{(see Figure 9.9)}$$

and

$$P_c \propto \hat{B}_e^n \quad \text{(see Figure 3.19)}$$

$$= \frac{K_1 U_1^n}{N_1^n} \quad \text{from Eqn (9.16), } K_1 \text{ being a constant of proportionality}$$

$$\therefore P_c + P_w = K_1 \left(\frac{U_1}{N_1}\right)^n + \left(\frac{U_1}{R_b'}\right)^2 R_s$$

From Eqn (11.11) $R_s = K_2 N_1^2$ where K_2 is another constant of proportionality.

$$\therefore P_c + P_w = K_1 \left(\frac{U_1}{N_1}\right)^n + K_2 \left(\frac{U_1 N_1}{R_b'}\right)^2$$

Differentiating with respect to N and equating to zero, the following condition for minimum total power loss is obtained

$$K_1 \left(\frac{U_1}{N_1}\right)^n = \frac{2K_2}{n} \left(\frac{U_1 N_1}{R_b'}\right)^2$$

i.e.
$$P_c = \frac{2}{n} P_w \quad (9.24)$$

It has been seen (Figure 3.19) that the value of n lies between about 2 and 3; only when it is 2 does the equating of the core and winding losses give minimum total loss.

In this analysis, it is implied that n is the Steinmetz exponent and therefore P_m is the hysteresis component of the total core loss. Since the eddy current loss component is proportional to $\hat{B}_e^2$ and it is usually small compared to the hysteresis loss, the error involved in the assumption is negligible.

It is not always possible or desirable to apply this result. As described at the beginning of this section, low-frequency power transformer designs are often regulation or saturation limited; under these conditions the winding loss predominates. At higher frequencies where the design is limited by core loss (temperature rise) it is possible to choose the most suitable ratio of core and winding loss. When small size is more important than efficiency it is usual for transformers with laminated iron cores to be designed to have high core loss relative to the winding loss. The resulting heat may be removed from the core by conduction to a heat sink, so the core temperature is prevented from being excessive.

As indicated in the previous section, the flux density is a major parameter in the design of power transformers. Given a limit value for the flux density, determined by saturation or core loss, the design of a power transformer on a given core can be very straightforward. The number of turns on the lowest voltage winding may be calculated from the appropriate induction equation, see Eqns (9.12) to (9.16) and, from the specified voltage ratios, the numbers of turns on the other windings may be derived. The conductor diameters and resistances may readily be obtained from the equations given in Sections 11.2 and 11.3, taking account of the operating temperature. The internal voltage drops due to winding resistances and, if significant, the leakage inductance may then be calculated, and the number of turns may be adjusted if necessary to obtain the specified full load secondary voltages.

At this stage of the design it may be found that the winding resistance is too high, leading either to excessive regulation or to excessive power loss in the windings. If so, a larger core must be selected and the design repeated.

When the design satisfies the flux density and winding loss requirements, the shunt inductance should be calculated to ensure that it is adequate to meet the specification for magnetizing current or possibly return loss.

This trial and error approach can, with experience, lead quite quickly to a satisfactory design. However it is helpful to have some advance information of the power-handling capacity of available cores. In the following section the various design approaches are analyzed with the object of deriving parameters that will facilitate core selection.

9.4.2 Design methods for low-frequency transformers

The object of this section is to establish some parameters that will provide guidance as to the power handling capacity of given ferrite transformer assemblies.

If the only source of power dissipation in the transformer is that due to the d.c. winding resistance, then a relation between the input power and the winding loss may be readily obtained. Assuming that the transformer has two windings and that each has the same cross-sectional area (see Section 11.3), then from Eqn (11.11), the d.c. resistance of the primary winding is

$$R_1 = \frac{2m\rho_c N_1^2 l_w}{A_a F_a} \quad \Omega \qquad (9.25)$$

where ρ_c is the conductor resistivity at 20°C,
A_a is the total window area,
F_a is the corresponding copper factor and
m is the fractional increase of conductor resistivity above the value at 20°C*.

$$\therefore R_1 + R_2' = \frac{4m\rho_c N_1^2 l_w}{A_a F_a} \quad \Omega$$

where R_2' is the secondary winding resistance referred to the primary, see Figure 9.9. The total winding loss $P_w = I^2(R_1 + R_2')$, therefore

$$P_w = \frac{4I_1^2 m\rho_c N_1^2 l_w}{A_a F_a} \quad W \qquad (9.26)$$

Assuming sinusoidal excitation, the number of turns may be eliminated by substitution from Eqn. 9.16, putting $U_1 = E$. Thus

$$\left.\begin{array}{l} P_w = \dfrac{8 I_1^2 U_1^2 m\rho_c l_w}{A_e^2 A_a F_a \omega^2 \hat{B}_e^2} \quad W \\[2mm] = \dfrac{P_i^2 m k_1}{f^2 \hat{B}_e^2} \quad W \\[2mm] \text{where } P_i \text{ is the input power and} \\[2mm] k_1 = \dfrac{2\rho_c l_w}{\pi^2 A_e^2 A_a F_a} \quad \Omega\,\text{m}^{-4} \\[2mm] = \dfrac{2\rho_c l_w}{\pi^2 A_e^2 A_w F_w} \quad \Omega\,\text{m}^{-4} \end{array}\right\} \qquad (9.27)$$

where A_w is the cross-sectional area occupied by the windings and F_w is the winding copper factor (see Section 11.2)
Transposing,

$$P_i^2 = \frac{P_w f^2 \hat{B}_e^2}{m k_1} \qquad (9.28)$$

(9.27) $\quad P_w = \dfrac{P_i^2 m k_1 \times 10^8}{f^2 \hat{B}_e^2} \quad W$

where $k_1 = \dfrac{2\rho_c l_w \times 10^8}{\pi^2 A_e^2 A_w F_w} \quad \Omega\,\text{cm}^{-4}$

$\phantom{\text{where }k_1} = $ the SI values given in Table 9.3

(9.28) $\quad P_i^2 = \dfrac{P_w f^2 \hat{B}_e \times 10^{-8}}{m k_1}$

If P_w is limited by temperature rise or some other constraint, then input power is related to the operating frequency and flux density by the constant, k_1, which depends on the geometry of the core and winding.

One constraint that can directly limit P_w at the lower frequencies is a limit on the voltage regulation. Such a requirement is not often encountered in ferrite transformer applications but the analysis is included here for the sake of completeness.

The voltage regulation may be defined as the difference between the output voltage on open circuit and that on full load, expressed as a fraction or percentage of the open circuit voltage. From Figure 9.9 the voltage regulation due to the winding resistance, referred to the primary side, is given by

$$\text{Voltage regulation} \simeq \frac{I_1(R_1 + R_2')}{U_1} = \frac{I_1^2(R_1 + R_2')}{\text{input power}}$$

If P_i and P_o denote the input and output power respectively, then for a voltage regulation of x per cent:

$$\frac{x}{100} = \frac{P_w}{P_i}$$

or $\quad P_w = P_i \dfrac{x}{100} \qquad (9.29)$

Substituting this in Eqn (9.28),

$$P_i = \frac{f^2 \hat{B}_e^2}{m k_1} \frac{x}{100}$$

In a regulation limited transformer the core loss is negligible compared to P_w, so from Eqn (9.29),

$$P_o = P_i - P_w = P_i \left(1 - \frac{x}{100}\right) \qquad (9.30)$$

$$= \frac{f^2 \hat{B}_e^2}{m k_1} \frac{x}{100} \left(1 - \frac{x}{100}\right)$$

$$\therefore \quad P_o = \frac{f^2 \hat{B}_e^2 (x - 0.01 x^2)}{m k_1 \; 100} \quad W \qquad (9.31)$$

or $\quad k_1 = \dfrac{f^2 \hat{B}_e^2 (x - 0.01 x^2)}{m P_o \; 100} \quad \Omega\,\text{m}^{-4} \qquad (9.32)$

If the regulation is 10%

$$k_1 = 0.09 \frac{f^2 \hat{B}_e^2}{m P_o}$$

---

*Copper temperature:	20	30	40	50	60	70	80	90	100
m	1	1.039	1.079	1.118	1.157	1.197	1.236	1.275	1.314

For a given core material, $\hat{B}_e^2$ is limited to a value determined by the saturation flux density; the frequency, the required output power and the permissible regulation are given parameters in the design. Therefore the design value of k_1 may be obtained from Eqn (9.32). It is then necessary to find a core having a value of k_1 (from Eqn (9.27)) that is equal to or smaller than the design value. The core selection is facilitated if the available cores are tabulated with their k_1 values. Table 9.3, which is introduced in Section 9.4.4, gives data for several ranges of ferrite cores suitable for power transformers and includes values of k_1 evaluated on the basis of an assumed value of copper factor.

Finally the possibility of poor regulation due to excessive leakage inductance must again be mentioned. At the low frequencies used in power distribution the regulation is almost invariably due to winding resistance. However, at higher frequencies it is advisable to check the contribution of the leakage inductance, the voltage drop across it being $jI\omega L_l$. If it is too large the winding must be sectionalized as described in Section 11.7 or a larger core must be used.

If the transformer is not regulation limited, and this will be true for most ferrite-cored power transformers, then the next limitation encountered is that imposed by the saturation of the core.

As the flux density rises in the course of the magnetization cycle, the field strength and the magnetizing current will rise in accordance with the shape of the magnetization curve, as indicated in Figure 9.10(a). As the flux density approaches the saturation value the magnetizing current will begin to rise sharply; Figure 9.10(b) and (c) shows this for both square and sine waves. The electrical circuit design will generally set a limit on the excess rise of current that can be tolerated; for example, in SMPS circuits it is essential to limit the current peaks to prevent damage to the semiconductor switches.

For the guidance of the designer, the manufacturer sets a limit to the flux density which should not be exceeded in any part of the core under any operational conditions. This may be regarded as the design saturation flux density; its value depends on the type of ferrite and the operating temperature. For a power ferrite operating at a temperature of 100°C, a typical value would be 320 mT. Figure 3.3 shows the temperature dependence of the actual saturation flux density for a number of ferrites.

The maximum design flux density, $\hat{B}$, i.e. the flux density at the minimum core cross-section, is set at a lower value, providing a margin that depends on the upper limit of the supply voltage and transient conditions that could otherwise drive the core into saturation. Thus it depends in part on the circuit and is not a core parameter.

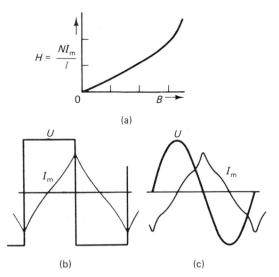

Figure 9.10 (a) Representation of an H–B curve. (b) The effect on the magnetizing current of approaching saturation due to a square voltage pulse. (c) The same for a sinusoidal voltage wave. (Courtesy Research Studies Press)

As indicated in Section 9.3, some manufacturers prefer to give data relating to specific core types in terms of flux, Φ, rather than flux density. This is because the problem of non-uniform core cross-section is thereby avoided, and, for square waves, the value of flux in terms of volt.seconds/turn follows directly from equations such as Eqn (9.14). In Figure 9.8(a) the operating limit is shown in terms of total flux for a typical ferrite power transformer core as presented in a manufacturer's data book.[26]

As a guide to core selection, the constant k_1 defined in Eqn (9.27) is also relevant to the saturation limited case. Eqn (9.28) gives the relation between total winding loss and the input power, P_i. Since the efficiency of a ferrite cored power transformer may be typically 97%, the total loss in the transformer is small compared to the throughput power so in the following analyses the input power may be put equal to the output power. Eqn (9.28) may then be rewritten:

$$P_o^2 = \frac{P_w f^2 \hat{B}_e^2}{mk_1} = \frac{P_w f^2 \hat{\Phi}^2}{mk_2} \qquad (9.33)$$

where $k_2 = k_1 A_e^2$ and could be used also in Eqns (9.27) and (9.31) if it were wished to express these equations also in terms of flux rather than flux density.

Therefore, from Eqn (9.21), the output power for

sine wave excitation is given by

$$P_o = (P_{tot} - P_c)^{1/2} \frac{f\hat{B}_e}{\sqrt{(mk_1)}} = (P_{tot} - P_c)^{1/2} \frac{f\hat{\Phi}}{\sqrt{(mk_2)}} \text{ W} \quad (9.34)$$

The equivalent equation for symmetrical square waves is

$$P_o = (P_{tot} - P_c)^{1/2} \left(\frac{A_w F_w}{mp_c l_w}\right)^{1/2} f\Phi_{p-p} \text{ W}$$

$$= (P_{tot} - P_c)^{1/2} k_3 F_w^{1/2} f\Phi_{p-p} \text{ W} \quad (9.35)$$

where $k_3 = (A_w/mp_c l_w)^{1/2}$

This is a form that is convenient for the design of SMPS transformers.[27]

The constants k_2 and k_3 for a wide range of commercially available ferrite power transformer cores are listed, together with k_1, in Table 9.3. These constants, and the value of P_{tot} which is also quoted in Table 9.3, provide a guide to the power handling capacity of a given core. The core will have a specified limit of flux density or flux from which, for a given circuit configuration, the design value of $\hat{B}_e$ or Φ_{p-p} for these equations may be derived, allowance being made for a suitable margin as described above. If the permitted temperature rise differs from the 40°C assumed for Table 9.3 then a revised value can be readily obtained from the thermal resistance, R_{th}, also given in Table 9.3 (see also Section 9.2.6).

A value cannot, of course, be assigned to P_c at this stage. However, in a saturation limited design, the core loss will generally be significantly less than half of the total loss. So, due to the influence of the square root, P_c may be put equal to zero without appreciable loss of accuracy, bearing in mind that the object is to estimate core performance at the outset of a design.

At low frequencies, while the core is saturation limited and the core loss is a rather small fraction of P_{tot}, it can be seen from Eqn (9.34) and (9.35) that the output power will be approximately proportional to frequency. If the frequency were standardized then, by assigning typical values to parameters such as temperature rise, copper factor, etc., a value for the power-handling capacity, P_o, could be given for any specific core. In practice, ferrite power cores are used over a wide range of frequencies, so it is not practicable to assign explicit values of P_o. Some manufacturers publish a graph of P_o as a function of frequency for each core in their range, based on a set of assumed parameters.

As the design frequency is increased from the region where the design is saturation limited, a point is reached at which the core loss equals the optimum proportion of the total loss. From Eqn (9.24), this occurs when $P_c = (2/n)P_w$.

Since $P_{tot} = P_c + P_w$, it follows that for the optimum division of losses:

$$P_w = \frac{P_{tot}}{1 + 2/n} \text{ W} \quad (9.36)$$

$$P_c = \frac{P_{tot}}{1 + n/2} \text{ W} \quad (9.37)$$

Beyond this point, at higher frequencies, the design flux density should be decreased as the frequency rises in order to limit the core loss to the optimum fraction of the total loss. From Eqns (9.34) and (9.35), the output power for the core loss limited design (taken to be equal to the input power for the purpose of this analysis) is given, for sine waves, by:

$$\left. \begin{aligned} P_o &= \left(\frac{P_{tot}}{1 + 2/n}\right)^{1/2} \times \frac{f\hat{B}_e}{\sqrt{(mk_1)}} \text{ W} \\ P_o &= \left(\frac{P_{tot}}{1 + 2/n}\right)^{1/2} \times \frac{f\hat{\Phi}}{\sqrt{(mk_2)}} \text{ W} \end{aligned} \right\} \quad (9.38)$$

and for symmetrical square waves by

$$P_o = \left(\frac{P_{tot}}{1 + 2/n}\right)^{1/2} \times k_3 F_w^{1/2} f\Phi_{p-p} \text{ W} \quad (9.39)$$

where $k_3 = (A_w/mp_c l_w)^{1/2}$

Given the values of P_{tot} and k_1, k_2 or k_3 from Table 9.3, these equations enable the power throughput of a given core to be estimated for the case of a loss limited transformer. The flux density or the flux must first be evaluated. This may be done by reference to the manufacturer's data in which graphs of core loss as a

---

(9.34) $P_o = (P_{tot} - P_c)^{1/2} \dfrac{f\hat{B}_e \times 10^{-4}}{\sqrt{(mk_1)}} = (P_{tot} - P_c)^{1/2} \dfrac{f\hat{\Phi} \times 10^{-8}}{\sqrt{(mk_2)}}$

where k_1 is defined in Eqn (9.27) and $k_2 = k_1 A_c^2 \times 10^{-8}$ and equals the SI values given in Table 9.3

(9.35) $P_o = (P_{tot} - P_c)^{1/2} k_3 F_w^{1/2} f\Phi_{p-p} \times 10^{-8}$ W

where $k_3 = (A_w/mp_c l_w)^{1/2}$ and equals the SI values given in Table 9.3

(9.38) $P_o = \left(\dfrac{P_{tot}}{1 + 2/n}\right)^{1/2} \times \dfrac{f\hat{B}_e \times 10^{-4}}{\sqrt{(mk_1)}}$ W

or $P_o = \left(\dfrac{P_{tot}}{1 + 2/n}\right)^{1/2} \times \dfrac{f\hat{\Phi} \times 10^{-8}}{\sqrt{(mk_2)}}$ W

(9.39) $P_o = \left(\dfrac{P_{tot}}{1 + 2/n}\right)^{1/2} \times k_3 F_w^{1/2} f\Phi_{p-p} \times 10^{-8}$ W

function of frequency and flux density or flux can usually be found. Since P_{tot} is known, the optimum value of P_c follows from Eqn (9.37); for most estimating purposes P_c may be put equal to $\frac{1}{2}P_{tot}$. The core loss graphs will then yield the appropriate value of flux density or flux corresponding to the operating frequency.

Although the constants k_1 to k_3 enable the performance of a given core to be estimated, they are of limited use in the selection of a core to satisfy a given performance. This need can be met to some extent by a further manipulation of Eqns (9.38) and (9.39).

For the case of sine wave excitation, Eqn (9.38) may be rewritten, substituting for k_1 from Eqn (9.27),

$$\frac{P_o(1+2/n)^{1/2}}{f\hat{B}_e} = \left(\frac{P_{tot}}{mk_1}\right)^{1/2}$$

$$\therefore \frac{\sqrt{2}}{\pi} \times \frac{P_o(1+2/n)^{1/2}}{f\hat{B}_e F_w^{1/2}} = \left(\frac{P_{tot}A_w}{m\rho_c l_w}\right)^{1/2} \times A_e$$

or, since it is now preferable to work in terms of the flux density at the minimum core cross-section,

$$\frac{\sqrt{2}}{\pi} \times \frac{P_o(1+2/n)^{1/2}}{f\hat{B}F_w^{1/2}} = \left(\frac{P_{tot}A_w}{m\rho_c l_w}\right)^{1/2} \times A_{min} = k_4 \quad (9.40)$$

From Eqn (9.39), the corresponding equation for symmetrical square waves is

$$\frac{P_o(1+2/n)^{1/2}}{2f\hat{B}F_w^{1/2}} = \left(\frac{P_{tot}A_w}{m\rho_c l_w}\right)^{1/2} \times A_{min} = k_4 \quad (9.41)$$

Assuming that the same temperature rise and maximum temperature used in compiling Table 9.3 applies, the constant k_4 may be evaluated as a design parameter. Table 9.3 lists values for a wide range of commercially available cores. For versions of these equations applicable to higher frequency designs see Eqns (9.54) and (9.55).

The parameters on the l.h.s. of these equations are either part of the given design specification or may be decided by the designer. P_o and f are usually specified, and the maximum flux density at the minimum core cross-section may initially be put at the maximum design value, which depends on the proposed grade of ferrite and the margin required by the circuit configuration.

For optimum division of core and winding loss, n may be put equal to 2.5 as a typical value. For a fully wound two-winding transformer, F_w may be put equal to 0.5. If these values are assumed then the equations reduce to the following very simple forms. For sine waves

$$0.854 P_o / f\hat{B} = k_4 \quad (9.42\text{(a)})$$

and for symmetrical square waves

$$0.949 P_o / f\hat{B} = k_4 \quad (9.42\text{(b)})$$

In all these equations f is in Hz and B is in tesla. The corresponding versions for higher frequency designs are Eqns (9.56) and (9.57).

For core selection, the l.h.s. is evaluated and a core having the next larger value of k_4 is chosen. This core should yield a winding loss that is equal to or less than the chosen fraction of the P_{tot} value quoted in Table 9.3. However, the core loss will in general not be equal to $(P_{tot} - P_w)$, since it depends arbitrarily on the chosen core material, frequency and flux density. So, having provisionally selected a core on the basis of k_4, the manufacturer's data should be consulted to see whether, at the given frequency and flux density, the core loss is within the limit implied by Eqn (9.37), i.e. $P_{tot}/2.25$ if $n = 2.5$.

If the core loss is smaller than this limit then the simplest course is to accept that the design is saturation limited. A move to a smaller core is unlikely to be satisfactory since this will, for the same flux density, substantially increase the winding loss.

If the initial selection suggests a core for which the core loss is excessive when a move to a larger core or a higher operating frequency is probably necessary. In the design example given in Section 9.4.5, these and other options are considered. The aim should always be to obtain the optimum loss balance if possible and the use of the constant k_4 should help in this selection process.

The equations and constants derived in this section are applicable to most ferrite transformer designs and particularly to the design of SMPS transformers.[27] In Section 9.4.4 the characteristics and design parameters of a number of commercially-available ranges of ferrite cores are reviewed, and in Section 9.4.5 there is a detailed design example based on the above approach.

9.4.3 Chokes and energy storage inductors

Chokes are inductors used to limit short term variations of current and voltage in power supply circuits, particularly at the output of d.c. power supplies. Magnetic energy storage (and recovery) is a technique

---

(9.40) $\quad \dfrac{\sqrt{2}}{\pi} \times \dfrac{P_o(1+2/n)^{1/2}}{f\hat{B}F_w^{1/2} \times 10^{-4}} = \left(\dfrac{P_{tot}A_w}{m\rho_c l_w}\right)^{1/2} \times A_{min} \times 10^{-4} = k_4$

(9.41) $\quad \dfrac{P_o(1+2/n)^{1/2}}{2f\hat{B}F_w^{1/2} \times 10^{-4}} = \left(\dfrac{P_{tot}A_w}{m\rho_c l_w}\right)^{1/2} \times A_{min} \times 10^{-4} = k_4$

(9.42(a)) $\quad 0.854 \times 10^4 P_o / f\hat{B} = k_4$

(9.42(b)) $\quad 0.949 \times 10^4 P_o / f\hat{B} = k_4$

where k_4 equals the SI values given in Table 9.3

used in the process of energy management in a variety of electronic equipment, particularly in power supplies using circuits such as flyback converters of the ringing choke type. Both categories are characterized by the need to achieve adequate inductance in the presence of large currents with a substantial d.c. component and both involve the use of gapped magnetic circuits. Much of the foregoing discussion concerning heat transfer, magnetization and losses is relevant to the design of chokes and energy storage inductors. An introduction to the basic relations expressing the properties of gapped magnetic circuits is given in Section 4.2.2. In the present section the design of chokes will be considered first.

The system or circuit design usually requires that the inductance of the choke should not fall below a specified value when the current in the winding rises to a maximum value. The current is composed of a maximum static component I_o with a superimposed ripple current of peak value $\hat{I}$, see Figure 9.11. Typically, $\hat{I}$ will be about 10% of I_o and there may be a safety margin of 10% on I_o to guard against transients. So the maximum current, I_m, for which the choke would be required to maintain a specified inductance would be typically $1.2I_o$.

Given a required minimum inductance and a maximum current, the conventional procedure is to use the Hanna curves[28] for a suitable material, e.g. a power ferrite, as described in the caption to Figure 3.7. The use of this material data gives rise to two difficulties. First, the input parameter is LI^2/V (where I would be put equal to I_m). However, the core volume, V, is not known until a suitable core has been selected, so an iterative process is inevitable. This is not, in practice, a serious obstacle. The second difficulty is that the scale of fractional air gaps on a Hanna curve, for a given material, corresponds to the physical gaps in the arbitrary core used in the experimental determination of the curve. Since the effective gap depends on the fringing flux, which in turn depends on the core

geometry, the gap length obtained from the Hanna curve will be subject to uncertainty, particularly at the larger values. Further, many of the practical core types, for example the RM cores, have non-uniform cross sectional area. So the reluctance of the gap depends not only on the fractional air gap, i.e. (air gap)/(effective path length), obtained from the Hanna curve for the material but also on the ratio of the area of the pole face embraced by the winding to the total pole face area in the return magnetic path.

As observed in connection with other design procedures, it is better to use data relating to specific cores made in a specific grade of ferrite. Most manufacturers provide such Hanna curves, obtained by direct measurement, for the appropriate cores in their product range. The vertical scale is in terms of LI^2 and the horizontal scale is in terms of NI. The volume and effective length of the core, and the effects of fringing and non-uniform core cross-section, are now incorporated into the presentation. It is only necessary to match the required value of LI_m^2 to such curves to find the smallest core that will give an acceptable design. Figure 9.12 shows a collection of such curves relating to a number of the core ranges listed in Table 9.3, the air gap being given in terms of spacer thickness.

How small the core can be depends on the design criteria. If a limit is imposed on the volt-drop across the choke and therefore on the power dissipated in the winding then the minimum size of the core will be determined by the winding resistance. Assuming that the optimum inductance factor does not change significantly with core size then, by analogy with inductor design (see Eqn (5.4) et seq.), the resistance for a given inductance will vary approximately as the inverse square of the linear dimensions. The procedure would be to make a trial design and then, depending on the winding resistance obtained, deduce the size of the smallest adequate core from this proportionality.

In most cases the design criterion is the minimum core size irrespective of the winding resistance. Then the design is constrained by two factors, saturation and heating. In principle, saturation can be held off indefinitely by making the gap larger and thus increasing the shearing of the magnetization curve. The practical limit is self-evident on a Hanna curve; towards the top of the curve the required gaps become impracticably large and the curve is terminated. The heating constraint is not so self-evident.

In most choke designs the core loss is negligible, so only the heating of the winding needs to be considered. If the ripple current is a significant fraction of the steady current, the r.m.s. current is given by:

$$I_{rms} = \sqrt{(I_o^2 + \hat{I}^2/2)} \text{ for sinusoidal ripple, or}$$
$$I_{rms} = \sqrt{(I_o^2 + \hat{I}^2/3)} \text{ for triangular ripple} \quad (9.43)$$

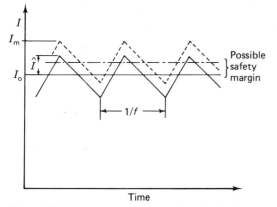

Figure 9.11 Representation of currents in a d.c. choke

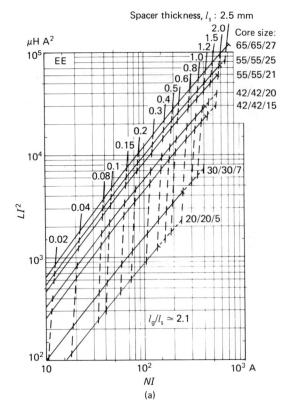
(a)

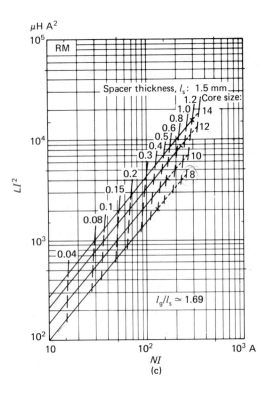
(c)

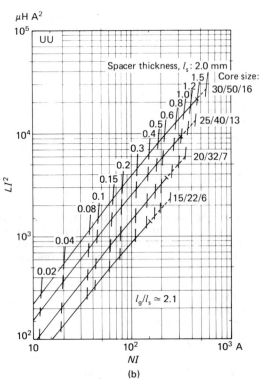
(b)

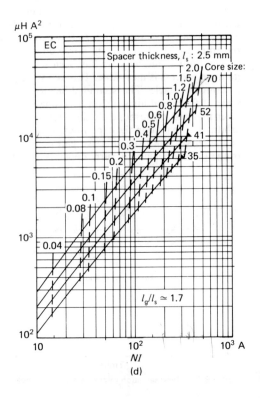
(d)

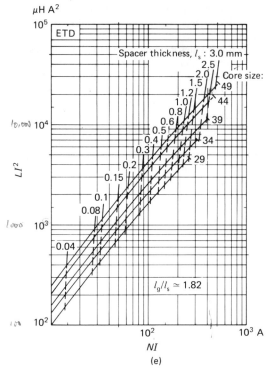

Figure 9.12 Hanna curves (at a temperature of 100°C) for a number of the ranges of ferrite cores included in Table 9.3.

All the air gap scales are in terms of spacer thickness in mm, l_s (total gap=$2l_s$). If l_g is the length of a single gap in the centre pole that would result in the same inductance as spacers of thickness l_s, then the ratio l_g/l_s depends mainly on the ratio of the pole face areas of the centre and combined outer limbs, respectively. The approximate value of l_g/l_s is indicated on each graph. The curves relate to a typical MnZn power ferrite

Where it occurs within the range of any curve, the thermal limit of NI for the core, taken from Table 9.1, is indicated by a marker and the curve is continued as a broken line

Normally the ripple current is only about 10% of the steady current, but even if $\hat{I} = 0.3 I_o$, I_{rms} is not more than 3% greater than I_o so, for most smoothing choke designs, I_o can in practice be used instead of the true r.m.s. value. The power dissipated in the winding is then $I_o^2 R$ where, in the present context, R is the d.c. resistance. If the ripple current were a large fraction of the total and if its frequency were high then it would be necessary to calculate separately the a.c. component of the winding loss using the r.m.s. value of the ripple current and the a.c. resistance, see Section 11.4.3.

For a given core and winding assembly, the power dissipation is limited by the permissible temperature rise and the thermal resistance as in the case of transformer design. It is usual to use the same temperature criteria, namely, an ambient temperature of 60°C and a core temperature limited to 100°C giving a permissible temperature rise of 40°C. If P_w is the winding loss corresponding to this temperature rise, then from Eqn (11.11),

$$P_w = I_o^2 R = \frac{m\rho_c(NI_o)^2 l_w}{A_w F_w} \quad \text{W} \qquad (9.44)$$

where ρ_c is the resistivity of copper at 20°C and
       m is the fractional increase of resistivity with temperature (see footnote to Eqn (9.25)).

So the maximum ampere turns in the winding if the winding loss is not to exceed P_w is

$$NI_o = \left(\frac{P_w A_w F_w}{m\rho_c l_w}\right)^{1/2} \quad \text{A} \qquad (9.45)$$

For a given core, the value of NI_o is clearly independent of the grade of material and the length of the air gap. Values of NI_o for a 40°C rise in temperature are listed in Table 9.3 for a large selection of currently used ferrite cores.

The design procedure is somewhat complicated by the fact that it is the maximum current, I_m (to which the required inductance must be maintained) that determines the gap length. This is best followed by considering the graph of L/N^2 as a function of NI for a given core in a specified grade of ferrite, see Figure 9.13. An upper boundary line can be drawn at the maximum value of NI_o based on heating considerations. If I_m is significantly larger than I_o, so that $I_m = k I_o$ where k would be typically 1.2, then a design utilizing the maximum permissible temperature rise would put the maximum amp.turns equal to kNI_o. This value would yield the required optimum gap length and, since kI_o is known, the maximum number of turns and the maximum inductance for this core can be obtained. L/N^2 would be marginally lower than the value corresponding to the NI_o limit.

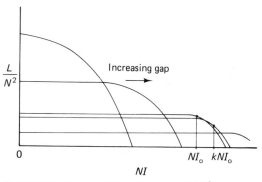

Figure 9.13 Representation of the graph of L/N^2 as a function of NI for a gapped core, with gap length as a parameter, to illustrate the significance of NI_o and NI_m ($=kNI_o$) on the design of a choke. See also Figure 3.6

In practice the Hanna curves for specific cores are more convenient. In Figure 9.12, a marker is drawn on each curve at the value of NI_o (taken from Table 9.3) corresponding to the maximum permissible power dissipation based on a temperature rise of 40°C (in some cases NI_o is beyond the end of the curve so the marker is omitted). The practical design procedure would be to evaluate the required LI_m^2 and establish a horizontal line on the appropriate graph of Figure 9.12 at that value. The optimum design would use the smallest core having a Hanna curve which is intercepted by the LI_m^2 line at, or a little above, kNI_o, i.e. typically 20% to the right of the NI_o marker. The gap length can be read from the curve and the required number of turns evaluated from kNI_o or NI_m. This design will provide the required inductance at I_m and will have the maximum permissible temperature rise of 40°C and therefore the least volume.

It may be argued that to operate a choke core at less than its maximum performance, i.e. its maximum temperature, is wasteful. Therefore, for the d.c. choke application, it would be sufficient to make a given core type with only one predetermined value of gap length,[29] namely that corresponding to the maximum permissible value of kNI_o; it would, however, be necessary to assume a typical value of k.

If the permitted temperature rise is not 40°C, either because the maximum temperature is to be lower than 100°C or the ambient is not 60°C, then P_w must be multiplied by a factor of $\theta_r/40$, where θ_r is the new temperature rise. This assumes that the thermal resistance is constant within the relevant temperature range. NI_o is then proportional to the square root of $\theta_r/40$ and this correction would have to be applied in the above procedure.

Another approach is described in References 30 and 31. In this, LI^2 and A_L, as functions of airgap, are presented graphically for each core. For both functions there are three curves, each representing a different application category. This approach allows the designer to select an air gap suitable for a particular application rather than accept the air gap predetermined by the maximum permissible temperature rise.

It is useful to note that $LI^2 = A_L N^2 I^2$ where L and the inductance factor, A_L, are in consistent units, so

$$NI = \sqrt{(LI^2/A_L)} \tag{9.46}$$

In general A_L depends on the air gap, the current and the ferrite material.

Some manufacturers feature cores with the optimum gap appropriate to the d.c. choke application while others offer a range of standard gaps to cover a variety of applications. Predetermined gaps are invariably provided during manufacture by cutting back one or both of the pole faces in the limb that is intended to carry the winding. Of course, the user always has the option of assembling ungapped cores with suitable spacers and thus retaining freedom of design.

When using design data such as Hanna curves, it is important to be clear whether the air gap values refer to a single gap of length l_g in the centre limb of the core or to a spacer of thickness l_s separating the core halves. If the core cross section were uniform and there were no fringing, the spacer thickness would be half the single air gap length; in practice this is usually not the case. Figure 9.12 gives the approximate ratios of (gap)/(spacer thickness) for the core shapes represented in each graph.

Moving on now to devices in which energy storage and recovery is the main function, such devices are essentially inductors but often, for the purpose of isolation or transformation, additional windings may be required.

The energy stored in a magnetic circuit of volume V_e is given by

$$W = V_e \int_{B_1}^{B_2} H\,dB \quad \text{J} \tag{9.47}$$

For the purpose of the present discussion it is convenient to consider only unidirectional magnetization and put $B_1 = 0$ and $B_2 = \Delta B$. This is often the case in practice and the resulting simplification does not affect the conclusions. Then

$$W = V_e \int_0^{\Delta B} H\,dB \quad \text{J} \tag{9.48}$$

or $$W = V_e \int_0^{\Delta \Phi} \frac{Nid\Phi}{l_e A_e} = \int_0^{\Delta \Phi} Nid\Phi \quad \text{J} \tag{9.49}$$

In Figure 9.14, the thick line represents an idealised magnetization curve of an ungapped core in which the hysteresis is neglected. The integral in Eqn (9.49) represents the area under sloping region of this magnetization curve, i.e. the area OAC when the maximum value of i is ΔI corresponding to $(N\Delta I)_o$, the subscript denoting the no-gap condition. Then

$$W = \tfrac{1}{2}(N\Delta I)_o \Delta \Phi \quad \text{J} \tag{9.50}$$

Assuming saturation is to be avoided, this is the maximum energy that can be stored in this particular

---

(9.47) $W = \dfrac{V_e}{4\pi} \displaystyle\int_{B_1}^{B_2} H\,dB \times 10^{-7}$ J

(9.48) $W = \dfrac{V_e}{4\pi} \displaystyle\int_0^{\Delta B} H\,dB \times 10^{-7}$ J

(9.49) $W = V_e \displaystyle\int_0^{\Delta \Phi} \dfrac{Nid\Phi}{l_e A_e} \times 10^{-8} = \displaystyle\int_0^{\Delta \Phi} Nid\Phi \times 10^{-8}$ J

(9.50) $W = \tfrac{1}{2}(N\Delta I)_o \Delta \Phi \times 10^{-8}$ J

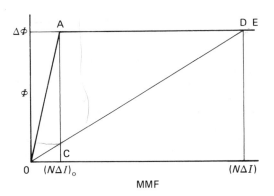

Figure 9.14 Idealized magnetization curve; OAE is the curve for an ungapped core and ODE is the sheared curve for the same core with an air gap

ungapped core. In fact, the introduction of a gap will shear the curve (see also Section 4.2.2) so the magnetomotive force may be increased to $(N\Delta I)$ and the stored energy increases to

$$W = \tfrac{1}{2}(N\Delta I)\Delta\Phi \quad \text{J} \qquad (9.51)$$

As shown in the discussion on choke design, $(N\Delta I)$ is proportional to $\{(\Delta I)^2 R\}^{1/2}$ and for a given waveform $(\Delta I)^2 R$ is proportional to $I_{rms}^2 R$, i.e. the winding loss, P_w. Again, the permissible temperature rise places a limit on P_w and therefore on $(N\Delta I)$.

It is of interest to take Eqn (9.51) one stage further by substituting $L = N\Delta\Phi/\Delta I$ so that

$$W = \tfrac{1}{2}L(\Delta I)^2 \quad \text{J} \qquad (9.52)$$

This is very similar to the corresponding LI^2 parameter used in the design of d.c. chokes.

In the design of an inductor to store the maximum amount of energy, the flux excursion should be as large as the saturation limit (and any safety margins) will allow. For a given applied voltage an equation such as Eqn (9.14) will determine the number of turns. The maximum temperature rise will then place a limit on I_{rms} and therefore, depending on the wave form, on ΔI. Having established the flux excursion, the number of turns and ΔI, the required inductance follows from

$$L = N\Delta\Phi/\Delta I \quad \text{H} \qquad (9.53)$$

Design data on suitable cores generally includes either graphs of inductance factor, A_L, as a function of air gap length or values of A_L for cores with predetermined air gaps; the required gap length can therefore be chosen.

Energy storage inductors, subject to large flux density excursions, are frequently used for energy conversion in switched mode power supplies, particularly those using the flyback (ringing-choke) principle. With additional windings to provide isolated outputs, they may be regarded as transformers having substantially unidirectional excitation. Depending on the switching frequency, their design may be saturation or loss limited. The above discussion has been concerned with the energy, current and air gap aspects of design and is based on the assumption that only the winding loss is significant. Section 9.2 and the foregoing parts of this section are also directly relevant to the design of these transformers, particularly where frequency and the amplitude of the flux excursion result in appreciable core loss.

Finally reference should be made to two less conventional types of gap that are sometimes used in smoothing chokes.

The first is the stepped or tapered gap.[32-34] In the case of a power supply in which the duty cycle is fixed, consideration of Figure 9.11 shows that, as the load current is reduced, the ripple current is an increasing fraction of the load current. This may, to some extent, also be true even when the duty cycle is regulated by the load current. In the limit, the troughs in the choke current will intercept the zero current axis and the choke current will become intermittent, possibly causing undesirable effects in the electronic circuit. This situation could be avoided if the choke inductance were to be increased as the load current is reduced. In a conventional choke, the inductance is virtually unaffected by current variations below I_m. However, if the gap is stepped or tapered so that over a fraction of its area the gap length is very small or zero, this region will shunt the main gap at very low currents. This will result in an inductance that is much higher than the value corresponding to the full gap. As the load current increases, the shunt region will progressively saturate, the gap will assume its total (uniform) length and the inductance will decrease towards its full load value. Figure 9.15 shows the characteristic of such a choke.

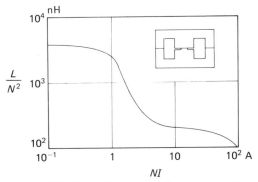

Figure 9.15 Characteristic of a non-linear choke having a stepped air gap

---

$(9.51) \quad W = \tfrac{1}{2} N\Delta I \Delta\Phi \times 10^{-8} \quad \text{J}$

The second type of unconventional gap is one that contains a thin slice of permanent magnet ferrite to premagnetize the core in a direction opposing the magnetization due to the load current.[35-38] This greatly increases the current that can be passed through the choke before saturation occurs.

The required premagnetization flux in the core determines the area of the permanent magnet slice. The inductance requirement determines the reluctance of the slice ($\simeq l_g/\mu_o A_g$, since the relative permeability of the magnet is substantially unity) and this in turn determines the thickness of the slice.

In calculating the area of the slice, care must be taken that, under any possible condition of current overload of the choke and at the maximum operating temperature of the permanent magnet, the latter cannot be demagnetized. Such demagnetization could have catastrophic consequences on the associated circuit. In practice, the area of the permanent magnet slice has to be much larger than that of the main core cross-section so the limb carrying the slice must be enlarged. Under some circumstances the use of permanent magnet premagnetization can result in size and weight reduction but its use is not common.

9.4.4 Core types and performance

The preceding sections have been concerned with the basic relations that express the performance of ferrite cores used in power applications. In this section attention is turned to the characteristics and performance of some specific core types that are in current use. The core performance data are mainly in terms of design parameters that are independent of the properties of the core material. Given these design parameters and the magnetic and electrical properties of the core (characteristic of a given grade of ferrite), the design can follow the procedures outlined in the foregoing sections.

Earlier in this chapter, and particularly during the development of the basic design relations, some of the desirable attributes of a good power ferrite have been identified. Before moving on to consider specific core geometries, it will be helpful to summarize the desirable material properties in qualitative terms. Material properties in quantitative terms are set out in Table 3.4, column IV and in the appropriate figures of Chapter 3.

For maximum power throughput the core should be operated at the highest permissible temperature which, for present day ferrites, lies in the region of 85 to 100°C and for most design purposes is taken to be 100°C. So the desirable properties listed below should relate to the operation at such temperatures.

(1) High saturation flux density, to enable a large flux excursion to be obtained and so minimize the required number of turns.
(2) The saturation should occur at a low field strength as indicated for example by the limit marks in Figure 9.8(a). This is equivalent to placing a lower limit on the amplitude permeability and minimizes the magnetizing current for a given flux density excursion.
(3) Low magnetic losses over the working range of frequencies to maximize the throughput for a given temperature rise. If the anisotropy compensation temperature (see Section 1.3.2) is placed at a suitably high value, the minimum in the magnetic loss/temperature curve can be situated near the maximum operating temperature, see Figure 9.16 (corresponding curves for a number of power ferrites are given in Figure 3.20).
(4) High resistivity, to minimize the eddy current core loss. This can be important at high operating frequencies.

With the great increase in the use of power ferrites as cores for SMPS transformers and the strong trend towards higher operating frequencies now apparent in this application, there have been strong incentives for the improvement of these material properties and a number of significant advances have been announced in recent years, see References 39–42. The values quoted in Table 3.4 take account of these improvements.

The same incentives that have led to improvements in power ferrite properties have encouraged the development of core shapes that can more fully exploit

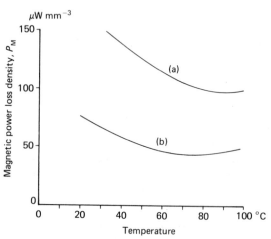

Figure 9.16 Magnetic loss density P_M at $f=100$ kHz and $B=100$ mT as a function of temperature for power ferrites, showing minima near the operating temperature; curve (a) is for a MnZn ferrite with improved processing and curve (b) is for a Ti-substituted MnZn ferrite[41]

these properties.[43] Some of the more important factors affecting the geometry of practical ferrite core assemblies and their suitability for power applications are:

(1) Physical constraints on transformer height or the board area occupied: such constraints may dictate whether the core should be mounted with its major axis perpendicular or parallel to the board.
(2) Optimum proportions for maximum performance; here the criteria depend somewhat on the actual application. Relevant considerations are the provision of adequate magnetic screening, the minimization of thermal resistance, the choice between round and rectangular cross section for the limb carrying the winding, the provision of adequate clearance for the winding terminations, and production cost.
(3) A given geometric design should be available in a suitable range of sizes and, for some applications, air gaps.
(4) The availability of convenient methods for clamping and mounting the cores.
(5) The availability of suitable coil formers having good provision for the termination of the windings, including those consisting of heavy round or strip conductors. Where mains isolation is required, the recommendations set out in IEC Standard Nos. 742,[44] 435[45] and 950[46] should be met; in terms of 240 V a.c. mains isolation the creepage requirement is 8 mm, see Figure 9.17. Also the relevant requirements concerning the flammability of the material and the solderability of the terminals should be observed.

Initially, there were very few available ferrite core shapes specifically intended for power applications so designers were often faced with using cores borrowed from other applications, e.g. the larger communication transformer E cores and the television U cores. In the first edition of this book the author, confronted with this lack of suitable ranges of ferrite cores, postulated a range based on various assemblies of building bricks consisting of two basic U core types, identical except that one had longer legs than the other. Such cores were, and still are, manufactured and provide a useful means of building up a range of ferrite cores, particularly for the higher powers.

Later, the demands of the developing SMPS application created a need for cores specifically designed for this purpose. The speed at which this demand materialized led to a proliferation of core types in spite of attempts by the various standards organizations to contain it. One of the reasons for such a proliferation was that, as the SMPS circuit techniques evolved and the procedures for design optimization improved, the perceived requirements of the ferrite assembly, as outlined above, changed. While this trend has provided the transformer designer with a wide choice of cores it has limited the economies of scale available to the manufacturer.

A brief survey of the currently available core types and their main characteristics now follows.

TV U cores. These were developed, in the early years of ferrite manufacture, for the transformers providing the line time base output and e.h.t. generation functions in domestic television receivers. This U core shape, with its wide aperture and round legs, allows good clearance for the e.h.t. winding and its termination. The use of these cores, while on a very large scale, is virtually confined to the TV receiver.

E cores. These cores, used in pairs or with an I core, have rectangular section limbs and consequently they can be pressed perpendicular to their principal plane. As this makes them relatively easy to manufacture there are many different national ranges and international standardization has been difficult. The DIN Standard 41 295[47] has become the basis for a generally accepted range and an IEC Standard is in the course of preparation. The E core shape enables a high pressed density, and therefore good uniform properties, to be obtained. The cross-sectional area is generally uniform and larger than the minimum area of the comparable round-legged E core, but the rectangular section leads to higher winding resistance and leakage inductance (for the same enclosed core area). They are particularly suitable for the higher powers.

U cores. These cores are equivalent to E cores folded along the principal axis and the same remarks apply.

Pot cores.[48–50] Inductor pot cores are often used for the smaller power, higher frequency, applications. In the past they have usually been available only in ferrite intended for low amplitude applications and have therefore offered a lower usable flux density than a power ferrite core. Also they have very restricted access for the winding terminations. However, the totally enclosed construction offers advantages where magnetic screening is important. The current trend towards higher frequencies has emphasised the need for improved magnetic screening and so the pot core

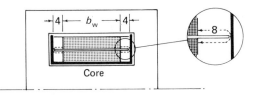

Figure 9.17 Cross-section of a transformer core with the winding breadth restricted to allow a 4 mm clearance at each side, thus complying with the requirements of IEC Standard No. 435 for a minimum creepage distance of 8 mm between windings of unimpregnated transformers

is being increasingly used in SMPS. To meet this demand, power ferrite pot cores are now becoming available.

RM cores.[51, 52] These were initially developed for low power inductor and transformer applications but, like the pot cores, tended to be used also for the smaller, higher frequency, power transformers and chokes. Subsequently the range was extended to larger sizes made in power ferrites to provide a useful range of cores specifically for power applications. For this purpose, RM cores can be supplied without the central hole thus offering significantly increased centre pole area. They retain much of the magnetic screening of the pot core but provide better access for the winding terminations. Alternative coil formers specially suited to power applications are also available.

EC cores.[53] This was probably the first range of cores to be specifically developed for SMPS applications. The geometry was based on a round centre limb and a wide, shallow, winding space to minimize winding resistance, leakage inductance and copper eddy current loss. The somewhat enlarged cross-sections of the back and outer limbs were designed to reduce thermal resistance.

PM cores.[54] These are very large pot cores specifically designed for power applications. They combine good magnetic screening and good access for winding termination with a large cross-section round centre pole.

PQ cores. This range has some similarity with the larger RM cores in that they have round centre poles and are designed to occupy minimum (square) area on the circuit board. The main difference is that the termination access is obtained by cutting away the core on two opposite sides of the square instead of opposite corners.

ETD cores.[55] The geometry of these cores is similar to that of the EC cores but the proportions have been optimised to a different set of criteria based on later trends in SMPS requirements. The round cross-section of the centre limb is nearer in area to that of the combined outer limbs. An IEC Standard is in the course of preparation.

LP cores. This design minimizes the height of the assembly thus providing a low profile component. It is evolved from the PQ core by increasing the terminal access on one side, cutting the ferrite back as far as the round centre pole and compensating the loss of outer limb area by reducing the aperture and dispensing with terminals on the opposite side. The core is mounted by the terminal pins with the core axis horizontal.

Figure 9.18 shows outline drawings of all these core types except the TV U core, and Figure 9.19 illustrates the cores assembled with their coil formers and mounting parts.

Table 9.3 sets out the geometry-dependent properties of cores representing most of the above ranges. The figures are based on information published in various manufacturers' catalogues and data books. For the physical and derived, e.g. effective, dimensions the published values often show small variations between manufacturers and in such cases some discretion has been used in choosing the values for this table. The design parameters have been computed from the dimensions actually used in the table.

The purpose of Table 9.3 is to facilitate core selection and to provide a guide to core performance; the figures are presented in good faith but no guarantee of accuracy is implied. Where questions of limits and guarantees are involved, reference should be made to manufacturers' data and specifications.

9.4.4.1 Parameters listed in Table 9.3

Core type. The core type designations are those commonly used in catalogues.

Assy. vol. This is an estimate of the volume of the smallest rectangular box that will enclose the core and winding assembly, ignoring the terminating pins or tags. It is based on approximate nominal dimensions.

Nominal core dimensions. The dimensions A, B and C are defined in Figure 9.18; the values given are generally the mean of the typical limit dimensions.

Dimensional parameters of the core. The effective dimensions, l_e, A_e, V_e and C_1, are generally those quoted in manufacturers' data books. In the case of diversity between manufacturers' figures an approximate mean value has been entered. A_{min} is the area of the smallest core cross-section and is calculated from the lower dimensional limits.

Winding dimensions of coil formers. The mean turn length, l_w, is generally the value quoted by the manufacturer and h_w is the minimum winding height based on limit dimensions of the coil former. In the next two columns, the winding breadth, b_w is the minimum dimension between flanges and $A_w = b_w \times h_w$. In the next column, b_w is the previous winding breadth minus 8 mm. This would be used where, in order to comply with the mains isolation requirements of IEC Standard Nos. 742,[44] 435[45] or 950,[46] it is necessary to allow 4 mm clearance at either side of the winding, see Figure 9.17. Where the net winding breadth would be less than 8 mm it is assumed that the coil former is too

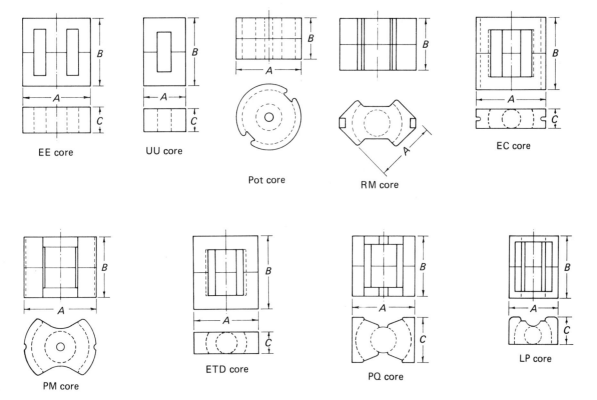

Figure 9.18 Outline diagrams of the transformer cores referred to in Table 9.3

narrow to accommodate creepage allowance. The next column gives the corresponding (curtailed) winding area.

Performance parameters, with no creepage allowance.
- k_1 This is defined in Eqn (9.27). It is calculated from the second version in which ρ_c is the resistivity of copper at 20°C, i.e. 1.709×10^{-8} Ωm, and F_w is put equal to 0.5. Its value relates to 20°C because Eqn (9.27) contains the temperature factor m.
- k_2 This is defined in Eqn (9.33). It is based on the same data as k_1 but relates to the total flux instead of the flux density
- k_3 This is defined in Eqn (9.35). Again ρ_c is the above 20°C value but m is put equal to 1.314 (see footnote to Eqn (9.25)) so this parameter relates to 100°C.
- k_4 This is defined in Eqn (9.40). The value of ρ_c is the same as that quoted for k_1, i.e. the 20°C value, m is put equal to 1.314 and P_{tot} is the value (for a 40°C temperature rise) quoted later in this table. Therefore these k_4 values refer to operation at 100°C.

- A_R This is the d.c. resistance of a full winding (see Eqn (11.11)), at a temperature of 100°C, divided by the square of the number of turns:

$$A_R = R/N^2 = m\rho_c l_w / A_w F_w$$

  where ρ_c is the above 20°C value, $m = 1.314$ and $F_w = 0.5$.
- R_{th} This is the thermal resistance of the core and winding assembly (see Section 9.2.6). It is based on manufacturers' data.
- P_{tot} This is the total power dissipation of the core and winding assembly based on a temperature rise of 40°C and calculated from Eqn (9.11).
- NI_o This is the number of (r.m.s.) ampere turns for a fully wound coil former corresponding to a temperature rise of 40°C in the complete (core and winding) assembly; it is the maximum value of NI_{rms} ($\simeq NI_o$) to be used in inductor and choke design, see Section 9.4.3 and in particular Eqn (9.45). It is based on the quoted value of P_{tot} ($\simeq P_w$ for chokes) and assumes a working temperature of 100°C and $A_w = 0.5$.

Table 9.3 Properties of ferrite power transformer cores

Core type	Assy. vol. (mm³ ×10³)	Nominal core dimensions			Dimensional parameters of the core					Winding dimensions of coil formers		No creepage allowance		With 8 mm creepage allowance		Performance param.	
		A (mm)	B (mm)	C (mm)	l_e (mm)	A_e (mm²)	V_e (mm³ ×10³)	C_1 (mm^{-1})	A_{min} (mm²)	l_w (mm)	h_w min (mm)	b_w min (mm)	A_w min (mm²)	b_w min (mm)	A_w min (mm²)	k_1 (Eqn. 9.27) ($\Omega\,m^{-4}$)	k_2 (Eqn. 9...) (Ω)
EE cores																	
20/20/5	5.3	20.2	20.0	5.1	42.8	31.2	1.34	1.37	23.5	36	2.6	10.6	27	—	—	9400	9.2 × 10
30/30/7	17.7	30.1	30.0	7.1	66.9	59.7	4.00	1.12	46	52	4.7	17.1	80	9.1	43	1250	4.5 × 10
42/42/15	62	42.2	42.0	15.0	98.0	182	17.6	0.534	172	87	6.9	26.0	179	18.0	124	100	3.35 × 10
42/42/20	70	42.2	42.0	19.7	98.0	236	23.1	0.415	227	96	6.9	26.0	179	18.0	124	66	3.70 × 10
55/55/21	135	55.2	55.0	20.7	123	354	43.7	0.348	341	114	8.4	33.2	279	25.2	212	22.5	2.80 × 10
55/55/25	150	55.2	55.0	24.7	123	420	52.0	0.293	407	122	8.4	33.2	279	25.2	212	17.0	3.00 × 10
65/65/27	230	65.1	65.0	27.0	147	532	78.2	0.275	513	138	10.3	39.5	407	31.5	324	8.2	2.35 × 10
VV cores																	
15/22/6	8.1	15.2	22.4	6.5	48.0	31.0	1.49	1.50	30	42	3.7	9.8	36	—	—	8300	8.0 × 10
20/32/7	17.4	20.8	31.2	7.5	68.0	56.0	3.80	1.24	52	54	5.1	14.5	74	—	—	1600	5.0 × 10
25/40/13	51	24.8	39	12.8	86	100	8.6	0.860	100	73	6.5	19.0	124	11	72	400	4.0 × 10
30/50/16	110	31.0	50.6	16.3	112	160	17.9	0.700	157	93	8.9	25.9	231	17.9	159	110	2.80 × 10
Pot cores																	
18/11	3.4	18.0	10.55	—	25.9	43	1.12	0.60	36.3	36.7	3.0	6.0	18	—	—	7600	14.0 × 10
22/13	6.3	21.6	13.4	—	31.6	63	2.00	0.50	51.3	44.5	3.5	7.75	27.1	—	—	2840	11.2 × 10
26/16	10.5	25.5	16.1	—	37.2	93	3.46	0.40	76.5	52.6	4.0	9.55	38.2	—	—	1100	9.5 × 10
30/19	16.9	30.0	18.8	—	45	136	6.1	0.33	115	62.0	4.8	11.3	55.4	—	—	415	7.7 × 10
36/22	27.5	35.5	21.7	—	52	202	10.6	0.264	172	74.3	5.8	12.7	73.7	—	—	171	7.0 × 10
42/29	53	42.4	29.4	—	69	265	18.2	0.259	214	86.2	7.8	17.7	138	9.7	76	61	4.3 × 10
RM cores*																	
RM8	9.1	19.3	16.4	—	38.0	64.0	2.43	0.590	53.4	42	3.4	8.7	29.6	—	—	2380	9.7 × 10
RM10	14.3	24.2	18.6	—	44.0	98	4.31	0.450	86.6	52	4.15	10.1	42	—	—	860	8.3 × 10
RM12	27	29.3	24.5	—	57.0	146	8.34	0.390	121	61	4.95	14.4	71	—	—	279	6.0 × 10
RM14	38	34.3	30.1	—	70.0	200	14.0	0.350	165	71.5	5.85	18.1	106	10.1	59	117	4.7 × 10
EC cores																	
EC35	34	34.5	34.6	9.5	77.4	84.3	6.53	0.918	66.5	52	4.6	21.2	97	13.2	61	520	3.7 × 10
EC41	101	40.6	39.0	11.6	89.3	121	10.8	0.735	100	62	5.7	24.0	137	16.0	91	210	3.1 × 10
EC52	148	52.2	48.4	13.4	105	180	18.8	0.581	134	72	7.5	27.8	209	19.8	149	73	2.35 × 10
EC70	322	70.0	69.0	16.4	144	279	40.1	0.514	201	96	11.2	41.4	464	33.4	374	18.3	1.42 × 1
PM cores																	
PM50/39	172	49.2	38.8	—	87	340	29.6	0.255	271	96.8	7.0	22.0	154	14	98	37	4.3 × 10
PM62/49	288	61.0	48.8	—	113	550	62.2	0.205	455	120	8.7	30.0	261	22	191	10.5	3.18 ×
PM74/59	464	72.8	58.7	—	133	740	98	0.180	612	140	11.9	37.5	446	29.5	351	4.0	2.15 × 1
PM87/70	744	85.5	69.6	—	153	915	140	0.167	679	158	15.1	43.4	655	35.4	535	2.0	1.65 ×
PM114/93	1000	111.8	92.5	—	208	1730	360	0.120	1333	210	19.0	56.3	1070	48.3	918	0.45	1.35 × 1
PQ cores																	
PQ20/16	11.4	20.5	16.2	14.0	37.4	62	2.31	0.605	58.1	44	3.05	7.8	23.8	—	—	3300	12.7 × 1
PQ20/20	13.6	20.5	20.2	14.0	45.4	62	2.79	0.738	58.1	44	3.05	11.7	35.7	—	—	2200	8.5 ×
PQ26/20	21.2	26.5	20.2	19.0	46.3	119	5.49	0.391	109.4	56.2	3.50	9.0	31.2	—	—	870	12.3 × 1
PQ26/25	25.1	26.5	24.8	19.0	55.5	118	6.53	0.472	109.4	56.2	3.50	13.6	47.6	—	—	580	8.1 ×
PQ32/20	29.8	32.0	20.6	22.0	55.5	170	9.42	0.326	136.8	67.1	5.15	8.7	44.8	—	—	360	10.3 ×
PQ32/30	41.3	31	30.4	22.0	74.6	161	11.97	0.464	136.8	67.1	5.15	18.5	95.3	10.5	54	186	4.8 ×
PQ35/35	59	35.1	34.8	26.0	87.9	196	17.26	0.448	156.1	75.2	6.95	22.1	154	14.1	98	88	3.35 ×
PQ40/40	80	40.5	39.8	28.0	101.9	201	20.45	0.508	167.4	83.9	9.10	26.2	238	18.2	166	60	2.40 ×
PQ50/50	195	50.0	50.1	32.0	113	328	37.24	0.346	303	104	9.65	32.6	315	24.6	237	21	2.25 ×
ETD cores																	
ETD29	31	29.8	31.6	9.5	72	76.	5.47	0.947	66.5	53	4.9	19.4	93	11.4	55	680	3.90 ×
ETD34	59	34.2	34.6	10.8	78.6	97.1	7.64	0.810	86.6	60	5.9	20.9	123	12.9	76	355	3.35 ×
ETD39	80	39.1	39.6	12.5	92.2	125	11.5	0.737	117	69	6.9	25.7	177	17.7	122	170	2.70 ×
ETD44	105	44.0	44.6	14.9	103	173	17.8	0.589	167	77	7.25	29.5	214	21.5	156	83	2.45 ×
ETD49	135	48.7	49.4	16.4	114	211	24.0	0.534	204	85	8.35	32.7	273	24.7	206	48	2.15 ×
LP cores																	
LP23/8	6.9	16.5	23.4	8.7	44.1	31.3	1.377	1.41	24.6	30.9	2.15	15.0	32.0	—	—	6800	6.6 ×
LP22/13	12	25.0	22.4	12.9	49.0	67.9	3.327	0.721	55.4	45.8	3.28	13.9	46.0	—	—	1480	6.8 ×
LP32/13	16	25.0	31.8	12.9	64.0	70.3	4.498	0.909	55.4	45.8	3.28	21.6	70.8	13.6	45	900	4.4 ×

*with no conetral hole; assembly volume assumes use of coil former with extended terminations.

	no creepage allowance					Performance parameters, with 8 mm creepage allowance							Standards
k_3 (Eqn. 9.35)	k_4 (Eqn. 9.40)	A_R $(=R/N^2)$	R_{th}	P_{tot} (for 40°C temp. rise)	NI_o	k_1 (Eqn. 9.27)	k_2 (Eqn. 9.33)	k_3 (Eqn. 9.35)	k_4 (Eqn. 9.40)	A_R $(=R/N^2)$	R_{th}	P_{tot} (for 40°C temp. rise)	
(Ω^{-1})	$(W^{\frac{1}{2}}\Omega^{-\frac{1}{2}}m^2)$	$(\mu\Omega)$	(°C/W)	(W)	(A)	(Ωm^{-4})	(Ω)	(Ω^{-1})	$(W^{\frac{1}{2}}\Omega^{-\frac{1}{2}}m^2)$	$(\mu\Omega)$	(°C/W)	(W)	
35	0.0046	59	35	1.1	140	—	—	—	—	—	—	—	DIN 41 295
55	0.0157	28.5	23.5	1.7	245	2350	8.3×10^{-6}	193	0.0107	54	27	1.5	
05	0.102	21.5	10.5	3.8	420	145	4.8×10^{-6}	255	0.079	31	12	3.3	
0	0.131	23.5	10	4.0	410	95	5.3×10^{-6}	240	0.102	34	11.5	3.5	
0	0.28	18.0	6.7	6.0	580	29.5	3.7×10^{-6}	290	0.23	23.5	7.4	5.4	
0	0.34	19.3	6.0	6.7	590	22.5	4.0×10^{-6}	280	0.27	25.5	6.8	5.9	
5	0.53	15.0	5.0	8.0	730	10.3	2.9×10^{-6}	330	0.42	18.7	6.1	6.6	
7	0.0065	50	33	1.2	155	—	—	—	—	—	—	—	DIN 41 296
0	0.0165	32	24	1.65	225	—	—	—	—	—	—	—	
0	0.044	26	15.7	2.55	310	700	7.0×10^{-6}	210	0.031	45.0	18	2.2	
0	0.103	18.0	10.2	3.9	470	160	4.1×10^{-6}	275	0.080	26.5	11.7	3.4	
0	0.0057	90	36	1.1	110	—	—	—	—	—	—	—	IEC 133
5	0.0100	72	29	1.4	138	—	—	—	—	—	—	—	
2	0.0192	61	21	1.9	177	—	—	—	—	—	—	—	DIN 41 293
0	0.036	49	16.5	2.4	220	—	—	—	—	—	—	—	
0	0.061	45	14	2.9	251	—	—	—	—	—	—	—	
0	0.110	28	11	3.6	365	112	7.9×10^{-6}	200	0.081	51	11	3.6	
0	0.0098	63	38	1.1	130	—	—	—	—	—	—	—	IEC 431
5	0.0195	53	29	1.4	160	—	—	—	—	—	—	—	
8	0.036	38.5	23	1.7	210	—	—	—	—	—	—	—	DIN 41 980
7	0.065	30.3	17	2.35	280	210	8.4×10^{-6}	190	0.047	54	18	2.2	
0	0.029	24	17.5	2.3	310	830	5.9×10^{-6}	230	0.0215	38	20	2.0	IEC 647
0	0.051	20.0	15.0	2.65	365	320	4.7×10^{-6}	260	0.039	30	17	2.4	
0	0.096	15.2	10.3	3.9	505	103	3.3×10^{-6}	305	0.074	21.5	12	3.4	
0	0.221	9.1	7.1	5.6	790	22.5	1.75×10^{-6}	420	0.19	11.3	7.8	5.1	
	0.118	27.5	15	2.7	310	59	6.80×10^{-6}	215	0.09	43.5	16.2	2.5	DIN 41 989
	0.26	20.5	11.5	3.5	420	14.4	4.30×10^{-6}	270	0.22	27.5	12.5	3.2	
	0.48	14.0	9.3	4.3	560	5.0	2.75×10^{-6}	340	0.41	17.5	10.0	4.0	
	0.66	10.8	7.7	5.2	690	2.4	2.05×10^{-6}	390	0.58	13.0	8.3	4.8	
	1.70	8.8	5.7	7.0	900	0.53	1.6×10^{-6}	440	1.50	10.3	6.2	6.5	
	0.0087	82	43	0.93	107	—	—	—	—	—	—	—	
	0.0115	55	36	1.1	142	—	—	—	—	—	—	—	
	0.022	79	24.5	1.6	145	—	—	—	—	—	—	—	
	0.027	52	24	1.65	178	—	—	—	—	—	—	—	
	0.031	66	22.5	1.8	165	—	—	—	—	—	—	—	
	0.051	31	18.3	2.2	265	330	8.5×10^{-6}	190	0.037	55	19.2	2.1	
	0.076	21.5	15.3	2.6	350	137	5.3×10^{-6}	245	0.059	34	16.1	2.5	
	0.11	15.5	12.1	3.3	460	86	3.5×10^{-6}	300	0.089	22.5	12.7	3.15	
	0.25	14.5	8.0	5.0	590	28	3.0×10^{-6}	320	0.211	19.3	8.4	4.8	
	0.025	25	22	1.8	270	1150	6.7×10^{-6}	215	0.0175	42.5	27	1.48	
	0.039	21.5	18	2.2	320	570	5.4×10^{-6}	240	0.027	35	23	1.75	
	0.065	17.2	15	2.7	395	250	3.9×10^{-6}	285	0.050	25	17	2.4	
	0.108	15.8	12	3.3	460	113	3.4×10^{-6}	300	0.089	22	13	3.1	
	0.147	13.7	11	3.6	520	64	2.85×10^{-6}	330	0.123	18.2	12	3.3	
	0.0052	42.5	42	0.96	150	—	—	—	—	—	—	—	
	0.0136	44	30	1.3	175	—	—	—	—	—	—	—	
	0.019	28.5	24	1.7	240	1430	7.1×10^{-6}	210	0.015	45.2	25	1.61	

Table 9.4 Summary of equations relating to design parameters k_1–k_4 listed in Table 9.3

Design limitation	Equation for obtaining the value of the parameter from the requirements of the transformer design		Expression for the numerical value of the parameter for a given core in Table 9.3		Conditions			
	SI units	Eqn. No.	SI units	Eqn. No.	Waveform	Temperature (P_{tot} in Table 9.3 assumes 40°C temp. rise)	Copper Factor F_w	Steinmetz exponent n
Voltage regulation ($x\%$)	$k_1 = \dfrac{f^2 \hat{B}_c^2}{mP_o} \dfrac{(x - 0.01x^2)}{100}$	9.32			sine			
Saturation	$k_1 = (P_{tot} - P_J)\dfrac{(f\hat{B}_c)^2}{mP_o^2}$	9.34	$\dfrac{2\rho_c l_w}{\pi^2 A_c^2 A_w F_w}$	9.27	sine	Optional: set m in Eqn (9.32)	$F_w = 0.5$ in Eqn. (9.27)	
	$k_2 = (P_{tot} - P_J)\dfrac{(f\hat{\Phi})^2}{mP_o^2}$	9.34	$\dfrac{2\rho_c l_w}{\pi^2 A^2 A_w F_w}$	9.33	sine	Optional: set m in Eqn (9.34)	$F_w = 0.5$ in Eqn (9.33)	
	$k_3 = \dfrac{P_o}{(P_{tot} - P_c)^{1/2} F_w^{1/2} \hat{\Phi}_{p\text{-}p}}$	9.35	$\left(\dfrac{A_w}{mp_c l_w}\right)^{1/2}$	9.35	symmetrical square		Optional: set F_w in Eqn (9.35)	

Core loss:					
Low frequency ($F_R = 1$)	$k_4 = \frac{\sqrt{2}}{\pi} \times \frac{P_o(1+2/n)^{1/2}}{f\hat{B}F_w^{1/2}}$	9.40	sine	Optional: set F_w in Eqn (9.40)	Optional: set n in Eqn (9.40)
	$\frac{P_o(1+2/n)^{1/2}}{2f\hat{B}F_w^{1/2}}$	9.41*	symmetrical square*	Optional: set F_w in Eqn (9.41)	Optional: set n in Eqn (9.41)
	$\frac{0.854 P_o}{f\hat{B}}$	9.42 (a)	sine	$F_w = 0.5$ assumed in Eqns (9.42) (a) & (b)	$n = 2.5$ assumed in Eqns (9.42) (a) & (b)
	$\frac{0.949 P_o}{f\hat{B}}$	9.42 (b)*	symmetrical square*		
				Operating temp. assumed = 100°C	
	$\left(\frac{P_{tot} A_w}{m p_e l_w}\right)^{1/2} \times A_{min}$ (Note: $\hat{B}$ is flux density at A_{min})				
High frequency ($F_R > 1$)	$\frac{\sqrt{2}}{\pi} \times \frac{P_o(1+2/n)^{1/2}}{f\hat{B}} \times \left(\frac{F_R}{F_w}\right)^{1/2}$	9.54	sine	Optional: set F_R/F_w in Eqn (9.54)	Optional: set n in Eqn (9.54)
	$\frac{P_o(1+2/n)^{1/2}}{2f\hat{B}} \times \left(\frac{F_R}{F_w}\right)^{1/2}$	9.55*	symmetrical square*	Optional: set F_R/F_w in Eqn (9.55)	Optional: set n in Eqn (9.55)
	$\frac{0.604 P_o}{f\hat{B}} \times \left(\frac{F_R}{F_w}\right)^{1/2}$	9.56	sine	$n = 2.5$ assumed in Eqn (9.56)	$n = 2.5$ assumed in Eqn (9.56)
	$\frac{0.671 P_o}{f\hat{B}} \times \left(\frac{F_R}{F_w}\right)^{1/2}$	9.57*	symmetrical square*	Optional: set F_R/F_w in Eqn (9.57)	$n = 2.5$ assumed in Eqn (9.57)

*For **SMPS** transformers other than those having the two-winding push-pull configuration, Eqns (9.41), (9.42(b)), (9.55) and (9.57) should be multiplied by the appropriate numerical factor listed on p. 293.

Figure 9.19 Ferrite power transformer cores assembled with their coil formers, clamping and mounting accessories.

(Samples by courtesy of Philips Components, Siemens and TDK)

| E42/42/20 | U25/40/13 | 36/22 Pot core | RM10 | EC41 |
| PM62/49 | PQ32/30 | ETD34 | | LP32/13 |

Performance parameters, with 8 mm creepage allowance. The parameters in the first five columns are based on the same assumptions as the corresponding parameters defined above except that the 8 mm creepage version of A_w has been used. The next column is R_{th} appropriate to a transformer where the 8 mm creepage allowance has been incorporated. The non-italic numbers relate to measured data supplied by one manufacturer while the italic numbers are the author's estimates which are consistent with the measured data. The corresponding values of P_{tot} are given in the following column.

Standards. This column indicates some of the relevant standards that specify the physical dimensions of the cores.

Table 9.3 contains all the physical parameters of the cores necessary to implement the design procedures described in foregoing sections. The presence of a number of different types and styles in one integrated presentation facilitates comparison. As indicated earlier, this table cannot include parameters that depend on the properties of the ferrite as these will differ between grades and manufacturers and will be subject to continuing development. So it is not possible for the table to provide a direct indication of core performance in terms power handling capability. However, using the design equations presented in Section 9.4.2 together with manufacturer's data on the magnetic and electric properties of their cores, an iterative route should lead rapidly to an optimum design.

Table 9.4 summarizes the more important design equations involving the parameters k_1, k_2, k_3, and k_4 thus providing a useful adjunct to Table 9.3. These two tables provide, in a compact form, most of the information required for the design process.

Some manufacturers are able, by making certain assumptions, to combine their core and material parameters into a direct presentation of core performance, e.g. in terms of throughput power as a function of frequency. Where this information is available the task of the transformer designer may be simplified, but often such presentations do not provide the numerical resolution necessary for actual designs. The equations and performance parameters presented in this chapter are directly relevant and applicable, particularly where it is wished to modify the design assumptions or to optimize a design.

9.4.5 Example power transformer design

To illustrate the use of the design guides developed in Section 9.4.2. and the data presented in Section 9.4.4., the following transformer specification will be used as an example. Since Section 9.4.2 is restricted to low frequency transformers, any eddy current loss in the windings is ignored.

Nominal output power	120 W
Input	200 V, 25 kHz symmetrical square wave
Output	55 V

It will be assumed that the primary and secondary windings nominally fill the net winding space after allowing an 8 mm margin to meet the IEC creepage requirement. It will also be assumed, initially, that the division of losses between core and windings will be near optimum and that the winding copper factor will be 0.5. Therefore the second of the equations (9.42) may be used to make the initial core selection, i.e.

$$\frac{0.949 P_o}{f \hat{B}} = k_4$$

The initial value of the peak flux density would normally be put equal to the maximum design value as defined in Section 9.4.2 (see text relating to Figure 9.10). This would be considerably less than the manufacturer's limit based on saturation (e.g. see Figure 9.8(a)) because it is necessary to allow for voltage tolerances and transient effects. A typical value would be 200 mT. However, as this design is expected to be core loss limited, a lower value is appropriate. So a peak flux density of 170 mT will be used. Therefore,

$$k_4 = \frac{0.949 \times 120}{25 \times 10^3 \times 0.17} = 26.8 \times 10^{-3}$$

An inspection of Table 9.3 shows that the EDT34 core has a k_4 value (with creepage allowance) of 0.0270 so this core will now be used for the design. It has the following parameters:

$A_e = 97.1$ mm^2, $V_e = 7640$ mm^3, $A_{min} = 86.6$ mm^2,
$l_w = 60$ mm, $h_w = 5.9$ mm,

and, providing 8 mm creepage allowance,

$b_w = 12.9$ mm, $A_w = 76$ mm^2, $P_{tot} = 1.75$ W

Referring to manufacturers' data sheets, a typical power loss for this core at $f = 25$ kHz and a peak flux density (at the minimum core cross-section) of 200 mT is 1.1 W. Assuming that $P_c \propto B^{2.5}$, the corresponding loss at 170 mT would be $1.1 \times (0.17/0.2)^{2.5} = 0.73$ W. See also Figure 9.8(b).

For the temperature rise assumed in Table 9.3, P_{tot} for the ETD34 core (with creepage allowance) is 1.75 W. If the division of losses between core and winding is to be optimum, the allowable core loss would be, from Eqn (9.37), $P_{tot}/2.25 = 0.78$ W. Had the above predicted loss exceeded this figure it would have been necessary to further reduce the flux density and recalculate k_4. However, with the above figures it is safe to proceed with the design of the windings.

Because of the need for rounding it is preferable to calculate the smaller number of turns first. The subscripts 1 and 2 refer to the primary and secondary windings respectively.

From Eqn (9.15),

$$N_2 = \frac{U_2 t_d}{2 \hat{B} A_{min}} = \frac{U_2}{4 \hat{B} A_{min} f} =$$

$$\frac{55}{4 \times 0.17 \times 86.6 \times 10^{-6} \times 25 \times 10^3} = 37.4$$

Therefore the number of turns can be provisionally assigned as follows:

$N_2 = 38$ turns, $N_1 = 38 \times 200/55 = 138$ turns

Assuming that each winding occupies half the available winding space and allowing a packing factor of 0.9 (see Table 11.1), the space available for each winding is $76 \times 0.9/2 = 34.2$ mm^2.

From Eqn (11.6),

$(d_o)_1 = (34.2/138)^{1/2} = 0.498$ mm
$(d_o)_2 = (34.2/38)^{1/2} = 0.949$ mm

The use of metric winding wire with Grade 2 (medium) thickness enamel insulation will be assumed. From Appendix B Table B.1.3, the next smallest standard diameters (using the principal rather than the intermediate sizes), are:

$(d_o)_1 = 0.462$ mm;
  bare copper diameter, $d_1 = 0.40$ mm
$(d_o)_2 = 0.885$ mm;
  bare copper diameter, $d_2 = 0.80$ mm

From Appendix B Table B.1.1, the corresponding values of resistance/metre, R_c, at 20°C are, for d_1, 0.136 Ω/m and, for d_2, 0.03401 Ω/m.

Therefore, multiplying R_c by the number of turns, the mean turn length, l_w, and $m \simeq 1.3$, the winding resistances at 100°C will be:

$R_1 = 0.136 \times 138 \times 0.06 \times 1.3 = 1.464$ Ω
$R_2 = 0.03401 \times 38 \times 0.06 \times 1.3 = 0.1008$ Ω

This calculation follows the usual practice of using the same mean turn length for both windings, the errors cancelling in the summation of the total winding loss. It should be noted that this example is concerned only with d.c. resistances, it having been postulated at the beginning of this section that, as the present subject is low frequency transformers, eddy current effects in the windings will be ignored.

From the specification, the r.m.s. currents are: $I_1 = 120/200 = 0.6$ A and $I_2 = 120/55 = 2.18$ A. Therefore the winding losses will be:

$(P_w)_1 = 0.6^2 \times 1.464 = 0.527$ W
$(P_w)_2 = 2.18^2 \times 0.1008 = 0.479$ W
$\therefore P_w = 1.006$ W

From the core parameters, $P_{tot} = 1.75$ W. So from Eqn (9.36), for optimum division of losses and assuming $n = 2.5$, the winding loss should be $P_{tot}/1.8 = 0.97$ W. The agreement is good.

As a check on the value of F_w (0.5) assumed in Eqn (9.42), the actual copper areas, using the conductor areas quoted in Table B.1.1, are for the primary $138 \times 0.1257 = 17.35$ mm² and for the secondary $38 \times 0.5027 = 19.1$, giving a total copper area of 36.45 mm². Dividing by the available winding area, 76 mm², the actual value of F_w realized is 0.48.

Staying with the winding design, it is necessary to check that the winding can indeed be accommodated within the available winding height, 5.9 mm.

Primary turns/layer = $b_w/d_o = 12.9/0.462 = 27.9$; number of layers = $138/27.9 = 4.95$. A satisfactory arrangement would be 5 layers of 28 turns, giving an adjusted number of primary turns, $N_1 = 140$

Secondary turns/layer = $12.9/0.885 = 14.6$; so 3 layers of 13 turns can be used, giving an adjusted number of secondary turns, $N_2 = 39$.

The turns ratio has now changed slightly from the initial 3.63 to 3.59.

This adjustment to obtain winding layers having the same number of turns, i.e. integral numbers of layers, is good practice particularly if, as will be seen later, it is necessary to minimize leakage inductance and eddy current winding loss.

The actual build-up of winding height would be as follows. Primary, assuming 0.1 mm thick interleaving:

$h_1 = 5 \times 0.462 + (4 \times 0.1) = 2.71$ mm

Secondary, assuming 0.2 interleaving:

$h_2 = 3 \times 0.885 + (2 \times 0.2) = 3.05$ mm

This gives a total build-up of 5.76 mm compared with an available h_w of 5.9 mm. This is a little tight considering the need for insulation between and over the windings but, as all the dimensions have been given the worst-case limits and some compaction of the windings is to be expected, the design appears to be satisfactory. Had the design called for auxiliary windings it would have been necessary to assume a smaller copper factor, F_w, at the outset, using Eqn (9.41) instead of Eqn (9.42) to obtain the design value of k_4. This would have led to the choice of a larger core.

The adjustment to the numbers of turns results in the following adjusted winding resistances; $R_1 = 1.49\,\Omega$, $R_2 = 0.103\,\Omega$. The corresponding winding losses are 0.54 W and 0.49 W respectively.

This slightly increased winding loss is offset by a lower adjusted core loss. The actual flux density is now

$$\hat{B} = \frac{U_1}{4A_{min}N_1 f}$$

$$= \frac{200}{4 \times 86.6 \times 10^{-6} \times 140 \times 25 \times 10^3} = 165\,\text{mT}$$

Assuming again that $P_c \propto \hat{B}^{2.5}$, the adjusted core loss will be $1.1(165/200)^{2.5} = 0.68$ W. So the final power budget in watts is:

$(P_w)_1 = 0.54$ W $\bigg\}$ $P_w = 1.03$ W
$(P_w)_2 = 0.49$ W

$P_c = 0.68$ W
$\therefore P_{tot} = 1.71$ W

This compares with the listed value of P_{tot} for the ETD34 core of 1.75 W. It is, of course, to be expected that the winding loss will approximate to the correct proportion of P_{tot} provided the design parameters are met, but the core loss depends on frequency, flux density and the quality of the ferrite and, as observed in Section 9.4.2., there is no reason arising from the design procedure why it should assume any particular value.

Had the core loss been less than the value that satisfied the power budget, then an overall improvement could be obtained by increasing the operating flux density towards the limit imposed by saturation considerations. This would reduce k_4, and the winding loss, so the core would be under utilized. A smaller core might then be tried.

If the core loss had exceeded the acceptable figure, then lowering the flux density would not be a remedy since this would increase k_4, increase the winding loss, and invalidate the choice of core. If a lower loss ferrite is not available, an obvious course is to move to the next larger core. This is, in fact, a rather large step in size if it is wished to stay in the ETD range, and the resulting transformer would have a poor utilization of the core.

A better course would be to raise the frequency if this is possible. A modest increase in frequency would, for a constant number of turns, reduce the flux density in inverse proportion and this would result in a substantial reduction in core loss. The winding loss, still ignoring eddy currents, would be unchanged. Another option may be to dispense with the 8 mm creepage penalty by meeting the isolation requirement in some other way, e.g. by using an earthed screen between windings or by impregnation of the windings (or both). If the excess loss is small, another possibility is to use thinner enamel insulation on the winding wire, i.e. Grade 1. This would allow a small increase in the number of turns.

Returning now to the design, it remains to check the voltage drops and the magnetizing current.

From the winding resistances and currents, the volt-drops in the primary and secondary windings are 0.9 and 0.22 V respectively; these are probably too small to merit correction. If this were not so, the primary turns could be appropriately adjusted to provide the specified secondary voltage on full load.

The value of the primary magnetizing current may be obtained from data such as that given in Figure 9.8(a). The peak flux is $BA_{min} = 0.17 \times 86.6 \text{ mm}^2 \times 10^{-6} = 14.7 \,\mu\text{Wb}$ so, using that data, the corresponding value of NI (typical) would be 3.0. Therefore the magnetizing current would be $3.0/140 = 21$ mA, which is very small compared with the load current of 0.6 A.

This worked example has resulted in a satisfactory design of a 120 W transformer and has illustrated some of the design procedures involved. The omission of eddy current loss in the windings has simplified the design. This aspect is considered in the following sections.

9.5 Higher frequency power transformers

9.5.1 General

The power transformers considered in this section are those in which eddy currents influence the design or performance of the windings. The actual frequency at which this occurs depends on the design requirements. For example, as noted earlier, a 1 kW transformer designed to have a low voltage output will have such thick conductors that the a.c. resistance may be much greater than the d.c. resistance at frequencies as low as a few kilohertz, while on the other hand a transformer delivering relatively little power into a high impedance load might have negligible eddy current winding loss at several hundred kilohertz.

In low power wide-band transformers the winding resistance is limited by the insertion loss requirements but in higher power transformers it is usually the heat dissipation that is the limiting factor. Thus the higher the power the more important is the effect of eddy currents; high power implies high currents, and heavy conductors imply relatively high a.c. resistance due to eddy currents.

As indicated in the introduction to this chapter, the major application of high frequency power transformers has become their use in Switched Mode Power Supplies, and the emphasis in the following sections will be on this application. However, there are other important applications such as wide band power transformers, for example in antenna matching units, and these will receive some attention.

The design procedures outlined in the foregoing sections may be adapted to take account of eddy current losses in the windings. It will be shown that, in the matter of core selection, the eddy current winding loss may be taken into account rather easily by modifying the evaluation of the core selection parameters, such as k_4.

Given adequate data, the higher frequency performance of the ferrite core can be predicted as easily as for the lower freqencies. However, the design of an optimum winding, in terms of eddy current loss, leakage inductance and self capacitance can involve considerable design effort. The following section discusses the various aspects of core selection and performance, and considers the problems of cooling in transformers required to operate at the higher end of the freqency and power range. Section 9.5.3. derives the basic criteria for the design of optimum windings and sets out guide lines for their implementation.

9.5.2 Core selection

This section considers the selection and performance of cores for transformers operating in the higher frequency range as defined above. It is divided into two parts; the first is concerned with transformers at the lower end of the high-frequency range and the second deals with the upper end of the range where the design approach can be radically different.

For the majority of power transformers operating in the higher frequency range the design will be core loss limited. Possible exceptions are wide-band transformers that have to meet a severe attenuation or return loss specification at the lower end of the transmission band; in such cases the shunt inductance may determine the number of turns.

9.5.2.1 The lower frequency range

At the lower end of this range the transformers will not be very different in application or appearance to the low-frequency transformers considered earlier in this chapter. They will probably be single frequency or narrow band devices, they will have windings that occupy a large proportion of the available winding space and the total cooling surface area will be approximately as expressed by R_{th} in Table 9.3. Under these circumstances the earlier sections of this chapter are entirely relevant, only requiring some modification to take account of the eddy current effects in the windings.

Anticipating the winding design criteria derived in Section 11.4.4 and developed in Section 9.5.3, the winding will have an a.c. resistance that is a factor F_R times the d.c. resistance. Except for windings having very few layers, the optimum value of F_R lies between 1.33 and 1.5 depending on the type of conductor. For the purpose of core selection it is reasonable to assume a nominal value for F_R.

If the operating frequency is relatively low and the circuit impedances are high, the eddy current effects in the largest conductor cross-section that can be accommodated may be so small that the value of F_R would be nearer to unity than 1.5. If, on the other hand, it is probable that eddy current loss will be such that optimum conductor cross-section can be used then it is reasonable to put $F_R = 1.5$.

This factor can be introduced into Eqn (9.25) with the result that ρ_c is replaced by $F_R \rho_c$ in subsequent equations, e.g. in the expression for k_1 in Eqn (9.27). In particular Eqns (9.40) and (9.41) become:

For sine waves,

$$\frac{\sqrt{2}}{\pi} \times \frac{P_o(1 + 2/n)^{1/2}}{f\hat{B}} \times \left(\frac{F_R}{F_w}\right)^{1/2} = k_4 \qquad (9.54)$$

and for symmetrical square waves,

$$\frac{P_o(1 + 2/n)^{1/2}}{2f\hat{B}} \times \left(\frac{F_R}{F_w}\right)^{1/2} = k_4 \qquad (9.55)$$

where $\hat{B}$ is the maximum allowable flux density in the minimum core cross-section, A_{min}. As before,

$$k_4 = \left(\frac{P_{tot} A_w}{m \rho_c l_w}\right)^{1/2} \times A_{min}$$

and is listed in Table 9.3 for a number of ranges of cores.

If, following the procedure used in Section 9.4.2, the value of n, the Steinmetz exponent, is put nominally equal to 2.5, the following simplified expressions, corresponding to Eqns (9.42), result.

For sine waves,

$$\frac{0.604 P_o}{f\hat{B}} \times \left(\frac{F_R}{F_w}\right)^{1/2} = k_4 \qquad (9.56)$$

and for symmetrical square waves,

$$\frac{0.671 P_o}{f\hat{B}} \times \left(\frac{F_R}{F_w}\right)^{1/2} = k_4 \qquad (9.57)$$

The winding copper factor, F_w (see Eqn (11.2)) has been retained in Eqns (9.56) and (9.57) to provide a factor F_R/F_w, the value of which may be chosen by the designer.

The values of k_4 are based on A_w, the net available winding area in the coil former (with or without creepage allowance). If the net winding area is fully occupied by close-packed windings of round wire, then $F_w \simeq 0.5$. If in addition the windings have conductors of optimum diameter so that $F_R \simeq 1.5$, then

$$\left(\frac{F_R}{F_w}\right)^{1/2} \simeq 1.73 \qquad (9.58)$$

At higher frequencies and higher currents the winding design may result in optimum conductor cross-sections that are significantly less than the maximum that can be accommodated; the optimum windings would then occupy less than the total winding space and so F_w will be reduced. The use of strip windings or bunched conductors will also adversely affect F_w.

Further, auxiliary windings for feedback, control, energy recovery or voltage limitation may be required. Assuming that the power windings would otherwise occupy all the available area, F_w would be reduced in proportion to the cross-sectional area required for these auxiliary windings. However, at the higher frequencies where the use of optimum conductor cross-sections results in the power windings occupying significantly less than the available winding space, the auxiliary windings may possibly be accommodated without penalty in the free space.

With a little experience, a reasonable estimate of F_R/F_w can be made.

In Eqns (9.54)–(9.57) it is assumed that the windings carry a continuous sinusoidal or symmetrical square wave current, the r.m.s. value of which is used in the derivation. In Switched Mode Power Supplies there are many circuit arrangements where this assumption is not valid. Jansson in Reference 27 analyzed a number of commonly-used circuits and derived factors to modify the design equations to allow for the actual current waveforms. The effect is to multiply the l.h.s. of Eqns (9.55) and (9.57) (i.e. the equations relating to symmetrical square wave excitation) by a numerical factor. A few of the more important results are:

(9.54) $\quad \dfrac{\sqrt{2}}{\pi} \times \dfrac{P_o(1 + 2/n)^{1/2}}{f\hat{B} \times 10^{-4}} \times \left(\dfrac{F_R}{F_w}\right)^{1/2} = k_4$

(9.55) $\quad \dfrac{P_o(1 + 2/n)^{1/2}}{2f\hat{B} \times 10^{-4}} \times \left(\dfrac{F_R}{F_w}\right)^{1/2} = k_4$

(9.56) $\quad \dfrac{0.604 \times 10^4 P_o}{f\hat{B}} \times \left(\dfrac{F_R}{F_w}\right)^{1/2} = k_4$

(9.57) $\quad \dfrac{0.671 \times 10^4 P_o}{f\hat{B}} \times \left(\dfrac{F_R}{F_w}\right)^{1/2} = k_4$

where k_4 equals the SI values given in Table 9.3.

Circuit	Factor*
A two-winding push-pull converter transformer:	1
A four-winding push-pull converter transformer where, during each half-cycle, one of the primary and one of the secondary windings are inactive:	$\sqrt{2}$
A two-winding forward converter transformer:	$2\sqrt{2}$
A ringing-choke transformer:	$4\sqrt{(2/3)}$

Reference 27 gives a more complete list of such factors together with a comprehensive survey of SMPS circuits. The literature on SMPS circuits is very extensive; References 56–66 represent a broad selection of publications in most of which there is reference to the transformer design.

The initial design for the transformer may now proceed, following the route outlined towards the end of Section 9.4.2. The required throughput power and the frequency are known. The maximum design flux density is set at a value that will avoid saturation by a suitable margin. Using these values and an estimated value of (F_R/F_w), Eqn (9.57) (for SMPS) or Eqn (9.56) (for sinusoidal excitation) can be used to obtain the value of the parameter k_4. In the case of SMPS transformers, the above circuit-dependent factor must be first applied to Eqn (9.57). From Table 9.3 a suitable core type may then be selected. The selected core should then be able to accommodate a winding design that has a total winding loss that is not greater than $P_{tot}/(1 + 2/n) \simeq 0.56 P_{tot}$, see Eqn (9.36). P_{tot} is the total transformer loss corresponding to the assumed temperature rise and is quoted for the selected core in Table 9.3.

The core loss will depend on the core performance and this will need to be evaluated from the manufacturer's data using the limit values. In general it will not equal $P_{tot}/(1 + n/2) \simeq 0.44 P_{tot}$, see Eqn (9.37), since it will depend on the chosen core material, the frequency and the flux density.

As in the case of the low-frequency design, if the core loss is lower than $0.44 P_{tot}$, the design will be safe; it will be saturation limited. Unless the core loss is excessively low there would probably be no advantage in moving to a smaller core. If the core loss data shows that P_c will exceed $0.44 P_{tot}$ then the total dissipation will lead to an excessive temperature rise and it would probably be necessary to choose a core with a lower loss, or a larger core, or move to a higher frequency. Some of the options are examined in Section 9.4.5.

Having selected a suitable core, the numbers of turns may be evaluated, starting with the lower voltage winding, using the appropriate induction formula (Eqns (9.13)–(9.16)). Sufficient information is now available to begin the winding design, starting with the calculation of the optimum conductor cross-sections and then following the guidelines in Section 9.5.3.

If the transformer is to operate over a band of frequencies, the design should be carried out initially using the lowest frequency and then checked at the highest frequency using the same numbers of turns and the corresponding (lower) flux density.

If there is a minimum inductance limit, the design should be checked to ensure that this limit is not infringed. In some cases, e.g. wide-band transformers operating at relatively low power, the inductance consideration may override that of core loss in determining the number of turns, and the design procedure approaches that described for wide-band transformers in Section 7.3.2.

9.5.2.2 The higher frequency range

As observed at the beginning of this chapter, the trend in SMPS development is towards the use of higher switching frequencies. This has been made possible by advances in the performance of semiconductors and capacitors and results in size and weight reduction, not only in the transformers and chokes but also in the capacitors. As far as the transformers are concerned, the implications of higher frequency operation have been examined in Reference 67. That analysis is relevant to the present discussion and will therefore now be summarized.

Referring to Eqn (9.20), it has been seen that the ratio of eddy current to magnetic loss (per unit volume) in a core has the following approximate proportionality

$$P_F/P_M \propto A_e f^{0.7}/\rho \hat{B}^{1/2} \qquad (9.59)$$

where ρ is the core resistivity.

For typical power ferrites, experimental results suggest that this holds for frequencies up to about 200 kHz, above which the exponent of frequency will decrease.

If it is assumed that the temperature rise of the core is to be fixed at the maximum permissible value and, for the purpose of this analysis, the core may be considered thermally separate from the winding, then P_c/A_c is a constant, where A_c is the effective cooling surface area of the core. Since P_F is generally small compared with P_M, the flux density in the core is determined mainly by the latter. Therefore from Eqn (9.17),

*Note. These factors are adapted from the Jansson factors by dividing the latter by 2 and taking the reciprocal.

$P_c/A_c = P_M V_e/A_c \propto f^{1.3} \hat{B}_e^{2.5} V_e/A_c$

= a constant for a given core temperature rise.

$\therefore \hat{B}_e \propto f^{-0.52}(A_c/V_e)^{0.4}$

Putting this in Eqn (9.59),

$P_F/P_M \propto f A_e (V_e/A_c)^{0.2}/\rho$ approx.

Therefore if the core shape is constant,

$P_F/P_M \propto f L^{2.2}/\rho$ approx. (9.60)

where L is a linear dimension of the core.

It follows that for the proportion of eddy current core loss to remain constant with frequency in a given core size and shape, the core resistivity must increase approximately in proportion to the frequency. In practice, as the frequency rises it is likely that the size of core will be reduced thus decreasing the proportion of eddy current loss if ρ/f is constant.

The core resistivity can in principle be increased by changes to the composition or the processing parameters, see Section 1.3. This is often a major objective in the development of power ferrite materials intended for higher frequency applications.[39] In this connection it is important that it is the resistivity at the operating temperature, e.g. 100°C, that is considered since this can be more than five times lower than the 20°C value. Inasmuch as the ferrite resistivity can be increased in proportion to the application frequency, the eddy current contribution to the core loss will remain small and it can be neglected in the further development of this analysis.

Therefore, from Eqn (9.17), the total core loss may be expressed by

$P_c = P_M V_e = k f^m \hat{B}_e^n V_e$

and, as before, for constant temperature rise

$P_c/A_c = k f^m \hat{B}_e^n V_e/A_c = k f^m \hat{B}_e^n L =$ a constant (9.61)

where L is a linear dimension of the core assuming a constant core shape.

From the induction equation, $\hat{B}_e \propto U/NA_e f \propto U/NL^2 f$ so

$k f^m (U/Nf)^n L^{1-2n} =$ a constant

$\therefore N L^{2-1/n} \propto k^{1/n} U f^{(m/n)-1}$ (9.62)

Inserting typical experimental values, e.g. $m = 1.3$ and $n = 2.5$

$N L^{1.6} \propto k^{0.4} U f^{-0.48}$

$\propto k^{0.4} U/f^{1/2}$ approx.

The parameter k represents the magnetic loss coefficient of the core material and its reduction has always been an objective of ferrite development. References 39–42 and 68–70 provide information on recent advances in the development of power ferrites

with particular reference to loss reduction. Further improvements may be expected but for the purpose of this analysis k may be assumed constant. Then

$N^2 L^{3.2} \propto N^2 V_e \propto U^2/f$ approx (9.63)

This expresses the proportional relations between the number of turns, the core volume, the throughput power ($P_o \propto U^2$) and the frequency, given a constant temperature rise in the core and no influence from the winding. If the proportionalities are substituted into the induction equation, the relation can be extended to the flux density.

The results may be conveniently summarized as follows.*

For a constant core shape and a given temperature rise, doubling the frequency would result in the following related changes:

	Power	Core volume	Turns	Flux density
a)	const.	const.	$N/\sqrt{2}$	$\hat{B}/\sqrt{2}$
or b)	const.	$V_e/2$	const.	$\hat{B}/2^{1/3}$
or c)	$2P_o$	const.	const.	$\hat{B}/\sqrt{2}$

Thus for a given power ferrite of typical loss performance, and having a sufficiently high resistivity to make the eddy current loss negligible over the applicable frequency range, the performance of a transformer will be improved by increasing the operating frequency. This result is, of course, well known in qualitative terms but the above analysis indicates the possible trade-offs between the throughput power, core volume, number of turns and the operating flux density. The achievement of a lower hysteresis loss in the ferrite would be an additional advantage but would not influence the conclusions.

It has been noted elsewhere that the exponent of f rises slowly with frequency and this will limit the validity of the above analysis. However, the results are probably valid at frequencies up to about 200 kHz. The caption to Figure 3.19 comments further on the behaviour of m.

A similar analysis may be applied to an optimized winding. Again thermal isolation is assumed, i.e. the winding temperature rise is independent of the conditions in the core. Assuming close-packed windings of round conductors occupying a given coil former, then from Eqns (11.31) and (11.33),

$P_w = 1.5 P_{dc} = \dfrac{6 I^2 N l_w \rho_c}{\pi d^2}$ (9.64)

From Eqn (11.36) the optimum conductor diameter is proportional to $(b_w/fN)^{1/3}$. Substituting this for d in Eqn (9.64),

$P_w \propto \dfrac{I^2 N^{5/3} l_w f^{2/3}}{b_w^{2/3}}$

*Courtesy of Research Studies Press, Letchworth, Hertfordshire, England (see Reference 67, p 149).

From the above core loss analysis, $N \propto f^{-1/2}$ (Eqn (9.63)), other parameters being constant, so

$$P_w \propto \frac{I^2 l_w}{f^{1/6} b_w^{2/3}} \quad (9.65)$$

This suggests that for a given load current and a given coil former wound with optimum diameter wire, the winding loss changes only weakly with frequency, mainly because with increasing frequency the required number of turns decreases. In practice, some of the assumptions on which this conlusion is based may be invalidated under some design conditions. However, it indicates that in general winding losses need not rise strongly as the design frequency is increased.

The overall conclusion is that the increase of operating frequency will substantially improve the core performance, e.g. leading to a smaller core, lower loss or higher throughput, and the winding loss need not rise unduly. This analysis provides theoretical support to the current trend and quantifies some of the options.

Attention will now be turned to transformers operating at much higher frequencies, e.g. > 5 MHz. Such transformers will differ appreciably in application and construction to the those hitherto considered.[71] They will usually be required to transmit a broad band of frequencies, a typical bandwidth ratio being 10. They will consequently need to have low leakage inductance and low self capacitance, so the windings will usually be designed as single layers, well spaced from the core and from each other. The cooling area will not therefore correspond to that used for low frequency transformers. To ease the problem of designing the winding to meet the high frequency specification it is necessary to restrict the size of the transformer. This usually entails a more effective means of heat transfer than free air convection; oil cooling is often used.

It follows from these considerations that h.f. power transformers tend to have an open construction in which the entire core surface can transfer heat directly to the convection medium. If this type of construction is assumed, the thermal balance of the core is almost entirely isolated from that of the winding and so the design of the core may be largely independent of the winding design.

It is assumed that the frequency and core size are such that, to avoid eddy current losses in the core, the choice of ferrite must be confined to the nickel zinc grades. To avoid an excessive core loss density at high frequencies it is usually necessary to restrict the flux density to quite low values, e.g. less than 10 mT (100 Gs). Under these conditions the residual loss is the main component of the magnetic loss in the core (see Section 2.2.5). Therefore the magnetic core loss density may be related to the residual loss factor by transposing Eqn (2.54):

$$P_m = \left(\frac{\tan \delta_r}{\mu}\right) \frac{\pi f \hat{B}_e^2}{\mu_o} \quad \text{W m}^{-3} \quad (9.66)$$

The residual loss factor $(\tan \delta_r)/\mu$ is a function of frequency and temperature, see Figures 3.12 and 3.13. The ferrite with the lowest residual loss factor under the appropriate conditions will be the most suitable. Given a grade of ferrite, Eqn (9.66) relates the core loss density to the flux density.

In the problem of core selection the starting parameters depend to some extent on the circumstances. One approach is to start with the core temperatures and base the selection on thermal equilibrium. To establish the relevant relations, a hypothetical core having a uniform circular cross-section of diameter d and a volume V may be considered. The temperature difference between the centre and the surface of the core is designated θ_{cs} and is related to the core loss density by Eqn (9.9):

$$\theta_{cs} = \frac{P_m d^2}{16\lambda} \quad °C \quad (9.67)$$

Assuming the flux density is uniform the total power dissipated in the core is $P_m V$ and at equilibrium this equals the total rate of heat transfer, i.e.

$$P_m V = P_{conv} A_c \quad \text{W} \quad (9.68)$$

where A_c is the surface area of the core

P_{conv} is the heat transfer per unit area

Combining this in Eqn (9.67):

$$P_{conv} = \frac{16 V \lambda \theta_{cs}}{A_c d^2}$$

$$= \frac{4 \lambda \theta_{cs}}{d} \quad \text{W m}^{-2} \quad (9.69)$$

If d is in mm and λ for ferrite is 4000 µW mm^{-1} °C^{-1}:

$$P_{conv} = \frac{16 \times 10^3 \times \theta_{cs}}{d} \quad \mu\text{W mm}^{-2} \quad (9.70)$$

If thermal considerations enable the designer to decide the temperature of the core surface and the temperature of the core centre these equations enable the corresponding diameter and core loss density to be determined. The method of heat transfer and the permissible surface temperature together determine the rate of heat transfer P_{conv} (e.g. see Figure 9.3). The

$$(9.66) \quad P_m = \left(\frac{\tan \delta_r}{\mu}\right) \frac{f \hat{B}_e^2 10^{-7}}{4} \quad \text{W cm}^{-3}$$

difference between the surface and centre temperatures equals θ_{cs}. Thus Eqn (9.70) yields d and Eqn (9.67) yields P_m and hence the flux density.

These parameters are not dependent on the total power to be transmitted because the only design criterion was temperature distribution. If the transformer is required to transmit a band of frequencies, the flux density calculation must be made at several frequencies within the band so that when the corresponding numbers of turns are subsequently calculated the limiting case, i.e. that yielding the largest number of turns, may be determined.

It is interesting to note that if, for the purpose of facilitating the h.f. aspects of the design, it is desired to minimize the length of the winding conductor then it is necessary to minimize πdN, $\propto d/\hat{B}A$ from the induction equation (9.16), i.e. to maximize $\hat{B}d$ since $A = \pi d^2/4$, or $\sqrt{P_m d}$ since $P_m \propto \hat{B}^2$ approx. Applying this to Eqn (9.67) it is clear that θ_{cs} should be made as high as thermal considerations will allow; this implies a high core loss density and a high rate of heat removal from the core.

From the induction equation, the number of turns may be calculated and the subsequent winding design will determine the winding breadth required. A single layer spaced winding of tape supported on a skeletal former of low permittivity and dielectric loss is sometimes used to minimize both the eddy current losses and self capacitance.

The cross-sectional area of the core and the winding breadth having been evaluated, a core of appropriate ferrite grade and dimensions may be selected. In practice a suitable core with a cylindrical section may not be available, in which case a rectangular section could be used, the diameter, d, now referring to a circle of equivalent cross-sectional area.

The above approach will lead to the selection of a suitable core providing the design is core loss limited. This is not always true; if the bandwidth is wide the minimum shunt inductance will determine the number of turns. Therefore when the core has been selected on magnetic loss considerations and the corresponding number of turns has been calculated, the shunt inductance must be checked using Eqn (4.19) and the value of C_1 for the selected core. If it is found that the inductance is insufficient to meet the attenuation or return loss specification a number of courses are open. It would, for example, be possible simply to increase the number of turns but this would probably make it more difficult to meet the leakage inductance and self capacitance requirements, and since it would lower the flux density it would result in the core being operated at temperatures lower than the permitted values. A better course is to repeat the design on a ferrite having a higher permeability. This will probably have a higher value of loss factor $(\tan\delta_r)/\mu$ so the design approach starting from core temperatures will, in fact, result in a higher number of turns.

However, the number of turns required to obtain the specified inductance will be less than for the previous design because the permeability is higher. In principle it would be possible to balance the design so that the specified inductance and the specified core temperatures were simultaneously achieved, giving the most economical use of the core material.

Sometimes, for the larger transformers, it may be necessary to stack E cores or U cores side-by-side to obtain sufficient cross-section. This may cause excessive temperature rise in the centre of the stack. In such a case, cooling ducts could be provided by spacing the cores in the stack thus allowing convective cooling of the inner sections of the core. If oil cooling is used, the spacing should not be less than 3 mm (see Section 9.2.3.3).

So far all the core loss has been assumed as magnetic in origin. Where there are high frequency, high voltage terminals adjacent to the ferrite the intense electric fields can give rise to high dielectric losses in the core. The resultant local heating may fracture the ferrite. The remedy is to screen those parts of the core that are exposed to intense h.f. electric fields. The screens should be applied in such a way as to avoid excessive interception of the magnetic fields, e.g. a cylindrical core might be covered by an earthed screen which is divided into narrow strips parallel to the axis. In the application of such screens the cooling of the core must not be impeded and the clearance to the winding must be adequate to prevent excessive capacitance or the chance of voltage breakdown.

Summarizing the problem of core selection for the higher frequency transformer, the temperature limitations will narrow the choice to cores of particular cross sectional dimensions and the inductance considerations may in addition determine the core factor C_1. The final selection will depend on the winding; in particular the core must provide sufficient winding breadth. Some of the aspects of winding design will be considered in the next section.

9.5.3 Windings

The foregoing section considered the problem of core selection for transformers operating in the higher frequency range and, in doing so, anticipated some of the results concerning the a.c. resistance of windings derived in Section 11.4.4. It was shown that if, as is often the case, the fractional increase in winding resistance due to eddy currents, F_R ($= R_{AC}/R_{dc}$), can be estimated in advance, then Eqns (9.54)–(9.58) may be used to evaluate the parameter k_4 on which an initial core selection may be based.

Assuming that a suitable core has been selected and the required numbers of turns have been calculated, this section considers the practical design of optimized windings, i.e. windings that have minimum a.c. resistance. It draws on the general principles set out in Section 11.4 and in particular the optimization criteria in Section 11.4.4. Reference should be made to these sections for the derivation of the design equations and the assumptions that qualify the results.

Sometimes optimization procedures may not provide a clear choice between alternative winding arrangements. In such cases the alternatives can be assessed by evaluating the a.c. resistances using Figure 11.14 and the parameters of the Dowell analysis described in Section 11.4.4.2. The lower, more relevant, region of Figure 11.14 is reproduced, with an expanded scale, in Figure 9.20.

Before considering a procedure for winding optimization, it is instructive to illustrate the method of calculating the a.c. resistance of a given set of windings by applying the Dowell method and Figures 9.20 or 11.14 to a specific example.

The low-frequency transformer example worked in Section 9.4.5 will be used. The simple winding arrangement shown in Figure 11.10(a) will be assumed, so each winding consists of one portion having p layers. The low frequency design may be summarized as follows.

$I_1 = 0.6$ A, $N_1 = 140$, $d_1 = 0.4$ mm, $R_1 = 1.49\,\Omega$,
$N_l = 28$, $p = 5$

$I_2 = 2.18$ A, $N_2 = 39$, $d_2 = 0.8$ mm, $R_2 = 0.104\,\Omega$,
$N_l = 13$, $p = 3$

From Table 11.2, the penetration depth Δ for a frequency of 25 kHz and a temperature of 100°C is $75(25\,000)^{-1/2} = 0.474$ mm.
The core type is ETD34; from Table 9.3, $b_w = 12.9$ assuming that an 8 mm creepage allowance is required.

Primary
The conductor height $h\ (=b) = 0.886 \times 0.4 = 0.354$ mm (see Figure 11.13).
The layer copper factor, $F_l\ (=N_l b/b_w)$
$= 28 \times 0.354/12.9 = 0.768$.
$\varphi = (h\sqrt{F_l})/\Delta = 0.354 \times 0.877/0.474 = 0.655$.
From Figure 9.20, for $\varphi = 0.655$ and $p = 5$, $F_R = 1.5$.
Therefore $R_{ac} = 1.49 \times 1.5 = 2.24\,\Omega$.

Secondary
$h = 0.886 \times 0.8 = 0.709$ mm.
$F_l = 13 \times 0.709/12.9 = 0.715$.
$\varphi = 0.709 \times 0.845/0.474 = 1.26$.
The value of F_R corresponding to $\varphi = 1.26$ and $p = 3$ lies outside the range of Figure 9.20 but Figure 11.14 gives $F_R = 3.24$.
Therefore $R_{AC} = 0.104 \times 3.24 = 0.337\,\Omega$.

This result, of course, completely invalidates the original design because in that design the core selection deliberately ignored eddy current losses in the windings. Some improvement would result from rearranging the windings into the sandwich configuration shown in Figure 11.10(b). However, while the secondary may be divided into two portions each having $p = 1\frac{1}{2}$, the two primary portions (on either side of the secondary) must have integral numbers of layers. As the existing primary design has five layers the primary winding cannot therefore be simply divided. This illustrates a practical constraint in the design of sandwich windings; the total number of layers in the outer windings must be even. Although a redesign of this transformer with six primary layers and two secondary layers would give a lower overall a.c. winding resistance, there is no possibility of meeting the specification with the size of core selected on the basis of d.c. resistance. However, the above example does illustrate the ease with which trial winding configurations may be evaluated.

As a preliminary to considering the practical aspects of winding optimization, it is important to note that the analyses set out in Section 11.4.4 depend on the assumption that the leakage flux in the winding space lies wholly parallel to the winding interface(s). The mutual repulsion of the leakage flux on either side of an interface (see Figure 11.20) ensures that the flux is indeed parallel in the vicinity of an interface.

However, over the winding area in general there is, in reality, considerable departure from parallelism. Figure 11.11 shows computer derived flux plots for the more important configurations. Two important conclusions may be drawn from these plots. The first is that, while it is well established that the theoretical predictions in Section 11.4.4 are sufficiently accurate for most practical design purposes, they should not be

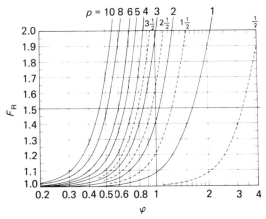

Figure 9.20 Resistance factor F_R as a function of $\varphi = (h\sqrt{F_l})/\Delta$ with the number of layers per portion p as a parameter

pursued into the realms of high resolution. The second conclusion is that the leakage flux is much more parallel in a sandwich winding, i.e. a winding located between the split halves of the associated winding as in Figure 11.10(b). Therefore if a winding design leads to a low voltage (high current) winding consisting of tape, strip or parallel connected conductors then this winding should be sandwiched between the equal halves of the higher voltage winding. This will minimize any additional eddy currents in such windings due to the components of the leakage field that are perpendicular to the surface of the winding.

A practical approach to optimum winding design can now be outlined. Reference 72 gives a more detailed procedure. For convenience, the more important equations in Section 11.4.4 will be repeated here with Chapter 9 numbering and dual references.

Assuming that the guide lines and data given in the earlier sections of this chapter have resulted in the selection of a suitable core, and the required numbers of turns have been calculated, then N, b_w and h_w are known. The penetration depth corresponding to the operating frequency may be obtained from Eqn (11.15), Table 11.2 or Figure 11.7.

If either of the main power windings require a relatively few turns, commensurate with only one or two layers, then a sandwich arrangement may be assumed at the outset. In this case, as noted in the foregoing example, the other winding must be designed to have an even number of total layers so that it can be divided into equal portions.

However, for the purpose of following an orderly development of the design process, it will be supposed initially that the windings are simply arranged, one over the other, as in Figure 11.10(a) and that they are composed of close-packed round wires. Each winding then constitutes a winding portion for which, if p is 2 or more, the optimum conductor diameter may be obtained from Eqn (9.71), (11.36).

$$d_{opt} = 2.6 \left(\frac{b_w}{fN}\right)^{1/3} \text{ mm} \qquad (9.71)$$

where b_w is in mm and f is in kHz.

If the transformer has multiple secondary windings, then the actual m.m.f. diagram (see Figures 11.10 and 11.12) must be established and a value of ϑ obtained for each secondary. For windings where $\vartheta \neq 0$, the optimum conductor diameter is given by Eqn (9.72), (11.37).

$$d_{opt} = 2.6 \left(\frac{b_w}{fN}\right)^{1/3} \times \frac{(1-\vartheta)^{1/2}}{(1-\vartheta^3)^{1/6}} \text{ mm} \qquad (9.72)$$

where the units are the same as in Eqn (9.71) and N is the number of turns between B_{min} and B_{max} in Figure 11.12.

A standard wire, having a bare diameter nearest to d_{opt} and a suitable grade of insulation, may be selected from wire tables such as those in Appendix B. From these tables, the overall diameter, d_o, and the resistance per meter, R_c, are noted.

If $p < 2$, then the above expressions for d_{opt} are invalid and the wire diameter should be the largest that can be accommodated within the available winding breadth.

For each winding, a trial number of turns per layer, N_l ($= b_w/d_o$), and the number of layers per portion, p ($= N/N_l$) may now be obtained. It is emphasized in Section 11.4.4 that for the assumption of parallel flux to be reasonably true, all winding portions and all layers should have the same winding breadth, b_w, and this should be the maximum that is available (taking account of creepage distance if required). Therefore the values of N_l and p should now be adjusted so that for each winding, p has a minimum integral value and N_l is the same (within one turn) for each layer. Generally this will require a minor adjustment to the number of turns. Preferably this should be a small increase to avoid raising the flux density in the core; the disturbance to the turns ratio should be minimal. As a result of this adjustment, the pitch of the turns in the layer will generally be somewhat larger than d_o, i.e. the turns will be spaced. This will tend to reduce the a.c. resistance since F_l is then reduced.

Having provisionally decided on a winding design based on optimum choice of conductor cross-section, the next step is to calculate the total build-up of the winding, allowing for insulation, screens and any auxiliary windings that may be required. Where the core is of rectangular cross-section allowance must be made for any bowing of the winding, see Figure 11.4.

If the windings can be adequately accommodated within the winding height available (h_w in Table 9.3) then all that remains is to check the design by using Figure 9.20 as previously described or by applying Eqn (9.73), (11.38), which is introduced below. This check is to ensure that the resulting values of F_R are near optimum (i.e. 1.5) and that the total a.c. resistance (referred to one winding) is within the limits imposed by the winding loss requirements. This involves the calculation of the d.c. resistances using Eqn (11.9). For initial designs, it is usual to use the mean turn length, l_w, quoted for the coil former (see Table 9.3) for each winding, the errors generally cancelling in the final power summation. If in practice F_R is very different for the different windings then it may be advisable to use mean turn lengths evaluated separately for each winding. The use of such specific mean turn lengths would be essential for definitive designs, where manufacturing limits must be given.

For windings where $\vartheta \neq 0$, Figure 9.20 does not apply, so having obtained d_{opt} from Eqn (9.72), the resistance factor F_R should be derived from Eqn (9.73), (11.38).

If windings having optimum or near-optimum conductor cross-section cannot be accommodated but the resistances of such windings would be well within limits, then conductors of smaller cross-section should be considered. In winding portions consisting of round wire, the criterion should be to decrease the wire diameter by an amount that, if possible, will lead to a reduction in the number of (whole) layers since this will result in a stepwise reduction in both F_R and winding height. The need to allow creepage distance (see Figure 9.17), if applicable, is clearly a significant disadvantage in attempting to obtain low a.c. resistance because it not only reduces the available winding space but it does so by reducing b_w and therefore increasing the number of layers.

Having obtained new values of d that allow the windings to be accommodated, new values of F_R may be obtained (for winding portions where $p \geq 2$) from Eqn (9.73), (11.38):

$$F_R = 1 + \tfrac{1}{2}\left(\frac{d}{d_{opt}}\right)^6 \qquad (9.73)$$

In fact this equation can be used to estimate by how much the wire diameters may be reduced while keeping the a.c. resistances within limits.

For winding portions where $p < 2$ (for a simple two-winding arrangement p would therefore equal unity), the best conductor diameter is, as previously stated, the largest that can be accommodated. F_R would then be obtained from Figure 9.20, using $\varphi = (h\sqrt{F_l})/\Delta$. Alternatively, for close-packed layers of round wire operating at a temperature of 100°C, Eqn (9.74), (11.49), may be applied:

$$\varphi = 0.352 d \sqrt{(fd/d_o)} \qquad (9.74)$$

where the diameters are in mm and f is in kHz. In general, this equation may be used also for spaced windings, in which case d_o becomes the pitch of the turns.

If it is found that there is ample room for windings of optimum conductors but the resistances are too high, a sandwich configuration should be tried. Assuming that the winding that is to provide the outer portions has an even number of layers, the reconfiguration will halve the number of turns and layers per portion, thus increasing the optimum conductor diameters and therefore reducing the a.c. resistances at the expense of increased winding height.

Also, as indicated at the beginning of this account of design procedure, the sandwich arrangement will probably be the best solution whenever one of the power windings has few turns resulting in less than four layers, i.e. $p < 2$.

If, in a sandwich winding, round wire is to be used and $p < 2$ then, as for the simple winding, the wire diameter should be made as large as possible and the paragraph leading to Eqn (9.74) applies.

When the required conductor cross-section becomes large, it will usually be preferable to use tape or strip conductors rather than round wire. In the case of tape, where the layer may consist of N_l turns of width b, the cross-section is $b \times h$, see Figure 11.13(a). The tape width should be made as large as the layer width will allow, the best compromise between b, N_l and p being obtained by trial. The extreme option is $b = b_w$; a split or sandwich winding will then consist of a total of $N = 2p$ spiral turns of full-width strip.

For strip conductors of width b_w, the optimum thickness when $p > 2$ is given by Eqn (9.75), (11.44),

$$h_{opt} = \frac{3.14}{\sqrt{(fp)}} \text{ mm} \qquad (9.75)$$

where f is in kHz

The optimum strip thickness when $p \leq 2$ is given in Table 9.5 together with the corresponding values of φ and F_R. These results were obtained by computation from Dowell's basic equations, see Section 11.4.4.2.

For a winding of tape of width b, having N_l turns in a layer of width b_w, the h_{opt} values should be divided by $\sqrt{F_l} = (N_l b/b_w)^{1/2}$; generally this will be near enough to unity to be ignored.

Proceeding as for the simple winding arrangement, the winding height should be checked, using d_{opt} for the windings of round wire and h_{opt} for tape or strip windings. In the case of spiral strip, each turn will be interleaved by insulation and this must be taken into account; it can result in a poor copper factor. If the winding build-up exceeds h_w, then reduced conductor cross-sections should be considered. The increase in resistance may be estimated, for round wires, from Eqn (9.73) and from Eqn (9.76), (11.45), for tape or strip.

$$F_R = 1 + \tfrac{1}{3}\left(\frac{h}{h_{opt}}\right)^4 \text{ provided } \begin{cases} h/h_{opt} < 1.4 \\ \text{and } p > 2 \end{cases} \qquad (9.76)$$

$$= 1.33 \text{ when } h = h_{opt}$$

The d.c. resistance of full-width spiral strip is given by

$$R_{dc} = \frac{\rho_c N l_w}{h b_w}$$

$$= 22.5 \times 10^{-6} \frac{N l_w}{h b_w} \qquad (9.77)$$

when the dimensions are in mm and the resistivity of copper at 100°C is used.

When a winding consists of only very few turns, the length of conductor between the actual turns and the terminals of the transformer, i.e. the 'leadouts', can contribute an appreciable addition to the calculated winding resistance. This would be particularly im-

Table 9.5 Optimum parameters for tape and strip when $p \leqslant 2$

p	2	$1\frac{1}{2}$	1	$\frac{1}{2}$
φ	0.96	1.14	1.57	3.14
F_R	1.35	1.37	1.44	1.44
h_{opt}	$2.28/f^{\frac{1}{2}}$	$2.70/f^{\frac{1}{2}}$	$3.72/f^{\frac{1}{2}}$	$7.45/f^{\frac{1}{2}}$

Notes (1) h_{opt} is in mm when f is in kHz; it is the physical thickness of the strip, i.e. twice $h/2$ in the case of $p = \frac{1}{2}$.
(2) Eqn (11.43) gives h_{opt} for a strip winding when $p > 2$ and $\theta \neq 0$; this of course does not apply to a sandwich winding.

portant, for example, in the case of a spiral winding consisting of 2 or 3 turns terminated by folding the ends at right angles. Inasmuch as it is possible to isolate these leadout sections from the bulk of the transformer assembly by allowing airspace around them, their contribution to the transformer temperature rise may be minimized; their contribution to the winding loss is unavoidable and must be taken into account when the voltage drop is being estimated.

Returning to the general consideration of high current windings consisting of relatively few turns, other winding constructions are possible. A tape winding may be simulated by using a bi- or trifilar winding of parallel wires connected in parallel.[73] This option should only be used in a sandwich winding, where the number of layers per portion will probably be less than two. In that case, the optimum wire diameter is the maximum that can be accommodated within the available winding breadth. F_R may be obtained from Figure 9.20 as for single round wire, using, for example, Eqn (9.74) to obtain φ. R_{dc} is the dc resistance of the parallel connection of the two or three strands of wire. The copper factor will be inferior to that of tape but the winding operation may be a little easier.

Another possibility, which need not be confined to sandwich windings, is to use bunched conductors, see Section 11.4.3. Because bunched conductors become very expensive as the number of strands increases and the strand diameter is reduced, the choice will probably be confined to rather few strands of relatively large diameter, contrasting with the usage for low-loss inductors, see Section 5.7.3. The optimum strand diameter will be given by Eqn (9.71) when N is replaced by nN, where n is the number of strands (cp Eqn (5.22)). The approximate overall diameter and the resistance per unit length is obtainable from wire tables such as those in Appendix B, Tables B.4. Although bunched conductors have rather poor copper factors they can under some conditions provide the lowest loss windings and they are easy to apply; they may be the best solution when performance has priority over cost.

Finally it should be noted that eddy current loss will occur in any non-conducting or non-power winding that is exposed to leakage field. Such a situation can occur in SMPS transformers where a winding may be inactive for part of the switching cycle, or where a winding is providing a low-power auxiliary function. In terms of power, the loss (assuming round wire) is expressed by Eqn (11.29). The value of $\bar{B}^2$ now depends on the leakage field due to the other windings and must be evaluated from the actual m.m.f. diagram that applies to the particular situation (see, for example, Figure 11.12 and the method used to obtain Eqn (11.27)). Often it will become apparent that a non-power winding is exposed to little or no leakage field in which case it can be ignored. Where applicable, allowance must be made for the proportion of the switching period for which a winding is inactive. Any significant loss of power due to eddy currents in non-active windings should be included in the summation of the total winding losses

If, in the space available, it is not possible to hold the total winding loss within the required limits with any practical winding arrangement, then the initial core selection was at fault. Either a larger core must be used or the operating frequency raised.

This outline of winding design procedure should be adequate for most applications. Jongsma gives a more comprehensive, step-by-step, procedure in Reference 73 (see also Reference 72).

Other winding properties such as self capacitance and leakage inductance may be calculated from the expressions given in Sections 11.6 and 11.7 respectively. Concerning leakage inductance at high frequencies, it should be remembered that the current flows on the conductor surface nearest the winding interface, i.e. nearest the maximum m.m.f. surface (see Figure 11.10). When there is only one or very few layers this effect reduces the leakage inductance. It may easily be taken into account by making the dimensions Σx and Σx_Δ in Figure 11.21 apply to the current configuration rather than to that of the conductors. Reference 74 includes formulae for the precise evaluation of the effect of eddy currents on leakage inductance.

Often there will be a need for screens between windings. When these are exposed to significant leakage fields, eddy current losses will be induced. Obviously, screens should be made as thin as possible. Reference 73 recommends the use of high resistivity metal foil and suggests phosphor bronze which has a resistivity of $(10 \text{ to } 16) \times 10^{-8} \, \Omega\text{m}$ at $100°\text{C}$. The overlap of the ends of the screen should be minimized to avoid the effects of capacitive currents in the screen.

Finally the specification of the winding and insulation should be carefully considered in relation to the operating environment, with particular attention to the temperature, frequency and voltages.

9.6 References and bibliography

Section 9.1

1. BOULTON, P. G., GANDER, M. C. and GENT, D. Transistor 90° colour line deflection and e.h t. circuit, *Mullard Tech. Commun.* No. 112, 44, Oct. (1971)
2. BARROW, C. Cores for switched-mode power supply transformers, *Mullard tech. Commun.* No. 124, 170, Oct. (1974)
3. ROESS, E. Soft magnetic ferrites and applications in telecommunication and power converters, *I.E.E.E Trans. Magn.*, **MAG-18**, 1529 (1982)
4. SNELLING, E. C. and GILES, A. D. *Ferrites for inductors and transformers*, pp. 158–160, Research Studies Press, Letchworth, Hertfordshire, England; distrib. by John Wiley & Sons, New York (1983)
5. GOLDMAN, A. Competing nonferrite materials for switched-mode power supplies, *Fourth Int. Conference on Ferrites, Part II*, edited by F. F. Y. Wang, *Advances in Ceramics*, **16**, 421 (1985)
   MACFADYAN, K. A. *Small transformers and inductors.* Chapman and Hall, London (1953)
   GROSSNER, N. R. *Transformers for electronic circuits*, 2nd Edn, McGraw-Hill Book Co., New York (1983)
   BRISBANE, J. A. and THOMPSON, J. W. The use of ferrite-cored transformers in high power broadcast transmitters, *Int. Broadcasting Convention, I.E.E. Conf, Publication*, 159, I.E.E., London (1970)

Section 9.2.1

6. KING, W. J. The basic laws and data of heat transfer, *Mech. Engrg*, **54**, 190 (Part 1), 275 (Part 2), 347 (Part 3), 410 (Part 4) and 492 (Part 5) (1932)
7. KARLEKAR, B. V. and DESMOND, R. M. *Heat Transfer*, 2nd Edn, West Publishing Co., St. Paul, Minnesota, USA (1977)
8. PITTS, D. R. and SISSON, L. E. *Theory and problems of heat transfer*, McGraw-Hill Book Co., New York (1977)
9. GROSSNER, N. R. *Transformers for electronic circuits*, 2nd Edn, Chap. 4, McGraw-Hill Book Co., New York (1983)
10. POLAK, S. J., SCHROOTEN, J. and BARNVELD BINKHUYSEN, C. TEDDY 2, a program package for parabolic composite region problems, *ACM Trans. on Math. Software*, 4, 209 (1978)
11. TURNER, W. D., ELROD, D. C. and SIMAN-TOV, I. I. HEATING-5, An IBM360 heat conduction program, *ORNL/CSD/TM-15*, March (1977) (*see also* NEA Data Bank Abstract 0517)
12. GROSSNER, N. R. Analysis of transient loading and heating of the electronic transformer, *Trans. I.E.E.E*, **PMP-3**, 30 (1967)

Section 9.2.2

13. KING, W. J. *loc cit.*, Table 1, p. 494

Section 9.2.3

14. BLUME, L. F., BOYAJIAN, A., CAMILLI, G., LENNOX, T. C., MINNECI, S. and MONTSINGER, V. M. *Transformer Engineering*, John Wiley & Sons, New York (1951)
15. KING, W. J. *loc cit.*, p. 351
16. KING, W. J. *loc cit.*, p. 350
17. KING, W. J. *loc cit.*, p. 351
18. KING, W. J. *loc cit.*, p. 426
19. KING, W. J. *loc cit.*, p. 353

Section 9.2.6

20. BRACKE, L. P. M. Optimizing the power density of ferrite-cored transformers, *Powerconvers. Int.*, **8**, 19 (1982)
21. HESS, J. and ZENGER, M. The influence of material properties of ferrite cores on the temperature rise of transformers, *Fourth Int. Conference on Ferrites, Part II*, edited by F. F. Y. Wang, *Advances in Ceramics*, **16**, 501 (1985)
22. BRACKE, L. P. M. High-frequency ferrite power transformer and choke design, Part 2; Switched-mode power supply magnetic considerations and core selection, *Philips Elcoma Publication*, No. 9398 923 30011, Sept. (1982)
23. ROESPEL, G. and ZENGER, M. Ferrite core types for power electronics, *Components Rep.*, **14**, 209 (1979)

Section 9.3

24. ANNIS, A. D. Philips Research Laboratories, Redhill, Surrey, England, unpublished report (1979)
25. SNELLING, E. C. and GILES, A. D. *Ferrites for inductors and transformers*, p. 136, Research Studies Press, Letchworth, Hertfordshire, England; distrib. by John Wiley & Sons, New York (1983)
26. *Mullard Technical Handbook*, Book 3, Part 2a, Mullard, London (1986)

Section 9.4.4

27. JANSSON, L. E. Power-handling capability of ferrite transformers and chokes for switched-mode power supplies, *Mullard Tech. Note*, No. 31, Mullard, London (1975)

Section 9.4.3

28. HANNA, C. R. Design of reactances and transformers which carry direct current, *J. Am. Inst. elect. Engrs*, **46**, 128 (1927)
29. D.C. choke core selection for SMPS, *Mullard Technical Publication*, No. M81-0035, Mullard, London (1981)
30. JONGSMA, J. and BRACKE, L. P. M. Improved method of power-choke design, *Electronic Components and Applications*, **4**, 66 (1982)
31. *Philips Elcoma Data Handbook*, C5 (1986)
32. ROESSLER, W. Non-linear storage chokes improve operating performance of switched-mode power supplies, *Siemens Components*, **20**, 24 (1985)
33. SIBILLE, R. Smoothing choke materials for switched mode power supplies, *J. Magn. and Magn. Mater.*, **26**, 281 (1982)
34. RUSCHMEYER, K. and WIETERS, H. Stufendrosseln mit weichmagnetischen Einlagen im Luftspalt, *Valvo Berichte*, Hamburg, 10, Nov (1986)

35. The application of Ferroxdure for premagnetizing cores in power carrying coils, *Matronics*, No. 9, Philips Tech. Inf. Bull. (Elcoma), Eindhoven (1955)
36. BRUIJNING, H. G. and RADEMAKERS, A. Pre-magnetization of the core of a pulse transformer by means of Ferroxdure, *Philips tech. Rev.*, **19**, 28 (1957/58)
37. SIBILLE, R. and BEUZELIN, P. New ferrite and new cores for switched mode power supplies, *Powerconvers. Int.*, **8**, 32 (1982)
38. RUSCHMEYER, K. Höher ausnutzbare Spulen durch permanentmagnetische Vormagnetisierung, *Valvo Berichte*, Hamburg, 1, Nov (1986)

Section 9.4.4
39. OCHIAI, T. and OKUTANI, K. Ferrites for high-frequency power supplies, *Fourth Int. Conference on Ferrites, Part II*, edited by F. F. Y. Wang, *Advances in Ceramics*, **16**, 447 (1985)
40. MOCHIZUKI, T., SASAKI, I. and TORRI, M. Mn-Zn ferrite for 400–600 kHz switching power supplies, *ibid.*, 487 (1985)
41. STIJNTJES, TH. G. W. and ROELOFSMA, J. J. Low-loss power ferrites for frequencies up to 500 kHz, *ibid.*, 493 (1985)
42. KAMADA, A. and SUZUKI, K. Optimum ferrite core characteristics for a 500 kHz switching mode converter transformer, *ibid.*, 507 (1985)
43. ZENGER, M. and ROESPEL, G. Ferrite core types for power electronics', *Components Report*, **14**, No. 5 (1979)
44. Isolating transformers and safety isolating transformers: Requirements, *Int. Electrotechnical Commission, Publication 742*, Geneva (1983)
45. Safety of data processing equipment, *Int. Electrotechnical Commission, Publication 435*, Geneva (1983)
46. Safety of information technology equipment including electrical business equipment, *Int. Electrotechnical Commission, Publication 950*, Geneva (1986)
47. Cores of soft magnetic ferrite; E-cores, *German Standard DIN 41 295*, Berlin, May (1984)
48. Dimensions of pot-cores made of ferromagnetic oxides and associated parts *Int. Electrotechnical Commission, Publication 133*, Geneva (1985)
49. Dimensions of pot-cores made of ferromagnetic oxides for use in telecommunications and allied electronic equipment, *BS4061*, British Standards Institution, London (1966)
50. Pot cores of soft magnetic ferrite, *German Standard DIN 41 293*, Berlin, August (1971)
51. Dimensions of square cores (RM-cores) made of magnetic oxides and associated parts, *Int. Electrotechnical Commission, Publication 431*, Geneva (1983)
52. Cores of soft magnetic ferrites; RM-cores, *German Standard DIN 41 980*, Berlin, February (1982)
53. Dimensions for magnetic oxide cores intended for use in power supplies (EC-cores), *Int. Electrotechnical Commission, Publication 647*, Geneva (1979)
54. Cores made of soft magnetic oxides; PM-cores, *German Standard DIN 41 989*, Berlin, August (1985)
55. Dimensions of magnetic oxide cores intended for use in power supply applications, *Int. Electrotechnical Commission*, Geneva, to be published

Section 9.5.2
56. JANSSON, L. E. A survey of converter circuits for switched-mode power supplies, *Mullard Tech. Commun.*, **12**, No. 119, 271 (1973)
57. BURGUM, F. Switched-mode power supply transformer design nomograms, *Philips Elcoma tech Information*, 005, Dec. (1975)
58. Transformer core selection for SMPS, *Mullard tech. Publication*, M81-0032 (1981)
59. CATTERMOLE, P. A. Optimizing flyback transformer design, *Powerconvers. Int.*, **7**, 74 (1981)
60. BRACKE, L. M. P. and GEERLINGS, F. C. High-frequency ferrite power transformer and choke design, Pt. 1, Switched-mode power supply magnetic component requirements, *Philips Elcoma Publication*, 9398 923 20011, Sept. (1982)
61. BRACKE, L. M. P. High-frequency ferrite power transformer and choke design, Pt. 2, Switched-mode power supply magnetic considerations and core selection, *Philips Elcoma Publication*, 9398 923 30011, Sept. (1982)
62. JANSSON, L. E. and HARPER, D. J. The application of POWERMOS transistors in switched-mode power supplies, *Proc. 5th Int. PCI Conf., Geneva*, 86 (1982)
63. 44 to 80 V d.c. input 5 V 100 W output SMPS using a BUZ36 POWERMOS transistor and ETD cores, *Mullard tech. Publication*, March (1984)
64. 100 W mains-input SMPS using BUT11 bipolar transistors and ETD cores, *Mullard tech. Publication*, July (1984)
65. COONROD, N. R. Transformer computer design aid for higher frequency switching power supplies, *PESC '84 Record*, 257 (1984)
66. Mains-input 5 V 100 W output SMPS using a BUZ80 POWERMOS transistor and ETD cores, *Mullard tech. Publication*, Jan. (1986)
67. SNELLING, E. C. and GILES, A. D. *Ferrites for inductors and transformers*, Section 5.3.6, Research Studies Press, Letchworth, Hertfordshire, England; distrib. by John Wiley & Sons, New York (1983)
68. ZENGER, M. Ferrites and core forms for power electronics at frequencies up to 1 MHz, *Siemens Components*, **17**, 113 (1982)
69. LOEHN, K. Modelling and optimization of inductive power components for high switching frequencies, *Proc. Int. Macroelectronics Conf.*, Munich, 239, Nov. (1982)
70. BUETHKER, C. and HARPER, D. J. Improved ferrite materials for high frequency power supplies, *Proc. High Freq. Power Convers. Conf., Virginia Springs*, 186 (1986)
71. BRISBANE, J. A. and THOMPSON, J. W. The use of ferrite cored transformers in high power broadcast transmitters, *Int. Broadcasting Convention, I.E.E. Conf, Publication*, 159, I.E.E., London (1970)

Section 9.5.3
72. JONGSMA, J. Minimum loss transformer windings for ultrasonic frequencies, *Electron. Appl. Bull.*, **35**, No. 3, 146, and No. 4, 221 (1978)
73. JONGSMA, J. High-frequency ferrite power transformer and choke design; Pt. 3, Transformer winding design, *Philips Elcoma Publication*, 9398 923 40011, Sept. (1982)
74. DOWELL, P. L. Effects of eddy currents in transformer windings, *Proc. Inst. elect. Engrs*, **113**, 1387 (1966)

10 Ferrite antennas

10.1 Introduction

As domestic radio receivers became more portable and self-contained there was a trend towards the use of built-in antennas. At first these were frame windings consisting of a number of turns enclosing the largest available area within the dimensions of the receiver. The received signal was the e.m.f. induced in the winding by the magnetic component of the electromagnetic field. As receivers became smaller it became increasingly difficult to achieve sufficient sensitivity with frame windings. The difficulty was relieved by the introduction of ferrite cores. The dimensions of the winding could then be reduced to those of a small r.f. tuning coil and it could be provided with a high permeability core which could magnify by at least a hundred times the actual area enclosed by the winding. In this way sufficient sensitivity could be obtained within the dimensions of a very small receiver. Almost all a.m. broadcast receivers now manufactured are provided with ferrite antennas; indeed this is true not only of personal receivers but also of larger systems such as music centres where the ferrite antenna is a convenient means of avoiding external antenna connections.

The use of ferrite antennas to replace the telescopic aerial in small v.h.f. receivers is an attractive proposition but it has so far proved impracticable. The appropriate NiZn ferrites for operation at 100 MHz have low permeability and relatively high losses so in order to obtain a useful performance the required volume of ferrite is unacceptably high for use in a personal receiver. The criteria for such antennas are not considered here; Reference 1 gives analytical and design details.

The theory of cylindrical ferrite cores is considered in detail in Section 4.3. The basic relations are established and data concerning the magnetization of cylinders are given. These results will now be applied to the use of ferrite cores specifically for antennas.

In practice the majority of antenna cores are cylindrical, but the dimensions of the receiver may be such that other shapes, notably a slab or plate, may be preferable. Although the following treatment is based on the use of cylindrical cores it applies in principle to cores of any cross-section. This is also true of the definitions of μ_{rod} and μ_{coil}, but the data concerning these properties given in Section 4.3 will not in general apply to cores having cross-sections departing appreciably from circular.

10.2 Antenna circuit theory

The approach used in this section is based on that used by Maanders and van der Vleuten[2].

10.2.1 Induced e.m.f., effective height

If a short circular winding of N turns enclosing an effective area A_N is placed in a uniform alternating magnetic field with the axis of the winding parallel to the field strength vector H, then the e.m.f., E_s, induced in it will, by the induction equation, be:

$$E_s = \mu_o \omega H A_N N \quad \text{V} \qquad (10.1)$$

If the aperture of the winding is now filled with a long ferrite cylinder coaxial with the winding and having its centre coinciding with the centre of the winding, then the flux density at the centre of the winding will, by the definition given in Section 4.3.1, be increased by the factor μ_{rod}. If the cross-sectional area of the rod is A then the total flux linking the winding will be $\mu_o \mu_{rod} H A$ (neglecting the air flux that passes through any area difference between A_N and A). Therefore

$$E_s = \mu_o \mu_{rod} \omega H A N \quad \text{V} \qquad (10.2)$$

The strength of an electromagnetic field is usually expressed in terms of the electric field strength E. Using the fundamental relation:

$$H = \sqrt{\left(\frac{\varepsilon_o}{\mu_o}\right)} E \quad \text{A m}^{-1}$$

where E is in V m^{-1}

(10.1) $E_s = \omega H A N 10^{-8}$ V

(10.2) $E_s = \mu_{rod} \omega H A N 10^{-8}$ V

$$\mu_o H = \sqrt{(\mu_o \varepsilon_o)} \, E = \frac{E}{c_o}$$

where $c_o = 1/\sqrt{(\mu_o \varepsilon_o)}$ = velocity of electromagnetic waves in vacuo $\simeq 3 \times 10^8$ m s^{-1}

This may be substituted into the previous equation:

$$E_s = \mu_{rod} \omega E A N / c_o \quad \text{V} \qquad (10.3)$$

The effective height, h_e, of an antenna is defined as the ratio of the induced e.m.f. to the electric field strength:

$$h_e = \frac{E_s}{E} \quad \text{m} \qquad (10.4)$$

Therefore for a ferrite rod antenna

$$h_e = \mu_{rod} \omega A N / c_o$$
$$= \mu_{rod} 2 \pi A N / \lambda \quad \text{m}$$

where λ is the wavelength in m

If the winding is not short compared with the rod or if it is not central on the rod, the induced e.m.f. will be less than the value given by Eqn (10.3). The reduction is proportional to the e.m.f. averaging factor F_A; this factor is defined in Section 4.3.2 and data are given which enable it to be estimated in certain cases. The foregoing equations may now be rewritten in a more general form:

$$E_s = \mu_{rod} \omega E A N F_A / c_o \quad \text{V} \qquad (10.5)$$

and $h_e = \mu_{rod} \omega A N F_A / c_o$
$= \mu_{rod} 2 \pi A N F_A / \lambda$ m $\qquad (10.6)$

10.2.2 Signal strength

The magnetic antenna described above behaves as a signal source having an e.m.f., E_s, and a reactance ωL. Associated with this reactance there will be a loss tangent, tan δ, which will arise from the radiation resistance and the losses in the winding and the core. The radiation resistance of a ferrite antenna is usually negligible compared with the losses. The estimation of losses will be considered in Section 10.3.3 but for the purpose of the present discussion it is sufficient to represent the losses, whatever their origin, by a shunt resistance, R_p, so that the unloaded Q-factor of the antenna is given by

$$Q = R_p / \omega L$$

The inductance of the antenna is usually resonated, at the frequency of the signal to be detected, by a capacitance, C. This resonant circuit is in general loaded by the first amplification stage of the receiver. Figure 10.1(a) shows the equivalent circuit; for convenience the loss element R_p is shown in parallel with

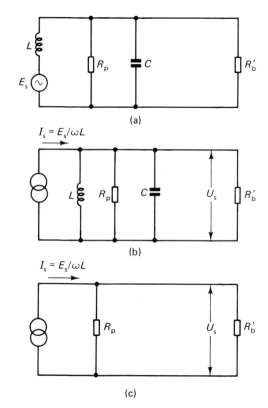

Figure 10.1 Equivalent circuits of a rod antenna connected to a signal source

C. There will, in most practical circuits, be an impedance transformation between the capacitance and the load and this is usually achieved by connecting the load to a tap on the winding. However, in the following discussion this transformation will be ignored and R'_b will be defined as the load resistance referred to the primary circuit. If the amplification stage is voltage operated the value of R'_b is high so the antenna operates substantially on open circuit. In a conventional circuit using bipolar semiconductors, R'_b is usually comparable to R_p and the circuit design is governed by power transfer, noise or bandwidth considerations.

The presence of the load lowers the Q-factor to a loaded value denoted by

$$Q_L = \frac{R_p R'_b}{\omega L (R_p + R'_b)} \qquad (10.7)$$

Thus

$$\frac{Q_L}{Q} = \frac{R'_b}{R_p + R'_b} \qquad (10.8)$$

The above circuit may be converted to the constant

current equivalent (see Figure 10.1(b)), by applying Thevenin's theorem. Since at resonance $\omega L = -(1/\omega C)$ the circuit reduces to that shown in Figure 10.1(c).

The signal voltage, U_s, developed across the load, R'_b, is given by

$$U_s = I_s \frac{R_p R'_b}{R_p + R'_b}$$

$$= \frac{E_s}{\omega L} \frac{R_p R'_b}{(R_p + R'_b)} = E_s Q_L \quad \text{V} \quad (10.9)$$

This result is rather obvious; the e.m.f. induced in the antenna is magnified by the value of the loaded Q-factor. Thus the voltage appearing across the load referred to the primary will be Q_L times the e.m.f. given by Eqn (10.5):

$$U_s = \mu_{rod} \omega E Q_L A N F_A / c_o = E h_e Q_L \quad \text{V} \quad (10.10)$$

The voltage appearing at the terminals of the first amplifier stage will of course differ from U_s if there is an impedance transformation.

If the first stage of amplification is voltage operated then the maximum sensitivity will be obtained when the r.h.s. of Eqn (10.10) is a maximum, i.e. when $h_e Q_L$ is maximum. For a given antenna loss, Q_L will be maximum when the circuit has negligible load ($R'_b \gg R_p$). If, on the other hand, the first stage is current operated, as in the case of a bipolar transistor, it is the power in the load that is of interest.

From Eqn (10.9)

$$P_s = \frac{U_s^2}{R'_b} = \frac{E_s^2}{\omega^2 L^2} \frac{R_p^2 R'_b}{(R_p + R'_b)^2} = \frac{E_s^2 Q R_p R'_b}{\omega L (R_p + R'_b)^2} \quad \text{W} \quad (10.11)$$

where P_s is the signal power delivered to the load. This will be a maximum when $R'_b = R_p$, i.e. when $Q_L = Q/2$. The maximum signal power that may be delivered to the load is:

$$(P_s)_{max} = \frac{E_s^2 Q}{4 \omega L} = \frac{E^2 h_e^2 Q}{4 \omega L}$$

In practice, bandwidth and signal/noise ratio considerations usually lead to $R'_b > R_p$. In either case the signal power is proportional to $h_e^2 Q$.

10.2.3 Noise power

Figure 10.2 shows a source, having resistance R_p, connected to an active 4-pole having input resistance R'_b referred to the primary side of any transformer that may be in the circuit. A noise generator having an e.m.f. of r.m.s. value E_n is shown connected in series with R_p to represent the thermal noise due to the source resistance. The noise voltage is given by

$$E_n = \sqrt{(4kT\Delta f R_p)} \quad \text{V} \quad (10.12)$$

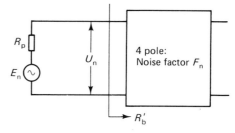

Figure 10.2 An active 4-pole connected to a noise source

where

k = Boltzmann's constant
 = 1.380×10^{-23} J K^{-1}
T = temperature in K
Δf = bandwith at the 3 dB points

If the 4-pole consisted of only a shunt resistor R'_b then the noise voltage U_n at the input terminals would simply correspond to that due to the thermal noise of a resistance equal to the parallel combination of R_p and R'_b:

$$U_n = \sqrt{\left(\frac{4kT\Delta f R_p R'_b}{R_p + R'_b}\right)} = E_n \sqrt{\left(\frac{R'_b}{R_p + R'_b}\right)} \quad (10.13)$$

However, as the 4-pole in practice is an active device, e.g. an amplifier, the input resistance R'_b itself is not regarded as a noise source. Instead the 4-pole is said to have a noise factor F_n which may be defined by:

$$F_n = \left(\frac{\text{Output noise power due to source and 4-pole}}{\text{Output noise power if source was only noise contribution}}\right) \quad (10.14)$$

This ratio may also be expressed in decibels. Because R'_b is not regarded as a source of noise, the input noise voltage, U_n, due to the source E_n is given by

$$U_n = \frac{E_n R'_b}{R_p + R'_b} \quad \text{V} \quad (10.15)$$

and the input noise power is

$$P_n = \frac{U_n^2}{R'_b} = \frac{E_n^2 R'_b}{(R_p + R'_b)^2} \quad \text{W} \quad (10.16)$$

If the 4-pole contributes no noise and has a gain G, the output noise power is

$$\frac{G E_n^2 R'_b}{(R_p + R'_b)^2} \quad \text{W} \quad (10.17)$$

The noise factor, F_n, is the factor by which the output noise power of an actual 4-pole exceeds this value.

It is interesting to note that if the only source of

noise in the 4-pole were a resistance corresponding to R'_b then from Eqn (10.13) the output noise power would be

$$\frac{GE_n^2}{R_p + R'_b}$$

and then

$$F_n = \frac{GE_n^2}{R_p + R'_b} \frac{(R_p + R'_b)^2}{GE_n^2 R'_b} = \frac{R_p + R'_b}{R'_b} \quad (10.18)$$

If, under these conditions, the source is matched to the 4-pole, i.e. $R_p = R'_b$, then $F_n = 2$ or alternatively this may be expressed as 3 dB.

In practice F_n may be greater or smaller than this value even if the input is matched since in an active 4-pole R'_b does not normally exist as a physical resistance and there will in general be gain and noise generation distributed in a complex manner throughout the network. For any active 4-pole, F_n is an approximately quadratic function of the source resistance R_p and has a minimum at a particular value of this resistance.

It follows from Eqns (10.14) and (10.17) that the noise power at the output of the 4-pole is

$$P_{no} = \frac{GE_n^2 R'_b}{(R_p + R'_b)^2} \times F_n$$

$$= GF_n \frac{4kT\Delta f R_p R'_b}{(R_p + R'_b)^2} \quad W \quad (10.19)$$

It has been assumed that the only noise arising from the antenna is due to the loss resistance R_p. There will of course be noise induced in the aerial from the noise components of the electromagnetic field (due to natural causes and man-made electromagnetic interference). However, this noise is transmitted through the antenna circuit in the same way as the signal, so its amplitude in relation to that of the signal is unaffected by the circuit design. Its only significance in the present context is that it can provide a measure of the relative importance of the thermal noise generated in the circuit.

10.2.4 Signal-to-noise ratio

It is readily seen from Eqn (10.11) that the signal power in the output of the 4-pole is

$$P_{so} = \frac{GE^2 h_e^2 Q R_p R'_b}{\omega L (R_p + R'_b)^2} \quad W \quad (10.20)$$

Therefore the signal-to-noise ratio is

$$\frac{P_{so}}{P_{no}} = \frac{E^2 h_e^2 Q}{4kT\Delta f F_n \omega L} \quad (10.21)$$

This expression applies to those stages of the receiver up to but not including the detector. The bandwidth Δf is the total bandwidth of the part of the receiver being considered.

After detection the signal-to-noise ratio may be shown to be[3]:

$$\left(\frac{P_{so}}{P_{no}}\right)_{af} = \frac{m^2 E^2 h_e^2 Q}{4kT\Delta f F_n \omega L} \quad (10.22)$$

where m is the modulation factor of the signal, and Δf is the effective noise bandwidth of the receiver. For an input signal/noise ratio greater than 4, this bandwidth may be taken as twice the audio frequency bandwidth.

For the maximum signal-to-noise ratio the antenna circuit should have the highest possible value of $h_e^2 Q/F_n$. The effective height is limited by the permissible size of the antenna, the noise factor may be minimized by proper choice of source resistance referred to the input of the 4-pole, and the unloaded Q-factor is limited by the bandwidth requirements or the impedance requirements or the losses in the antenna.

10.2.5 Bandwidth and impedance requirements

The bandwidth at the 3 dB points is given by the well-known relation:

$$\Delta f = \frac{f}{Q_L} \quad (10.23)$$

so from Eqn (10.8)

$$\Delta f = \frac{f(R_p + R'_b)}{QR'_b} \quad (10.24)$$

To obtain the best compromise between the various requirements, an impedance transformation is required between the antenna and the input of the 4-pole. This is shown in Figure 10.3, where the transformation, of voltage ratio r, is obtained by tapping into the antenna winding L (tight coupling is assumed). C is the resonating capacitor; the choice of a convenient value for C will usually determine the value of L and hence the number of primary turns. R_b is determined by the design of the 4-pole (the 4-pole being an amplifier or frequency changer), R'_p is the value of source impedance that will make the noise factor minimum and R_p is determined by the losses in antenna and the number of turns on the antenna winding.

Rewriting Eqn (10.8) for the secondary side of the transformer:

$$Q_L = \frac{QR_b}{r^2 R_p + R_b}$$

since $R_b = r^2 R'_b$

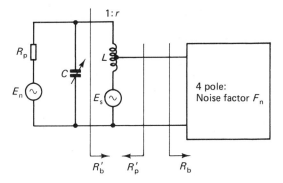

Figure 10.3 The equivalent circuit of a rod antenna having both signal and noise sources, and connected to an active 4-pole

$$\therefore Q_L = \frac{QR_b}{R'_p + R_b}$$

$$\therefore Q = \frac{Q_L(R'_p + R_b)}{R_b} \quad (10.25)$$

All the quantities on the r.h.s. of this equation are determined by bandwidth or impedance requirements, so the unloaded Q-factor may be calculated. From this value, R_p may be obtained:

$$R_p = \omega L Q \quad \Omega \quad (10.26)$$

where L is determined by the tuning requirements. Finally the transformation ratio is calculated:

$$r = \left(\frac{R'_p}{R_p}\right)^{1/2} \quad \text{assuming tight coupling.} \quad (10.27)$$

These conditions may be met, in general, only at one frequency since r cannot be varied with frequency within a given band. It is usual to obtain the best design at the middle of a broadcast band and accept some departure from optimum as the frequency departs from mid-band. For another broadcast band a separate antenna circuit is designed, the additional winding usually being placed on the same antenna rod.

10.2.6 A design example

To illustrate the antenna circuit theory, a basic design for a medium wave ferrite antenna will be derived. It is assumed that the broadcast band extends from 0.5 to 1.5 MHz and that at 1 MHz the bandwidth of the antenna circuit is 10 kHz. From Eqn (10.23),

$$Q_L = 100.$$

The 4-pole is assumed to be a mixer circuit having a typical h.f. bipolar transistor at the input. At 1 MHz a typical input conductance would be 0.9 mA/V and it is assumed that the noise factor has a minimum of 2.5 dB when the source conductance is 1.5 mA/V. Therefore $R_b = 1/0.009 = 1100 \, \Omega$ and for minimum noise $F_n = 1.8$ and $R'_p = 1/0.0015 = 670 \, \Omega$.

If the design were to to be based on minimum noise, then from Eqn (10.25) the required unloaded Q-factor would be

$$Q = 100 \frac{670 + 1100}{1100} = 161$$

From Eqn (10.26), assuming $L = 200 \, \mu H$

$$R_p = 6.28 \times 10^6 \times 200 \times 10^{-6} \times 161 \simeq 2 \times 10^5 \quad \Omega$$

Therefore from Eqn (10.27)

$$r = \left(\frac{670}{2 \times 10^5}\right)^{1/2} = 1/17.3$$

A better strategy would be to optimize the signal-to-noise ratio. Figure 10.4 shows a graph of F_n against source resistance (R'_p) for a typical h.f. bipolar transistor at a signal frequency of 1 MHz. R'_p and Q are related by Eqn (10.25); putting $Q_L = 100$ (10 kHz bandwidth) and $R_b = 1100 \, \Omega$, the values of Q corresponding to R'_p may be calculated and these have been marked off along the abscissa. Since the signal-to-noise ratio is proportional to Q/F_n (from Eqn (10.21)) this quantity has been plotted in the same figure.

For this example, Figure 10.4 shows that the maximum signal-to-noise ratio is obtained when the unloaded Q-factor is about 190. This, fortuitously, almost coincides with the condition for maximum

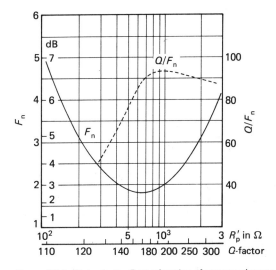

Figure 10.4 Noise factor F_n as a function of source resistance R'_p for a typical bipolar transistor operating as a mixer at 1 MHz. The abscissa has been scaled in terms of Q-factor using Eqn (10.25), putting $Q_L=100$ (10 kHz bandwidth at 1 MHz) and $R_b=1100 \, \Omega$. A curve of Q/F_n (proportional to signal-to-noise ratio) has been added

signal (i.e. that $Q = 2Q_L$) deduced from Eqn (10.11). Putting Q equal to a design value of 200 yields a new value of $R'_p = 1100\,\Omega$ (from Figure 10.4). The new value of $R_p = \omega LQ \simeq 250\,\text{k}\Omega$ and the transformation ratio, r, becomes 1/15.

In general the Q-factor for maximum signal-to-noise ratio will not coincide with the value for maximum signal power and a compromise may be necessary. Further, a design optimized at mid-band will not have optimum performance at other frequencies.

An overriding design limitation is the unloaded Q-factor that can be achieved in practice. Experimental results given in Section 10.3.3 show that while values over 200 may be readily obtained at 1 MHz with an isolated ferrite rod, when the antenna is assembled in a small receiver the values may fall to 120–150. Thus the *in situ* unloaded Q-factor will generally limit the signal-to-noise ratio that can be achieved when low-noise transitors are used.

Considering only the bandwidth, adjacent channel selectivity requirements usually limit the bandwidth of the i.f. amplifier in small receivers to about 8 kHz. The bandwidth of the antenna circuit is limited by the need for image channel rejection. Thus for best overall performance the antenna circuit bandwidth, f/Q_L, should be constant over the broadcast band, having a value of about 10 kHz. Therefore from Eqns (10.8) and (10.26)

$$Q = Q_L\left(\frac{\omega LQ}{R'_b} + 1\right) = \frac{Q_L}{1 - (Q_L\omega L)/R'_b}$$

For $L = 200\,\mu\text{H}$, $R_b = 1100\,\Omega$ (assumed const.), $Q_L = f/10^4$ and $r = 1/15$

f(MHz)	Q_L	Q
0.5	50	57
1.0	100	200
1.5	150	—

At 1.5 MHz the value of R'_b limits Q_L to less than 150 even if the unloaded Q-factor were infinite, so the ferrite antenna inevitably has excessive bandwidth at the upper end of the medium wave broadcast band.

In the long wave band, e.g. at 200 kHz, a bandwidth of 10 kHz requires a Q-factor of not more than 20. In practice a value nearer 30 is used and usually the antenna circuit has to be artificially degraded to restrict the Q-factor to this value.

If the amplifier or frequency changer is voltage operated, as in the case of a circuit using MOS devices, the situation is a little different. Since such a circuit imposes negligible load on the antenna, $Q \simeq Q_L$. Thus for the same bandwidth the value of Q will need to be lower than for the loaded case.

10.3 Practical aspects of design

10.3.1 General

It has been seen (Eqn (10.21)) that for the greatest signal-to-noise ratio, $h_e^2 Q/L$ should be as large as possible. It was subsequently shown that Q and L are often determined by requirements of bandwidth and tuning respectively. Thus the primary considerations in ferrite antenna design are:

(1) The effective height should be as large as possible within the permissible dimensions.
(2) The unloaded Q-factor should be reasonably near the design value.
(3) The inductance of the winding should be as close as possible to the design value.

In addition there may be a secondary consideration, e.g. the temperature coefficient of inductance will, in general, have to be within specified limits.

These factors will now be studied from the design point of view.

10.3.2 Maximum effective height

From Eqn (10.6), the effect height:

$$h_e = \mu_{\text{rod}}\omega ANF_A/c_o \quad \text{or} \quad \mu_{\text{rod}}2\pi ANF_A/\lambda \quad \text{m}$$

The number of turns N and the e.m.f. averaging factor F_A are usually determined by the inductance and temperature coefficient requirements and are not variables in the attainment of maximum effective height. Therefore:

$$h_e \propto \mu_{\text{rod}}A \tag{10.28}$$

It is shown in Section 4.3.1 that μ_{rod} depends on the permeability of the rod material and the length/diameter ratio, m, of the rod. Figure 4.11 shows the relations. For maximum μ_{rod} the ferrite permeability should be high and the rod should be long compared with its diameter, d. However, considering only μ_{rod}, the graph shows that for a given ferrite permeability there is a limit to the useful increase in m while for a given value of m there is a limit to the useful increase in ferrite permeability.

From Eqn (10.28), it is the product $\mu_{\text{rod}}A$ that should be maximized. Referring again to Fig. 4.11 it is seen that none of the curves has a slope as steep as 2, therefore the quotient μ_{rod}/m^2 ($=\mu_{\text{rod}}d^2/l^2$) increases monotonically as m decreases. It follows that, for a given length, l, the products $\mu_{\text{rod}}d^2$ and $\mu_{\text{rod}}A$ increase as the diameter is increased.

The maximum effective height is therefore obtained when the ferrite material has the highest permeability consistent with loss limitations, the length of the rod is the maximum permitted by the space limitations and the diameter is as large as weight and cost considerations will allow.

If the maximum effective height per unit core volume is required it is necessary to maximize $\mu_{rod}A/ld^2$, i.e. maximize μ_{rod}/l. From Eqn (4.70)

$$\frac{\mu_{rod}}{l} \simeq \frac{1}{l([1/\mu] + N)}$$

Figure 4.10 shows N as a function of m for various magnetic bodies. For rods and similar bodies N decreases as m increases, therefore μ_{rod}/l will increase as m becomes larger. It follows that for a given rod length the maximum effective height per unit core volume will be obtained when the rod is as thin as possible. Alternatively if the value of m is fixed the effective height per unit core volume will increase as the rod becomes shorter.

Some of these relations are illustrated numerically in Table 10.1 in which the calculated properties of some typical antenna rods are shown.

Brief mention may be made at this point about the possible use of tubular cores. If a given ferrite rod is made into a tube of the same overall dimensions, the value of μ_{rod} does not change nor does the flux density in the central portion of the ferrite tube. Therefore the total flux due to a given field is diminished in the ratio of the ferrite cross-sectional areas. It follows that in the expression for the effective height the value of A must be that of the cross-sectional area of the tube.

Thus for the same overall dimensions the tube will give a smaller effective height. The volume of material would be better utilized in a rod of the same length and cross-sectional area as this would have a larger value of m and therefore a higher μ_{rod}. If the rod becomes too slender then mechanical considerations would invalidate this conclusion.

10.3.3 Unloaded Q-factor

The Q-factor of a resonant circuit is the reciprocal of the sum of all the contributory loss tangents. If the inductive element is a rod antenna the principal loss tangents to be considered are:

$(\tan \delta_m)_{rod}$ = loss tangent due to magnetic losses in the rod,
$\tan \delta_{AC}$ = loss tangent due to total (a.c.) copper losses,
$\tan \delta_{rad}$ = loss tangent due to radiation,
$\tan \delta_{mc}$ = loss tangent due to the magnetic coupling of the rod to adjacent conductive parts, e.g. metal supports or chassis.

All of these losses depend on the distribution of the flux density along the rod due to current in the winding. This distribution is in general different from and independent of the distribution of flux density due to the magnetic component of the incident elec-

Table 10.1 Properties of some typical ferrite antenna rods

This table illustrates the influence of factors that affect the effective height. The effective height is based on a frequency of 1 MHz and a short central winding of 40 turns ($F_A = 1$). Such a winding would have an inductance between 200 μH and 400 μH depending on which of the rods is used.

μ	l (mm/in)	d (mm/in)	m (l/d)	N	μ_{rod}	A (m²)	h_e (m)	Vol (m³)	h_e/Vol (m⁻²)
175	200 / 8	9.5 / 0.375	21	0.0043	100	71×10^{-6}	5.9×10^{-3}	14.2×10^{-6}	420
175	200 / 8	12.7 / 0.50	15.8	0.0070	79	127×10^{-6}	8.4×10^{-3}	25.4×10^{-6}	330
175	150 / 6	9.5 / 0.375	15.8	0.0070	79	71×10^{-6}	4.7×10^{-3}	10.7×10^{-6}	440
175	150 / 6	12.7 / 0.50	11.8	0.0112	59	127×10^{-6}	6.3×10^{-3}	19.1×10^{-6}	330
500	200 / 8	9.5 / 0.375	21	0.0045	159	71×10^{-6}	9.5×10^{-3}	14.2×10^{-6}	670
500	200 / 8	12.7 / 0.50	15.8	0.0073	111	127×10^{-6}	11.8×10^{-3}	25.4×10^{-6}	460
500	150 / 6	9.5 / 0.375	15.8	0.0073	111	71×10^{-6}	6.6×10^{-3}	10.7×10^{-6}	620
500	150 / 6	12.7 / 0.50	11.8	0.0115	76	127×10^{-6}	8.1×10^{-3}	19.1×10^{-6}	420

tromagnetic field. This is because the e.m.f. induced in the winding by the incident field has a phase that is 90° in advance of that of the incident field. If the winding is at resonance the current is in phase with this e.m.f. so the resultant secondary or self-induced field is in quadrature with the incident field. Thus the e.m.f. that energizes the circuit arises wholly from the incident field, while the secondary or self-induced field determines the inductance and magnetic loss associated with the winding.

The calculation of $(\tan \delta_m)_{rod}$ has been discussed in some detail in Section 4.3.5; its value may readily be obtained from a knowledge of the loss tangent of the material and the geometry of the rod. The loss tangent of the material of a given rod may be obtained by measurement in the following way. The rod is closely wound along its entire length with a bunched conductor composed of many strands of fine wire, so chosen that the copper loss may be neglected. This may be checked by subsequently replacing the winding with another using a conductor having a somewhat higher h.f. resistance and ensuring that the measured loss tangent is unaffected. Provided $\tan \delta_{rad}$ and $\tan \delta_{mc}$ are negligible the measured loss tangent will equal $(\tan \delta_m)_{rod}$. The material loss tangent may then be calculated from Eqn (4.78). This value may then be used in evaluating $(\tan \delta_m)_{rod}$ for more practical winding configurations.

The prediction of $\tan \delta_{AC}$ is difficult. This loss tangent depends on the h.f. resistance of the winding which in turn depends on the type of conductor, the shape of the winding and the configuration of the secondary magnetic field. An adequate method of evaluating this loss would require a large amount of empirical data and the results would probably be too unwieldy for convenient use. As a guide, the discussion of copper losses in the chapter on inductors (Section 5.7.3) may be consulted in conjunction with Section 11.4. The principles established in these sections will indicate in a general way how a given copper loss will be influenced by a given change in conductor composition or configuration. A practical difficulty is the isolation of the copper loss for the purpose of measurement. This may be overcome by using a rod of known material loss angle and calculating the value of $(\tan \delta_m)_{rod}$. If $\tan \delta_{rad}$ and $\tan \delta_{mc}$ are made negligible, the value of $\tan \delta_{AC}$ may be obtained by subtracting $(\tan \delta_m)_{rod}$ from the measured loss tangent. In practice it is not difficult to make the winding loss small compared with the core loss, although economy may dictate a design or type of conductor for which the winding loss is by no means negligible.

The value of $\tan \delta_{rad}$ is always negligibly small. This follows from the fact that the effective height (see Section 10.2.1 and Table 10.1) is always very much smaller than that of the equivalent dipole antenna.

In practice the loss due to the proximity of conductive parts is usually quite large. In the lay-out of a portable radio receiver it is always essential to make a compact assembly, indeed the main advantage of a ferrite antenna is that it may be contained physically in quite a small volume. It is therefore inevitably placed near other components and metal parts. The resultant degradation of performance compared with that obtained when the rod antenna is isolated may amount to a doubling of the total loss tangent, i.e. a halving of Q-factor.

Because of this last consideration there seems to be little point in elaborating on methods of calculating Q-factor. The usual procedure is to make a trial winding giving the required inductance. For a.m. broadcast bands the conductor usually consists of a number of strands, e.g. 24, of 0.04 mm diameter enamelled wire. The Q-factor may then be measured in isolation and *in situ*; if a higher Q-factor is required the number of strands may be increased or the length of the winding reduced. The magnetic loss in the rod eventually limits the Q-factor that may be obtained.

The effect on the Q-factor of varying the length of the winding may be deduced from Eqns (4.78), (4.79) and (4.81). The longer winding gives the higher value of $(\tan \delta_m)_{rod}$ because the flux density distribution is more uniform or, putting it in a different way, μ_{coil} is higher.

As examples of unloaded Q-factors, Table 10.2 shows results quoted by Maanders and van der Vleuten[2] for various rod antennas in both isolated and *in situ* positions.

It was seen in the example of Section 10.2.6 that it may be desirable to obtain a high Q-factor in order to maximize the signal-to-noise ratio, so it is clearly worth while making an effort to reduce the degradation of the Q-factor when the receiver layout is being considered.

10.3.4 Inductance

The inductance of a winding centrally placed on a ferrite cylinder has been considered in Section 4.3.4. From Figure 4.14 the inductance of any central winding on any rod may be estimated with reasonable accuracy. An experimental winding based on such an estimate may be made and its inductance measured; the number of turns may then be altered if necessary. Final adjustment is often made by moving the winding along the rod until the correct inductance is obtained.

It is quite common for one rod to be used for two frequency bands. In such cases it is usual to place one winding towards one end of the rod and the other winding towards the other end of the rod. A winding that is midway between the centre and the end of the

Table 10.2 Unloaded Q-factors of typical antenna rods

Q is the value measured with the rod in isolation,
Q' is the value measured with the rod *in situ* in a receiver.

Winding details:
$L = 200\,\mu\text{H}$, $N = 44$ turns for 4B1 ferrite or 37 turns for 3D3 ferrite
Winding was short and central

Rod details:
Length = 200 mm (8 in)
Diameter = 9.5 mm (0.375 in)

Ferrite grade	Frequency (MHz)	Bunched conductors									
		45		28		20		12		8	
		\multicolumn{10}{c}{strands of 0.04 mm (0.0016 in) dia. copper}									
		Q	Q'	Q	Q'	Q	Q'	Q	Q'	Q	Q'
4B1	0.5	282	125	234	122	221	117	180	108	142	92
	0.75	272	133	238	131	231	128	199	120	165	107
	1.0	246	130	222	128	215	128	191	120	164	111
	1.2	206	120	193	118	191	118	172	112	150	107
	1.5	170	105	175	105	166	105	154	98	140	96
3D3	0.5	376	136	316	128	274	125	228	112	166	98
	0.75	332	145	300	140	260	137	238	127	188	114
	1.0	256	145	246	142	232	140	212	131	179	121
	1.2	195	140	191	135	182	135	172	127	153	119
	1.5	135	128	135	125	128	125	125	119	120	113

Ferrite grade	Frequency (MHz)	Solid copper wire (approx. dia.)									
		0.4 mm 0.0164 in		0.3 mm 0.0116 in		0.2 mm 0.0076 in		0.16 mm 0.0060 in		0.12 mm 0.0048 in	
		Q	Q'	Q	Q'	Q	Q'	Q	Q'	Q	Q'
4B1	0.5	163	87	180	106	204	112	190	109	153	97
	0.75	149	89	172	109	200	115	197	118	172	110
	1.0	140	88	159	107	185	112	186	118	172	112
	1.2	126	82	143	100	164	104	168	110	158	108
	1.5	112	73	126	89	140	93	145	99	140	97
3D3	0.5	195	94	240	109	260	115	238	113	187	101
	0.75	183	100	216	112	252	121	241	124	205	115
	1.0	163	100	186	112	212	121	210	126	192	122
	1.2	141	97	155	107	181	117	179	122	165	118
	1.5	114	92	124	101	137	110	137	113	133	112

rod will have about 10% less inductance than the same winding placed centrally.

10.3.5 Temperature coefficient

From Eqn (4.82) the temperature coefficient of inductance of a winding on a ferrite rod is given approximately by:

(T.F.) × μ_{coil}

where T.F. is the temperature factor, $\Delta\mu/\mu^2\Delta\theta$, of the ferrite and is a material parameter. If μ_{coil} is 20 and T.F. is $10 \times 10^{-6}\,°\text{C}^{-1}$ then the temperature coefficient

$$\frac{\Delta L}{L\Delta\theta} = 200\text{ ppm/°C}$$

The temperature coefficient of frequency, assuming that the temperature coefficient of the resonating capacitance is negligible,

$$\frac{\Delta f}{f\Delta\theta} \simeq -\tfrac{1}{2}\frac{\Delta L}{L\Delta\theta} \quad \text{since } f \propto L^{-1/2}$$

Its value is 100 ppm/°C for the above example. If $\Delta\theta$ is 20 °C and f is 1 MHz, Δf is -2 kHz.

The temperature coefficient may be somewhat reduced by reducing the length of the winding in relation to the length of the rod; this reduces μ_{coil}.

10.4 References and bibliography

Section 10.1
1. SHIEFER, G. A small ferroxcube aerial for VHF reception, *Philips tech. Rev.*, **24**, 332 (1962–63)

Section 10.2
2. MAANDERS, E. J., *et al.*, see General

Section 10.2.4
3. BLOCK, H. and RIETVELD, J. J. Inductive aerials in modern broadcast receivers, *Philips tech. Rev.*, **16**, 181 (1955)
   BAELDE, A. Theory and experiments on the noise of transistors, *Philips Res. Rep. Supplement No.* 4 (1965)
   MEYER, R. G. Noise in transistor mixers at low frequencies, *Proc. Instn elect. Engrs*, **114**, 611 (1967)

Section 10.2.6
2. MAANDERS, E. J., *et al.*, see General

General
VAN SUCHTELEN, H. Ferroxcube aerial rods, *Electron Applic. Bull.*, **13**, 88 (1952)
CRAWTHORNE, M. J. Ferroxcube aerial rods for medium and long wave reception, *Mullard tech. Commun.*, **1**, 181 (1954)
GRIMMETT, C. A. Ferrite cored antennae, *Conv. Rec. Inst. Radio Engrs*, **2**, Part 7, 3 (1954), also *Proc. Instn Radio Engrs Aust.*, **16**, 31 (1955)
Ferroxcube aerial rods for medium wave reception, *Matronics*, no. 3, 41 (1953)
DUPUIS, J. Cadres utilisant des ferrites, *Onde élect.*, **35**, 379 (1955)
WRIGHT, C. M. Ferrite rods for broadcast receiver antenna coils, *Proc. Instn Radio Engrs Aust.*, **21**, 410 (1960)
MAANDERS, E. J. and VAN DER VLEUTEN, H. Ferrite aerials for transistor receivers, *Matronics*, no. 18, 354 (1961)
NILSSEN, O. K. A nondirectional ferrite rod antenna arrangement suitable for A.M. radios, *Proc. Inst. Radio Engrs*, **49**, 1222 (1961)
POLYDOROFF, W. J. Recent advances in ferromagnetics, *Electronic Inds*, **20**, no. 9, 102 (1961)
LAURENT, H. J. and CARVALHO, C. A. B. Ferrite antennas for A.M. broadcast receivers, *Trans. Inst. Radio Engrs*, **BTR-8,** no. 2, 50 (1962)
HUMPHREY, L. C. Design of ferrite-cored antennas, *Solid St. Des.*, **5**, no. 1, 24 (1964)
BITTERLICH, W. Magnetische Dipolantennen für Feldstärkemessungen im LF- und VLF-Bereich, *Ekektron. Rdsch.*, **21**, 225 (1967)
PETTENGILL, R. C., GARLAND, H. T. and MEINDL, J. D. Receiving antenna design for miniature receivers, *IEEE Trans. Antennas and Propag.*, **AP-25,** 528 (1977)

11 Properties of windings

11.1 Introduction

The ferrite manufacturer usually supplies his product as a set of ferrite parts and possibly ancillary parts from which the user may construct a wound component to meet specific requirements. The winding, with all its possible variety in design and execution, is usually the concern of the user. In this the magnetic component differs from other electrical components. Whereas a resistor or a capacitor leaves the manufacturer as a finished and tested component, the quality of a wound magnetic component depends on both the core manufacturer and the user.

The core manufacturer normally provides data on the magnetic properties of the core; these are the properties that he can test and control. Where the application of the core involves a well-established design technique, e.g. cores for inductors or transformers, the manufacturer will usually give some additional design information to enable the properties of the core to be fully exploited. This information, e.g. winding data, Q-curves, etc., may usually be taken as a guide only, since so much depends on the techniques of winding design and construction that will be used. Also, it usually applies only to specific cores. Often such data are not available and the winding must be designed from first principles.

In most of the preceding chapters on the applications of ferrites, the relations between the winding properties and the overall design have been established for each type of application. The purpose of the present chapter is to consider these winding properties and the way in which they may be related to or calculated from the winding design.

In any calculations involving windings the first essential is adequate data on the wires or conductors which make up the winding. The wire tables in Appendix B give numerical information on copper winding wires conforming to the metric and AWG standards. The tables cover the properties of bare round wire, enamelled solid wire and enamelled bunched conductors; the quoted properties include the dimensional limits, the resistance per unit length and parameters such as the copper factor.

11.2 Number of turns in a given winding area

The number of turns of a given conductor that may be wound into a given winding area depends not only on the overall diameter of the conductor and the dimensions of the winding area but also on the following factors:

(1) whether the winding is in regular layers,
(2) how much insulation is used to interleave or cover the windings,
(3) the winding tension,
(4) how many separate windings and connections there are, and the technique used for bringing out the connections.

The possible number of turns predicted from calculation will depend on a proper allowance for these factors; the result will also depend on whether dimensions of the conductor and winding area are the nominal values or whether they are respectively maximum and minimum limit values. Figure 11.1 defines

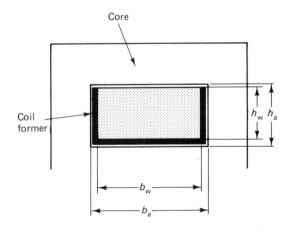

A_a = window area = $b_a h_a$
A_w = net winding cross-sectional area = $b_w h_w$

Figure 11.1 Some symbols used in winding calculations

some of the symbols used in calculations relating to the winding.

Two ideal arrangements of conductors are shown in cross-section in Figure 11.2; these will be referred to as the square and hexagonal arrangement respectively. If the overall conductor diameter, d_o, is much smaller than the dimensions of the available winding space, then the number of turns, N, in a winding area, A_w, is given for the square arrangement by:

$$N = \frac{A_w}{d_o^2} \quad (11.1)$$

If the winding copper (space) factor is defined as

$$F_w = \frac{\text{total cross-sectional area of copper in winding}}{\text{actual cross-sectional area of the winding space}}$$

then for the perfect square arrangement:

$$F_w = \frac{\pi}{4} \left(\frac{d}{d_o}\right)^2 \quad (11.2)$$

where d is the bare diameter of the conductor.

This factor is given in the wire tables in Appendix B for all the commonly used insulated conductors. It is calculated from Eqn (11.2) using the nominal value of d and the maximum overall value of d_o. When $d/d_o \to 1$, $F_w \to 0.785$.

The square arrangement is one extreme of close-packed ordering; the other is the hexagonal arrangement. In this case the winding copper factor is given, for a large number of turns, by

$$F_w = \frac{\pi}{2\sqrt{3}} \left(\frac{d}{d_o}\right)^2 \quad (11.3)$$

When $d/d_o \to 1$, $F_w \to 0.907$.

In a practical winding the conductor pattern will correspond to neither ideal. However, in a carefully wound coil, without interlayer insulation, the hexagonal arrangement will occur over large regions of the winding. Although successive layers approximate to helices of opposite lay, in practice the conductor will tend to lay in the groove formed by the previous layer for the majority of the turn before being forced to cross over into the next groove. In the region of the crossover the square configuration will occur.

From the above definition of winding copper factor, its value for a practical winding is given by

$$F_w = \frac{N\pi d^2}{4} \frac{1}{A_w} \quad (11.4)$$

and will usually be somewhat less than the value of F_w given in the Appendix B wire tables for a given value of d.

Another useful factor is the overall copper factor F_a defined by

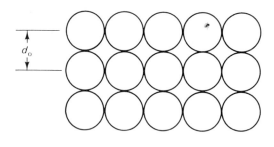

(a) Square

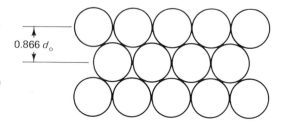

(b) Hexagonal

Figure 11.2 Cross-sections of ideal arrangements of conductors in a winding

$$F_a = \frac{\text{total cross-sectional area of copper in winding}}{\text{actual window area in the core}}$$

$$= \frac{N\pi d^2}{4} \frac{1}{A_a} \quad (11.5)$$

where A_a is defined in Figure 11.1.

Figure 11.3 shows sections of some cylindrical windings, obtained by encapsulating, cutting and photographing. The most common type of winding is the scramble or random winding. The wire is guided by hand or by mechanical traverse but because the diameter is small and there is no interleaving between layers a fully-ordered arrangement is not often possible. In the first cross-section the winding has been hand controlled by a skilled winder; almost perfect ordering has been achieved. In the second, the traverse was mechanical and some disordering has occurred. The third photograph, (c) shows the cross-section of a random winding, hand-controlled without much care. The first few layers are well-ordered but then disordering sets in and spreads from the flanges of the coil former. If the wire is fed onto the coil former with the same traverse or pitch that would be used for close-wound layers, the winding area will be well filled, varying from the theoretical maximum filling of the hexagonal arrangement in some places to a somewhat spaced version of the square arrangement

Table 11.1

Overall conductor diameter		F_p
(mm)	(in)	
0.03–0.07	$(1.2–2.8) \times 10^{-3}$	0.77–0.86
0.10–0.15	$(4–6) \times 10^{-3}$	0.81–0.90
0.20–0.4	$(8–16) \times 10^{-3}$	0.86–0.95
0.50–2.0	$(20–80) \times 10^{-3}$	0.81–0.90

in others. In practice the number of turns that may be randomly wound into a given area may be calculated from Eqn (11.1) provided a suitable packing factor, F_p, is used:

$$N = \frac{A_w}{d_o^2} \times F_p \qquad (11.6)$$

The packing factor depends on the overall diameter of the conductor, the winding technique (wire tension, traverse, skill of the operator, etc.) and to some extent on the dimensions of the winding area. Table 11.1 gives some experimentally-determined values applying to cylindrical windings on all but the smallest coil formers.

The higher figure in the F_p range, combined with nominal values of A_w and d_o, will give a typical number of turns; the lower figure combined with a minimum value of A_w and a maximum value of d_o will give a fairly safe limit value for N, i.e. the number of turns that may be specified with reasonable certainty that they can be wound in the given winding area. The value of F_p has been calculated for each of the cross-sections in Figure 11.3 and the results are quoted in the caption. If the winding area dimensions are very small the above values of F_p must be reduced, e.g. by 10%.

If the overall conductor diameter is not very small compared to the dimensions of the winding area, then it becomes relatively easy to wind the whole coil in regular layers. Such a coil is shown in Figure 11.3(d). The conductor arrangement is close-packed and varies from the square to the hexagonal pattern. For most purposes the number of turns may be calculated from Eqn (11.6). However, when the conductor is so large that only a few turns per layer are possible then it is necessary to calculate the number of turns per layer and the possible number of layers to arrive at the number of turns. The possible number of layers, assuming a square pattern of conductors, is given by h_w/d_o rounded to the next lower integer.

Another type of winding is that in which the layers are insulated with paper or plastic foil. Figure 11.3(e) shows an actual cross-section. Normally this type of winding uses a coil former without flanges. The winding breath, b_w, must be less than the width of the interleaving to prevent the end turns from collapsing onto the layer below; the size of the margin depends on the conductor diameter and the stiffness of the interleaving. If h_w is the permissible winding height and the layers are in tight contact (see also last paragraph of this section), the number of turns is given by

$$N = p \times \frac{b_w}{d_o} \times F_p \qquad (11.7)$$

where p is the number of layers = $h_w/(d_o + t)$ rounded to the next lower integer,
t is the thickness of the interleaving,
F_p is a packing factor.

A good winder will attain this number of turns with unity packing factor if d_o used in the calculation is the maximum value. If a safety margin is required F_p may be put equal to 0.95.

Where a self-supporting, low-capacitance winding is required, wave-winding is often used. In this the conductor is made to traverse across the whole winding breadth and back one or more times during each revolution. Its main use is for h.f. coils in radio receivers, antenna rod coils, etc. Since wave windings rarely have to fill the available winding space the design is not usually critical. A successful winding depends on the proper choice of gear ratios on the winding machine so that the correct traverse is obtained. Instructions for gear wheel selection are usually available with the winding machine.

Returning to layer or random winding it has been assumed in calculations involving winding height that successive layers are in close contact. This is justified in the case of round coils but when the coil former is rectangular as in Figure 11.4(a) the actual winding height in the region covered by the core will be greater than the calculated winding height, h_w, due to the bowing of the windings. The amount of the bowing depends on the proportions of the coil former and the height of the winding. Usually the nett available winding height should be reduced by 15 to 20% to obtain the value of h_w to be used in the above calculations.

11.3 The d.c. winding resistance

The d.c. resistance of a winding consisting of a total length, l, of conductor of cross-sectional area A is

$$R_{dc} = \frac{\rho_c l}{A} \quad \Omega \qquad (11.8)$$

where ρ_c is the resistivity of the conductor material.
For copper at a temperature of 20 °C the nominal

316 Properties of windings

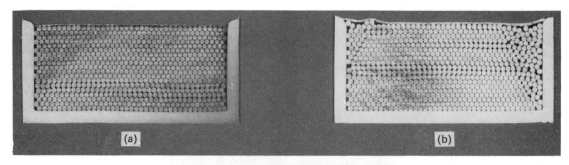

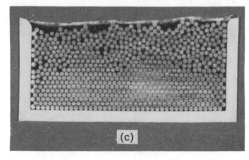

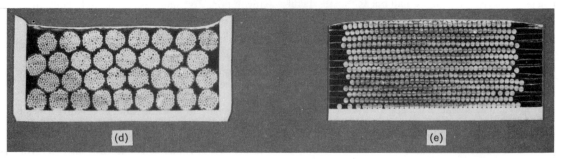

Figure 11.3 Cross-sections of some actual windings. Nominal winding area dimensions = 9.18 × 4.0 = 36.72 mm².

Type of winding		Conductor	Turns	F_w	F_p
(a)	Layer wound, hand traverse	0.190 mm Grade 1 enamelled	740	0.59	0.93
(b)	Layer wound, machine traverse	0.190 mm Grade 1 enamelled	733	0.58	0.92
(c)	Random wound by hand	0.190 mm Grade 1 enamelled	690	0.55	0.87
(d)	Layer wound by hand	81 × 0.071 mm SSC bunched conductors	32	0.28	0.87
(e)	Interleaved and layer wound by hand	0.190 mm Grade 1 enamelled	490	0.39	0.61

F_w and F_p are based on nominal overall conductor diameter except for F_p in (d) where conductor was appreciably oversize.

value of ρ_c is $1.709 \times 10^{-8}\,\Omega\text{m}$. The temperature coefficient of this resistivity is 0.00393 per °C*.

If d is the bare conductor diameter, l_w is the mean turn length and N is the number of turns on the winding, then Eqn (11.8) becomes

$$R_{dc} = \frac{4\rho_c N l_w}{\pi d^2} = N l_w R_c \quad \Omega \qquad (11.9)$$

where R_c is the resistance per unit length of conductor $= 4\rho_c/\pi d^2$.

Values of R_c are given in the wire tables in Appendix B. For a circular winding the mean turn length is readily calculated from the mean diameter. When the winding is wound on a rectangular coil former as

* At a temperature, θ, the resistivity of copper is $m\rho_c$ where:

$\theta =$	20	30	40	50	60
$m =$	1	1.039	1.079	1.118	1.157
$\theta =$	70	80	90	100	
$m =$	1.197	1.236	1.275	1.314	

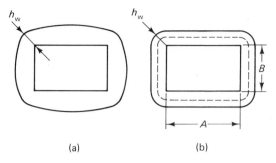

Figure 11.4 Geometry of a winding on a rectangular coil former

shown in Figure 11.4, the bowing of the winding may usually be ignored in its effect on the mean turn length and this may be calculated, by reference to Figure 11.4(b), from

$$l_w = 2(A + B) + \pi h_w \tag{11.10}$$

Combining Eqns (11.9) and (11.4) or (11.5)

$$R_{dc} = \frac{\rho_c N^2 l_w}{A_w F_w} = \frac{\rho_c N^2 l_w}{A_a F_a} \quad \Omega \tag{11.11}$$

Thus for a winding fully occupying a given winding space, $R_{dc} \propto N^2$. If the winding embraces a magnetic core, the inductance, L, is also proportional to N^2 (see Eqn (4.24) for example). It follows that a given core may be characterized by a particular value of R_{dc}/L, i.e. the ratio of the resistance to the inductance for a winding that occupies the full winding space. From the expressions for R_{dc} and L

$$\left. \begin{array}{l} R_{dc}/L = \dfrac{\rho_c}{\mu_0 \mu_e} \times \dfrac{l_w C_1}{F_w A_w} \quad \Omega\,H^{-1} \\[1em] \text{or} \quad \dfrac{\rho_c l_w 10^9}{F_w A_w A_L} \quad \Omega\,H^{-1} \end{array} \right\} \tag{11.12}$$

from Eqn (5.6).

Inasmuch as F_w is not greatly affected by the conductor diameter, R_{dc}/L is approximately independent of the number of turns and depends mainly on the core geometry. This ratio is often used to compare the merits of alternative core designs.

If the winding is made with bunched conductors consisting of n strands, each of diameter d, then provided the twist of the bunch is neglected, the resistance of the conductor or winding may be calculated from Eqns (11.8) or (11.9) by using d or R_c for a single strand and dividing the result by n.

So far the discussion has been confined to a single winding occupying substantially all the available winding space. Another important case is the transformer which, in its simplest form, consists of two windings having in general unequal numbers of turns, e.g. N_1 and N_2, and carry currents I_1 and I_2 respectively such that

$$I_1 N_1 = I_2 N_2$$

Continuing to use the subscripts to distinguish the properties of the windings, the power dissipated in the windings is, using Eqn (11.11)

$$P = I_1^2 \frac{\rho_c N_1^2 l_{w1}}{A_{w1} F_{w1}} + I_2^2 \frac{\rho_c N_2^2 l_{w2}}{A_{w2} F_{w2}}$$

if $l_{w1} \simeq l_{w2} = l_w$, $F_{w1} = F_{w2} = F_w$ and $A_{w1} + A_{w2} =$ the total winding area, A_w, then

$$P = \frac{\rho_c I_1^2 N_1^2 l_w}{F_w} \left(\frac{1}{A_{w1}} + \frac{1}{A_w - A_{w1}} \right)$$

Differentiating w.r.t. A_{w1} and equating to zero:

$$\frac{1}{A_{w1}^2} = \frac{1}{(A_w - A_{w1})^2}$$

or $\quad A_{w1} = \dfrac{A_w}{2} \tag{11.13}$

Therefore for minimum power loss in the windings the winding areas should be approximately equal. If l_{w1} is significantly less than l_{w2} then A_{w2} should be somewhat greater than A_{w1}. An exact result may be obtained in specific cases by expressing l_{w1} and l_{w2} in terms of A_{w1} before differentiating. In practice the minimum is fairly flat, so equal or nearly equal winding areas may be used for the first trial calculation. If, as a result, the winding resistance R_1 differs significantly from $R_2(N_1/N_2)^2$ (this may be due to unequal copper factors as well as a poor division of the winding space) then the ratio of winding areas should be suitably adjusted until the resistances referred to one side of the transformer are approximately equal.

11.4 Power loss due to eddy currents in the winding

11.4.1 General

A conductor immersed in an alternating magnetic field will have e.m.f.s induced in it and these will give rise to eddy currents and associated power losses[1]. The magnetic fields may be due to currents flowing in other conductors or due to current flowing in the conductor in question. The losses may be represented as an increase in resistance of the current-carrying conductors above the value measured with direct current; this increase of resistance becomes larger as the frequency increases.

A conductor which forms part of a current-carrying winding will in general experience the magnetic field due to its own alternating current and also the field due to all the other current-carrying conductors in the vicinity. Usually these two contributions are considered separately and the resultant losses or resistance increments are added together. This approach is considered in the two sections immediately following. However, it is only valid if the field due to the winding as a whole is approximately uniform across the diameter of the individual conductors. If the winding consists, for example, of a single layer of large diameter wire, the transverse field will vary greatly across the thickness of the layer and the simple approach might lead to appreciable error in the estimation of eddy current losses. Instead it is preferable to analyse the eddy current losses as a whole. The results of such analysis, applicable particularly to transformer design, are given in Section 11.4.4.

11.4.2 Skin effect

A single straight isolated conductor carrying an alternating current, see Fig. 11.5(a), will be surrounded by, and at low frequencies permeated by, a concentric magnetic field. This field will induce opposing eddy currents within the conductor itself as shown in the centre diagram. These currents tend to oppose the main current in the vicinity of the axis of the conductor and to enhance it at the surface. Thus the current distribution tends to become non-uniform across the section, the current being least at the centre and greatest at the surface of the conductor. As the frequency increases, the induced e.m.f.s increase and the non-uniformity becomes more pronounced until the current is virtually confined to a thin skin at the surface, and the inner region plays no part in the conduction.

The a.c. resistance of a straight, isolated, conductor of circular cross-section is given by

$$R_{ac} = R_{dc} + R_{se} = R_{dc}(1 + F) \quad \Omega \qquad (11.14)$$

where R_{se} is the increase in resistance due to skin effect, and F is the skin effect factor.

The skin effect factor is a function of d/Δ where Δ is the penetration depth. This is a property of a conducting material and is strictly the depth beneath an infinite plane surface at which an incident plane e.m. wave is attenuated to $1/\varepsilon$ or 37% of its surface value due to the effect of eddy currents[2].

$$\Delta = \sqrt{(\rho_c/\pi\mu_o\mu_c f)} \quad \text{m} \qquad (11.15)$$

where ρ_c is the conductor resistivity, and μ_c is the relative permeability of the conductor.

When the current in a round conductor is confined by skin effect to a depth beneath the surface that is much less than the diameter, this depth, measured as the distance over which the current falls to $1/\varepsilon$ of its surface value, tends to the value Δ. Some values of Δ

Table 11.2 Penetration depth. $\Delta = kf^{-1/2}$

Conductor material	Temperature	k	
	°C	m. Hz$^{1/2}$	mm Hz$^{1/2}$
Copper	20	0.0658	65.8
	70	0.072	72
	100	0.075	75
Aluminium	20	0.084	84
	70	0.092	92
	100	0.095	95
Silver	20	0.064	64
	70	0.070	70
	100	0.073	73

in terms of frequency are given in Table 11.2 for three common conducting materials at several temperatures.

Figure 11.6 gives F and $1 + F$ as functions of d/Δ. For convenience in calculations involving copper conductors Figure 11.7 gives Δ for copper as a function of frequency, and also the threshold diameter, i.e. the diameter at which the skin effect has increased the resistance of the conductor to 1.2 times the d.c. value. In using these graphs and applying the following equations it should be remembered that the skin effect factor, F, as given in Figure 11.6 applies only to a straight isolated conductor. In a close-packed winding the circular fields of adjacent conductors will tend to cancel and this significantly reduces the skin effect in inductor windings.

Referring to Figure 11.6 it is seen that when d/Δ for a round conductor is less than 2 the skin effect is negligible. It increases rapidly as d/Δ increases and when d/Δ is greater than 5:

$$1 + F \simeq \frac{1}{4}\left(\frac{d}{\Delta} + 1\right) \qquad (11.16)$$

When d/Δ is very large, $1 + F$ approximates to $d/4\Delta$ and the a.c. resistance,

$$R_{ac} = R_{dc} + R_{se} \simeq R_{dc}d/4\Delta = \frac{l}{d}\sqrt{\left(\frac{\mu_o\mu_c\rho_c f}{\pi}\right)} \quad \Omega \quad (11.17)$$

from Eqn (11.15) and $R_{dc} = 4\rho_c l/\pi d^2$.

This equation shows that, although an increase in diameter will increase the ratio of a.c. resistance to d.c. resistance, it will always reduce the actual value

(11.15) $\quad \Delta = \dfrac{1}{2\pi}\left(\dfrac{\rho_c 10^9}{\mu_c f}\right)^{1/2} \text{cm}$

(11.17) $\quad R_{ac} = R_{dc} + R_{se} \simeq R_{dc}d/4\Delta = \dfrac{2l\sqrt{(\mu_c f \rho_c 10^{-9})}}{d} \quad \Omega$

Properties of windings 319

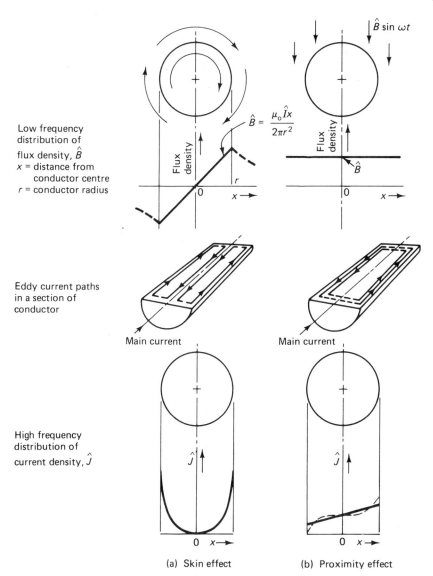

Figure 11.5 Skin effect and proximity effect in round conductors

of a.c. resistance due to skin effect at any frequency. At the higher frequencies at which Eqn (11.17) holds, the conductor current is equivalent to a current of uniform density flowing in a surface layer of depth Δ. This follows from Eqn (11.17) since

$$R_{dc} + R_{se} \simeq R_{dc}d/4\Delta = \frac{4\rho_c l}{\pi d^2} \times \frac{d}{4\Delta} = \frac{\rho_c l}{\pi d \Delta} \quad (11.18)$$

and $\pi d \Delta$ is the cross-sectional area of a surface layer of depth Δ.

Skin effect may be virtually eliminated by using conductors consisting of thin insulated strands so composed that individual strands weave cyclically from the centre of the conductor to the outside and back as they run along the length of the conductor. Such a conductor is referred to as Litz wire; the stranding and transposition makes the current density uniform. However, since skin effect is not usually the most important form of eddy current loss in winding conductors, it is not usual to use this special stranding. To combat the proximity effect loss described in the next section bunched conductor is often used. This consists of a number of thin insulated strands

320 Properties of windings

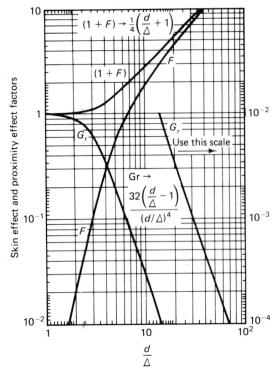

Figure 11.6 Skin effect factor F and proximity effect factor G_r as functions of d/Δ for round conductors, based on figures given by Butterworth[3]

simply twisted into the form of a rope. If the strands formed perfect helical paths, keeping at a constant distance from the axis, the skin effect would be the same as for a solid conductor having the same copper cross-section, assuming that all the strands are connected together at each end of the conductor. In practice, however, such bunched conductors are usually made up of groups of strands and appreciable transposition of the strands occurs. Careful measurements have shown that most bunched conductors behave as though the strands are transposed almost perfectly and the skin effect can be ignored.

In isolation, a perfectly transposed bunched conductor will be subject to a further eddy current loss due to its own field traversing the strands, each of which carries the same current. This may be called internal proximity effect. It may be comparable to, and at high frequencies may be larger than, the skin effect in an equivalent conductor consisting of perfectly untransposed strands. However, when a perfectly transposed bunched conductor is in a winding, the field due to the bunch, as such, largely disappears due to the proximity of the other turns; the winding behaves as though it consists only of strands, all carrying equal current. Under these conditions only the normal proximity effect described in the next section is important.

Returning to the consideration of skin effect loss in general, the loss tangent due to skin effect in an inductor of L henries follows from Eqn (11.14)

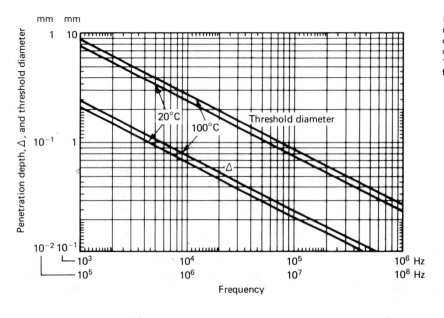

Figure 11.7 Skin effect in round conductors. Penetration depth Δ and the threshold diameter (diameter at which $1+F=1.2$) as functions of frequency

$$\tan \delta_{se} = \frac{R_{dc}F}{\omega L} \qquad (11.19)$$

At low values of d/Δ, F is proportional to $(d/\Delta)^4$ and is therefore proportional to f^2. Thus at low frequencies $\tan \delta_{se}$ is proportional to f. It may easily be shown that it reaches a peak when $d/\Delta \simeq 6$. Thereafter it falls; as F becomes approximately proportional to $f^{1/2}$, $\tan \delta_{se}$ approaches proportionality to $f^{-1/2}$.

11.4.3 Proximity effect

This is the eddy current effect in a conductor due to the alternating magnetic field of other conductors in the vicinity. In practice this may usually be interpreted as the eddy current effect in the conductors of a winding due to the field of the winding as a whole. The field will, in general, depend on the geometry of the core, if any. It must be remembered that any additional windings or conductors in the same field will have eddy currents induced in them whether or not they are carrying a main current; the resultant energy loss will simply add to the corresponding loss in the current-carrying winding and will be apparent as an additional resistance in that winding.

The field of a current-carrying winding will normally cut the conductors of that winding, or associated windings, perpendicular to the conductor axis. The resultant eddy current paths and current distribution are shown in Figure 11.5(b) for a round wire for which d/Δ is not large.

When d/Δ is less than unity the effect of the magnetic fields of the eddy currents themselves may be ignored and the calculations of the eddy current loss are quite simple; the following derivation of proximity effect in a thin tape illustrates the principles.

Figure 11.8 shows the cross-section of a tape having width b and thickness d. An alternating magnetic flux density, $\hat{B} \sin \omega t$, is everywhere parallel to the plane of the tape. The e.m.f. induced in a loop consisting of the two elementary laminae is given by

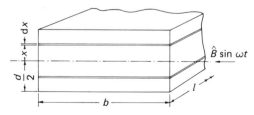

Figure 11.8 Calculation of eddy currents in a thin tape

$$E = \omega \hat{B} 2xl/\sqrt{2} \quad \text{V} \qquad (11.20)$$

where l is the length of tape being considered.

An eddy current will flow along the laminae parallel to the length of the tape, in one direction on one side of the axis and in the opposite direction on the other side. Neglecting the short paths connecting the laminae at the ends of the tape, the resistance of the elementary eddy current circuit is given by

$$R = \frac{2\rho_c l}{bdx} \quad \Omega$$

Therefore the power loss due to proximity effect is

$$dP_{pe} = E^2/R = \omega^2 \hat{B}^2 x^2 lb dx / \rho_c$$

$$P_{pe} = \frac{\omega^2 \hat{B}^2 lb}{\rho_c} \int_0^{d/2} x^2 dx$$

$$= \frac{\omega^2 \hat{B}^2 lb d^3}{24 \rho_c} \quad \text{W} \qquad (11.21)$$

It must be remembered that this result applies only when the flux density is parallel to the plane of the tape; a small inclination may result in much greater losses. An analogous expression may be obtained for a round conductor of diameter d, namely

$$P_{pe} = \frac{\pi \omega^2 \hat{B}^2 l d^4}{128 \rho_c} \quad \text{W} \qquad (11.22)$$

Since the latter case is the more common, the following discussion will be in terms of round conductors. If the transverse magnetic field is associated with a main current in the conductor, the eddy currents cause a distortion of the current density, enhancing it on one side of the axis and diminishing it on the other, as shown by the full line in the lower diagram of Figure 11.5(b).

At higher frequencies or with larger conductor diameters such that ratio d/Δ becomes larger than unity and field due to these eddy currents significantly reduces the flux density inside the conductor and the associated current distribution becomes non-linear as shown by the broken line.

The effect of the eddy current field and the associated non-linear current distribution is to reduce the proximity effect loss that would otherwise be present. Butterworth[3] calculated the eddy current loss in round conductors taking this eddy current screening

---

(11.20) $\quad E = \dfrac{\omega \hat{B} 2xl}{\sqrt{2}} 10^{-8} \quad$ V

(11.21) $\quad P_{pe} = \dfrac{\omega^2 \hat{B}^2 lb d^3}{24 \rho_c} 10^{-16} \quad$ W

(11.22) $\quad P_{pe} = \dfrac{\pi \omega^2 \hat{B}^2 l d^4}{128 \rho_c} 10^{-16} \quad$ W

into account. His results are most conveniently expressed by introducing a proximity effect factor, G_r, into Eqn (11.22):

$$P_{pe} = \frac{\pi\omega^2 \bar{B}^2 l d^4 G_r}{128\rho_c} \quad \text{W} \tag{11.23}$$

This factor, which is a dimensionless function of d/Δ is plotted in Figure 11.6; it is valid only for round conductors. When d/Δ decreases to unity, $G_r \to 1$; when d/Δ is increasing beyond 4,

$$G_r \to \frac{32}{(d/\Delta)^4}\left(\frac{d}{\Delta} - 1\right) \tag{11.24}$$

It is often necessary to express this loss as the tangent of the proximity effect loss angle, δ_{pe}, referred to the current-carrying winding. For simplicity it will be assumed that there is only one winding concerned and that it is embraced by a magnetic core, e.g. a ferrite pot core. The flux density in the winding space, although non-uniform, is everywhere assumed proportional to the ampere-turns in the winding. Eqn (11.23) may be combined with the following relations:

$$P_{pe} = I^2 R_{pe}$$

where $\overline{B^2}$ is B^2 ($=\hat{B}^2/2$) averaged over the winding space and k is a constant;

$$\overline{B^2} = kN^2I^2$$

where $\overline{B^2}$ is $\hat{B}^2$ ($=B^2/2$) averaged over the winding space and k is a constant;

$$\tan \delta_{pe} = R_{pe}/\omega L$$

and $\quad L = \mu_o\mu_e N^2/C_1$

It follows that

$$\tan \delta_{pe} = \frac{P_{pe}}{\omega L I^2} = \frac{P_{pe}C_1}{I^2\omega\mu_o\mu_e N^2}$$

$$= \frac{\pi\omega^2 kN^2 I^2 l d^4 G_r}{64\rho_c} \times \frac{C_1}{I^2\omega\mu_o\mu_e N^2}$$

$$= \frac{\pi^2 k C_1}{32\rho_c\mu_o} \times \frac{f l d^4 G_r}{\mu_e} \tag{11.25}$$

When d/Δ is less than unity, e.g. at low frequencies, $G_r \simeq 1$ and $\tan \delta_{pe}$ is proportional to the frequency. At higher frequencies the proximity effect usually makes an appreciable contribution to the total loss of a wound component. When $d/\Delta \simeq 3.5$ the value of

---

(11.23) $\quad P_{pe} = \dfrac{\pi\omega^2 \hat{B}^2 l d^4 G_r}{128\rho_c} 10^{-16} \quad \text{W}$

(11.25) $\quad \tan \delta_{pe} = \dfrac{\pi k C_1}{128\rho_c} \times \dfrac{f l d^4 G_r}{\mu_e} \times 10^{-7}$

$\tan \delta_{pe}$ reaches a maximum. At higher frequencies G_r approaches proportionality to $(d/\Delta)^{-3}$ or $f^{-3/2}$, so $\tan \delta_{pe}$ approaches proportionality to $f^{-1/2}$, i.e. it falls with increasing frequency.

A reduction of conductor diameter, d, will be very effective in reducing proximity effect but, by itself, this reduction will rapidly increase the value of R_{dc} and may result in a greater overall loss. The usual method of combating proximity effect is to use bunched conductors consisting of n insulated strands of diameter d. At each end of the winding the strands are soldered together; at low frequencies the conductor has an effective area of $n\pi d^2/4$. The conductor is twisted so that, in its simplest form, each strand follows a helical path. Figure 11.9 shows a side view of two strands in such a bunched conductor; the e.m.f.s induced in each half-twist are cancelled by those in the next. If the bunched conductor has a more complicated composition, e.g. seven twisted bunches each of seven twisted strands, cancellation will still occur. Provided the number of twists in a winding is greater than about ten the eddy currents circulating between the strands may be ignored and only those circulating within the strands need to be considered. Thus Eqn (11.25) may be applied to the strands of the conductor; d then refers to the bare diameter of the strand and l refers to the total length of strands, i.e. Nnl_w, where l_w is the mean turn length. The above equation may now be written in more general form

$$\left.\begin{array}{l}\tan \delta_{pe} = \dfrac{\pi^2 k l_w C_1}{32\rho_c\mu_o} \times \dfrac{fNnd^4 G_r}{\mu_e} \\[2mm] \quad\quad= \dfrac{k_e fNnd^4 G_r}{\mu_e} \\[2mm] \text{or} \\[2mm] \quad\quad \dfrac{k_E fNnd^4 G_r}{A_L}\end{array}\right\} \tag{11.26}$$

where k_e and k_E are alternative forms of proximity effect constant; $k_e = 10^{-9} k_E C_1/\mu_o$ when A_L is in nH per turn^2 and C_1 is in m^{-1}. The constant k_e has the basic units [s m^{-4}] but may be more conveniently expressed in [s mm^{-4}] in which case d is in mm. The constant k_E has the basic units [s m^{-4} H] but may similarly be expressed in terms of mm and nanohenries.

If solid conductors are being considered the same equations may be used but then $n = 1$.

Section 5.7.3 examines, in greater detail, the effects of eddy currents in inductor windings. It is shown that, at frequencies where d/Δ is less than about 1.5 and $G_r \simeq 1$, the combined d.c. and proximity effect loss resistance is minimum when it approximately equals one and a half times the d.c. resistance. Expressions are derived for the optimum conductor diameter to meet this condition. The way in which the proxim-

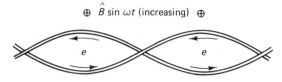

Figure 11.9 The cancellation of eddy-current e.m.f.s induced in twisted strands of bunched conductors by a transverse flux.

ity effect loss tangent varies in relation to the d.c. resistance loss tangent over a wide range of frequencies is also examined.

Since the flux density in the winding space is not uniform and is not easily calculated, the value of k_e or k_E is usually determined experimentally for a given core. It will depend on the geometry of the core, air gap and the winding. The experimental method is basically as follows. Two coils are wound, each having the overall cross-section for which the constant is to be determined, e.g. a full coil former. One is wound with solid conductor of such a diameter that, at the measuring frequency, d/Δ is about 3.5 so that $\tan \delta_{pe}$ is near maximum. The diameter should be preferably on the high side so that the number of turns is not excessive, bearing in mind the need to make self capacitance loss negligible. Further, the frequency should not be so high that the core loss is large compared with the proximity effect loss. The other coil is wound with the same number of turns of a bunched conductor of the same overall diameter. The strand diameter is chosen so that the proximity effect loss is negligible and if the strands are well transposed the skin effect loss may also be ignored. After the d.c. resistance has been measured, each coil is placed in turn in the given core and the overall loss tangent is determined. The inductances are also measured.

For each winding, the loss tangent due to the respective d.c. resistance is subtracted from the measured loss tangent. This leaves only the core loss and the proximity loss tangents in the case of the winding of solid conductor, and just the core loss tangent in the case of the winding of bunched conductor. The resulting loss tangents are subtracted to yield the proximity effect loss tangent of the solid conductor winding.

Thus k_e and k_E may be obtained from Eqn (11.26) since all the other quantities are known (G_r may be obtained by reference to Figure 11.6). The proximity effect constant, having been determined for a given core (with a given air gap), may be used to calculate the proximity effect loss tangent of any winding on that core type (with the same air gap) provided the overall geometry of the winding is similar to that used for the experimental determination. Typical values of k_e and k_E for inductor cores are given in Table 5.5.

The graphs of Figure 5.14 show the relative magnitudes of the winding resistance and eddy current loss tangents calculated for comparable windings of solid and bunched conductors placed in a typical ferrite pot core.

With modern CAD techniques[4] it is possible to compute the leakage flux density at any number of mesh points covering the winding area. Using a suitable post-processing package, the proximity effect loss can then be evaluated from Eqn (11.23) by numerical integration and the proximity effect loss tangent can be calculated from $P_{pe}/\omega L I^2$. This procedure yields results that agree closely with those obtained experimentally.

11.4.4 Total eddy current loss in transformer windings

Within the winding space of an inductor, the leakage flux lines assume a very curved distribution. As seen in the previous section, their effect in inducing eddy currents can only be quantified by an experimentally-measured constant such as k_e or by computer-aided field analysis.

In a transformer, the mutual repulsion of the primary and secondary leakage fields tends to make the fields lie parallel to the winding interface. An assumed parallel field distribution is the basis of the long-established formulae for leakage inductance (see Section 11.7), and has been successfully used for the prediction of eddy current loss in transformer windings.

The ampere turn product, or m.m.f., due to the primary winding in a transformer is very nearly balanced by that of the secondary. In Section 11.7 the m.m.f. diagram is introduced (see Figure 11.20). Moving from the innermost or outermost surface of the winding assembly, the m.m.f. increases linearly from zero until it reaches a maximum at a primary/secondary interface after which it begins to fall linearly. Figure 11.10 shows two common winding arrangements and the corresponding m.m.f. diagrams.

Since the basis of the analyses in this section is the assumption of parallel leakage field it is instructive to see actual flux patterns as computed by a magnetic field analysis program such as MAGGY[4].

Such plots have been computed for both simple and sandwich windings on a pair of E cores. Figure 11.11(a) and (b) show the leakage flux lines within the winding window of the core. However, a large proportion of the winding lies outside this window, e.g. in a direction perpendicular to the plane of the core, so Figure 11.11(c) and (d) show the flux pattern for this region.

It can be seen in Figure 11.11(a) that for a simple two-winding arrangement there is considerable diver-

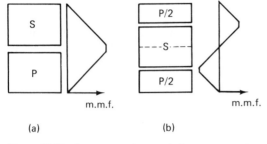

Figure 11.10 Common transformer winding arrangements and the corresponding m.m.f. diagrams. (a) Simple primary/secondary windings; (b) secondary winding sandwiched between a split primary winding

gence of the flux in the corners of the total winding cross-section. This reduces the flux density in these regions so the calculation of eddy currents in a winding composed of round wire, based on parallel flux distribution, will be pessimistic. However, if these windings were to consist of strip conductor, then the radial component of the flux would introduce additional components of eddy current loss and the calculation based on parallel flux would underestimate the loss.

Figure 11.11(c) shows a similar flux pattern in the cross-section perpendicular to the plane of the core. Here the divergence is greater in the winding nearer the core but less in the other winding.

Figure 11.11(b) shows that in a sandwich winding, e.g. a secondary sandwiched between two equal primary portions, as in Figure 11.10(b), the flux is constrained to be substantially parallel over most of the middle winding. Again, the orthogonal cross-section, Figure 11.11(d), shows a similar flux pattern.

These flux plots are useful in showing the reality of the flux paths. However, for the purpose of analysis, parallel flux is assumed and the results have been shown to predict the a.c. resistance with reasonable accuracy.

It must be emphasized that the approximately parallel flux pattern will only hold provided that the widths of the various windings, and layers within windings, are equal. This constraint is assumed in the following analyses and is a condition for obtaining low eddy current losses in practice.

11.4.4.1 A simple analysis

Figure 11.12 shows a schematic cross-section of a winding of N turns carrying a current of r.m.s. amplitude I amps. Usually, in a simple transformer, the minimum flux density within a winding is zero and this occurs at the level remote from the winding interface, see Figure 11.10. However, in a multi-output transformer this condition will not apply for the individual secondary windings, so for the purpose of generality, Figure 11.12 shows the r.m.s. leakage flux density varying linearly from B_{min} at h_1 to B_{max} at h_2, where $B_{min} = \vartheta B_{max}$ and therefore $h_1 = \vartheta h_2$.

From $x = h_1$ to $x = h_2$,

$$B = \frac{x}{h_2} B_{max}$$

or $\quad B^2 = \left(\frac{x}{h_2} B_{max}\right)^2$

The average value of B^2 from h_1 to h_2 is

$$\overline{B^2} = \frac{1}{h_2 - h_1} \int_{h_1}^{h_2} B^2 \, dx = \frac{1}{h_2(1 - \vartheta)} \int_{h_1}^{h_2} \frac{B_{max}^2 x^2}{h_2^2} \, dx$$

$$\therefore \overline{B^2} = \frac{B_{max}^2}{h_2(1 - \vartheta)} \left(\frac{h_2}{3} - \frac{h_1^3}{3h_2^2}\right) = \frac{B_{max}^2}{3} \times \frac{(1 - \vartheta^3)}{(1 - \vartheta)}$$

(11.27)

The return magnetic path outside the winding cross-section has relatively low reluctance due to the presence of the core and the rapid divergence of flux beyond the edges of the winding. It can therefore be assumed that the flux density at a winding height embracing NI amp turns is given by $B = \mu_0 NI/b_w$. Therefore from Figure 11.12,

$$B_{max} = B_{min} + \mu_0 NI/b_w = \vartheta B_{max} + \mu_0 NI/b_w$$
$$= \mu_0 NI/b_w (1 - \vartheta)$$

and from Eqn (11.27),

$$\overline{B^2} = \frac{1}{3} \left(\frac{\mu_0 NI}{b_w}\right)^2 \times k \qquad (11.28)$$

where $k = (1 - \vartheta^3)/(1 - \vartheta)^3$

$= 1$ when $B_{min} = 0$

The following analysis applies only to a multi-layer winding, e.g. a winding having three or more layers, where it can be assumed that the flux density is substantially uniform over a distance Δx comensurate with the conductor diameter.

The proximity effect loss in a round conductor subject to a transverse magnetic field of flux density $\hat{B}$ ($= \sqrt{2}B$) is given by Eqn (11.22). This equation, rather than Eqn (11.23), is used because, as it will be seen later, the optimum conductor diameter for a multi-layer winding is such that $d/\Delta < 2$, so eddy current screening can be ignored and G_r can be taken as unity, see Figure 11.6.

Therefore the total proximity effect loss in a wind-

(11.28) $\quad \overline{B^2} = \frac{1}{3}\left(\frac{4\pi NI}{10 b_w}\right)^2 \times k \quad (Gs)^2$

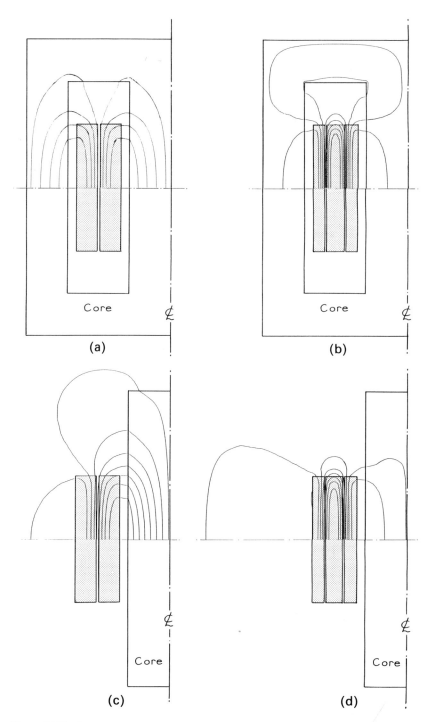

Figure 11.11 Computed plots of the leakage flux in the winding space of a transformer wound on a pair of E cores. The l.h. plots are for a simple two-winding arrangement, while the r.h. plots are for an arrangement in which the secondary is sandwiched between the split primary; (a) and (b) for the region enclosed by the window of the core, (c) and (d) for the region outside the core, in a plane perpendicular to the plane of the core

326 Properties of windings

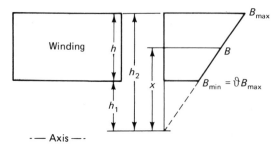

Figure 11.12 The generalized linear variation of the leakage flux density B over the height of a winding h

ing consisting of a number of layers of closely-wound round conductor (of total length $l = Nl_w$) will be given by

$$P_{pe} = \frac{\pi\omega^2 \times 2\overline{B}^2 l d^4}{128\rho_c} \tag{11.29}$$

$$= \frac{2\pi(\omega\mu_o NI)^2 Nl_w d^4 k}{3 \times 128\rho_c b_w^2} \quad W \tag{11.30}$$

from Eqn (11.28).

The power loss due to the d.c. resistance of the same winding is, from Eqn (11.9),

$$P_{dc} = I^2 R_{dc} = \frac{4I^2 Nl_w \rho_c}{\pi d^2} \tag{11.31}$$

The total winding loss is $P_{pe} + P_{dc}$. Differentiating the sum of the expressions in Eqns (11.30) and (11.31) with respect to d and equating to zero gives the condition for minimum winding loss as

$$\frac{4 \times 2\pi(\omega\mu_o NI)^2 Nl_w d^3 k}{3 \times 128\rho_c b_w^2} = \frac{2 \times 4I^2 Nl_w \rho_c}{\pi d^3} \tag{11.32}$$

i.e.
or
$$\begin{aligned} 2P_{pe} &= P_{dc} \\ P_{dc} + P_{pe} &= 1.5 P_{dc} \\ R_{AC} &= 1.5 R_{dc} \\ F_R &= 1.5 \end{aligned} \tag{11.33}$$
and

where F_R is the resistance factor $= R_{AC}/R_{dc}$.

This result, which is independent of k, is the same as that obtained in Section 5.7.3 (Eqn (5.20)). From Eqn (11.32), the conductor diameter, d_{opt}, for minimum winding loss is given by

$$d_{opt}^6 = \frac{3 \times 32}{\pi^4 \mu_o^2} \left(\frac{b_w \rho_c}{fN} \right)^2 \times \frac{1}{k} \tag{11.34}$$

(11.29) $P_{pe} = \dfrac{\pi\omega^2 2\overline{B}^2 l d^4}{128\rho_c} \times 10^{-16}$ W

(11.30) $P_{pe} = \dfrac{32\pi^3(\omega NI)^2 Nl_w d^4 k}{300 \times 128\rho_c b_w^2} \times 10^{-16}$ W

$$\therefore d_{opt} = 92.45 \left(\frac{b_w \rho_c}{fN} \right)^{1/3} \times \frac{1}{k^{1/6}} \quad m \tag{11.35}$$

Therefore at 100 °C, when $\rho_c = 2.246 \times 10^{-8} \, \Omega m$, and for a normal winding in which $B_{min} = 0$,

$$d_{opt} = 2.6 \left(\frac{b_w}{fN} \right)^{1/3} \quad mm \tag{11.36}$$

where b_w is in mm and f is in kHz.

For a winding forming a part of a multi-output transformer where $B_{min} \neq 0$

$$d_{opt} = 2.6 \left(\frac{b_w}{fN} \right)^{1/3} \times \frac{(1-\vartheta)^{1/2}}{(1-\vartheta^3)^{1/6}} \quad mm \tag{11.37}$$

where the units are the same as in Eqn (11.36).

It is of interest to note that if $\vartheta = \frac{1}{2}$, the correction factor is approximately 0.72.

The expression for d_{opt} can be used to obtain a simple relation between F_R and any chosen conductor diameter, d, (provided that it is within the range of validity of the above optimization):

$$F_R = \frac{R_{dc} + R_{pe}}{R_{dc}} = 1 + \frac{P_{pe}}{P_{dc}}$$

If the expressions for P_{pe} and P_{dc} are substituted from Eqns (11.30) and (11.31) respectively and then a substitution made from Eqn (11.34) to introduce d_{opt}, the following result is obtained,

$$F_R = 1 + \tfrac{1}{2} \left(\frac{d}{d_{opt}} \right)^6 \tag{11.38}$$

Reference 5 states that this expression is valid for multi-layer windings provided d/d_{opt} is less than 1.25 (or less than 1.15 when $p = 1.5$, where p is the number of layers per winding portion, defined later in the Dowell analysis). These restrictions are not serious in practice since the equation clearly shows the penalty if the optimum diameter is substantially exceeded.

The same method, with the same conditions assumed, gives the corresponding result for a winding consisting of a spiral coil of strip conductor of width b_w, thickness h and $N \gg 1$. In this case the starting point is Eqn (11.21) in which $b = b_w$ and $d = h$, so

$$P_{pe} = \frac{\omega^2 \times 2\overline{B}^2 Nl_w b_w h^3}{24\rho_c} \quad W \tag{11.39}$$

Again, for the general winding shown in Figure 11.12, $\overline{B}^2$ is given by Eqn (11.28) and P_{dc} is given by Eqn (11.31) in which the conductor cross-sectional area, $\pi d^2/4$, is replaced by hb_w. So the total winding loss is given by

---

(11.39) $P_{pe} = \dfrac{\omega^2 \times 2\overline{B}^2 Nl_w b_w h^3}{24\rho_c} \times 10^{-16}$ W

$$P_{pe} + P_{dc} = \frac{2(\omega\mu_o NI)^2 Nl_w h^3 k}{3 \times 24\rho_c b_w} + \frac{I^2 Nl_w \rho_c}{hb_w} \quad (11.40)$$

Differentiating with respect to h yields the conditions for minimum winding loss as

or
$$\left.\begin{array}{c} 3P_{pe} = P_{dc} \\ P_{pe} + P_{dc} = (4/3)P_{dc} \\ R_{AC} = 1.33 R_{dc} \\ F_R = 1.33 \end{array}\right\} \quad (11.41)$$

and

again independent of k,

and
$$h_{opt}^4 = 12 \left(\frac{\rho_c}{\omega N \mu_o}\right)^2 \times \frac{1}{k} \quad (11.42)$$

$$\therefore \quad h_{opt} = \frac{3.14}{\sqrt{(fN)}} \left(\frac{(1-\vartheta)^3}{1-\vartheta^3}\right)^{1/4} \quad \text{mm} \quad (11.43)$$

or if $B_{min} = 0$

$$h_{opt} = \frac{3.14}{\sqrt{(fN)}} \quad \text{mm} \quad (11.44)$$

where, in the last two equations, f is in kHz and ρ_c is given the value of the resistivity of copper at 100 °C.

Again, using the same method that led to Eqn (11.38), a corresponding expression for F_R in terms of h and h_{opt} (valid[5] for $h/h_{opt} < 1.4$) can be obtained:

$$F_R = 1 + \frac{1}{3}\left(\frac{h}{h_{opt}}\right)^4 \quad (11.45)$$

It should be noted that where a winding is sectioned into portions, e.g. in a split or sandwich winding as in Figure 11.10(b), then N in the above equations is the number of turns in the portion being considered, not the total number of turns in the complete winding.

The above results provide a simple basis for the optimization of transformer windings consisting of a number of layers and enables the a.c. resistance to be calculated with reasonable accuracy provided the conductor thickness does not substantially exceed the optimum and all the winding layers have the same width.

11.4.4.2 The Dowell analysis

A more general analysis was carried out by Dowell[6] who adapted the calculation of eddy current losses in armature conductors[7] to the case of transformer windings. His analysis allows for the non-uniform flux density over the thickness of a layer and also takes into account eddy current screening. Qualitatively, the effect of the latter is that the combination of the transverse and conductor fields normally causes the current density to increase towards the surface of the conductor that faces the nearest winding interface. This not only results in an increase in winding resistance but also a small decrease in the leakage inductance (see Section 11.7) as the frequency rises.

The results of the Dowell analysis may also be expressed in terms of the resistance factor F_R. As in the above simpler analysis, F_R is a function of the winding geometry and a parameter analogous to d/Δ.

The assembly of windings is divided into portions in each of which the low frequency m.m.f. varies linearly from zero. A portion may be a whole winding as in Figure 11.10(a) or it may be half a winding as in secondary of Figure 11.10(b). This analysis does not consider the case of multi-secondaries, in which the m.m.f. does not vary from zero but from some intermediate value (see Figure 11.12).

The parallel flux assumption requires that all the layers should have the same width. Fractional layers must be avoided by dividing the total number of turns per portion equally between the layers, marginally adjusting the total turns as necessary to achieve this. So all winding portions consisting of complete windings, such as S and P in Figure 11.10(a) will have integral numbers of layers. Split windings such as $P/2$ must consist of equal portions each having an integral number of layers; as previously noted, this requires that the complete winding must have an even number of layers.

Sandwich windings such as S in Figure 11.10(b) are also divided into two equal portions. If the complete sandwich winding has an even number of layers then each portion will have an integral number of layers. If, however, the winding has an odd number of layers, then the centre layer is notionally divided into two half-layers, each of full width but having a height of $h/2$. The number of layers per portion is then taken to be an integral number plus a half. If a sandwich winding consists of only one layer then $p = \frac{1}{2}$; the half-height is $h/2$ so the total physical thickness of the winding is h.

For each winding portion the following parameters are required:

The number of layers	p
The number of turns per layer	N_l
The effective conductor height	h
The copper layer factor	F_l

Figure 11.13(a) shows a typical layer of rectangular conductors. The layer copper factor is defined by

$$F_l = \frac{N_l b}{b_w} \quad (11.46)$$

where b_w is the overall winding breadth and b is the conductor breadth. If the conductors are round, the equivalent square conductors must first be derived as in Figure 11.13(b) and then the above parameters may be obtained by putting $h = b = 0.886d$.

From the above parameters the variable, φ, corres-

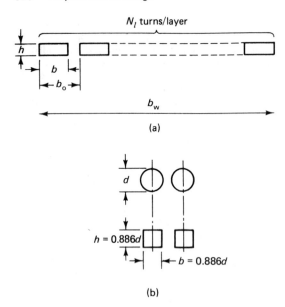

Figure 11.13 Winding layer parameters (note: h, b and d refer to bare conductors). (a) A layer of rectangular conductors, (b) conversion of round conductors to square

ponding to d/Δ in the previous section must be calculated:

$$\varphi = \frac{h\sqrt{F_l}}{\Delta} \quad (11.47)$$

where Δ is the penetration depth, having units consistent with h. Relevant values can be obtained from Eqn (11.15), Table 11.2 or Figure 11.7.

Figure 11.14 shows F_R as a function of φ with the number of layers per portion as a parameter. Given the d.c. resistance of a winding portion, see Section 11.3, the a.c. resistance is $F_R R_{dc}$.

Where two winding portions are the halves of a split or sandwich winding, it is necessary only to obtain F_R for one portion; it will be the same for both. Further, for the purpose of estimation, the mean turn length (for the coil former) may be used to obtain R_{dc} for both portions; the over-estimate for one portion will be approximately cancelled by the under-estimate for the other. However, for a simple winding arrangement as shown in Figure 11.10(a), the error arising from assuming a common mean turn length will not cancel if the two portions have significantly different values of F_R. In such cases and also where precise results are required, the mean turn length should be evaluated for each winding portion.

For a close-packed winding consisting of round conductor of diameter d and overall diameter d_o:

$$h = 0.886d \text{ and } F_l = 0.886d/d_o$$

$$\therefore \varphi = \frac{0.886d \times \sqrt{(0.886d/d_o)}}{\Delta}$$

$$\therefore \varphi = \frac{0.834d \times \sqrt{(d/d_o)}}{\Delta} \quad (11.48)$$

where $\sqrt{(d/d_o)}$ ranges from 0.9 to 0.97 over the practical range of Grade 2 (medium) enamel covered wires. For spaced windings, d_o becomes the pitch of the turns.

Substituting for Δ from Table 11.2, using the value for copper at 100 °C

$$\varphi = 0.352d\sqrt{(fd/d_o)} \quad (11.49)$$

where d is in mm and f is in kHz.

In the region where $F_R < 2$, the curves in Figure 11.14 are closely approximated by the equation:

$$F_R = 1 + \frac{(5p^2 - 1)}{45} \varphi^4 \quad (11.50)$$

For a close-packed winding of round wire, R_{dc} is proportional to d^{-2} (see Section 11.3), and is therefore, from Eqn (11.48), closely proportional to φ^{-2}. So R_{AC} ($= F_R R_{dc}$) will be proportional to φ^{-2} when $\varphi \ll 1$ (from Eqn 11.50)), i.e. where $F_R \to 1$ in Figure 11.14.

Inspection of the curves for $p > 2$ shows that as φ increases beyond the lower region of curvature, F_R approaches proportionality to φ^4 (the steep straight part of these curves). So in this region R_{dc} approaches proportionality to φ^2. It follows that R_{AC}, having been proportional to φ^{-2} for low values of φ, will have a minimum in the region of maximum curvature that characterizes the lower end of the curves. In fact, this can be demonstrated using Eqn (11.50). For a close-packed multilayer winding of round wire, the second term is closely proportional to $p^2\varphi^4$. Since $p = N/N_l \propto Nd/b_w F_l$ this second term is proportional to d^6. If the expression is divided by d^2 so that it is proportional to $F_R R_{dc}$ ($= R_{AC}$) and it is differentiated with respect to d, the minimum R_{AC} occurs when $F_R = 1.5$. This is, of course, in agreement with the result obtained earlier by the direct method and provides a useful connection between the two approaches. Similarly, for a multi-turn winding of strip of given width, where R_{dc} is proportional to $1/h$, division by h yields an optimum when $F_R = 1.33$.

The main advantage of the more general analysis is that the results, expressed in Figure 11.14, apply for values of p down to $\frac{1}{2}$ and also for values to φ extending into the region of eddy current screening represented by the upper straight region where $F_R \propto \varphi$.

As seen from the previous analysis, when $p > 2$ there is no advantage, indeed there is generally a substantial loss penalty, in using a cross-section that is larger than optimum, so the upper region of Figure

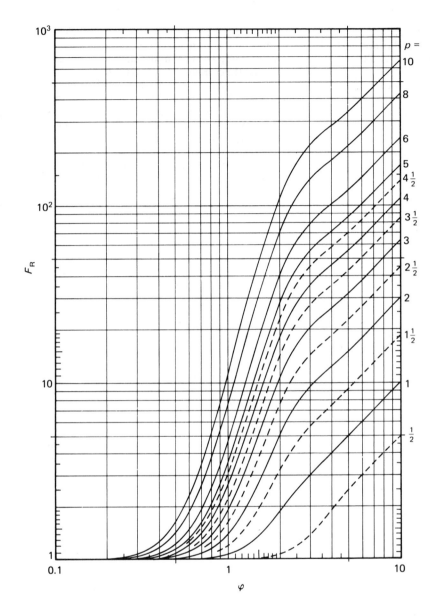

Figure 11.14 Resistance factor F_R as a function of $\varphi = (h\sqrt{F_l})/\Delta$, with the numbers of layers per portion p as a parameter

11.14 is not relevant to winding design except when p is very small.

When $p = 1.5$ or less it is seen from Figure 11.14 that the lower curvature is less pronounced and when $\varphi > 3$ it merges with the unity slope region, indicating that eddy current screening is becoming dominant.

For round conductors there is no optimum diameter as such when $p < 2$; the lowest a.c. resistance will be obtained when the conductor diameter is the largest that can be accommodated within the available winding breadth.

For windings made with tape or full-width strip there is an optimum strip thickness for all values of p. When $p > 2$ the optimum thickness has the value derived in the foregoing simple analysis, see Eqn (11.44), but as p tends to the very low values it becomes larger. For $p \leqslant 2$, Table 9.5 gives the optimum tape or strip thicknesses and the corresponding values of F_R and φ. These results were obtained by computing F_R/h using Dowell's basic equations and finding the minima.

Tape or strip conductor should only be used in a sandwich winding, such as the secondary in Figure 11.10(b). This is because, as observed earlier, the leakage flux is more nearly parallel to the layers in such a winding; any components of flux normal to the surface of the tape or strip will induce additional eddy current loss.

In practice, Figure 11.14 can be used to assess quite rapidly any proposed winding arrangement and so test it for optimum. For this purpose it is useful to have the region below $F_R = 2$ expanded to give a better resolution, and this is done in Figure 9.20.

The practical application of the results presented in this section are considered in Section 9.5.3. Reference 5 provides a more detailed discussion of the subject.

11.5 The inductance of an air-cored coil

The inductance of a cylindrical air-cored solenoid having N turns is given by Nagaoka's formula[8].

$$L = \frac{\mu_o \pi D^2 N^2 K}{4l_c} \times 10^6 \quad \mu H \quad (11.51)$$

where D is the mean diameter of the coil in m
l_c is the axial length of the coil in m
K is a constant depending on l_c/D; as l_c/D becomes larger $K \to 1$.

If K is modified to take account of the radial thickness of the coil, the above equation may be used to calculate the inductance of any cylindrical coil having a rectangular winding section. Since the constant will depend among other things on l_c/D the equation may be simplified to:

$$L = D_o N^2 k \quad \mu H \quad (11.52)$$

where D_o is the outside diameter of the coil in mm,
k is a function of $l_c D_o$ and $D_i D_o$,
D_i is the internal diameter of the coil in mm,
l_c is in mm.

The function k has been calculated from data given by Grover[9] and it is plotted in Figure 11.15. Using this graph and Eqn (11.52) the inductance of any cylindrical air-cored coil may easily be obtained with reasonable accuracy. For greater precision reference should be made to the data by Grover, mentioned above.

11.6 The self capacitance of a winding

11.6.1 Introduction

This subject has been extensively discussed in the literature[10,11,12]; in the present context it is only neces-

sary to describe briefly the basic approach and to summarize the more important results.

It is usually assumed in the analysis of self capacitance that the current distribution in the winding is not significantly different from that which exists at low frequency. More specifically, it is assumed that the voltage between any two points on a winding is proportional to the number of turns separating them. This assumption is adopted in the analytical results quoted here.

The distributed capacitance between turns of a winding, and between the winding and the surroundings, may for most purposes be represented by a self capacitance C_s connected across the winding. The end capacitances shown in Figure 5.13(a) may or may not be included in C_s depending on the circumstances (see also Eqn (5.13)). The general method of calculating C_s is to equate the energy stored in it, i.e. $\frac{1}{2}C_s U^2$, to the sum of all the energy in the distributed capacitance. Assuming that the winding consists of one or more regular layers wound over a screen or a conductive core, three main categories of distributed capacitance may be distinguished, namely, that between adjacent layers, that between layers and screen (or core) and that between adjacent turns. The self capacitance reflected across the terminals of a winding having N turns, due to the capacitance between a pair of adjacent turns, is $(1/N)^2 \times$ the adjacent-turn capacitance so the total contribution of the adjacent-turn capacitance is only significant when there are few turns. Thus it is in single-layer windings that it becomes important. In such windings the capacitance between non-adjacent turns may also contribute significantly. The problem of the isolated single-layer winding is considered at the end of this section.

11.6.2 A general method of analysis

It follows from the above discussion that the problem is to find an expression for C_s in terms of the inter-layer and/or the layer-to-screen capacitance. In the following treatment, which is based on a paper by Duerdoth[10], the term 'screen' is taken to include any conductive surface, such as a core, adjacent to the first layer.

Figure 11.16 shows two adjacent surfaces representing winding layers or a layer and a screen. In general the potential of surface A may be taken as varying linearly from U_{AO} at the lower end to U_{AD} at the upper end. Similarly, the potential of the surface B varies linearly from U_{BO} to U_{BD}. If surface A is a screen, the surface is at uniform potential and $U_{AO} = U_{AD}$. The potential difference between the two surfaces will in general vary linearly along the length from $U_O = U_{BO} - U_{AO}$ at the bottom to $U_D = U_{BD} - U_{AD}$ at the top, possibly becoming zero at some

---

(11.51) $\quad L = \dfrac{\pi^2 D^2 N^2 K 10^{-3}}{l_c} \quad \mu H$

Properties of windings 331

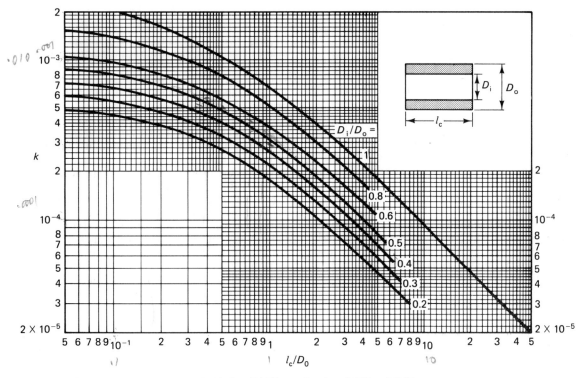

Figure 11.15 Inductance of air-cored coils: k in Eqn (11.52) as a function of l_c/D_o and D_i/D_o

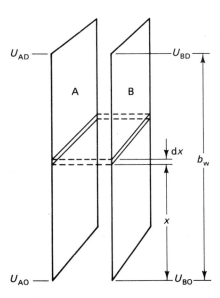

Figure 11.16 Calculation of self capacitance due to winding layers. Representation of two adjacent conductive layers each having a linear potential distribution. (After Duerdoth[10])

point depending on the polarity and distribution of the surface potentials.

The potential difference at the element dx is

$$U_o + (U_D - U_o)\frac{x}{b_w} \quad \text{V}$$

where b_w is the width of the layer. The associated electrical energy is

$$dW = \tfrac{1}{2}C_l \frac{dx}{b_w} \left\{ U_o + (U_D - U_o)\frac{x}{b_w} \right\}^2 \quad \text{J}$$

where C_l is the capacitance between the two surfaces considered as a parallel-plate capacitor, i.e. when there is a uniform voltage distribution over each surface. Integrating over the surface, the total energy is

$$W = \frac{C_l}{6}(U_o^2 + U_o U_D + U_D^2) \quad \text{J}$$

U_o and U_D must be of consistent sign convention. If one of the surfaces represents a layer of a winding having a total voltage U_p across it then the self capacitance, C_s, appearing across the terminals of that winding due to the distributed capacitance of that layer may be obtained by equating energies.

$$\tfrac{1}{2}C_s U_p^2 = \frac{C_l}{6}(U_o^2 + U_o U_D + U_D^2)$$

$$\therefore \quad C_s = \frac{C_l}{3 U_p^2}(U_o^2 + U_o U_D + U_D^2) \quad \text{F} \quad (11.53)$$

If there are several layers of distributed capacitance contributing to the total self capacitance of a winding, the contributions of these layers must be added to the r.h.s. of this equation to obtain the total self capacitance, observing that C_l may not necessarily be the same for each layer.

As an example, an expression will be derived for the self capacitance of the two-layer winding over a screen illustrated in Figure 11.17. Let the total applied voltage be U_p and let C_m and C_l represent the parallel-plate capacitance between layer and screen, and between layers, respectively. Applying Eqn (11.53) to each part:

$$C_s = \frac{C_m}{3U_p^2}\left(0 + 0 + \frac{U_p^2}{4}\right) + \frac{C_l}{3U_p^2}(U_p^2 + 0 + 0)$$

$$= \frac{C_m}{12} + \frac{C_l}{3} \qquad (11.54)$$

Table 11.3 gives expressions for the self capacitance of a number of commonly used winding arrangements. The first part of this table is concerned with single windings varying in configuration from one layer to a multi-layer winding having several sections. In most cases the screen is also shown and this is connected to some point on the winding. In practice, even if no physical connection is made, the screen or core may often be considered to be at a potential equal to that at some point on the winding. It has been convenient in the table to consider the windings as wound *over* the screen but the results are, of course, true for the inverted arrangement.

The second part of Table 11.3 is concerned with the winding arrangements commonly used in wide-band transformers. These are all simple single-layer arrangements and there is usually a screen which is connected to, or considered to be equal in potential to, a point on one of the windings. The windings are connected together at points which are assumed to be of equal potential. For transformers having many layers it may be possible to neglect the winding-winding or the winding-screen capacitances and calculate the self capacitance of each winding considered separately as in the first part of the table. For the intermediate cases of transformers having a few layers the self capacitance must be calculated from first principles by the method described above.

It should be noted that in the second section of the table a number of additional variations are implied. For example, for item 8 the variations may be set out as in Table 11.4.

To use the expressions for self capacitance it is necessary to determine C_l and C_m. This may be done experimentally by measuring the direct capacitance between two single layers or between a layer and a screen; the winding conductor and interleaving insulation must correspond with the proposed winding but any difference in superficial area may be corrected by proportion.

Alternatively the values may be calculated by treating the adjacent surfaces as a parallel plate capacitor in which

$$C = \frac{\varepsilon_o \varepsilon A}{S_e} 10^{12} \quad \text{pF} \qquad (11.55)$$

where $\varepsilon_o = 8.854 \times 10^{-12}$
ε is the effective permittivity of the dialectric
A is the area of the surface $(= b_w l_w)$ in m²
S_e is the effective thickness of the dialectric in m.

If the dimensions are in mm

$$C = 0.008854 \frac{\varepsilon A}{S_e} \quad \text{pF} \qquad (11.55a)$$

Since one or both of the surfaces consist of a layer of round conductors the effective thickness of the dialectric is not defined.

Empirical formulae for calculating S_e were proposed by Zuhrt[13] and are quoted in Figure 11.18. These are claimed to be accurate to 2% provided d is between $0.5c$ and $0.9c$ and S is between $1.2c$ and $1.8c$.

In the absence of more specific data the value of the effective permittivity of the dialectric may be taken as 4 if the winding is impregnated with wax and 3.5 if it is unimpregnated.

A useful guide to the self capacitance of a simple multi-layer winding may be obtained from expression 5 of Table 11.3(a) and Eqn (11.55). Considering only the inter-layer capacitance and assuming that there

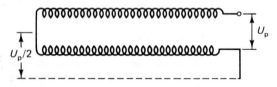

Figure 11.17 Representation of two layers wound over a screen. (After Duerdoth[10])

$$(11.55) \quad C = 0.08854 \frac{\varepsilon A}{S_e} \quad \text{pF}$$

Properties of windings

Table 11.3(a) Self capacitance of various arrangements of a single winding and screen.

C_m is the low-frequency capacitance that would be measured between layer and screen when disconnected, (see Eqn 11.55).
C_l is the low frequency capacitance that would be measured between adjacent layers when disconnected, (see Eqn (11.55)).
C_m and C_l refer to layers extending across the full winding width except in case 6 where they refer to the layers in each section.
p is the number of layers wound one above another.
q is the number of sections.
N is the number of turns.

Winding arrangement	Description	Self capacitance
1	Unbalanced single layer winding and screen	$\dfrac{C_m}{3}$
2	Balanced single layer winding and screen	$\dfrac{C_m}{12}$
3	Balanced double layer winding and screen	$\dfrac{1}{12}\left(\dfrac{C_m}{4}+C_l\right)$
4	Double layer winding, both layers starting at r.h.s. Start connected to screen	$\dfrac{C_m}{12}+\dfrac{C_l}{4}$
5	Normal winding of p layers, start connected to screen	$\dfrac{C_m}{3p^2}+\dfrac{4C_l(p-1)}{3p^2}$
6	Normal winding with q sections, each with p layers: without screen	$\dfrac{4C_l(p-1)}{3p^2 q}$
	Additional self capacitance due to screen:	
	$q=2$	$\dfrac{C_m}{3}\cdot\dfrac{3p^2+3p+2}{4p^2}$
	$q=3$	$\dfrac{C_m}{3}\cdot\dfrac{5p^2+3p+1}{3p^2}$
	$q=4$	$\dfrac{C_m}{3}\cdot\dfrac{42p^2+18p+4}{16p^2}$
7	Single spiral of flat strip	$C_l\cdot\dfrac{N-1}{N^2}$

are a large number of layers

$$C_s \propto \frac{C_l}{p} \propto \frac{A}{pS_e} \propto \frac{b_w l_w}{pS_e}$$

where b_w = width of the winding
l_w = mean turn length.

For a given winding cross-sectional area $A_w (= b_w h_w)$ and a given number of turns it follows

Table 11.3(b) Self capacitance of various arrangements of transformer windings and screen.

The self capacitances are referred to the primary winding, i.e. the lower winding in each diagram.
r = Secondary e.m.f./primary e.m.f. = turns ratio. The expressions given apply when both primary and secondary are wound in the same direction. If the winding directions are opposite, r must be treated as a negative number.
The connection shown between primary and secondary need not be a physical one: it is intended to indicate points of common potential.
Other symbols as in Table 11.3(a).

Winding arrangement	Description	Self capacitance referred to primary
8	Unbalanced single layer windings, both connected to the screen at one end	$\dfrac{C_m}{3}+\dfrac{C_l}{3}(1-r)^2$
9	Unbalanced single layer windings, opposite ends connected to the screen	$\dfrac{C_m}{3}+\dfrac{C_l}{3}(1+r+r^2)$
10	Unbalanced single layer windings connected to interposed screen. Winding to screen capacitances equal	$\dfrac{C_m}{3}(1+r^2)$
11	Balanced to unbalanced single layer windings without screen	$\dfrac{C_l}{3}\left(1-\dfrac{r}{2}+\dfrac{r^2}{4}\right)$
12	Balanced to unbalanced single layer windings with interposed screen. Winding to screen capacitances equal	$\dfrac{C_m}{3}\left(1+\dfrac{r^2}{4}\right)$
13	Balanced to balanced single layer windings without screen	$\dfrac{C_l}{12}(1-r)^2$

Table 11.4 Variations of expression in Table 11.3(b) using expression 8 as example

	Self capacitance referred to primary, C_s	Self capacitance referred to secondary, $C'_s = C_s/r^2$
r very large	$\dfrac{C_m}{3}+\dfrac{C_l r^2}{3}$	$\dfrac{C_m}{3r^2}+\dfrac{C_l}{3}$
r very small	$\dfrac{C_m}{3}+\dfrac{C_l}{3}$	$\dfrac{C_m}{3r^2}+\dfrac{C_l}{3r^2}$
$r=1$	$\dfrac{C_m}{3}$	$\dfrac{C_m}{3}$

334 Properties of windings

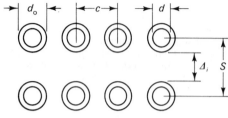

$S_e = S - 1.15d + 0.26c$
$ = \Delta_i + d_o - 1.15d + 0.26c$

For an unspaced winding, $c = d_o$,
then $S_e = \Delta_i + 1.26d_o - 1.15d$

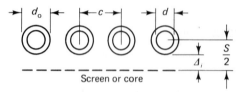

$S_e = \frac{1}{2}(S - 1.15d + 0.26c)$
$ = \Delta_i + \frac{1}{2}(d_o - 1.15d + 0.26c)$

For an unspaced winding, $c = d_o$,
then $S_e = \Delta_i + \frac{1}{2}(1.26d_o - 1.15d)$

Figure 11.18 Effective thickness of dielectric between adjacent winding layers or between a winding layer and a screen, (after Zuhr[13]). See wire tables in Appendix B for values of d and d_o

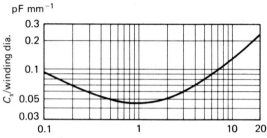

Figure 11.19 Self capacitance of a physically isolated single layer winding, according to Medhurst.[14] One end of the winding is earthed and the capacitance due to the connection leads has not been included

from Eqn (11.1) that

$$p = \frac{h_w}{d_o} = h_w\sqrt{(N/A_w)} \propto h_w$$

S_e will be constant because if the winding cross-sectional area and the number of turns are both constant the wire diameter is constant,

$$\therefore \quad C_s \propto \frac{b_w l_w}{h_w} \qquad (11.56)$$

This expression shows the relation between self capacitance and the geometry of the winding. Clearly it is an advantage to use a winding area which is tall in comparison with its breadth. Sectionalizing is an artificial method of achieving this if the given winding area is unfavourable.

For random or scramble wound coils the calculation of self capacitance will obviously be less reliable than the corresponding calculation for multi-layer coils. However, provided the winding is laid evenly using a traversing pitch appropriate to a layer winding then a reasonable guide to the self capacitance of a cylindrical winding may be obtained by assuming that it is built up as shown in Figure 11.2(a) and using expression 5 of Table 11.3(a). The number of layers may be taken as the winding height divided by the overall wire diameter, and C_l may be calculated from Eqn (11.55) and Figure 11.18 by putting $\Delta_i = 0$.

As an example, consider a 25 mm diameter ferrite pot core. The winding height h_w is 4 mm, the winding breadth is 9.0 mm and the mean turn length is 52.7 mm. Therefore the area of the mean layer is 474 mm². Assuming an unimpregnated winding consisting of 0.2 mm Grade 1 enamel covered wire ($d_o = 0.23$ mm)

$\varepsilon = 3.5$
$S_e = 1.26 d_o - 1.15 d = 0.060$ mm

Therefore from Eqn (11.55(a))

$$C_l = \frac{0.008854 \times 3.5 \times 474}{0.06}$$

$$= 245 \text{ pF}$$

The number of layers, $p = 4.0/0.23 = 17$. Therefore from expression 5 of Table 11.3(a)

$$C_s = \frac{4}{3} \times \frac{16}{289} \times 245 = 18 \text{ pF}$$

This corresponds to the element designated C_a in Figure 5.13 and is in close agreement with the results quoted in Table 5.4. The value of C_a obtained on a given coil former wound to a given winding height depends very little on the number of turns because as the wire diameter gets smaller, the number of layers increases (tending to reduce C_a) while S_e decreases (tending to increase C_a).

To calculate the capacitance between the start of a multi-layer winding and the core, assuming that the core is isolated from the winding and that the number of layers is large, Eqn (11.55(a)) may again be used, S_e

being obtained from the lower part of Figure 11.18 by putting Δ_i equal to the separation of the first layer from the core and choosing an appropriate value of ε. If Δ_i is large compared with the wire diameter then S_e may be taken as equal to Δ_i.

The discussion so far has not applied to physically isolated single-layer windings. The single-layer windings considered have been in the proximity of a conductive layer so that the capacitance between the turns and that layer have made the major contribution to the self capacitance. For an isolated single-layer winding the capacitance between turns and the capacitance between the winding as a whole and its (distant) surroundings are both important.

Medhurst[14] studied this subject and came to the conclusion that the self capacitance is independent of the conductor diameter and the spacing of the turns; it depends only on the diameter and axial length of the winding. His experimental results, based on measurements of a wide variety of single-layer windings, are summarized in Figure 11.19. These results may also be used to estimate the self capacitance of a single-layer winding located approximately half-way up the winding area of a ferrite core, e.g. a pot core, particularly if the ferrite is of a high resistivity grade.

11.7 Leakage inductance in transformers

11.7.1 General

The flux linkage beteeen two windings or parts of the same winding is never complete. In addition to the mutual flux, which does link both of the windings, there are various leakage fluxes. Some are those which fail to link all the turns of the winding which generates them, while others, having completely linked the generating winding, fail to link all the turns of an adjacent winding. This situation is obvious in the case of an air-cored mutual inductance, but, to a lesser degree, it also applies to windings surrounded by a high permeability core as shown in Figure 11.20.

If the primary winding, P, is energized by a current flowing in the direction shown then, if the secondary winding is loaded, the secondary current will flow in the reverse direction. The instantaneous difference between the primary and secondary ampere turns will equal the instantaneous magnetizing ampere turns and determine the instantaneous mutual flux. In addition there will be leakage fluxes shown diagrammatically by broken lines in the l.h.s. of Figure 11.20. In practice these fluxes combine, within the core, with the main flux and so lose their individual identities. The resultant flux pattern not only depends on the geometry of the winding and core but also varies with time over the cycle.

If the secondary is short-circuited, the main flux which links both windings will be negligible because the primary and secondary ampere turns almost cancel. The flux pattern will then be as shown in the l.h.s. of Figure 11.20. The fluxes contributed by each winding are in the same direction in the winding space and mutually repel. Figure 11.11 shows computer plots of the leakage flux for two common transformer configurations. It is seen that within the winding area the mutual repulsion causes the leakage flux to lie approximately parallel to the winding interface. For the purpose of analysis it is assumed to be parallel. The residual e.m.f. due to the leakage flux will appear as inductance in series with the primary terminals and is termed the leakage inductance referred to the primary. The symbol L_l will be used to denote leakage inductance. Its magnitude may be calculated by equating its energy, $\frac{1}{2}L_l I^2$, to the magnetic energy of the leakage flux.

The r.h.s. of Figure 11.20 shows an elementary layer of winding, of thickness dx, situated at a distance x from the inner surface of the primary winding. The field strength along the flux path which includes this layer depends on the number of ampere turns linked by the path, i.e.

$$\oint H ds = N_1 I_1 \frac{x}{h_1} \quad \text{A}$$

where s is the distance along the flux path
N_1 is the number of primary turns
I_1 is the primary current.

Since the reluctance of the path within the magnetic core is negligible compared with that of the path in the winding, H may be taken as the field strength in the winding layer dx. It is assumed to be constant along the layer and this is supported by the experimental results. As previously indicated, the uniformity arises mainly from the mutual repulsion of the leakage fluxes which ensures that the flux direction is substantially parallel to the interface between the windings (if the primary and secondary were wound side-by-side, the leakage flux would be perpendicular to the direction shown in Figure 11.20). Thus

$$H = \frac{N_1 I_1}{b_w} \frac{x}{h_1} \quad \text{A m}^{-1} \qquad (11.57)$$

It is usual to use the winding breadth, b_w, rather than the window breadth, b_a, because the flux disperses rapidly on leaving the winding and the associated energy is much reduced. This is particularly so at

---

(11.57) $\quad H = \dfrac{4\pi N_1 I_1}{10 b_w} \dfrac{x}{h_1} \quad$ Oe

336 Properties of windings

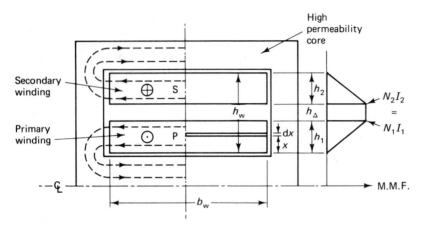

Figure 11.20 Leakage inductance in a simple transformer. The l.h.s. of the diagram represents the leakage flux paths; the r.h.s. shows how the magnetomotive force varies with the distance x from the inner surface of the winding

winding edges that may be outside the confines of the core, e.g. an E or U core (see also Figure 11.11).

The magnetomotive force will vary linearly from zero when $x = 0$ to $N_1 I_1$ when $x = h_1$, as shown in the diagram on the r.h.s. of Figure 11.20. It will be constant across the interwinding space because the ampere turns embraced by the line integral in this region are constant. As the secondary winding space is traversed the magnitude of the magnetomotive force will fall linearly to zero since $N_1 I_1 = -N_2 I_2$.

The volume of the elementary layer is $l_w b_w \, dx$, therefore the energy stored in the field is

$$\frac{\mu_0}{2} \int_0^h H^2 dx l_w b_w$$

The mean length, l_w, is usually taken as a constant for the windings as a whole, i.e. it is based on $h_w/2$. The energy in the total winding space is then

$$\mu_0 \frac{l_w b_w}{2} \left\{ \int_0^{h_1} \left(\frac{N_1 I_1 x}{b_w h_1} \right)^2 dx + \left(\frac{N_1 I_1}{b_w} \right)^2 h_\Delta + \int_0^{h_2} \left(\frac{N_2 I_2 x}{b_w h_2} \right)^2 dx \right\}$$

$$= \tfrac{1}{2} L_l I_1^2 \text{ by definition}$$

Since $|N_1 I_1| = |N_2 I_2|$ this equation reduces to

$$\tfrac{1}{2} L_l I_1^2 = \frac{\mu_0 l_w}{2 b_w} \left(\frac{h_1 + h_2}{3} + h_\Delta \right) N_1^2 I_1^2$$

$$\therefore \quad L_l = \mu_0 N_1^2 \frac{l_w}{b_w} \left(\frac{h_1 + h_2}{3} + h_\Delta \right) \quad \text{H} \quad (11.58)$$

This equation gives the value of the leakage inductance referred to the primary. The value referred to the secondary may be obtained by simply substituting N_2 for N_1.

If the physical dimensions are in millimetres:

$$L_l = 4\pi 10^{-4} N_1^2 \frac{l_w}{b_w} \left(\frac{h_1 + h_2}{3} + h_\Delta \right) \quad \mu\text{H} \quad (11.59)$$

This method of derivation may be extended to a variety of winding arrangements. Figure 11.21 gives results for windings sectionalized various ways. Other arrangements are considered in the literature.

It often happens that Σx_Δ is negligible compared with Σx so that $(\Sigma x/3 + \Sigma x_\Delta) \propto X$, the total winding dimension perpendicular to the inter-section layers (see Figure 11.21). Therefore, for a given winding arrangement $L_l \propto N^2 l_w X/Y$. For a transformer having winding sections wound one above another $X = h_w$, the winding height, and $Y = b_w$, the winding breadth, so

$$L_l \propto N_1^2 \frac{l_w h_w}{b_w} \quad (11.60a)$$

while for a transformer having sections wound side-by-side $X = b_w$ and $Y = h_w$ so

$$L_l \propto N_1^2 \frac{l_w b_w}{h_w} \quad (11.60b)$$

The latter arrangement requires a winding area which is tall and narrow if the leakage inductance is to be made as small as possible. Some of the advantage would be lost because such a winding shape would lead to a larger value of l_w. For these reasons the side-by-side arrangement is not often used. The more conventional method of winding the sections one above another leads to a leakage inductance which is related to the winding area geometry as shown in Eqn (11.60a), i.e. for minimum leakage inductance the winding area is wide and shallow. This is the reverse of the requirement for least self capacitance (see Eqn (11.56)).

(11.58) $\quad L_l = 4\pi 10^{-9} N_1^2 \frac{l_w}{b_w} \left(\frac{h_1 + h_2}{3} + h_\Delta \right) \quad$ H

Properties of windings 337

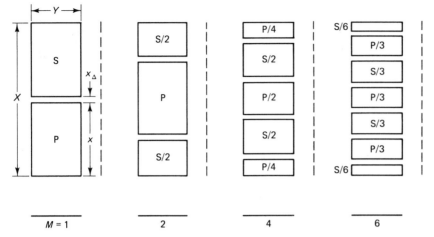

Figure 11.21 The leakage inductance of commonly used winding arrangements. The diagram and the equation below may be used for transformers in which the primary and secondary sections are wound one above the other, i.e. where $X=h_w$, $Y=b_w$ and the solid line represents the winding axis, or for transformers in which the sections are wound side-by-side, i.e. where $X=b_w$, $Y=h_w$ and the broken line represents the winding axis.

$$L_l = 4\pi 10^{-4} \frac{N^2 l_w}{M^2 Y} \left(\frac{\Sigma x}{3} + \Sigma x_\Delta \right) \quad \mu H$$

where N is the number of turns on the winding to which the leakage inductance is to be referred
M is the number of section interfaces
l_w is the mean turn length in mm
X is the overall winding dimension perpendicular to the section interfaces in mm
Y is the winding dimension parallel to the section interfaces, in mm
Σx is the sum of all section dimensions perpendicular to the section interfaces, in mm, i.e. $X - \Sigma x_\Delta$
Σx_Δ is the sum of all inter-section layer thicknesses, in mm

The presence of an air gap in the core usually has little effect on the leakage inductance provided it is not so large as to cause appreciable distortion of the field in the winding space. Since the air gap reduces the primary open-circuit inductance, L_p, the ratio L_p/L_l is reduced and this, as discussed in Chapter 7, reduces the possible bandwidth of the transformer.

At high frequencies the current distribution in the conductors will be disturbed due to eddy currents. The effect is to cause the current to concentrate at the surface of the conductor that faces the nearest winding interface. This results in a lower leakage inductance, although the reduction is only significant when the windings consist of few layers of fairly thick conductors. Dowell[6] has analysed this effect and has given data for the calculation of high frequency leakage inductance. It is related to the high frequency resistance of transformer windings discussed in Section 11.4.4 and is only significant when there is an appreciable high frequency contribution to the winding resistance.

11.7.2 Auto-transformer windings and bifilar windings

A diagram of a typical auto-transformer is shown in Figure 11.22. Between the start of the primary and the secondary tap the magnetomotive force rises linearly as before, reaching $I_1(N_1 - N_2)$. The secondary ampere-turns are then superimposed on those of the primary so that viewed from the other end of the winding the magnetomotive force at the tap is

$$N_2(I_2 - I_1) = N_2(I_1 N_1/N_2 - I_1) = I_1(N_1 - N_2)$$

Integrating as before, the leakage inductance referred to the primary winding is found to be

$$L_l = \frac{\mu_0 l_w}{b_w}(N_1 - N_2)^2 \frac{h_w}{3} \quad H \qquad (11.61)$$

If the physical dimensions are in mm

$$L_l = 4\pi 10^{-4} \frac{l_w}{b_w}(N_1 - N_2)^2 \frac{h_w}{3} \quad \mu H \qquad (11.62)$$

Thus an auto-transformer has lower leakage inductance than the corresponding two-winding transformer.

A bifilar wound transformer is one in which the primary and secondary windings are wound together by feeding on the two winding conductors side-by-side so that each layer consists of alternate primary and secondary conductors. This represents the ultimate intermingling of the windings and if the primary and secondary have equal numbers of turns then the leakage inductance will be extremely small; the self

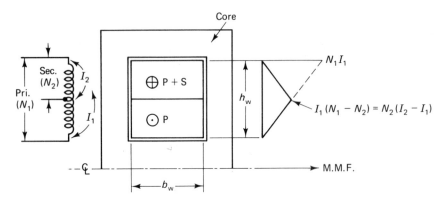

Figure 11.22 Leakage inductance in an auto-transformer. The r.h.s. of the diagram shows how the magnetomotive force varies with the distance from the inner surface of the winding

capacitance may, however, be prohibitively large. If the numbers of turns are not equal then there will be a part of the larger winding in which there is no bifilar cancellation of the field. The situation is then similar to the auto-transformer and the result derived above may be used.

The case of the multi-secondary transformer has been considered in relation to eddy current losses in windings in Section 11.4.4. Here the leakage flux does not necessarily vary from zero over the winding height. Figure 11.12 shows the general case and an expression for the mean square leakage flux density, which is proportional to the magnetic energy, is derived in the associated text. This may be applied to obtain an expression for the corresponding leakage inductance.

11.8 References and bibliography

Section 11.4.1
1. WELSBY, V. G. *The theory and design of inductance coils*, Macdonald and Co., London, 2nd Edn (1960)

Sections 11.4.2 and 11.4.3
2. HOWE, G. W. O. The application of telephone transmission formulae to skin effect problems, *J. Instn elect. Engrs.*, **54**, 473 (1916)
3. BUTTERWORTH, S. Effective resistance of inductance coils at radio frequency, *Expl. Wireless*, **3**, 203 (1926)
   HOWE, G. W. O. The high frequency resistance of wires and coils, *J. Instn elect. Engrs*, **58**, 152 (1920)
   BUTTERWORTH, S. Eddy current losses in cylindrical conductors with special applications to the alternating current resistance of short coils, *Phil. Trans. R. Soc.*, **A222**, 57 (1922)
   CASIMIR, H. B. G. and UBBINK, J. The skin effect, *Philips tech. Rev.*, **28**, 271 (Part 1) and 300 (Part 2) (1967)
4. DE BEER, A., POLACK, S. J., WACHTERS, A. J. H. and WELIJ, J. V. *MAGGY User manual*, Publ. UDV-DSA/SCA/AdB/SP/JvW/77/067/ap, Cat. No.

4322 270 20661, Philips Gloeilampenfabrieken, Eindhoven, The Netherlands (1985)

Section 11.4.4
SNELLING, E. C. Ferrites for power applications, *Paper No. 6*, in *Colloq. on trends in forced commutation components*, Instn. Elect. Engrs., Digest No. 1978/3 (1978)
5. JONGSMA, J. High-frequency ferrite power transformer and choke design, Part 3; Transformer winding design, *Philips Elcoma Publication*, 9398 923 40011, Eindhoven (1982)
6. DOWELL, P. L. Effects of eddy currents in transformer windings, *Proc. Instn elect. Engrs*, **113**, 1387 (1966)
7. FIELD, A. B. Eddy currents in large slot-wound conductors, *Proc. Am. Inst. elect. Engrs*, **24**, 659 (1905)

Section 11.5
8. NAGAOKA, H. The inductance coefficients of solenoids, *J. Coll. Sci. imp. Univ. Tokyo*, **27**, art 6 (1909)
9. GROVER, F. W. *Inductance calculations*, van Nostrand Company (1946)

Section 11.6
10. DUERDOTH, W. T. Equivalent capacitances of transformer windings, *Wireless Engr*, **23**, 161 (1946)
11. MAURICE, D. and MINNS, R. H. Very-wide band radio frequency transformers, *Wireless Engr*, **24**, 168 (Part 1) and 209 (Part 2) (1947)
12. MACFADYAN, K. A. *Small transformers and inductors*, Chapman and Hall (1953)
13. ZUHRT, H. Einfache Näherungsformeln für die Eigenkapazität Mehrlagiger-Spulen, *Elektrotech. Z.*, **55**, 662 (1934)
14. MEDHURST, R. G. H.F. resistance and self-capacitance of single-layer solenoids, *Wireless Engr*, **24**, 35 (1947)

Section 11.7
STAFF OF DEPT. OF ELECT. ENGNG, M.I.T. *Magnetic circuits and transformers*, John Wiley & Sons, New York (1943)
MAURICE, D., *et al.*, see Section 11.6
MATSCH, W. M. *Capacitors, magnetic circuits and transformers*, Prentice-Hall, Englewood Cliffs, N.J. (1964)
GROSSNER, N. R. *Transformers for electronic circuits*, McGraw-Hill Book Co. (1983)

Appendix A
Calculation of the core factors of a pot core

Two core factors are used in the calculation of the effective magnetic dimensions of a magnetic circuit. They are introduced and defined in Section 4.2.1:

$$C_1 = \Sigma l/A, \qquad C_2 = \Sigma l/A^2$$

where A is the cross-sectional area of an element, of length l, in the magnetic path.

The procedures for calculating these core factors have been laid down in IEC Publication 205 (1966). Thus the following procedure is in accordance with that recommendation.

Figure A.1 represents the essential geometry of a pot core and defines the symbols used. The object is to derive expressions for $\Sigma l/A$ and $\Sigma l/A^2$ for each section of the magnetic path; the core factors C_1 and C_2 are then the sum of the respective expressions. The mean magnetic path is assumed to divide the main cross-sections into equal areas, thus the dimensions S_1 and S_2 are not identically equal to half the radial thickness of the inner and outer rings respectively.

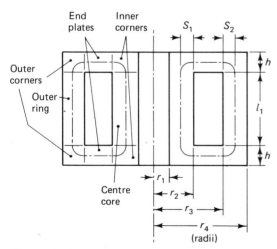

Figure A.1 Cross-section of a simple pot core, defining symbols and indicating sections of the magnetic path

A.1 The outer ring

Cross-sectional area of the ring $= \pi(r_4^2 - r_3^2)$

$$\therefore \quad \Sigma l/A = \frac{l_1}{\pi(r_4^2 - r_3^2)}$$

$$\Sigma l/A^2 = \frac{l_1}{\pi^2(r_4^2 - r_3^2)^2}$$

A.2 The end plates

For an elementary ring at a radius x in the end plate

$$\frac{l}{A} = \frac{dx}{2\pi x h}$$

$$\therefore \quad \Sigma l/A = \frac{1}{2\pi h}\int_{r_2}^{r_3}\frac{dx}{x} = \frac{1}{2\pi h}\log_e\frac{r_3}{r_2}$$

and $\Sigma l/A^2 = \dfrac{1}{4\pi^2 h^2}\int_{r_2}^{r_3}\dfrac{dx}{x^2} = \dfrac{1}{4\pi^2 h^2}\left(\dfrac{1}{r_2} - \dfrac{1}{r_3}\right)$

$$\therefore \quad \text{For both plates } \Sigma l/A = \frac{1}{\pi h}\log_e\frac{r_3}{r_2}$$

$$\Sigma l/A^2 = \frac{1}{2\pi^2 h^2}\left(\frac{r_3 - r_2}{r_2 r_3}\right)$$

A.3 The centre core

Cross-sectional area of the centre core $= \pi(r_2^2 - r_1^2)$

$$\therefore \quad \Sigma l/A = \frac{l_1}{\pi(r_2^2 - r_1^2)}$$

$$\Sigma l/A^2 = \frac{l_1}{\pi^2(r_2^2 - r_1^2)^2}$$

A.4 The outer corners

The magnetic path divides the cross-sectional areas of the end plate and outer ring into equal parts. The magnetic path is taken as a quadrant having a mean radius and the cross-sectional area is taken as the mean of the two terminating areas.

First it is necessary to evaluate an expression for S_2. By the equal-area definition

$$\pi(r_3 + S_2)^2 - \pi r_3^2 = \frac{\pi}{2}(r_4^2 - r_3^2)$$

$$\therefore\ S_2^2 + 2r_3 S_2 - \frac{1}{2}(r_4^2 - r_3^2) = 0$$

$$\therefore\ S_2 = -r_3 + \sqrt{\left(\frac{r_3^2 + r_4^2}{2}\right)}$$

The length of one magnetic path $= \frac{\pi}{4}(S_2 + h/2)$

The mean cross-sectional area

$$= \frac{1}{2}(2\pi r_3 h + \pi(r_4^2 - r_3^2))$$

$$= \frac{\pi}{2}(2r_3 h + r_4^2 - r_3^2)$$

Therefore, for the combination of both outer corners:

$$\Sigma l/A = \frac{(S_2 + h/2)}{2r_3 h + r_4^2 - r_3^2}$$

$$\Sigma l/A^2 = \frac{2(S_2 + h/2)}{\pi(2r_3 h + r_4^2 - r_3^2)^2}$$

A.5 The inner corners

The same procedure is followed. To find an expression for S_1:

$$\pi r_2^2 - \pi(r_2 - S_1)^2 = \frac{\pi}{2}(r_2^2 - r_1^2)$$

$$\therefore\ S_1^2 - 2r_2 S_1 + \frac{1}{2}(r_2^2 - r_1^2) = 0$$

$$\therefore\ S_1 = r_2 - \sqrt{\left(\frac{r_1^2 + r_2^2}{2}\right)}$$

The length of one magnetic path $= \frac{\pi}{4}(S_1 + h/2)$

The mean cross-sectional area

$$= \frac{1}{2}(2\pi r_2 h + \pi(r_2^2 - r_1^2))$$

$$= \frac{\pi}{2}(2r_2 h + r_2^2 - r_1^2)$$

Therefore, for the combination of both inner corners

$$\Sigma l/A = \frac{(S_1 + h/2)}{(2r_2 h + r_2^2 - r_1^2)}$$

$$\Sigma l/A^2 = \frac{2(S_1 + h/2)}{\pi(2r_2 h + r_2^2 - r_1^2)^2}$$

These expressions ignore any slots that might be provided for winding connections, e.g. see Figure 5.1(b). IEC Publication 205 includes correction expressions to take such slots into account; in effect the cross-sectional areas of the outer ring, the outer corners and the end plates are reduced in proportion to the total angle of the slots subtended at the centre.

It must be appreciated that these calculations are standardized approximations for the purposes of comparing similar cores; the need for relatively simple expressions limits the accuracy.

Appendix B
Wire tables

B.1.1	Metric Bare	B.3.1	AWG Bare
B.1.2	Metric Grade 1 enamelled	B.3.2	AWG Fine enamelled
B.1.3	Metric Grade 2 enamelled	B.3.3	AWG Medium enamelled
B.2	SWG Bare	B.4.1–5	Bunched enamelled wires

Appendix B

METRIC Bare

Table B.1.1 Properties of bare copper wires—Metric (based on BS 6811: 1987)

Extracts from BS 6811: 1987 in this and the following two tables are reproduced with permission of BSI. Complete copies of these (and related) standards can be purchased from BSI Sales Dept., Linford Wood, Milton Keynes, MK14 6LE.

Copper diameter (d)			Nominal copper area	Resistance per unit length (R_c)		Approx. weight
Nom. mm	Max. mm	Min. mm	mm²	Nom. Ω/m	Max. Ω/m	g/m
0.020			0.314×10^{-3}	54.41	59.85	0.00279
0.022			0.380×10^{-3}	44.97	49.47	0.00338
0.025			0.491×10^{-3}	34.82	38.31	0.00365
0.028			0.616×10^{-3}	27.76	30.54	0.00548
0.030			0.707×10^{-3}	24.18	26.60	0.00629
0.032			0.804×10^{-3}	21.25	23.38	0.00718
0.036			1.018×10^{-3}	16.79	18.46	0.00905
0.040			1.257×10^{-3}	13.60	14.92	0.0112
0.045			1.590×10^{-3}	10.75	11.76	0.0141
0.050			1.963×10^{-3}	8.706	9.489	0.0175
0.056			2.463×10^{-3}	6.940	7.530	0.0219
0.060			2.827×10^{-3}	6.046	6.542	0.0251
0.063			3.117×10^{-3}	5.484	5.922	0.0277
0.071	0.074	0.068	3.959×10^{-3}	4.318	4.747	0.0352
0.080	0.083	0.077	5.027×10^{-3}	3.401	3.703	0.0447
0.090	0.093	0.087	6.362×10^{-3}	2.687	2.900	0.0566
0.100	0.103	0.097	7.854×10^{-3}	2.176	2.333	0.0698
0.112	0.115	0.109	9.852×10^{-3}	1.735	1.848	0.0876
0.125	0.128	0.122	12.27×10^{-3}	1.393	1.475	0.109
0.132	0.135	0.129	13.68×10^{-3}	1.249	1.319	0.122
0.140	0.143	0.137	15.39×10^{-3}	1.110	1.170	0.137
0.150	0.153	0.147	17.67×10^{-3}	0.9673	1.016	0.157
0.160	0.163	0.157	20.11×10^{-3}	0.8502	0.8906	0.179
0.170	0.173	0.167	22.70×10^{-3}	0.7531	0.7871	0.202
0.180	0.183	0.177	25.45×10^{-3}	0.6718	0.7007	0.226
0.190	0.193	0.187	28.35×10^{-3}	0.6029	0.6278	0.252
0.200	0.203	0.197	31.42×10^{-3}	0.5441	0.5657	0.279
0.212	0.215	0.209	35.30×10^{-3}	0.4843	0.5026	0.314
0.224	0.227	0.221	39.41×10^{-3}	0.4338	0.4495	0.350
0.236	0.240	0.232	43.74×10^{-3}	0.3908	0.4079	0.389
0.250	0.254	0.246	49.09×10^{-3}	0.3482	0.3628	0.436
0.265	0.269	0.261	55.15×10^{-3}	0.3099	0.3223	0.490
0.280	0.284	0.276	61.58×10^{-3}	0.2776	0.2882	0.547
0.300	0.304	0.296	70.69×10^{-3}	0.2418	0.2506	0.628
0.315	0.319	0.311	77.93×10^{-3}	0.2193	0.2270	0.693
0.335	0.339	0.331	0.08814	0.1939	0.2004	0.784
0.355	0.359	0.351	0.09898	0.1727	0.1782	0.880
0.375	0.380	0.370	0.1104	0.1548	0.1604	0.981
0.400	0.405	0.395	0.1257	0.1360	0.1407	1.12
0.425	0.430	0.420	0.1419	0.1205	0.1244	1.26
0.450	0.455	0.445	0.1590	0.1075	0.1109	1.41
0.475	0.480	0.470	0.1772	0.09646	0.09938	1.58
0.500	0.505	0.495	0.1963	0.08706	0.08959	1.75
0.530	0.536	0.524	0.2206	0.07748	0.07995	1.96
0.560	0.566	0.554	0.2463	0.06940	0.07153	2.19

Copper diameter (d)			Nominal copper area	Resistance per unit length (R_c)		Approx. weight
Nom. mm	Max. mm	Min. mm	mm²	Nom. Ω/m	Max. Ω/m	g/m
0.600	0.606	0.594	0.2827	0.06046	0.06222	2.51
0.630	0.636	0.624	0.3117	0.05484	0.05638	2.77
0.670	0.677	0.663	0.3526	0.04848	0.04994	3.13
0.710	0.717	0.703	0.3959	0.04318	0.04442	3.52
0.750	0.758	0.742	0.4418	0.03869	0.03987	3.93
0.800	0.808	0.792	0.5027	0.03401	0.03500	4.47
0.850	0.859	0.841	0.5675	0.03012	0.03104	5.05
0.900	0.909	0.891	0.6362	0.02687	0.02765	5.66
0.950	0.960	0.940	0.7088	0.02412	0.02484	6.30
1.000	1.010	0.990	0.7854	0.02176	0.02240	6.98
1.060	1.071	1.049	0.8825	0.01937		7.85
1.120	1.131	1.109	0.9852	0.01735		8.76
1.180	1.192	1.168	1.094	0.01563		9.73
1.250	1.263	1.237	1.227	0.01393		10.9
1.320	1.333	1.307	1.368	0.01249		12.2
1.400	1.414	1.386	1.539	0.01110		13.7
1.500	1.515	1.485	1.767	0.009673		15.7
1.600	1.616	1.584	2.011	0.008502		17.9
1.700	1.717	1.683	2.270	0.007531		20.2
1.800	1.818	1.782	2.545	0.006718		22.6
1.900	1.919	1.881	2.835	0.006029		25.2
2.000	2.020	1.980	3.142	0.005441		27.9
2.120	2.141	2.099	3.530	0.004845		31.4
2.240	2.262	2.218	3.941	0.004338		35.0
2.360	2.384	2.336	4.374	0.003908		38.9
2.500	2.525	2.475	4.909	0.003482		43.6
2.650	2.677	2.623	5.515	0.003099		49.0
2.800	2.828	2.772	6.158	0.002776		54.7
3.000	3.030	2.970	7.069	0.002418		62.8
3.150	3.182	3.118	7.793	0.002193		69.3
3.350	3.384	3.316	8.814	0.001939		78.4
3.550	3.586	3.514	9.898	0.001727		88.0
3.750	3.788	3.712	11.04	0.001548		98.1
4.000	4.040	3.960	12.57	0.001360		112
4.250	4.293	4.207	14.19	0.001205		126
4.500	4.545	4.455	15.90	0.001075		141
4.750	4.798	4.702	17.72	0.0009646		158
5.000	5.050	4.950	19.63	0.0008706		175

Notes
The principal (preferred) diameters are indicated by bold type in the first column.
For $d < 0.071$ mm, the diameter is assumed to be within limits if the resistance is within limits.

METRIC Grade 1

Table B.1.2 Properties of enamelled copper wires, Grade 1—Metric (based on BS 6811: 1987)

Nom. copper dia (d)	Overall dia. (d_o) (max.)	Increase in dia. due to insulation (min.)	Turns per unit winding breadth ($1/d_o$)	Turns per unit winding area ($1/d_o^2$)	Winding copper space factor $F_w = \frac{1}{4}\pi(d/d_o)^2$
mm	mm	mm	mm^{-1}	mm^{-2}	
0.020	0.025	0.002	40.0	1600	0.503
0.022	0.028	0.002	35.7	1276	0.485
0.025	0.031	0.003	32.3	1041	0.511
0.028	0.035	0.003	28.6	816	0.503
0.030	0.038	0.003	26.3	693	0.490
0.032	0.040	0.004	25.0	625	0.503
0.036	0.045	0.004	22.2	494	0.503
0.040	0.050	0.004	20.0	400	0.503
0.045	0.056	0.005	17.9	319	0.507
0.050	0.062	0.005	16.1	260	0.511
0.056	0.070	0.006	14.3	204	0.503
0.060	0.075	0.006	13.3	178	0.503
0.063	0.078	0.007	12.8	164	0.512
0.071	0.088	0.008	11.4	129	0.511
0.080	0.098	0.009	10.2	104	0.523
0.090	0.110	0.010	9.09	82.6	0.526
0.100	0.121	0.011	8.26	68.3	0.536
0.112	0.134	0.012	7.46	55.7	0.549
0.125	0.149	0.013	6.71	45.0	0.553
0.132	0.157	0.014	6.37	40.6	0.555
0.140	0.166	0.015	6.02	36.3	0.559
0.150	0.177	0.015	5.65	31.9	0.564
0.160	0.187	0.016	5.35	28.6	0.575
0.170	0.198	0.016	5.05	25.5	0.579
0.180	0.209	0.017	4.78	22.9	0.583
0.190	0.220	0.017	4.55	20.7	0.586
0.200	0.230	0.018	4.35	18.9	0.594
0.212	0.243	0.018	4.12	16.9	0.598
0.224	0.256	0.019	3.91	15.3	0.601
0.236	0.269	0.019	3.72	13.8	0.605
0.250	0.284	0.020	3.52	12.4	0.609
0.265	0.300	0.020	3.33	11.1	0.613
0.280	0.315	0.021	3.17	10.1	0.621
0.300	0.337	0.021	2.97	8.81	0.622
0.315	0.352	0.022	2.84	8.07	0.629
0.335	0.374	0.023	2.67	7.15	0.630
0.355	0.395	0.024	2.53	6.41	0.634
0.375	0.417	0.024	2.40	5.75	0.635
0.400	0.442	0.025	2.26	5.12	0.643
0.425	0.469	0.026	2.13	4.55	0.645
0.450	0.495	0.027	2.02	4.08	0.649
0.475	0.522	0.028	1.92	3.67	0.650
0.500	0.548	0.029	1.82	3.33	0.654
0.530	0.579	0.029	1.73	2.98	0.658
0.560	0.611	0.030	1.64	2.68	0.660
0.600	0.653	0.031	1.53	2.35	0.663
0.630	0.684	0.032	1.46	2.14	0.666
0.670	0.726	0.033	1.38	1.90	0.669
0.710	0.767	0.034	1.30	1.70	0.673
0.750	0.809	0.035	1.24	1.53	0.675
0.800	0.861	0.036	1.16	1.35	0.678
0.850	0.913	0.037	1.10	1.20	0.681
0.900	0.965	0.038	1.04	1.07	0.683
0.950	1.017	0.039	0.983	0.967	0.685
1.000	1.068	0.039	0.936	0.877	0.689
1.060	1.130	0.040	0.885	0.783	0.691
1.120	1.192	0.041	0.839	0.704	0.693
1.180	1.254	0.042	0.797	0.636	0.695
1.250	1.325	0.042	0.755	0.570	0.699
1.320	1.397	0.043	0.716	0.512	0.701
1.400	1.479	0.044	0.676	0.457	0.704
1.500	1.581	0.045	0.633	0.400	0.707
1.600	1.683	0.046	0.594	0.353	0.710

Note
The principal (preferred) diameters are indicated by bold type in the first column.

Table B.1.3 Properties of enamelled copper wires, Grade 2—Metric (based on BS 6811: 1987)

Nom. copper dia (d)	Overall dia. (d_o) (max.)	Increase in dia. due to insulation (min.)	Turns per unit winding breadth ($1/d_o$)	Turns per unit winding area ($1/d_o^2$)	Winding copper space factor $F_w = \frac{1}{4}\pi(d/d_o)^2$	Nom. copper dia (d)	Overall dia. (d_o) (max.)	Increase in dia. due to insulation (min.)	Turns per unit winding breadth ($1/d_o$)	Turns per unit winding area ($1/d_o^2$)	Winding copper space factor $F_w = \frac{1}{4}\pi(d/d_o)^2$
mm	mm	mm	mm^{-1}	mm^{-2}		mm	mm	mm	mm^{-1}	mm^{-2}	
0.050	0.068	0.010	14.7	216	0.425	0.750	0.832	0.052	1.20	1.44	0.638
0.056	0.076	0.012	13.2	173	0.426	**0.800**	0.885	0.054	1.13	1.28	0.642
0.060	0.081	0.012	12.3	152	0.431	0.850	0.937	0.055	1.07	1.14	0.646
0.063	0.085	0.013	11.8	138	0.431	**0.900**	0.990	0.057	1.01	1.02	0.649
0.071	0.095	0.015	10.5	111	0.439	0.950	1.041	0.058	0.961	0.923	0.654
0.080	0.105	0.016	9.52	90.7	0.456	**1.000**	1.093	0.059	0.915	0.837	0.657
0.090	0.117	0.017	8.55	73.1	0.465	1.060	1.155	0.060	0.866	0.750	0.662
0.100	0.129	0.019	7.75	60.1	0.472	**1.120**	1.217	0.061	0.822	0.675	0.665
0.112	0.143	0.020	6.99	48.9	0.482	1.180	1.279	0.062	0.782	0.611	0.669
0.125	0.159	0.022	6.29	39.6	0.485	**1.250**	1.351	0.063	0.740	0.548	0.672
0.132	0.167	0.023	5.99	35.9	0.491	1.320	1.423	0.064	0.703	0.494	0.676
0.140	0.176	0.024	5.68	32.3	0.497	**1.400**	1.506	0.066	0.664	0.441	0.679
0.150	0.188	0.024	5.32	28.3	0.500	1.500	1.608	0.067	0.622	0.387	0.683
0.160	0.199	0.025	5.03	25.3	0.508	**1.600**	1.711	0.068	0.584	0.342	0.687
0.170	0.211	0.026	4.74	22.5	0.510	1.700	1.813	0.069	0.552	0.304	0.691
0.180	0.222	0.027	4.50	20.3	0.516	**1.800**	1.916	0.071	0.522	0.272	0.693
0.190	0.234	0.027	4.27	18.3	0.518	1.900	2.018	0.072	0.496	0.246	0.696
0.200	0.245	0.028	4.08	16.7	0.523	**2.000**	2.120	0.073	0.472	0.223	0.699
0.212	0.258	0.029	3.88	15.0	0.530	2.120	2.243	0.074	0.446	0.199	0.702
0.224	0.272	0.030	3.68	13.5	0.533	**2.240**	2.366	0.075	0.423	0.179	0.704
0.236	0.285	0.030	3.51	12.3	0.539	2.360	2.488	0.076	0.402	0.162	0.707
0.250	0.301	0.031	3.32	11.0	0.542	**2.500**	2.631	0.077	0.380	0.144	0.709
0.265	0.317	0.031	3.15	9.95	0.549	2.650	2.784	0.078	0.359	0.129	0.712
0.280	0.334	0.032	2.99	8.96	0.552	**2.800**	2.938	0.080	0.340	0.116	0.713
0.300	0.355	0.033	2.82	7.93	0.561	3.000	3.142	0.081	0.318	0.101	0.716
0.315	0.371	0.034	2.70	7.27	0.566	**3.150**	3.294	0.082	0.304	0.0922	0.718
0.335	0.393	0.035	2.54	6.47	0.571	3.350	3.498	0.083	0.286	0.0817	0.720
0.355	0.414	0.036	2.42	5.83	0.577	**3.550**	3.702	0.085	0.270	0.0730	0.722
0.375	0.436	0.037	2.29	5.26	0.581	3.750	3.905	0.086	0.256	0.0656	0.724
0.400	0.462	0.038	2.16	4.69	0.589	**4.000**	4.160	0.087	0.240	0.0578	0.726
0.425	0.489	0.039	2.05	4.18	0.593	4.250	4.414	0.088	0.227	0.0513	0.728
0.450	0.516	0.041	1.94	3.76	0.597	**4.500**	4.668	0.090	0.214	0.0459	0.730
0.475	0.543	0.042	1.84	3.39	0.601	4.750	4.923	0.091	0.203	0.0413	0.731
0.500	0.569	0.044	1.76	3.09	0.606	**5.000**	5.177	0.092	0.193	0.0373	0.733
0.530	0.601	0.045	1.66	2.77	0.611						
0.560	0.632	0.046	1.58	2.50	0.617						
0.600	0.675	0.047	1.48	2.19	0.621						
0.630	0.706	0.049	1.42	2.01	0.625						
0.670	0.749	0.050	1.34	1.78	0.628						
0.710	0.790	0.051	1.27	1.60	0.634						

Note
The principal (preferred) diameters are indicated by bold type in the first column.

SWG Bare

Table B.2 Diameters of bare copper wires—SWG

As the SWG range of sizes has been superseded by the metric range, complete tables of SWG properties are no longer relevant. This table of diameters will provide correspondence with the metric wire tables.

Gauge No. SWG	Copper diameter (d) Nominal	
	in	mm
50	0.001 0	0.025 4
49	0.001 2	0.030 5
48	0.001 6	0.040 6
47	0.002 0	0.050 8
46	0.002 4	0.061 0
45	0.002 8	0.071 1
44	0.003 2	0.081 3
43	0.003 6	0.091 4
42	0.004 0	0.102
41	0.004 4	0.112
40	0.004 8	0.122
39	0.005 2	0.132
38	0.006 0	0.152
37	0.006 8	0.173
36	0.007 6	0.193
35	0.008 4	0.213
34	0.009 2	0.234
33	0.010 0	0.254
32	0.010 8	0.274
31	0.011 6	0.295
30	0.012 4	0.315
29	0.013 6	0.345
28	0.014 8	0.376
27	0.016 4	0.417
26	0.018 0	0.457
25	0.020	0.508
24	0.022	0.559
23	0.024	0.610
22	0.028	0.711
21	0.032	0.813
20	0.036	0.914
19	0.040	1.02
18	0.048	1.22
17	0.056	1.42
16	0.064	1.63
15	0.072	1.83
14	0.080	2.03
13	0.092	2.34
12	0.104	2.64
11	0.116	2.95
10	0.128	3.25

Table B.3.1 Properties of bare copper wires—AWG (B & S)

AWG (B & S)	Nominal in	Nominal mm	Maximum in	Maximum mm	Minimum in	Minimum mm	Nominal copper area in²	Nominal copper area mm²	Ω/100 in	Ω/m
50	0.000 99	0.0252	0.001 07	0.0271 8	0.000 91	0.023 11	0.769 8 × 10⁻⁶	0.497 × 10⁻³	88.18	34.72
49	0.001 11	0.0282	0.001 19	0.0302 3	0.001 03	0.026 16	0.967 7 × 10⁻⁶	0.624 × 10⁻³	70.15	27.62
48	0.001 24	0.0315	0.001 32	0.0335 3	0.001 16	0.029 46	1.208 × 10⁻⁶	0.779 × 10⁻³	56.21	22.13
47	0.001 40	0.0356	0.001 48	0.0375 9	0.001 32	0.033 53	1.539 × 10⁻⁶	0.993 × 10⁻³	44.09	17.36
46	0.001 57	0.0399	0.001 65	0.0419 1	0.001 49	0.037 85	1.936 × 10⁻⁶	1.249 × 10⁻³	35.06	13.80
45	0.001 76	0.0447	0.001 84	0.0467 4	0.001 68	0.042 67	2.433 × 10⁻⁶	1.570 × 10⁻³	27.90	10.98
44	0.001 98	0.0503	0.001 06	0.0523 2	0.001 90	0.048 26	3.079 × 10⁻⁶	1.987 × 10⁻³	22.04	8.679
43	0.002 22	0.0564	0.002 30	0.0584 2	0.002 14	0.054 36	3.871 × 10⁻⁶	2.497 × 10⁻³	17.54	6.904
42	0.002 49	0.0633	0.002 58	0.0655 3	0.002 40	0.060 96	4.870 × 10⁻⁶	3.142 × 10⁻³	13.94	5.488
41	0.002 80	0.0711	0.002 89	0.0734 1	0.002 71	0.068 83	6.158 × 10⁻⁶	3.973 × 10⁻³	11.03	4.341
40	0.003 14	0.0798	0.003 23	0.0820 4	0.003 05	0.077 47	7.744 × 10⁻⁶	4.996 × 10⁻³	8.767	3.451
39	0.003 53	0.0897	0.003 63	0.0922 0	0.003 43	0.087 12	9.787 × 10⁻⁶	6.314 × 10⁻³	6.936	2.731
38	0.003 97	0.1008	0.004 07	0.1034	0.003 87	0.098 30	12.38 × 10⁻⁶	7.986 × 10⁻³	5.484	2.159
37	0.004 45	0.1130	0.004 55	0.1156	0.004 35	0.1105	15.55 × 10⁻⁶	10.03 × 10⁻³	4.364	1.718
36	0.005 00	0.1270	0.005 10	0.1295	0.004 90	0.1245	19.63 × 10⁻⁶	12.67 × 10⁻³	3.457	1.361
35	0.0056	0.1422	0.0057	0.1448	0.0055	0.1397	24.63 × 10⁻⁶	15.89 × 10⁻³	2.756	1.085
34	0.0063	0.1600	0.0064	0.1626	0.0062	0.1575	31.17 × 10⁻⁶	20.11 × 10⁻³	2.178	0.8573
33	0.0071	0.1803	0.0072	0.1829	0.0070	0.1778	39.59 × 10⁻⁶	25.54 × 10⁻³	1.714	0.6750
32	0.0080	0.2032	0.0081	0.2057	0.0079	0.2007	50.27 × 10⁻⁶	32.43 × 10⁻³	1.350	0.5316
31	0.0089	0.2261	0.0090	0.2286	0.0088	0.2235	62.21 × 10⁻⁶	40.14 × 10⁻³	1.091	0.4296
30	0.0100	0.2540	0.0101	0.2565	0.0099	0.2515	78.54 × 10⁻⁶	50.67 × 10⁻³	0.8642	0.3402
29	0.0113	0.2870	0.0114	0.2896	0.0112	0.2845	100.3 × 10⁻⁶	64.70 × 10⁻³	0.6770	0.2665
28	0.0126	0.3200	0.0127	0.3226	0.0125	0.3175	124.7 × 10⁻⁶	80.44 × 10⁻³	0.5439	0.2142
27	0.0142	0.3607	0.0143	0.3632	0.0141	0.3581	158.4 × 10⁻⁶	0.1022	0.4286	0.1687
26	0.0159	0.4039	0.0161	0.4089	0.0157	0.3988	198.6 × 10⁻⁶	0.1281	0.3419	0.1346
25	0.0179	0.4547	0.0181	0.4597	0.0177	0.4496	251.6 × 10⁻⁶	0.1624	0.2697	0.1062
24	0.0201	0.5105	0.0203	0.5156	0.0199	0.5055	317.3 × 10⁻⁶	0.2047	0.2139	0.08422
23	0.0226	0.5740	0.0228	0.5791	0.0224	0.5690	401.1 × 10⁻⁶	0.2588	0.1692	0.06662
22	0.0253	0.6426	0.0256	0.6502	0.0250	0.6350	502.7 × 10⁻⁶	0.3243	0.1350	0.05316
21	0.0285	0.7239	0.0288	0.7315	0.0282	0.7163	637.9 × 10⁻⁶	0.4116	0.1064	0.04189
20	0.0320	0.8128	0.0323	0.8204	0.0317	0.8052	804.2 × 10⁻⁶	0.5189	0.08439	0.03322
19	0.0359	0.9119	0.0363	0.9220	0.0355	0.9017	1 012 × 10⁻⁶	0.6530	0.06706	0.02640
18	0.0403	1.024	0.0407	1.033	0.0399	1.013	1 276 × 10⁻⁶	0.8229	0.05321	0.02095
17	0.0453	1.151	0.0458	1.163	0.0448	1.138	1 612 × 10⁻⁶	1.040	0.04212	0.01658
16	0.0508	1.290	0.0513	1.303	0.0503	1.278	2 027 × 10⁻⁶	1.308	0.03349	0.01319
15	0.0571	1.450	0.0577	1.466	0.0565	1.435	2 561 × 10⁻⁶	1.652	0.02651	0.01044
14	0.0641	1.628	0.0647	1.643	0.0635	1.613	3 227 × 10⁻⁶	2.082	0.02103	0.008281
13	0.0720	1.829	0.0727	1.847	0.0713	1.811	4 072 × 10⁻⁶	2.627	0.01667	0.006564
12	0.0808	2.052	0.0816	2.073	0.0800	2.032	5 128 × 10⁻⁶	3.308	0.01324	0.005212
11	0.0907	2.304	0.0916	2.327	0.0898	2.281	6 461 × 10⁻⁶	4.168	0.01051	0.004136
10	0.1019	2.588	0.1029	2.614	0.1009	2.563	8 155 × 10⁻⁶	5.261	0.008322	0.003277
9	0.1144	2.906	0.1155	2.934	0.1133	2.878	10 280 × 10⁻⁶	6.631	0.006603	0.002600
8	0.1285	3.264	0.1298	3.297	0.1272	3.231	12 970 × 10⁻⁶	8.367	0.005234	0.002061
7	0.1443	3.665	0.1457	3.701	0.1429	3.630	16 350 × 10⁻⁶	10.55	0.004151	0.001634
6	0.1620	4.115	0.1636	4.155	0.1604	4.074	20 610 × 10⁻⁶	13.30	0.003293	0.001296

AWG Fine Appendix B 347

Table B.3.2 Properties of fine enamelled copper wires—AWG (B & S)

AWG (B & S)	Copper dia. (d) in	Overall dia. (max.) (d_o) in	Overall dia. (max.) (d_o) mm	Increase in dia. due to insulation (min.) in	Increase in dia. due to insulation (min.) mm	Approximate weight lb/100 in	Approximate weight g/m	Turns per unit winding breadth ($1/d_o$) in^{-1}	Turns per unit winding breadth ($1/d_o$) mm^{-1}	Turns per unit winding area ($1/d_o^2$) in^{-2}	Turns per unit winding area ($1/d_o^2$) mm^{-2}	Winding copper space factor $F_w = \frac{1}{4}\pi(d/d_o)^2$
50	0.00099	0.00140	0.03556	0.00012	0.00305	0.02695 × 10^{-3}	0.00481	714	28.1	510000	791	0.393
49	0.00111	0.00150	0.03810	0.00013	0.00330	0.03333 × 10^{-3}	0.00595	667	26.2	444000	689	0.430
48	0.00124	0.00160	0.04064	0.00014	0.00356	0.04111 × 10^{-3}	0.00734	625	24.6	391000	605	0.472
47	0.00140	0.00180	0.04572	0.00015	0.00381	0.05250 × 10^{-3}	0.00938	556	21.9	309000	478	0.475
46	0.00157	0.00200	0.05080	0.00017	0.00432	0.06583 × 10^{-3}	0.01118	500	19.7	250000	387	0.484
45	0.00176	0.00220	0.05588	0.00018	0.00457	0.08250 × 10^{-3}	0.01471	454	17.9	207000	320	0.503
44	0.00198	0.00240	0.06096	0.00020	0.00508	0.1042 × 10^{-3}	0.01866	417	16.4	174000	269	0.535
43	0.00222	0.00270	0.06858	0.00020	0.00508	0.1267 × 10^{-3}	0.02261	370	14.6	137000	213	0.531
42	0.00249	0.00300	0.07620	0.00025	0.00635	0.1633 × 10^{-3}	0.02919	333	13.1	111000	172	0.541
41	0.00280	0.00330	0.08382	0.00025	0.00635	0.2044 × 10^{-3}	0.03665	303	11.9	91800	142	0.565
40	0.00314	0.00370	0.09398	0.00030	0.00762	0.2611 × 10^{-3}	0.04666	270	10.6	73000	113	0.566
39	0.00353	0.00410	0.1041	0.00030	0.00762	0.3278 × 10^{-3}	0.05885	244	9.60	59500	92.2	0.582
38	0.00397	0.00470	0.1194	0.00040	0.01020	0.4167 × 10^{-3}	0.07432	213	8.38	45300	70.2	0.560
37	0.00445	0.00530	0.1346	0.00040	0.01020	0.5097 × 10^{-3}	0.09101	189	7.43	35600	55.2	0.554
36	0.00500	0.00590	0.1499	0.00050	0.01270	0.6528 × 10^{-3}	0.11628	169	6.67	28700	44.5	0.564
35	0.00560	0.00650	0.1651	0.00050	0.01270	0.8156 × 10^{-3}	0.1451	154	6.06	23700	36.7	0.583
34	0.00630	0.00730	0.1854	0.00060	0.01520	1.033 × 10^{-3}	0.1845	137	5.39	18800	29.1	0.585
33	0.00710	0.00820	0.2083	0.00070	0.01780	1.313 × 10^{-3}	0.2343	122	4.80	14900	23.1	0.589
32	0.00800	0.00920	0.2337	0.00070	0.01780	1.663 × 10^{-3}	0.2963	109	4.28	11800	18.3	0.594
31	0.00890	0.01010	0.2565	0.00070	0.01780	2.053 × 10^{-3}	0.3665	99.0	3.90	9800	15.2	0.610
30	0.01000	0.01130	0.2870	0.00080	0.02030	2.589 × 10^{-3}	0.4622	88.5	3.48	7830	12.1	0.615
29	0.01130	0.01270	0.3226	0.00080	0.02030	3.300 × 10^{-3}	0.5890	78.7	3.10	6200	9.61	0.622
28	0.01260	0.01410	0.3581	0.00090	0.02290	4.103 × 10^{-3}	0.7320	70.9	2.79	5030	7.80	0.627
27	0.01420	0.01580	0.4013	0.00100	0.02540	5.206 × 10^{-3}	0.9290	63.3	2.49	4010	6.21	0.634
26	0.01590	0.01770	0.4496	0.00100	0.02540	6.519 × 10^{-3}	1.164	56.5	2.22	3190	4.95	0.634
25	0.01790	0.01980	0.5029	0.00110	0.02770	8.247 × 10^{-3}	1.472	50.5	1.99	2550	3.95	0.642
24	0.02010	0.02210	0.5613	0.00120	0.03050	10.44 × 10^{-3}	1.854	45.2	1.78	2050	3.17	0.650
23	0.02260	0.02470	0.6274	0.00120	0.03050	13.11 × 10^{-3}	2.340	40.5	1.59	1640	2.54	0.658
22	0.02530	0.02750	0.6985	0.00120	0.03050	16.41 × 10^{-3}	2.925	36.4	1.43	1320	2.05	0.665
21	0.02850	0.03080	0.7823	0.00130	0.03300	20.80 × 10^{-3}	3.714	32.5	1.28	1050	1.63	0.673
20	0.03200	0.03440	0.8738	0.00130	0.03560	26.20 × 10^{-3}	4.677	29.1	1.14	845	1.31	0.680
19	0.03590	0.03850	0.9779	0.00140	0.03560	32.93 × 10^{-3}	5.880	26.0	1.02	675	1.05	0.683
18	0.04030	0.04300	1.092	0.00150	0.03810	41.47 × 10^{-3}	7.404	23.3	0.916	541	0.838	0.690
17	0.04530	0.04810	1.222	0.00150	0.03810	52.34 × 10^{-3}	9.348	20.8	0.819	432	0.670	0.697
16	0.05080	0.05370	1.364	0.00160	0.04060	65.75 × 10^{-3}	11.74	18.6	0.733	347	0.538	0.703
15	0.05710	0.06020	1.529	0.00170	0.04320	83.00 × 10^{-3}	14.78	16.6	0.654	276	0.428	0.707
14	0.06413	0.06730	1.709	0.00170	0.04320	104.6 × 10^{-3}	18.68	14.9	0.585	221	0.342	0.713
13	0.07200	0.07540	1.915	0.00170	0.04320	131.9 × 10^{-3}	23.54	13.3	0.522	176	0.273	0.716
12	0.08080	0.08440	2.144	0.00180	0.04570	166.7 × 10^{-3}	29	11.8	0.466	140	0.218	0.720
11	0.09070	0.09450	2.400	0.00190	0.04830	208.3 × 10^{-3}	37	10.6	0.417	112	0.174	0.724
10	0.10190	0.10590	2.690	0.00200	0.05080	263.9 × 10^{-3}	47	9.44	0.372	89.2	0.138	0.727
9	0.11440	0.11870	3.015	0.00210	0.05330	333.3 × 10^{-3}	59	8.43	0.332	71.0	0.110	0.730
8	0.12850	0.13300	3.378	0.00220	0.05590	419.4 × 10^{-3}	71	7.52	0.296	56.5	0.0876	0.733
7	0.14430	0.14900	3.785	0.00230	0.05840	527.8 × 10^{-3}	94	6.71	0.264	45.0	0.0698	0.737
6	0.16200	0.16700	4.242	0.00240	0.06100	669.4 × 10^{-3}	111	5.99	0.236	35.9	0.0556	0.739

Table B.3.3 Properties of medium enamelled copper wires—AWG (B & S)

AWG (B & S)	Copper dia. (d) in	Overall dia. (max.) (d_o) in	Overall dia. (max.) (d_o) mm	Increase in dia. due to insulation (min.) in	Increase in dia. due to insulation (min.) mm	Approximate weight lb/100 in	Approximate weight g/m	Turns per unit winding breadth ($1/d_o$) in^{-1}	Turns per unit winding breadth ($1/d_o$) mm^{-1}	Turns per unit winding area ($1/d_o^2$) in^{-2}	Turns per unit winding area ($1/d_o^2$) mm^{-2}	Winding copper space factor $F_w = \frac{1}{4}\pi(d/d_o)^2$
44	0.001 98	0.002 6	0.066 04	0.000 3	0.007 62	0.106 1 × 10^{-3}	0.018 9	385	15.1	148 000	229	0.456
43	0.002 22	0.002 9	0.073 66	0.000 4	0.010 2	0.128 9 × 10^{-3}	0.023 1	345	13.6	119 000	184	0.460
42	0.002 49	0.003 2	0.081 28	0.000 4	0.010 2	0.166 4 × 10^{-3}	0.029 7	313	12.3	97 700	151	0.476
41	0.002 80	0.003 6	0.091 44	0.000 5	0.012 7	0.209 4 × 10^{-3}	0.037 3	278	10.9	77 200	120	0.475
40	0.003 14	0.004 1	0.104 1	0.000 6	0.015 2	0.270 0 × 10^{-3}	0.048 2	244	9.60	59 500	92.2	0.461
39	0.003 53	0.004 5	0.114 3	0.000 6	0.015 2	0.336 1 × 10^{-3}	0.060 0	222	8.75	49 400	76.5	0.483
38	0.003 97	0.005 1	0.129 5	0.000 7	0.017 8	0.426 9 × 10^{-3}	0.076 2	196	7.72	38 400	59.6	0.476
37	0.004 45	0.005 7	0.144 8	0.000 8	0.020 3	0.523 3 × 10^{-3}	0.093 4	175	6.91	30 800	47.7	0.479
36	0.005 00	0.006 4	0.162 6	0.000 9	0.022 9	0.669 2 × 10^{-3}	0.119 5	156	6.15	24 400	37.8	0.479
35	0.005 6	0.007 0	0.177 8	0.000 9	0.022 9	0.833 9 × 10^{-3}	0.148 9	143	5.62	20 400	31.6	0.503
34	0.006 3	0.007 8	0.198 1	0.001 0	0.025 4	1.053 × 10^{-3}	0.188 1	128	5.05	16 400	25.5	0.512
33	0.007 1	0.008 8	0.223 5	0.001 1	0.027 9	1.339 × 10^{-3}	0.239 1	114	4.47	12 900	20.0	0.511
32	0.008 0	0.009 8	0.248 9	0.001 2	0.030 5	1.694 × 10^{-3}	0.302 1	102	4.02	10 400	16.1	0.523
31	0.008 9	0.010 8	0.274 3	0.001 2	0.030 5	2.091 × 10^{-3}	0.373 3	92.6	3.65	8 570	13.3	0.533
30	0.010 0	0.012 0	0.304 8	0.001 3	0.033 0	2.632 × 10^{-3}	0.466 0	83.3	3.28	6 940	10.8	0.545
29	0.011 3	0.013 4	0.340 4	0.001 3	0.033 0	3.347 × 10^{-3}	0.597 1	74.6	2.94	5 570	8.63	0.559
28	0.012 6	0.014 8	0.375 9	0.001 5	0.038 1	4.158 × 10^{-3}	0.742 2	67.6	2.66	4 570	7.08	0.569
27	0.014 2	0.016 5	0.419 1	0.001 6	0.040 6	5.269 × 10^{-3}	0.941 0	60.6	2.39	3 670	5.69	0.582
26	0.015 9	0.018 5	0.469 9	0.001 7	0.043 2	6.594 × 10^{-3}	1.177	54.1	2.13	2 920	4.53	0.580
25	0.017 9	0.020 6	0.523 2	0.001 8	0.045 7	8.342 × 10^{-3}	1.489	48.5	1.91	2 360	3.65	0.593
24	0.020 1	0.022 9	0.581 7	0.001 9	0.048 3	10.50 × 10^{-3}	1.874	43.7	1.72	1 910	2.96	0.605
23	0.022 6	0.025 6	0.650 2	0.002 0	0.050 8	13.25 × 10^{-3}	2.365	39.1	1.54	1 530	2.37	0.612
22	0.025 3	0.028 4	0.721 4	0.002 0	0.050 8	16.55 × 10^{-3}	2.955	35.2	1.39	1 240	1.92	0.623
21	0.028 5	0.031 7	0.805 2	0.002 1	0.053 3	20.96 × 10^{-3}	3.743	31.5	1.24	995	1.54	0.635
20	0.032 0	0.035 3	0.896 6	0.002 2	0.055 9	26.38 × 10^{-3}	4.710	28.3	1.12	802	1.24	0.645
19	0.035 9	0.039 5	1.003	0.002 3	0.058 4	33.16 × 10^{-3}	5.922	25.3	0.997	641	0.993	0.649
18	0.040 3	0.044 0	1.118	0.002 4	0.061 0	41.72 × 10^{-3}	7.450	22.7	0.895	517	0.801	0.659
17	0.045 3	0.049 1	1.247	0.002 4	0.061 0	52.60 × 10^{-3}	9.392	20.4	0.802	415	0.643	0.669
16	0.050 8	0.054 7	1.389	0.002 5	0.063 5	66.08 × 10^{-3}	11.80	18.3	0.720	334	0.518	0.677
15	0.057 1	0.061 3	1.557	0.002 6	0.066 0	83.42 × 10^{-3}	14.92	16.3	0.642	266	0.412	0.682
14	0.064 1	0.068 4	1.737	0.002 7	0.068 6	105.0 × 10^{-3}	18.75	14.6	0.576	214	0.331	0.690
13	0.072 0	0.076 5	1.943	0.002 8	0.071 1	132.4 × 10^{-3}	23.64	13.1	0.515	171	0.265	0.696
12	0.080 8	0.085 5	2.172	0.002 9	0.073 7	166.7 × 10^{-3}	29	11.7	0.460	137	0.212	0.701
11	0.090 7	0.095 7	2.431	0.003 0	0.076 2	211.1 × 10^{-3}	37	10.4	0.411	109	0.169	0.706
10	0.101 9	0.107 1	2.720	0.003 1	0.078 7	263.9 × 10^{-3}	47	9.34	0.368	87.2	0.135	0.711
9	0.114 4	0.119 9	3.045	0.003 2	0.081 3	333.3 × 10^{-3}	59	8.34	0.328	69.6	0.108	0.715
8	0.128 5	0.134 2	3.409	0.003 3	0.083 8	422.2 × 10^{-3}	75	7.45	0.293	55.5	0.086 1	0.720
7	0.144 3	0.150 2	3.815	0.003 4	0.086 4	530.6 × 10^{-3}	94	6.66	0.262	44.3	0.068 7	0.725
6	0.162 0	0.168 2	4.272	0.003 5	0.088 9	669.4 × 10^{-3}	119	5.95	0.234	35.3	0.054 8	0.729

Tables B.4.1 to B.4.6 are based on numerical data contained in IEC Publication 317-11 (1972) and reproduced by permission of the International Electrotechnical Commission, which retains the copyright. Complete copies of this publication, which specifies the requirements for bunched enamelled copper wires, can be purchased from the IEC, 1 rue de Varembé, Geneva, Switzerland, or through BSI Sales Dept., Linford Wood, Milton Keynes, MK14 6LE.

Table B.4.1 Properties of bunched enamelled copper wires with silk covering.
Strand diameter: 0.025 mm

Number of strands (n)	Nominal overall diameter of bunch	Nominal copper cross-sectional area	Resistance per unit length		Approx. weight	Winding copper space factor (F_w)
	mm	mm²	Nom. Ω/m	Max. Ω/m	g/m	
3	0.095	0.001 50	11.94	13.97	0.018 2	0.166
4	0.105	0.002 00	8.956	10.48	0.023 5	0.181
5	0.115	0.002 50	7.164	8.562	0.030 3	0.189
6	0.120	0.003 00	5.970	6.985	0.033 9	0.208
7	0.120	0.003 51	5.117	5.987	0.038 7	0.244
8	0.135	0.004 01	4.478	5.239	0.044 2	0.220
9	0.140	0.004 51	3.980	4.657	0.049 2	0.230
10	0.150	0.005 01	3.582	4.191	0.054 5	0.223
12	0.160	0.006 01	2.985	3.493	0.064 6	0.235
16	0.180	0.008 01	2.239	2.619	0.084 7	0.247
20	0.200	0.010 0	1.791	2.096	0.105	0.250
25	0.220	0.012 5	1.433	1.676	0.130	0.258
27	0.225	0.013 5	1.327	1.599	0.140	0.267
32	0.245	0.016 0	1.119	1.349	0.165	0.267
40	0.270	0.020 0	0.896	1.079	0.205	0.274
50	0.295	0.025 0	0.716	0.863	0.254	0.287
60	0.320	0.030 0	0.597	0.719	0.304	0.293
63	0.330	0.031 5	0.569	0.689	0.320	0.289
80	0.365	0.040 1	0.448	0.540	0.402	0.301
81	0.370	0.040 6	0.442	0.533	0.408	0.297
100	0.405	0.050 1	0.358	0.432	0.501	0.305
120	0.440	0.060 1	0.299	0.360	0.600	0.310
160	0.505	0.080 1	0.224	0.270	0.796	0.314
200	0.560	0.100	0.179	0.216	0.993	0.319
250	0.625	0.125	0.143	0.173	1.26	0.320
320	0.730	0.160	0.112	0.135	1.61	0.300
400	0.805	0.200	0.089 6	0.108	2.00	0.309

Table B.4.2 Properties of bunched enamelled copper wires with silk covering.
Strand diameter: 0.032 mm

Number of strands (n)	Nominal overall diameter of bunch	Nominal copper cross-sectional area	Resistance per unit length		Approx. weight	Winding copper space factor (F_w)
	mm	mm²	Nom. Ω/m	Max. Ω/m	g/m	
3	0.115	0.002 46	7.290	8.310	0.027 9	0.186
4	0.125	0.003 28	5.467	6.232	0.036 1	0.210
5	0.140	0.004 10	4.374	4.986	0.044 5	0.209
6	0.150	0.004 92	3.645	4.155	0.052 7	0.219
7	0.150	0.005 74	3.124	3.561	0.060 4	0.255
8	0.165	0.006 56	2.734	3.116	0.068 8	0.241
9	0.175	0.007 38	2.430	2.77	0.077 0	0.241
10	0.180	0.008 20	2.187	2.493	0.084 9	0.253
12	0.195	0.009 84	1.822	2.077	0.101	0.259
16	0.220	0.013 1	1.367	1.558	0.133	0.271
20	0.245	0.016 4	1.093	1.246	0.165	0.273
25	0.275	0.020 5	0.875	0.997	0.205	0.271
27	0.280	0.022 1	0.810	0.951	0.221	0.282
32	0.305	0.026 3	0.683	0.802	0.261	0.283
40	0.335	0.032 8	0.547	0.642	0.324	0.292
50	0.370	0.041 0	0.437	0.514	0.404	0.300
60	0.405	0.049 2	0.364	0.428	0.483	0.300
63	0.415	0.051 7	0.347	0.408	0.507	0.300
80	0.460	0.065 6	0.273	0.321	0.641	0.310
81	0.465	0.066 4	0.270	0.317	0.649	0.307
100	0.515	0.082 0	0.219	0.257	0.799	0.309
120	0.560	0.098 4	0.182	0.214	0.957	0.314
160	0.670	0.131	0.137	0.160	1.30	0.292
200	0.740	0.164	0.109	0.128	1.61	0.300
250	0.820	0.205	0.087 5	0.103	2.01	0.305
320	0.920	0.263	0.068 3	0.080 2	2.56	0.311
400	1.020	0.328	0.054 7	0.064 2	3.19	0.315

Notes
The preferred numbers of strands are indicated by bold type in the first column.
For conductors above the line the covering normally consists of one layer of silk; for those below the line two layers are normally applied.

Table B.4.3 Properties of bunched enamelled copper wires with silk covering. Strand diameter: 0.040 mm

Number of strands (n)	Nominal overall diameter of bunch	Nominal copper cross-sectional area	Resistance per unit length		Approx. weight	Winding copper space factor (F_w)
	mm	mm²	Nom. Ω/m	Max. Ω/m	g/m	
3	0.135	0.003 85	4.665	5.226	0.041 8	0.211
4	0.150	0.005 13	3.499	3.919	0.054 5	0.228
5	0.165	0.006 41	2.799	3.135	0.067 2	0.235
6	0.175	0.007 69	2.332	2.613	0.079 7	0.251
7	0.175	0.008 97	1.999	2.240	0.091 7	0.293
8	0.200	0.010 3	1.749	1.960	0.105	0.258
9	0.210	0.011 5	1.555	1.742	0.117	0.261
10	0.220	0.012 8	1.399	1.568	0.130	0.265
12	0.235	0.015 4	1.166	1.306	0.155	0.279
16	0.270	0.020 5	0.875	0.980	0.205	0.281
20	0.300	0.025 6	0.700	0.784	0.254	0.284
25	0.335	0.032 0	0.560	0.627	0.316	0.285
27	0.345	0.034 6	0.518	0.598	0.341	0.291
32	0.375	0.041 0	0.437	0.505	0.403	0.292
40	0.415	0.051 3	0.350	0.404	0.501	0.298
50	0.460	0.064 1	0.280	0.323	0.625	0.303
60	0.500	0.076 9	0.233	0.269	0.748	0.308
63	0.515	0.080 8	0.222	0.256	0.785	0.305
80	0.575	0.103	0.175	0.202	0.994	0.312
81	0.580	0.104	0.173	0.199	1.01	0.309
100	0.640	0.128	0.140	0.161	1.26	0.313
120	0.720	0.154	0.117	0.135	1.51	0.297
160	0.825	0.205	0.087 5	0.101	2.00	0.301
200	0.915	0.256	0.070 0	0.080 7	2.50	0.306
250	1.015	0.320	0.056 0	0.064 6	3.11	0.311
320	1.150	0.410	0.043 7	0.050 5	3.98	0.310
400	1.280	0.513	0.035 0	0.040 4	4.96	0.313

Table B.4.4 Properties of bunched anemelled copper wires with silk covering. Strand diameter: 0.050 mm

Number of strands (n)	Nominal overall diameter of bunch	Nominal copper cross-sectional area	Resistance per unit length		Approx. weight	Winding copper space factor (F_w)
	mm	mm²	Nom. Ω/m	Max. Ω/m	g/m	
3	0.160	0.006 01	2.986	3.284	0.062 2	0.235
4	0.180	0.008 01	2.239	2.463	0.081 5	0.247
5	0.195	0.0100	1.791	1.970	0.101	0.263
6	0.210	0.0120	1.493	1.642	0.120	0.272
7	0.210	0.0140	1.280	1.407	0.138	0.318
8	0.240	0.0160	1.120	1.232	0.158	0.278
9	0.250	0.0180	0.995	1.095	0.177	0.288
10	0.260	0.0200	0.896	0.985	0.196	0.296
12	0.285	0.0240	0.746	0.821	0.234	0.296
16	0.325	0.0320	0.560	0.616	0.310	0.303
20	0.360	0.0401	0.448	0.493	0.386	0.309
25	0.405	0.0501	0.358	0.394	0.481	0.305
27	0.420	0.0541	0.332	0.376	0.519	0.307
32	0.455	0.0641	0.280	0.317	0.613	0.310
40	0.505	0.0801	0.224	0.254	0.764	0.314
50	0.560	0.100	0.179	0.203	0.953	0.319
60	0.615	0.120	0.149	0.169	1.14	0.317
63	0.625	0.126	0.142	0.161	1.20	0.323
80	0.725	0.160	0.112	0.127	1.54	0.304
81	0.730	0.162	0.111	0.125	1.56	0.304
100	0.805	0.200	0.089 6	0.101	1.92	0.309
120	0.875	0.240	0.074 6	0.084 6	2.30	0.314
160	1.005	0.320	0.056 0	0.063 4	3.05	0.317
200	1.120	0.401	0.044 8	0.050 7	3.81	0.320
250	1.245	0.501	0.035 8	0.040 6	4.75	0.323
320	1.400	0.641	0.028 0	0.031 7	6.06	0.327
400	1.560	0.801	0.022 4	0.025 4	7.56	0.329

0.063 & 0.071 Bunched

Table B.4.5 Properties of bunched enamelled copper wires with silk covering. Strand diameter 0.063 mm

Number of strands (n)	Nominal overall diameter of bunch mm	Nominal copper cross-sectional area mm^2	Resistance per unit length Nom. Ω/m	Resistance per unit length Max. Ω/m	Approx. weight g/m	Winding copper space factor (F_w)
3	0.190	0.009 54	1.881	2.050	0.0966	0.264
4	0.215	0.0127	1.410	1.537	0.127	0.275
5	0.235	0.0159	1.128	1.230	0.158	0.288
6	0.255	0.0191	0.940	1.025	0.188	0.294
7	0.255	0.0223	0.806	0.879	0.217	0.343
8	0.290	0.0254	0.705	0.769	0.249	0.302
9	0.295	0.0286	0.627	0.683	0.278	0.329
10	0.320	0.0318	0.564	0.615	0.309	0.311
12	0.350	0.0382	0.470	0.512	0.370	0.312
16	0.400	0.0509	0.353	0.384	0.491	0.318
20	0.450	0.0636	0.282	0.307	0.612	0.314
25	0.505	0.0795	0.226	0.246	0.763	0.312
27	0.525	0.0858	0.209	0.235	0.823	0.311
32	0.565	0.102	0.176	0.198	0.992	0.320
40	0.630	0.127	0.141	0.158	1.23	0.320
50	0.725	0.159	0.113	0.127	1.54	0.303
60	0.785	0.191	0.0940	0.106	1.84	0.310
63	0.805	0.200	0.0895	0.101	1.93	0.309
80	0.900	0.254	0.0705	0.0792	2.45	0.314
81	0.905	0.258	0.0696	0.0782	2.48	0.315
100	1.000	0.318	0.0564	0.0633	3.05	0.318
120	1.095	0.382	0.0470	0.0528	3.65	0.319
160	1.255	0.509	0.0353	0.0396	4.85	0.323
200	1.395	0.636	0.0282	0.0317	6.04	0.327
250	1.550	0.795	0.0226	0.0253	7.53	0.331
320	1.750	1.017	0.0176	0.0198	9.62	0.332
400	1.940	1.272	0.0141	0.0158	12.00	0.338

Table B.4.6 Properties of bunched enamelled copper wires with silk covering. Strand diameter: 0.071 mm

Number of strands (n)	Nominal overall diameter of bunch mm	Nominal copper cross-sectional area mm^2	Resistance per unit length Nom. Ω/m	Resistance per unit length Max. Ω/m	Approx. weight g/m	Winding copper space factor (F_w)
3	0.210	0.0121	1.481	1.607	0.120	0.274
4	0.240	0.0162	1.111	1.205	0.159	0.281
5	0.260	0.0202	0.888	0.964	0.197	0.299
6	0.285	0.0242	0.740	0.803	0.235	0.298
7	0.285	0.0283	0.635	0.689	0.272	0.348
8	0.325	0.0323	0.555	0.602	0.312	0.306
9	0.340	0.0363	0.494	0.536	0.350	0.314
10	0.355	0.0404	0.444	0.482	0.388	0.321
12	0.390	0.0485	0.370	0.402	0.465	0.319
16	0.445	0.0646	0.278	0.301	0.616	0.326
20	0.500	0.0808	0.222	0.241	0.769	0.323
25	0.560	0.101	0.178	0.193	0.959	0.322
27	0.585	0.109	0.165	0.184	1.035	0.317
32	0.630	0.129	0.139	0.155	1.24	0.325
40	0.730	0.162	0.111	0.124	1.55	0.304
50	0.810	0.202	0.0888	0.0993	1.93	0.308
60	0.880	0.242	0.0740	0.0827	2.31	0.313
63	0.900	0.254	0.0705	0.0780	2.42	0.314
71	0.950	0.287	0.0644	0.0699	2.73	0.318
80	1.005	0.323	0.0555	0.0620	3.07	0.320
81	1.010	0.327	0.0548	0.0613	3.11	0.321
100	1.120	0.404	0.0444	0.0496	3.82	0.322
120	1.220	0.485	0.0370	0.0413	4.58	0.326
160	1.400	0.646	0.0278	0.0310	6.08	0.330
200	1.560	0.808	0.0222	0.0248	7.58	0.332
250	1.735	1.010	0.0178	0.0199	9.45	0.336
320	1.950	1.292	0.0139	0.0155	12.08	0.340
400	2.180	1.615	0.0111	0.0124	15.07	0.340

Appendix C

Publications relevant to ferrite cores prepared by Technical Committee No. 51 of the International Electrotechnical Commission

133 (1985)	Dimensions for pot-cores made of ferromagnetic oxides and associated parts.
205 (1966)	Calculation of the effective parameters of magnetic piece parts. Amendment No. 1 (1976). Amendment No. 2 (1981).
205A (1968)	First supplement.
205B (1974)	Second supplement.
220 (1966)	Dimensions of tubes, pins and rods of ferromagnetic oxides.
221 (1966)	Dimensions of screw cores made of ferromagnetic oxides. Amendment No. 2 (1976).
221A (1972)	First supplement.
223 (1966)	Dimensions of aerial rods and slabs of ferromagnetic oxides.
223A (1972)	First supplement.
223B (1977)	Second supplement.
226 (1967)	Dimensions of cross cores (X-cores) made of ferromagnetic oxides and associated parts. Amendment No. 1 (1982).
226A (1970)	First supplement.
281 (1969)	Magnetic cores for application in coincident current matrix stores having a nominal selection ratio of 2:1 and in linear select memory stores. Amendment No. 1 (1975).
281A (1973)	First supplement.
367:—Cores for inductors and transformers for telecommunications.	
367-1 (1982)	Part 1: Measuring methods. Amendment No. 1 (1984).
367-2 (1974)	Part 2: Guides for the drafting of performance specifications. Amendment No. 1 (1983).
367-2A (1976)	First supplement.
392 (1972)	Guide for the drafting of specifications for microwave ferrites.
401 (1972)	Information on ferrite materials appearing in manufacturers' catalogues of transformer and inductor cores.
424 (1973)	Guide to the specification of limits for physical imperfections of parts made from magnetic oxides.
431 (1983)	Dimensions of square cores (RM-cores) made of magnetic oxides and associated parts.
492 (1974)	Measuring methods for aerial rods.

525 (1976)	Dimensions of toroids made of magnetic oxides or iron powder. Amendment No. 1 (1980).
556 (1982)	Measuring methods for properties of gyromagnetic materials intended for application at microwave frequencies.
647 (1979)	Dimensions for magnetic oxide cores intended for use in power supplies (EC-cores).
701 (1981)	Axial lead cores made of magnetic oxides or iron powder.

723:—Inductor and transformer cores for telecommunications.

723-1 (1982)	Part 1: Generic specification.
723-2 (1983)	Part 2: Sectional specification: Magnetic oxide cores for inductor applications.
723-2-1 (1983)	Part 2: Blank detail specification: Magnetic oxide cores for inductor applications—Assessment level A.
732 (1982)	Measuring methods for cylinder cores, tube cores and screw cores of magnetic oxides.

852:—Outline dimensions of transformers and inductors for use in telecommunication and electronic equipment.

852-1 (1986)	Part 1: Transformers and inductors using YEI-1 laminations.

Note: The following standard prepared by Technical Committee No. 68 is also relevant:

404:—Magnetic materials.

404-1 (1979)	Part 1: Classification.

Copies may be obtained from IEC, 1 rue de Varembé, Geneva, Switzerland.

Appendix D
Symbols

Following each symbol and definition there is, in general, a reference to the equation (in parentheses), the figure or section in which the symbol is introduced or first used.

Symbol	Definition	Reference
a	a dimension	Figure 4.3
a	a parameter used in the leading edge analysis of a pulse transformer	(8.3), (8.5)
a	a parameter used in the trailing edge analysis of a pulse transformer	(8.13)
a	hysteresis loss coefficient due to Legg	(2.61)
a_{10}, a_{11}, a_{02}, etc.	Peterson coefficients	(2.35), (2.36)
A	area	(2.11)
A	a dimension	Table 7.1, Figures 9.18 and 11.4
A_a	window area of a core	Figure 11.1
A_c	exposed (cooling) area of core	(9.68)
A_e	effective cross-sectional area of a core	(4.3)
A_e	effective loss of a network	Figure 7.2
A_g	cross-sectional area of air gap	(4.20)
A_i	insertion loss of a network	Figure 7.2
A_L	inductance factor in nH/turn^2	(5.6)
A_m	cross-sectional area of a magnetic core	(4.20)
A_N	effective area enclosed by an air cored coil	(4.80)
A_r	return loss of a network	Figure 7.2
A_R	resistance factor, R/N^2, at 100 °C	Table 9.3
A_w	cross-sectional area of a winding	Figure 11.1
A_1, A_2	particular cross-sectional areas of a core	(6.1)
b	a dimension	Figure 4.3
b	conductor breadth	Figure 11.13
b	a parameter used in the leading edge analysis of a pulse transformer	(8.3), (8.5)
b	a parameter used in the trailing edge analysis of a pulse transformer	(8.13)
b_a	breadth of core window	Figure 11.1
b_w	actual winding breadth	Figure 11.1
b_0	pitch of turns	Figure 11.13
B	magnetic flux density	(2.2)
B	a dimension	Table 7.1, Figures 9.18 and 11.4
B_a, B_{3a}	flux densities at frequencies f_a and $3f_a$ respectively	(2.66)–(2.67)
B_{av}	flux density averaged over a length of core	Figure 4.12
B_c	flux density at the centre of a cylindrical core	(4.67)
B_e	effective flux density	(4.16)
B_r	remanence	Figure 2.1

B_{sat}	saturation flux density	Table 3.4, Figure 3.3
$(B)_0$	flux density in core with zero air gap	(4.34)
B_1, B_2	flux density at particular cross-sections of a core	(6.1)
c	a dimension	Figure 4.3
c	residual loss coefficient due to Legg	(2.61)
c_o	the velocity of electromagnetic waves *in vacuo*	(2.49)
c	pitch of turns in a winding layer	Figure 11.18
C	capacitance	Figure 5.15
C	$= C_1 + C_2'$	(7.18)
C	a dimension	Table 7.1, Figure 9.18
C_a, C_b, C_c	components of the self capacitance of a single section winding	Figure 5.13
C_a', C_b', C_c'	components of the self capacitance of a multi-section winding	Figure 5.13
C_l	capacitance between two adjacent winding layers considered as a parallel plate capacitor	Table 11.3
C_m	capacitance between a winding layer and a screen (or other conducting surface) considered as a parallel plate capacitor	Table 11.3
C_p	shunt capacitance	(2.41)
C_{res}	total capacitance required to resonate an inductance at a given frequency	(5.24)
C_s	total self capacitance of a winding or windings	(5.13)
C_x	inter-section winding capacitance	Figure 5.13
C_1	core factor $= l_e/A_e$	(4.9)
$(C_1)_1$	core factor, C_1, for a pot core of unit radius	(5.4)
C_2	core factor $= l_e/A_e^2$	(4.11)
C_1	total capacitance appearing across the primary winding	Figure 7.3
C_2	total capacitance appearing across the secondary winding	Figure 7.3
d	disaccommodation coefficient (of permeability)	(2.26)
d	dimension	(2.46), Figure 4.3
d	diameter	Figure 1.8
d	diameter of bare conductor	(11.2)
d_{hor}	horizontal dimension of a body subject to convection	(9.3)
d_o	diameter of conductor over the insulation	(11.1)
d_{opt}	optimum conductor diameter	(5.21), (11.35)
d_{vert}	vertical dimension of a body subject to convection	(9.3)
d_1	inner diameter	(7.54)
d_2	outer diameter	Figure 7.14
D	disaccommodation (of permeability)	(2.25)
D	droop of the top of a pulse, per cent	(8.12)
D_F	disaccommodation factor (of permeability)	(2.27)
D_i	inner diameter of air cored coil	Figure 11.15
D_o	outer diameter of air cored coil	Figure 11.15
e	instantaneous value of e.m.f.	(2.10)
e	eddy current loss coefficient due to Legg	(2.61)
E	e.m.f.	
E	emissivity of a surface	(9.1)
E	electric field strength	(10.3)
E_a	source e.m.f.	Figure 7.2
$E_a, E_{3a}, E_{2a} \pm b$	e.m.f.s at frequencies f_a, $3f_a$ and $2f_a \pm f_b$ respectively	(2.70)–(2.72)
E_c	e.m.f. induced in a short coil placed centrally on a cylindrical core	Section 4.3.2
E_n	noise e.m.f.	(10.12)
E_{na}	distortion e.m.f. at frequency nf_a	Figure 4.8
E_s	signal e.m.f.	(10.1)

E_ρ	activation energy	(3.5), Figure 3.21
f	frequency	
f_a, f_b	frequencies	(2.69)
f_c	cut-off frequency	(7.23)
f_{res}	ferromagnetic resonant frequency	(3.2), Figure 3.11
Δf	bandwidth at 3 dB points	(10.12)
f_1, f_2	frequency of the lower and upper ends respectively of a transformer pass band	Figure 7.1
F	skin effect factor	(11.14)
F	a function	(7.46), (7.50)
F_a	overall copper factor	(11.5)
F_A	averaging factor	Figure 4.13(b)
F_h	core hysteresis factor	(4.42)
F_l	layer copper factor	(11.46)
F_n	noise factor	(10.14)
F_p	packing factor	(11.6)
F_R	factor by which the a.c. resistance of a transformer winding exceeds the d.c. value	(11.33), Figure 11.14
F_w	winding copper factor	(11.2)
F_1	particular value of function F	(7.47), (7.51)
G	conductance	
G	gain	(10.17)
G_h	conductance due to hysteresis loss	(4.53)
G_r	factor expressing the screening effect of eddy currents in a round conductor	(5.18), (11.23)
G_s	conductance expressing the loss in the self capacitance of a winding	(5.23)
h	thickness of a toroidal core	Figure 4.1
h	dimension	Figure 4.3
h	effective conductor height	Figure 11.13
h_a	height of core window	Figure 11.1
h_e	effective height of an antenna	(10.4)
h_{opt}	optimum thickness of strip conductor	(11.43)
h_w	actual height of a winding perpendicular to layers	Figure 11.1
h_Δ	distance separating primary and secondary windings	Figure 11.20
h_1, h_2	height boundaries of a winding	Figure 11.12
h_1, h_2	height of primary and secondary windings respectively	Figure 11.20
H	magnetic field strength	(2.1)
H_a	field strength at the fundamental frequency	(2.66)
H_c	coercivity	Figure 2.1
H_c	field strength at centre of rod	Section 4.3.1
H_D	demagnetizing field strength	(4.66)
H_i	internal field strength	(2.3)
H_2	field strength of a particular part of a core	(6.4)
i	instantaneous current	(8.23)
I	current	
I_m	maximum current in choke	Figure 9.11
I_{na}	distortion current at frequency nf_a	(4.63)
I_s	signal current	(10.9)
I_o	static current in a winding	Figures 3.7 and 9.11
I_1, I_2	current in windings having N_1 and N_2 turns respectively	(7.4)
j	$\sqrt{-1}$	
J	magnetic polarization (intrinsic flux density)	(2.5)

J	current density	Figure 11.5
J_c	magnetic polarization at the centre of a cylindrical core	(4.66)
k	constant of proportionality	(9.61)
k	constant	Figure 9.13
k	function of geometry for an air-cored coil	(11.52), Figure 11.15
k	Boltzmann's constant	(3.5), Figure 3.21, (10.12)
k	coupling factor	(7.27)
k	a function of ϑ	(11.28)
k_e, k_E	proximity effect constants	(11.26)
k_1–k_3	magnetic loss coefficients	(2.60)
k_1–k_5	proportions of a pot core	Figure 5.1
k_1–k_4	parameters used in power transformer design	Table 9.4
K	constant	(11.51)
K_1, K_2	constants	(9.24)
l	length	(2.1)
l_A	length of magnetic shunt in pot core adjuster	Figure 5.6
l_c	length of a coil	Figure 4.13(a)
l_e	effective path length of a magnetic circuit	(4.3)
l_g	path length of an air gap	(4.20)
l_m	path length in a magnetic core	(4.20)
l_s	spacer thickness	Figure 9.12
l_w	mean turn length of a winding	(5.2)
l_1, l_2	path lengths of parts of a magnetic core	(6.2)
L	dimension of length	(9.60)
L	inductance	(2.14)
L_a	the air inductance of a coil	(4.71)
L_l	total leakage inductance	Figures 7.3 and 11.21
L_{l1}, L_{l2}	leakage inductance associated with the primary and secondary windings respectively	Figure 7.3
L_o	inductance/μ	(2.14)
L_p	parallel inductance	(2.17)
L_s	series inductance	(2.15)
m	exponent of frequency in magnetic loss equation	(2.74)
m	dimensional ratio of a cylindrical or ellipsoidal core	Figure 4.10
m	fractional increase of conductor resistivity above the value at 20 °C	(9.25), (11.8)
m	modulation factor	(10.22)
M	magnetization	(2.3)
M	number of section interfaces in a transformer	Figure 11.21
M_c	magnetization at centre of cylindrical core	(4.66)
M_{sat}	saturation magnetization	(3.2), Figure 3.11
n	Steinmetz exponent	(2.74), Figure 3.19
n	number of insulated strands in a bunched conductor	(5.18)
N	number of turns	(2.14)
N	demagnetization factor	(4.66)
N_a	normalized number of turns	(7.48)
N_l	number of turns in a winding layer	(11.46)
N_1, N_2	number of turns on primary and secondary windings respectively	(2.1), (2.11)
p	$2\pi f$	(2.65)
p	porosity	(3.1)
p	number of layers in a winding or portion of a winding	(11.7), Figure 11.14

P	power	
P_a	applied power	Figure 7.2
P_{av}	average power	(8.1)
P_b	power in load	Figure 7.2
P_c	total core loss	(9.21)
P_{cond}	rate of heat transfer due to conduction	(9.7)
P_{conv}	rate of heat transfer due to convection	(9.2)
P_F	eddy current power loss (volume) density	(2.46)
P_h	hysteresis power loss (volume) density	(2.31)
P_i	input power	(9.27)
P_m	power loss (volume) density in a magnetic material due to any cause	(2.54)
P_M	magnetic loss (volume) density	(2.74)
P_n	noise power	(10.16)
P_{no}	noise output power	(10.19)
P_o	output power	(9.23)
P_p	power during pulse	(8.1)
P_{pe}	power loss due to proximity effect in a winding	(11.21)
P_r	reflected power	Figure 7.2
P_{rad}	rate of heat transfer due to radiation	(9.1)
P_s	signal power	(10.11)
P_{so}	signal output power	(10.20)
P_{tot}	power loss in core plus winding	(9.21)
P_w	power loss in windings	(9.21)
q	number of side-by-side sections in a winding	Table 11.3
Q	quality (Q) factor	(5.14)
Q_L	Q-factor of a loaded circuit	(10.7)
r, r_1, r_2	radius and particular radii	Figure 4.1
r	transformation ratio	(7.3)
R	resistance	
R	$= R_a R_b'/(R_a + R_b')$	(7.12)
R_a	source resistance	Figure 7.2
R_{ac}	d.c. resistance of a conductor plus the resistance due to skin effect	(11.14)
R_{AC}	d.c. resistance of a winding plus the resistance due to general eddy current phenomena	(11.33)
R_b	load resistance	Figure 7.2
R_c	resistance per unit length of conductor	(11.9)
R_{dc}	d.c. resistance of a winding	(11.9)
$(R_{dc}/L)_1$	ratio of d.c. resistance to inductance for a winding half filling the available winding height	(7.36)
R_F	eddy current series loss resistance	(2.47)
R_h	series resistance representing hysteresis loss	(2.32)
R_p	parallel resistance representing magnetic and/or circuit losses	(2.17), Figure 2.2
R_{pe}	series resistance representing proximity effect in a winding	Section 11.4.3
R_s	series resistance representing magnetic and/or circuit losses	(2.15), Figure 2.2
R_s	total winding resistance of a transformer referred to the primary winding	(7.11)
R_{se}	increase in conductor resistance due to skin effect	(11.14)
R_T	thermal resistance	(9.8)
R_{th}	thermal resistance of a wound component	(9.11)
R_o	nominal resistance of a half-section filter	(7.23)
R_1, R_2	resistance of primary and secondary windings respectively	Figure 7.3
s	length along a path	(4.1)

Appendix D 359

S	pitch of adjacent winding layers	Figure 11.18
S_e	effective thickness of dielectric	(11.55), Figure 11.18
S_f	time factor associated with trailing edge of a pulse	(8.13)
$(S_f)_0$	value of S_f corresponding to the fall time to zero (t_f)	Figure 8.16
S_r	time factor associated with leading edge of a pulse	(8.3)
S_1, S_2	dimensions relating to the magnetic path in pot core	Figure A.1
t	time	
t	thickness of layer insulation	(11.7)
t_d	pulse duration	Figure 8.1
t_f	pulse fall time	Figure 8.1
t_P	time between leading edges of consecutive pulses	Figure 8.1
t_r	pulse rise time	Figure 8.1
t_1	pulse rise time of source	(8.2)
t_2	pulse rise time of oscilloscope	(8.2)
T	absolute temperature	(3.5), Figure 3.21
T_s	absolute temperature of a surface	(9.1)
T_o	absolute temperature of objects surrounding a radiating surface	(9.1)
U	voltage	
U_a	voltage at fundamental frequency f_a	Figure 4.8
U_b	voltage across load	Figure 7.2
U_{AO}, U_{AD} U_{BO}, U_{BD}	voltage potentials at each end of winding layers A and B respectively	Figure 11.16
U_D	voltage difference $U_{BD} - U_{AD}$	(11.53)
U_n	noise voltage	(10.13)
U_{na}	distortion voltage at frequency nf_a	Figure 4.8
U_p	voltage applied across a winding	(11.53)
U_s	signal voltage	(10.9)
U_o	voltage difference $U_{BO} - U_{AO}$	(11.53)
v	velocity of propagation of e.m. waves in material	(2.48)
v	air velocity	(9.5)
V	volume	
V_e	effective (or hysteresis) volume of a core	(4.7)
V_T	overall volume of a pot core	(5.5)
w_h	hysteresis energy loss (volume) density	(2.31)
W	energy stored in a magnetic circuit	(9.47)
W_h	hysteresis energy loss	(4.15)
x	a length or distance variable	
x	a parameter used in the leading edge analysis of a pulse transformer	(8.9)
x	voltage regulation	(9.29)
Σx	transformer winding dimension	Figure 11.21
Σx_Δ	transformer winding dimension	Figure 11.21
X	transformer winding dimension	Figure 11.21
Y	admittance	(2.17)
Y	transformer winding dimension	Figure 11.21
z	variable	Figure 7.6
Z	impedance	(2.15)
Z_a	source impedance	Figure 7.2
Z_A	source impedance	Figure 4.8

Z_b	load impedance	Figure 7.2
Z_B	load impedance	Figure 4.8
Z_i	input impedance	Figure 7.7
Z_m	impedance due to the magnetic properties of a core	Figure 4.8
a	coefficient of linear expansion	Section 1.4.2
a	ratio of boundary thickness to crystallite thickness	Figure 2.5
a	angle	(4.40), Figure 4.5
β	eddy current shape factor	(2.46)
γ	gyromagnetic ratio	(3.2), Figure 3.11
δ_{ac}	loss angle due to the d.c. resistance plus skin effect in a conductor	Figure 5.14(a)
δ_{AC}	loss angle due to the total (a.c.) resistance in a conductor	Section 10.3.3
δ_{cp}	loss angle of an inductor due to dielectric loss in its self capacitance	(5.24), Table 5.6
δ_{cs}	loss angle due to the self capacitance of a series resonated inductance	(5.27), Table 5.6
δ_d	dielectric loss angle $= \varepsilon_p''/\varepsilon_p'$	(2.43), Table 5.6
δ_{dc}	loss angle due to d.c. winding resistance	(5.15), Table 5.6
δ_F	loss angle due to eddy currents in core	(2.47), Table 5.6
δ_h	hysteresis loss angle	(2.33), Table 5.6
δ_H	permeability rise factor	(2.64)
δ_L	loss angle of an inductor in the absence of self capacitance	(5.27), Table 5.6
δ_m	loss angle due to any magnetic loss	(2.16)
δ_{mc}	loss angle of a rod antenna due to magnetic coupling to adjacent conductive parts	Section 10.3.3
δ_{pe}	loss angle due to proximity effect	(11.26), Table 5.6
δ_r	residual loss angle	(2.51), Table 5.6
δ_{rad}	loss angle due to antenna radiation	Section 10.3.3
δ_{r+F}	loss angle due to residual plus eddy current loss in core	Table 5.6
δ_{se}	loss angle due to skin effect	(11.19), Table 5.6
δ_{tot}	total loss angle of an inductor	(5.14)
Δ	penetration depth	(2.50), (11.15)
ε	permittivity	(2.41)
ε_p', ε_p''	real and imaginary components respectively of the complex permittivity expressed in parallel terms	(2.42)
ε_o	electric constant $= 8.854 \times 10^{-12} \, F \, m^{-1}$	(2.41)
ε_1, ε_2	real components of the permittivity of crystallite and boundary respectively	Figure 2.5
η	load-source mismatch factor	Figure 8.2
η_B	IEC hysteresis loss coefficient of a material	(2.63)
η_i	IEC hysteresis loss factor of a core	(4.44)
ϑ	ratio of flux densities at height boundaries of a winding	Figure 11.12
θ	temperature in °C	
θ_{cs}	temperature difference between the centre and the surface of a magnetic core	(9.9)
θ_1, θ_2	particular temperatures	(2.21)
κ	susceptibility	(2.7)
λ	wavelength	(2.48)

λ	thermal conductivity	(9.7)
μ	permeability	(2.8)
μ_a	amplitude permeability	(2.28)
μ_c	conductor permeability	(11.15)
μ_{coil}	ratio of inductance of a coil with and without a core	(4.71)
μ_e	effective permeability	(4.24)
μ_i	initial permeability	(2.28)
μ'_p, μ''_p	real and imaginary components respectively of the complex permeability expressed in parallel terms	(2.17)
μ_p	pulse permeability	(8.30)
μ_{rev}	reversible permeability	Section 2.1
μ_{rod}	rod permeability	(4.69)
μ'_s, μ''_s	real and imaginary components respectively of the complex permeability expressed in series terms	(2.15)
μ_Δ	incremental permeability	Section 2.1, Figure 3.6
μ_o	magnetic constant = $4\pi 10^{-7}$	(2.2)
μ_1, μ_2	permeability at particular places, temperatures or times	(2.21)
ν	Rayleigh hysteresis coefficient	(2.28)
ν	viscosity	(9.6)
ρ	resistivity	(2.41)
ρ	nominal mismatch factor	(7.24)
ρ_c	resistivity of winding conductor (at 20 °C when associated with m)	(11.8)
ρ_s	sintered (mass) density	(3.1)
ρ_x	x-ray (mass) density	(3.1)
ρ_1, ρ_2	resistivity of crystallite and boundary respectively	Figure 2.5
ρ_∞	resistivity of polycrystalline ferrite extrapolated to $T = \infty$	(3.5), Figure 3.21
σ_f	damping factor associated with trailing edge of a pulse	(8.13)
σ_r	damping factor associated with leading edge of a pulse	(8.3)
τ	relaxation time = $1/\omega_r$	(2.45)
φ	parameter used to determine a.c. resistance of a transformer winding	(11.47)
φ	phase shift	(7.16)
φ, Φ	magnetic flux	(2.10), (4.58)–(4.60)
χ	a parameter used in the trailing edge analysis of a pulse transformer	(8.13)
ω	$2\pi \times$ frequency	
ω_c	$2\pi \times$ cut-off frequency	(7.23)
ω_r	$2\pi \times$ relaxation frequency	(2.45)
ω_1	$2\pi \times$ frequency at lower end of transformer pass band	(8.19)
$\hat{x}$	maximum value of x	
$\bar{x}$	average value of x	
Δx	change of x	
δx	small change of x	
x'	real part of x	
x''	imaginary part of x	
x'	value of x referred to the winding of a transformer under consideration	

Index

Additives in ferrite manufacture, 9
Admittance related to complex permeability, 29
Air-cored coil inductance, 330
Air gaps
    effect on $B-H$ loop, 142–143
    effect on core losses, 145–147
    effect on variability, 141–142
    in design of cores carrying d.c., 74, 77, 215, 275–280
    in inductor cores, 163–168
    reasons for, 139
    reluctance/permeability, 139–143
    residual, effect in wide-band transformers, 214
Alloys for high-power transformers, 232
Ampere turn products in transformer windings, 323
Anisotropy, magnetocrystalline, 5
Antennas, 303–312
    circuit theory, 303–308
        bandwidth/impedance requirements, 306–307
        design example, 307–308
        induced e.m.f./effective height, 303–304
        noise power, 305–306
        signal strength, 304–305
        signal-to-noise ratio, 306
    design, practical, 308–111
        effective height, 308–309
        Q-factor, unloaded, 309–310
        temperature coefficient, 311
    tuning, transductors for, 192–194
Applied power, 201–203
Atmosphere control in kilns, 13–15
Auto-transformer windings, leakage inductance, 337

$B-H$ loops, 59
    definitions, 27–28
    for gapped cores, 142–143
    minor, 67
    in pulse operation, 249–251
    see also Hysteresis
Bifilar windings, leakage inductance, 337–338
Binders in ferrite manufacture, 11
Bloch walls and magnetization, 5–6
Bonding, see Cementing
Box kilns
    computer controlled, 15–16
    mechanically controlled, 14
Bunched conductor, 322

Calcining, 10

Capacitor/inductor combination
    requirements, 159
    temperature coefficient compensation, 169–170
    see also Self capacitance
Cementing/bonding of core parts, 19
    effect on inductance variability, 172
Choke design, 275–278
    gaps, unconventional, 279–280
    Hanna curves in, 275, 276
    power dissipation, 275, 277–278
Circuit theory, magnetic, 136–157
    assumptions, 136
    closed cores, 136–149
        air gaps and reluctance/permeability, 139–143
        distortion/intermodulation, 148–149
        effective dimensions, 136–139
        losses, 143–148
    cylindrical open cores, 149–157
        flux density and demagnetization, 149–152
        flux distribution, 152–154
        losses, 155–157
        rod permeability, 150–152
        temperature coefficient, 157
        winding inductance, 155
Classification of ferrites by application, 45–54
    survey of manufacturers and ferrite grades, 46–54
Coercivity, 28
    typical values of, 56–57
Conduction heat loss, 260–261
Convection heat loss, 257–260
Copper factors for winding close packing, 314
Core factors, 137–138
Cores for power applications
    higher frequency transformers, 291–296
        higher range, 293–296
        lower range, 291–293
    low frequency, 280–291
        current types, 281–282
        desirable properties, 280
        geometry considerations, 281
        parameters, 282–283, 288
Coupling factor, 213
Curie point
    initial permeability, 80
    saturation flux density, 69
    typical values of, 56–57
Cylindrical cores, 149–157
    flux density and demagnetization factor, 149–152
    flux distribution, 152–154
    losses, 155–157
    temperature coefficient, 157
    winding inductance, 155

Damping factor in pulse transformers, 234–236, 238–239
Demagnetization factor, 150–152
Diamond machining, 18
Dielectric loss
    in ferrite, 33–36, 296
    in stray capacitance, 179–180
Dielectric properties of ferrites, 126–129
Differential permeability, 28
Dilution ratio, 142
Dimensional resonance, 36, 146, 215, 228
Disaccommodation, 30–31, 88–89
    coefficient, 31
    factor, of gapped core, 142
    and inductance variability, 170
    and thermal disturbance, 172–173
    typical values of, 56
Distortion
    waveform, 39–40, 111, 148–149
    in inductors, 187
    in transformers, pulse, 232–239
        droop in relation to bandwidth, 247–248
        leading edge, 233–237
        top of pulse, 237
        trailing edge, 237–239
Domains in magnetization, 5–6, 26
Dowell analysis of a.c. winding losses, 327–330
Droop in pulse transformers, 237, 247–248

Eddy current phenomena
    in cores, 33–36, 41–42, 146–147, 215
    losses, 35, 146–148
        in power transformers, 267–268
        in pulse transformers, 251
        in wide-band transformers, 216
    losses in windings, 175–178, 317–330
        proximity effect, 176–178, 321–323
        skin effect, 175–176, 318–321
        total, 323–330
    resistivity/permittivity, 33–35
    resonance, dimensional, 36, 146–147
Effective
    dimensions, 136–138
    loss of a transformer, 202–204, 208
    loss tangent of a cylinder, 156–157
    permeability, 38, 140, 163, 214
E.m.f. averaging factor for a cylinder, 153–154
    equation, 27, 139, 265
Electric
    constant, 33
    field strength, 303
Electromagnetic wave propagation, 36
Electron spin, 4–5, 26–28
Emissivities of surfaces, 257
Encapsulation, 20–21
Extrusion, 12–13

364 Index

Ferrimagnetism, ferromagnetism, 5
Ferromagnetic resonance, 90, 99, 109, 114
Field strength, 26
Filter techniques in transformer design, 206–211
Finishing processes
   in core manufacture, 16–17
   available to the user, 18–21
Flux, expressions for, 147
Flux density, 26
   amplitude of third harmonic and intermodulation products, 39–40
   distribution in cylindrical cores, 149–150, 152–154
   intrinsic, 27
   effective, 138–139
   in power transformers, expressions for, 265
   at low amplitudes, expressions for, 31–32
   saturation, 28
   and temperature, 69
   see also B–H loops
Fringing flux at air gap, 141

Gapped cores, *see* Air gaps; Cylindrical cores
Grain growth, 13
Granulation, 11
Green state, 12
Grinding for ferrite finishing, 16–17

Hanna curves, 77
   in choke circuit design, 275, 276
Hanna procedure in transformer design, 215
Hardness of ferrites, 54
Heat transfer and transformer design, 256–264
   calculation, 261–263
   conduction, 260–261
   convection, 257–260
      in air, 258–259
      in oil, 259–260
   effects of heat, 256
   isolation, thermal, 264
   radiation, 257
   resistance, thermal
      of a conduction path, 260
      of a complete core, 263–264
   shunts, thermal, 264
   sources of heat, 256
Hot pressing, 13
Hysteresis, 27–28
   coefficient, 39
      typical values of, 56–57
   constant defined, 39
   expression of, quantitative, 31–33
      Peterson relations, 32–33
      Rayleigh relations, 31–32
   loss
      in inductors, 180–181
      in effective dimension calculations, 137, 138
      expression of, 31–33, 41–42, 144–145
      in non-uniform cross-section, 138–139
      in wide-band transformer design, 216–217
   loss factor
      and flux density, 107
      and frequency, 109
      and temperature, 110
   magnetic, 6–7
      in inductor design, 159
   see also Magnetic loss

IEC publications, list of, 352–353

Impact strength, 54
Impedance
   conversion from series to parallel, 29
   in terms of complex permeability, 29
   losses in closed cores
      current/series, 143–144
      voltage/parallel, 144–145
Incremental permeability, *see* Permeability, incremental
Inductance, 28–29
   of air-cored coil, 330
   factor, 163, 165
   in non-uniform cross-section core, 139
   of pulse transformers, 248–249
   ratio, 155
      of d.c. winding resistance to, 161, 217, 317
   requirements in network design, 159
   temperature coefficient of, 169–170
   in transductors, 189
   variability, 168–173
      from ferrite, 169–171
      from non-ferrite parts, 171–172
      overall, 172–173
   of wide-band transformers, 213–215
   in windings
      with cylindrical core, 155
      leakage, 335–338
Induction
   equations in transformer design, 264–265
   law of, 27
Inductors, 158–188
   adjustment, 165–168
      air gap shunt, 165–168
      rotation method, 165
   air gap and inductance calculation, 163–165
   core form, 160–163
      design, 161–163
      geometries, 160–161
      practical modifications, 163
   and digital techniques, 158
   energy storage, design, 278–279
   intermodulation, 187–188
   network design requirements, 158–160
      hysteresis effects, 159
      inductance, 159
      magnetic coupling, 160
      Q-factor, 159
   Q-factor, 175–187
      capacitance loss, 179–180
      combined losses, 181–182
      core loss, 180–181
      effect of design parameters on, 183–187
      eddy current loss in windings, 175–178
      presentation of, 182–183
      winding resistance loss, d.c., 175
      summary of loss expressions, 181
   rod-cored, inductance of, 155
   self capacitance, 173–174
   waveform distortion, 187
   see also Choke design
Injection moulding, 13
Insertion loss, 202–203
Intermodulation
   in ferrite, 39–40
   in gapped cores, 148–149
   in inductors, 187–188
Iron alloys for high-power transformers, 232

Kilns
   for calcining, 10
   for sintering, 14–16

Kilns, for sintering (*cont.*)
   box, 15–16
   continuous, 14–15
Leakage
   flux, 141, 323–325, 335–336
   inductance, 335–338
Legg hysteresis coefficients, 38, 145
Loss(es)
   angle, magnetic, 29
   of antenna, 304
   in inductors, 175–182
   insertion/return, in wide-band transformers, 201–211, 225–227
   magnetic, in ferrites, 28–42
      eddy current, 33–36
      factor, 38
      general expressions, 37–39
      high amplitude, 40–42
      hysteresis, 32–33
      loss factor and flux density, 107
      loss factor and frequency, 109
      loss factor and temperature, 110
      residual, 36–37
      loss factor spectrum, 99
      loss factor and temperature, 103
      loss tangent and frequency/static field, 105
   magnetic, in closed cores, 143–148
      eddy current, 146
      high amplitude, 147–148
      hysteresis, 144–145
      residual, 147
   magnetic, in cylindrical cores, 155–157
   magnetic, in power transformers, 267–269
   magnetic, in pulse transformers, 251–252
   magnetic, in wide-band transformers, 215–217, 224–228
   magnetic power loss and flux density/frequency, 114
   magnetic power loss and temperature, 123
   tangent, magnetic, 29
   typical values of, 56–57
   windings,
      eddy current,
         proximity effect, 321–323
         skin effect, 318–321
         total in transformer windings, 323–330
      of inductors, 175–178
      of power transformers, 269–274, 292, 296–300
      of wide-band transformers, 217
Loudspeakers, hexaferrites in, 2

Machining in ferrite finishing, 19–20
Magnetic
   area moment, 26
   circuit theory, 136–157
   constant, 26
   field, internal, 27
   loss, definition, 42
   polarization, 26–27
Magnetism in ferrites, 3–7
   magnetization, 4–7
      anisotropy, magnetocrystalline, 5
      domains of spin orientation, 5–6
      expression of, 26–28
      hysteresis, 6–7
   spinel lattice structure, 3
Magnetite, early use of, 1

Index

Magnetization, 4–7, 26–28
 current, 272
 curve, 268
 saturation, 90
Magnetostriction, static, 130
 in transductors, 199
Magnets, permanent, hexaferrites in, 2
Manganese zinc ferrites
 classification, 45
 development, 1–2
 typical values of properties, 56
 see also Properties of ferrites
Magnetomotive force, 323, 336
Manufacture of ferrites, 1, 7–18
 calcining, 10
 finishing, 16–18
 forming, 11–13
  dry pressing, 11–12
  extrusion, 12–13
 inspection, 18
 milling, 10
 mixing, 9
 processes, 7
 processing, post-calcining, 10–11
 raw materials, 7–9
  additions, 9
  ferrous iron, 9
  particle size, 8, 10
 sintering, 13–16
  chemistry of, 13–15
  kilns for, 14–16
 user finishing, 18–21
  cementing, 19–20
  encapsulation, 20–21
  machining, 18–19
Manufacturers, survey of ferrite grades, 46–54
Mechanical properties of ferrites, 54
Microwave devices, magnetic oxides in, 2
Mismatch
 in pulse transformer design, 234–236
 in wide-band transformer design, 208–211
Motors, d.c., hexaferrites in, 2

Networks, passive
 concepts in, 201–204
  transformer losses, 203–204
 design requirements, 158–160
Nickel zinc ferrites
 classification, 45
 development, 1–2
 typical values of properties, 57
 see also Properties of ferrites
Noise factor/power/voltage in antenna circuit theory, 305–306

Overshoot in pulsed transformers, 235–236
Oxygen, in ferrite manufacture, 14

Packing factor, 315
Penetration depth
 in conductors, 318–320
 in ferrite, 36
Permeability
 absolute, 27
 apparent, 155
 complex
  expression of, 28–29
  initial, spectrum, 90–91
 constancy, and inductance, 169–171
  and amplitude variations, 171
  disaccommodation, 170

Permeability, constancy, and inductance (*cont.*)
  temperature coefficient compensation, 169–170
 effective
  in gapped cores, 140–143
  standardization in inductors, 163
  and transformer design, 214
 expression of, 27
 and flux density, 71, 73
 and hysteresis, 28
 incremental, 28
  and static field strength, 74
  and temperature, 87
 initial
  and stress, 131
  and temperature, 80
 in pulse transformers, 248–252
 and reluctance in closed cores, 139–143
 rise factor defined, 39
 rod, 150–152
 in transformers, wide-band, 214–215
 variability
  expression of, 29–31
  with temperature, 29–30
  with time, 30–31
Permittivity, 33–35
 and frequency, 127
 and temperature, 129
Peterson relations, 32–33
Phase shift, 205, 208, 210–211
Porosity, 13, 54
Pot cores, 160, 219–221, 281–282
 core factor calculation, 339–340
 geometry, 160
Power conversion, 255–302
 choke design, 275–278
  Hanna curves in, 275
  power dissipation, 275, 277–278
 core types/performance, 280–288
  current types, 281–282
  desirable properties, 280
  geometry considerations, 281
  parameters, 282–283, 288
 development/early use, 255
 heat transfer, 256–264
  calculation, 261–263
  conduction, 260–261
  convection, 257–260
  effects of temperature, 256
  radiation, 257
  resistance, thermal, 260, 263–264
  shunts/isolation, thermal, 264
  sources of heat, 256
 inductors, energy storage, design, 278–279
 transformers, *see* Transformers, power
Power loss, magnetic, definition, 42
 and flux density/frequency, 114
 and temperature, 123
Pressing methods in ferrite manufacture, 11–12, 13
Properties of ferrites, 44–135
 classification by application, 45
 magnetic/electrical, 55
  see also by particular property
 mechanical, 54
 survey by manufacturer, 46–54
 thermal, 55
 typical values of, 56–57
Proximity effect in windings
 general theory, 321–323
 in inductors, 176–178

Proximity effect in windings (*cont.*)
 reduction of, 176–177, 322
 in transformers, 324–330

Q-factor
 in antenna design, 309–310
 in inductors, 175–187
  contributory losses, 181–182
  design parameters and, 183–187
  and network design, 159
  presentation of, Q-maps, 182–183
 in transductors, 192

Radiation heat loss, 257
Raster correction in television, transductors for, 195
Rayleigh relations in hysteresis, 31–32
Recording heads, 7
Reflection coefficient, 203
Relative permeability, 27
Relaxation frequency of composite dielectric, 35
Reluctance, 138–141
Remanence, 28
 typical values of, 56–57
Residual loss, 36–37, 147
 coefficient, 38–39
 factor, 147, 180
  spectrum, 99
 and temperature, 103
 typical values of, 56–57
 tangent, and frequency/static field, 105
Resins
 as adhesive, 19–20
 for encapsulation, 20–21
Resistance of power cores, thermal, 263–264
Resistivity, 33–35
 and frequency, 127
 measurement, 33
 in polycrystalline ferrites, 34–35
 and temperature, 126, 129
 typical values of, 56–57
Resonance
 dimensional, 36, 146–147
 and transformer design, 215
 ferromagnetic, 90
 precessional, 2
Return loss, 202–203
Reversible permeability, 28
Rise time of pulse transformers, 233, 234
Rod cores, *see* Cylindrical cores
Rod permeability, 150–152

Saturation flux density, 28
 and temperature, 69
 typical values of, 56–57
Saturation magnetization, 90
Screen, self capacitance to, 332–334
Self capacitance
 analysis, 330–334
 assumptions, 330
 in inductors, 173–174
 loss from, 179–180
  circulating current, 179–180
  dielectric, 179
 of winding, 330–334
  and core geometry, 218
Shearing of B–H loop, 142–143
Shrinkage during sintering, 11, 13–14, 16
Signal strength in antenna theory, 304–305
Signal-to-noise ratio in antenna theory, 306

Single crystal ferrites, 7
Sintering, 13–16
   and density changes, 54
   kilns for, 14–16
Skin effect in winding loss, 175–176, 318–321
Slip casting, 13
Spinel
   formula, 1–2
   lattice structure, 3–4
Spray drying, 10–11
Stefan–Boltzmann law on heat dissipation, 257
Steinmetz exponent, 42
Stress
   induced variability, 171–172
   and permeability, 131
Sub-lattices, 4–5
Susceptibility, 27
Switch Mode Power Supplies, *see*
      Transformers, power
Symbols, list of, 354–361
Synchroton
   ferrites in, 3
   field control, transducers for, 194

Telephony, early ferrite use, 2
Television
   deflection yokes, finishing, 17–18
   early ferrite use, 2
   power transformers in, 255
   raster correction in, transducers for, 195
   U cores in, 281
Temperature
   coefficient
      compensation, inductors/capacitors, 169–170
      of gapped core, 142, 157
   cycling, and inductor variability, 172–173
   and flux density, saturation, 69
   and losses
      hysteresis loss factor, 110
      power, magnetic, 123
      residual, 103
   and permeability
      incremental, and static field strength, 87
      initial, 80
      variability with, 29–30
   and permittivity, high frequency, 129
   and resistivity, 126
      high-frequency, 129
   rise in power transformers, 256–264, 295–296
   thermal properties of ferrites, 54–55
   see also Heat transfer and transformer design
Thermal
   conductivities, 261
   resistance of a conduction path, 260
   resistance of a transformer core, 263–264
   shunt, 264
Time factor in pulse analysis, 234
Tin substitution and disaccommodation, 89
Titanium substitution and disaccommodation, 89
Tolerances on dimensions of ferrite parts, 16
Toroidal cores
   effective dimensions of, 136–138

Toroidal cores (*cont.*)
   ideal, 28
   for transformers, 219, 228–230
Transductors, high frequency, 189–200
   applications, 189, 192–195
      antenna tuning, 192–194
      models, experimental, 192
      raster correction in television, 195
      synchrotron field control, 194
   design/operating techniques, 196–198
      ferrite choice, 197–198
   mode of operation, 189–192
      core loss, 192
      magnetic circuits, 189–191
      temperature dependence, 191–192
Transformers, power, 255–302
   design, electrical, 264–268
      core losses and flux density, 265
      excitation waveforms, 266–267
      induction equations, 264–265
      losses, core, 267–268
      shunt inductance, 266
      windings, 296–301
   design requirements, 255
      core materials, 256
   ferrites, early use in, 255
   heat transfer, 256–264
      calculation, 261–263
      conduction, 260–261
      convection, 257–260
      effects of temperature, 256
      radiation, 257
      resistance, thermal, 263–264
      shunts/isolation, thermal, 264
      sources of heat, 256
   higher frequency, 291–301
      applications, 291
      core selection, 291–296
      windings, 296–301
   low frequency, 268–274
      design example, 289, 291
      design methods, 270–274
      losses, 267–270
      thermal considerations, 269
Transformers, pulse, 232–254
   alloys for high-power, 232
   definitions/terms, 232–233
   design, 241–247
      complete procedure, 241–245
      example, 245–246
      rapid procedure, 246–247
   function, 232
   practical considerations, 252–254
   pulse distortion, 232–233, 247–248
      in relation to bandwidth, 247–248
      leading edge, 233–237
      top of pulse, 237
      trailing edge, 237–239
   pulse permeability, 248–252
      core losses, 251–252
      flux density excursions increase, 250–251
Transformers, wide-band, 201–231
   design, 201
   high-frequency, 224–231
      core contribution, 224–228

Transformers, high-frequency (*cont.*)
      core form, 228–229
      design example, 230–231
      geometry, 228
      windings, 229–230
   low/medium frequency design, 213–224
      core form, 219–222
      coupling, 213
      design procedure/example, 222–224
      geometry, 217–219
      Hanna procedure, 215
      and losses, 215–217
      and permeability, 214–215
      and resonance, dimensional, 215
   self-capacitance analysis, 332–333
   transmission/reflection, 201–213
      equivalent circuit, 204–205
      filter techniques, 211–213
      higher frequency region, 206–211
      lower frequency region, 205–206
      mid-band region, 205
      network concepts, 201–204

Vacancies, 4, 14
Variability
   of inductance, 168–173
      from ferrite material, 169–171
      from non-ferrite parts, 171–172
      overall, 172–173
   of permeability
      expression of, 29–31
      with temperature, 29–30
      with time, 30–31, 88

Waveform distortion, 39–40, 111, 148–149
Windings, 313–338
   inductance
      of air-cored coil, 330
      of cylindrical core, 155
      leakage, in transformers, 335–338
   losses, eddy current, 317–330
      inductors, 176–178
      proximity effect, 321–323
      skin effect, 318–321
      total, 323–330
   resistance, d.c., 315–317
      loss, 175
   self-capacitance, 330–334
      analysis, 330–334
      assumptions, 330
      and core geometry, 218
   for transformers
      power, higher frequency, 292, 296–301
      wide-band, high frequency, 229–230
   turns and area, 313–315
      copper factors in close packing, 314
      packing factor, 315
      winding ordering, 314–315
Wire tables, 341–351

Zinc content, effect on
   permeability/temperature relation, 80
   saturation flux density/temperature relation, 69